Recent Developments in Plant Genomics

Volume I

Recent Developments in Plant Genomics
Volume I

Edited by **Isabelle Nickel**

R CALLISTO REFERENCE

New York

Published by Callisto Reference,
106 Park Avenue, Suite 200,
New York, NY 10016, USA
www.callistoreference.com

Recent Developments in Plant Genomics: Volume I
Edited by Isabelle Nickel

International Standard Book Number: 978-1-63239-536-8 (Hardback)

Printed in the United States of America.

Contents

Preface

The field of genetic study and genomes is extremely vast with sub-divisions opening up on a regular basis due to new and exciting discoveries. A genome is basically the blueprint of an organism that contains in itself all the data needed for that organism to survive. It can be described as a set of rules and instructions that gives an organism its basic characteristics and behavior. Plant genomics is a part of this field and is an extensive and constantly evolving and changing field of study. It has gained much popularity in the past few years and focuses on the development of data management tools and research. There are many research scientists who are in continuous exploration of the methodologies and currents issues of this promising field. Many researches focus on the areas like functional analysis and discovery of genes with an emphasis on branches like mRNA, gene mapping, and protein profiling. There are many new and interesting methodologies that probe into plant breeding technology and gene analysis through mutation, transformation and gene function. There is now a growth in the investments in the arena of plant genomics that is essential in the advancement of our knowledge of plant biology and helping researchers transmute that knowledge and expertise into tangible benefits for mankind.

This book is an attempt to collate and compile all currently available data on plant genomics and advances in this field. I am thankful to all those whose hard work went into the research in this field. I particularly wish to thank the publishing team for their constant support as and when required. Lastly, I wish to thank my family for their faith in me and their encouragement at every step in my life.

Editor

Advances towards a Marker-Assisted Selection Breeding Program in Prairie Cordgrass, a Biomass Crop

K. R. Gedye, J. L. Gonzalez-Hernandez, V. Owens, and A. Boe

Department of Plant Sciences, South Dakota State University, Brookings, SD 57007, USA

Correspondence should be addressed to J. L. Gonzalez-Hernandez, jose.gonzalez@sdstate.edu

Academic Editor: Shizhong Xu

Prairie cordgrass (*Spartina pectinata* Bosc ex Link) is an indigenous, perennial grass of North America that is being developed into a cellulosic biomass crop suitable for biofuel production. Limited research has been performed into the breeding of prairie cordgrass; this research details an initial investigation into the development of a breeding program for this species. Genomic libraries enriched for four simple sequence repeat (SSR) motifs were developed, 25 clones from each library were sequenced, identifying 70 SSR regions, and primers were developed for these regions, 35 of which were amplified under standard PCR conditions. These SSR markers were used to validate the crossing methodology of prairie cordgrass and it was found that crosses between two plants occurred without the need for emasculation. The successful cross between two clones of prairie cordgrass indicates that this species is not self-incompatible. The results from this research will be used to instigate the production of a molecular map of prairie cordgrass which can be used to incorporate marker-assisted selection (MAS) protocols into a breeding program to improve this species for cellulosic biomass production.

1. Introduction

Recent world issues associated with fuel consumption and supply have turned attention towards biofuel production, especially cellulosic biofuel. Perennial grasses provide an optimal source of cellulosic biomass due to their high yield potential. Prairie cordgrass (*Spartina pectinata* Bosc ex Link) is a perennial indigenous grass of North America and can be found as a native from Texas to near the Arctic Circle [1]. Ongoing studies on prairie cordgrass in comparison with switchgrass (*Panicum virgatum* L.) indicate that prairie cordgrass could produce more biomass than switchgrass [2]. Furthermore, results from the comparison of prairie cordgrass and switchgrass performed by Boe and Lee in 2007 [2] indicated that prairie cordgrass has a wider environmental amplitude and is adapted to poorly drained wet areas which can have high salinity and be poorly aerated, regions not suitable for the production of conventional crops such as maize (*Zea mays*) [2, 3]. These results are indicative of the potential of prairie cordgrass as a source of biomass for

cellulosic biofuel production. A research program at South Dakota State University (SDSU) is underway to develop native prairie cordgrass into a viable cellulosic biomass crop. The development of a new crop species requires a multidisciplinary approach; examining and validating each step before commercialization can occur. These steps include, but are not limited to, the assembling of a germplasm collection, the development of an accelerated breeding program, and the optimization of cultivation practices. A fundamental step to accelerate the breeding of prairie cordgrass is to determine and optimize a crossing protocol. prairie cordgrass has been intrinsically believed to be a protogynous outcrossing species, based on the mode of reproduction of its maritime relative smooth cordgrass (*Spartina alterniflora*) [4, 5]. From work performed upon other members of the *Spartina* genus, it has been conjectured that prairie cordgrass will have a similar method of reproduction. In the majority of other Graminaceae species, breeding is performed via the initial emasculation of the floret to ensure that only cross pollination can occur. Prairie cordgrass has an inflorescence composed of between 0 and 31 short paracladia and 11–13 long paracladia [6], bearing a total of 10–80 fertile spikelets, of single florets [7]. Physical emasculation of prairie cordgrass is essentially impractical, but this technique may not be necessary as with protogyny and ascertaining the appropriate timing, directed cross pollination is feasible. Self pollination has not been identified previously in prairie cordgrass and assumptions have been made that prairie cordgrass may be self incompatible, if this is the case then the classic mapping technique of producing recombinant inbred lines will be fundamentally impossible and alternate strategies will need to be utilized.

An initial stage of the prairie cordgrass project is to develop a molecular map of the species. Only limited molecular analyses have been performed upon prairie cordgrass, notably in contrast to its maritime relative S. alterniflora. In the National Centre for Biotechnology Information (NCBI), only 57 prairie cordgrass sequences have been deposited (predominantly regions of the nuclear and organelle genomes utilized for diversity analysis, that is, Waxy (AY508655 and AF372461), ITS (AF019843, AJ489796, and EF153082) and the chloroplast trnL-trnF intergenic spacer region (EF137568, EU056305, and AF372625)). A recent publication on the analysis of the transcriptome of prairie cordgrass has increased the available knowledge on the genome of the species [8]. Constructed from the analysis of the expression sequence tags (ESTs) and identified from the transcriptome of prairie cordgrass, a total of 26,302 contigs and 71,103 singletons were assembled with all sequence information available as supplemental data [8]. Additional molecular analysis of genetic diversity of prairie cordgrass in natural populations in Minnesota with amplified fragment length polymorphism (AFLP) has been performed [9].

An optimal marker system for the development of a molecular map of prairie cordgrass are SSRs or microsatellites. Microsatellites are highly utilized molecular markers and have been developed for the majority of agronomically and economically important crop species [10]; their efficacy arises predominantly due to their reproducibility, codominant inheritance, and abundance [11]. Modern techniques allow the rapid identification of microsatellite regions in species which have limited sequence information, and by using proprietary genomic library screening techniques, microsatellites have been developed for numerous non-model organisms, encompassing insects [12], birds [13], fish [14], mammals [15], and plants, including other species in the Spartina genera [16]. Previous studies on other members of the Spartina genera have developed 35 markers for microsatellite loci in S. alterniflora [16, 17]. Of these S. alterniflora markers, three have been found to amplify in prairie cordgrass [16]. SSRs have been used extensively in numerous crop species, in wheat (Triticum aestivum L.), and another Poaceae species; SSRs have been used to enhance molecular maps begun with restriction fragment length polymorphism (RFLP) and to identify genes of interest [18]. Furthermore, Wilde et al. in 2007 [19] found that the incorporation of MAS with SSR markers into a traditional

breeding program can result in a substantial increase in the incorporation of Fusarium head blight resistance, a quantitative trait, in wheat with the shortest possible time. The results of Wilde et al. [19] are indicative of the potential of MAS for the breeding of prairie cordgrass.

This paper details the first investigation into initial breeding and crossing of prairie cordgrass, specifically the production of F_1 individuals and the validation of a crossing protocol. Furthermore, the characterization of 35 microsatellite loci in prairie cordgrass from genomic DNA is discussed. The validation of the microsatellite loci occurred in a prairie cordgrass germplasm collection and a reciprocal cross as an initial step in the development of a molecular map, which can be further utilized during marker-assisted selection of lines with traits desirable for the improvement of prairie cordgrass as a biofuel crop.

2. Materials and Methods

2.1. Genomic Library Construction. Genomic DNA was extracted from four lines of prairie cordgrass using the method as described by Karakousis and Langridge in 2003 [20], with minor modifications. A mixture of the four random prairie cordgrass clones DNA, totaling > 100 μg, was sent to Genetic Identification Services (GIS) (http://www.genetic-id-services.com/) for the development of genomic libraries. The genomic libraries were enriched for the four simple sequence repeat (SSR) motifs $(CA)_n$, $(GA)_n$, $(AAG)_n$, and $(CAG)_n$. Subsamples of 100 clones were sequenced and primers were designed to flank the SSR regions using DesignerPCR, version 1.03 (Research Genetics, Inc.) and PRIMER 3 [21]. The SSR regions were classified according to Jones et al. [22] as being pure repeats (i.e., $[N_1N_2]_X$), compound repeats (i.e., $[N_1N_2]_X[N_3N_4]_Y$), and interrupted repeats (i.e., $[N_1N_2]_XN_3[N_4N_5]_Y$).

2.2. Plant Material. Wild germplasm has been collected from throughout the mid-western states of the United States, creating a core germplasm collection to be utilized in prairie cordgrass breeding (unpublished data). A geographically diverse sample of sixteen plants, collected from South Dakota, North Dakota, Minnesota, and Iowa, was characterized using the identified SSR loci. A sample from the closely related species S. spartinae was also included to examine cross species amplification. The germplasm collection was grown in standard greenhouse conditions.

Crosses between and amongst these genotypes were performed. Crossing between two plants of prairie cordgrass was performed in the following manner; two inflorescence at the appropriate stage of development were placed inside a crossing bag for seven days, producing F_1 plants (Figure 1). A reciprocal cross between two genotypes (designated RR2 and RR21) from the Red River morphotype was produced. The reciprocal cross produced 45 F_1 plants from the RR21 × RR2 directed cross and 49 F_1 plants from the RR2 × RR21 directed cross, a total of 94 plants. Two genotypes identified as clones (designated SP1.1B and SP1.2A) from the SSR evaluation were crossed via the same technique, producing a total of

FIGURE 1: Two prairie cordgrass heads undergoing crossing, contained within a single bag.

46 F_1 individuals; a total of 20 F_1 plants were produced when SP1.1B was used as the maternal parent and 26 F_1 plants when SP1.2A was used as the maternal parent. All F_1 plants and their respective parents were grown in standard greenhouse conditions.

2.3. DNA Extraction and Evaluation of SSR Primers. DNA was extracted from all examined plants using the method as described by Karakousis and Langridge in 2003 [20], with minor modifications. Evaluation of the primers was performed in a PCR reaction consisting of 2 U *Taq* DNA polymerase (Promega GoTaq), 1 × PCR Buffer (Promega GoTaq), 2 mM $MgCl_2$, 0.6 mM dNTPs, and 0.6 mM of each primer, with a final volume of 20 μL [23]. PCR reactions were performed in a BioRad MyCycler (BioRad, Hercules, CA). Initially a gradient of annealing temperatures was used to determine optimal T_a, as these primers were designed to be utilized in a prairie cordgrass breeding program, a robust and high-throughput protocol was developed. Specific thermocycling conditions for the primers used in this study were as follows: an initial denaturation at 94°C for 5 minutes, followed by 35 cycles of 94°C for 1 minute, 53 or 55°C for 1 minute, 72°C for 1 minute, followed by an extension step of 72°C for 10 minutes, and a 10°C hold. PCR product was visualized on 8% nondenaturing PAGE; bands were scored and sized by comparison to a 100 bp ladder using AlphaEaseFC Software (Alpha Innotech, San Leandro, CA).

2.4. SSR Markers Analysis. Polymorphic bands in the two examined populations described above were scored on a presence (1) or absence (0) basis. An estimation of genetic distance was calculated with Nei and Li's algorithm [24] and the resulting matrix was clustered with unweighted pair group method with arithmetic mean (UPGMA) [25]. Analysis was performed using multivariate statistics package (MVSP) [26].

3. Results

3.1. Germplasm and Crossing. The crossing of the germplasm collection was successful with the production of a total of 14,813 putatively viable seeds from 110 crosses, performed in lines and collected from 48 distinct collection locations (Table 1). Numerous nonviable seeds were produced from each cross. A seed was determined to be viable if the actual seed (endosperm and embryo) was visible within the glumes (Figure 2). The number of viable seeds produced per head varied from 6 to 642, with an average of 134 seeds per cross. The crosses can be grouped into two, dependent upon the parents, those that were outcrossed and those that were selfed. Crosses that were designated "out" were between genotypes that were geographically diverse, while crosses designated "self" were produced from members of the population collected at the same location. The average number of seeds produced from outcrossed individuals was 143, with a range of 12 to 250, while selfed individuals produced an average of 112 seeds per cross, with a range of 6 to 642 (Figure 3).

3.2. Microsatellite Characterization. Of the total 100 genomic clones sequenced, 70 contained microsatellite motifs, 26 from the $(CA)_n$ enriched library, 25 from the $(GA)_n$ enriched library, 3 from the $(AAG)_n$ enriched library, and 16 from the $(CAG)_n$ enriched library. All 70 loci were examined and 35 were amplified with the standardized conditions (Table 2). The repeat SSR structure that occurred most frequently was the pure repeat (27), followed by the compound repeat (7) and only one interrupted repeat sequence was found (Table 2).

Only two primers produced monomorphic profiles from the analysis of the sixteen lines, SPSD004 from the $(GA)_n$ library and SPSD048 from the $(AAG)_n$ library. The remaining primers produced between 1 and 12 scorable bands. Of the examined primers, 11 were amplified between 1 and 7 bands in *S. spartinae* (Table 2).

All sequences were examined using BLASTn [27] on the National Center for Biotechnology Information (NCBI) server for similarities to other recorded sequences. Only two of the prairie cordgrass sequences displayed homology ($e \leq e^{-20}$) to recorded sequences, SPSD003 to six sequences from rice, wheat, and field mustard (*Brassica rapa*) (accession numbers: AP008214.1, AP005495.3, NM_001067901.1, AK111788.1, EU660901.1, and AC189183.2) and SPSD050 to five sequences all from rice (accession numbers: CR855236.1, AP008210.1, AL606607.3, NM_001059819.1, and AK108706.1) [25].

3.3. Crossing Population. A total of 94 individuals from the RR2 × RR21 and the RR21 × RR2 crosses were examined with 35 SSR primers. The amplified bands were scored based on presence or absence, the resulting data was analyzed with Nei and Li's coefficient producing a similarity matrix (data not shown), and the data was clustered with UPGMA producing a dendogram (Figure 4). The results of the dendogram indicate that each individual was a cross

TABLE 1: Summary of crosses performed amongst the core prairie cordgrass germplasm collection. The type of cross is designated "Out" for crosses between geographically diverse members and "Self" for members of the population collected at the same location.

Female parent	Male parent	Number of crosses	Type	Total no. of seeds
SP1	SP21	2	Out	126
SP1	SP71	1	Out	71
SP21	SP30	1	Out	51
SP21	SP31	6	Out	484
SP21	SP4	2	Out	248
SP21	SP42	1	Out	78
SP21	SP66	2	Out	233
SP21	SP7	1	Out	82
SP24	SP30	1	Out	180
SP29	SP22	1	Out	146
SP29	SP30	1	Out	42
SP3	SP77	1	Out	282
SP30	SP31	1	Out	164
SP30	SP52	1	Out	246
SP31	SP21	4	Out	408
SP31	SP30	1	Out	28
SP31	SP32	1	Out	38
SP31	SP46	3	Out	211
SP31	SP5	1	Out	81
SP31	SP66	1	Out	46
SP31	SP79	1	Out	198
SP32	SP21	1	Out	173
SP32	SP31	2	Out	310
SP32	SP45	1	Out	296
SP32	SP66	1	Out	240
SP32	SP67	1	Out	193
SP32	SP78	1	Out	94
SP35	SP78	1	Out	279
SP41	SP42	1	Out	6
SP41	SP45	1	Out	86
SP44	SP54	1	Out	312
SP44	SP59	1	Out	274
SP45	SP31	1	Out	282
SP45	SP49	1	Out	225
SP45	SP67	1	Out	138
SP49	SP40	1	Out	203
SP49	SP45	1	Out	176
SP49	SP66	2	Out	237
SP5	SP31	1	Out	189
SP5	SP78	1	Out	407
SP53	SP67	2	Out	43
SP54	SP44	1	Out	145
SP54	SP59	2	Out	81
SP56	SP58	1	Out	235
SP65	SP57	1	Out	70
SP66	SP31	1	Out	93
SP67	SP32	1	Out	99

TABLE 1: Continued.

Female parent	Male parent	Number of crosses	Type	Total no. of seeds
SP67	SP45	2	Out	583
SP67	SP53	1	Out	185
SP7	SP46	1	Out	160
SP72	SP1	1	Out	125
SP72	SP30	2	Out	459
SP75	SP13	1	Out	59
SP77	SP72	1	Out	163
SP77	SP79	1	Out	184
SP78	SP35	1	Out	94
SP78	SP79	1	Out	113
SP82	SP14	1	Out	642
SP21	SP21	13	Self	1470
SP22	SP22	1	Self	110
SP31	SP31	13	Self	1377
SP40	SP40	1	Self	126
SP41	SP41	2	Self	393
SP53	SP53	1	Self	12
SP74	SP74	2	Self	229

between the two parents. When the population derived from the cross between the two plants (SP1.1B and SP1.2A) and determined to be clones was examined, bands present in both parents were found to segregate in the progeny (data not shown), indicative of sexual recombination; this result provides validation that successful crossing occurred.

4. Discussion

The primary requirement of any breeding program is to ensure that accurate crosses are made; in many other members of the Poaceae this is achieved by physical or chemical emasculation. The prolific numbers of flowers per head in addition to the small size of the flowers make physical emasculation unfeasible. Furthermore, due to limited knowledge about the nature of the fertility of this species, chemical emasculation has not been developed for prairie cordgrass. The results from the SSR analysis indicate that utilizing the inherent protogyny of prairie cordgrass allows successful crossing between two individuals without the need for emasculation, confirming the validity of the breeding methodology used. The presence of individuals in the F_1 mapping populations which show limited genetic dissimilarity from the parents could be evidence of selfing; further investigations are required. The presence of these potential selfed individuals indicates that future breeding and/or mapping populations should be examined with the molecular markers devised in this research to remove suspect individuals. Subsequently, the successful crossing between two clones is indicative that prairie cordgrass may not be self-incompatible and that it may be possible to develop in this species conventional mapping populations, such as recombinant inbred lines. Further studies into self-compatibility

(a)

5mm

(b)

FIGURE 2: Prairie cordgrass seeds photographs taken over a light box. Illuminating prairie cordgrass with light box is used to distinguish the presence of a developed embryo in the seeds. (a) Non-viable seeds. (b) Viable seeds.

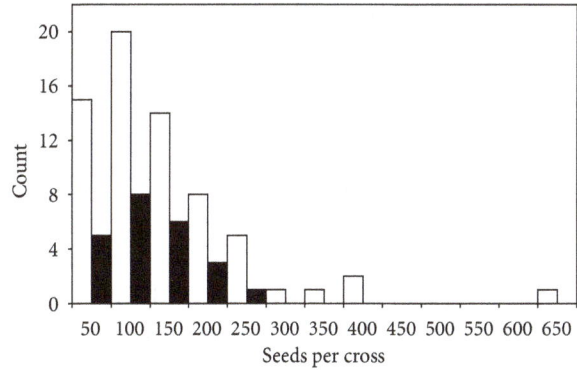

FIGURE 3: Summary graph of results of crosses between members of the core germplasm collection of prairie cordgrass. White bars indicate self crosses and black bars indicate out crosses.

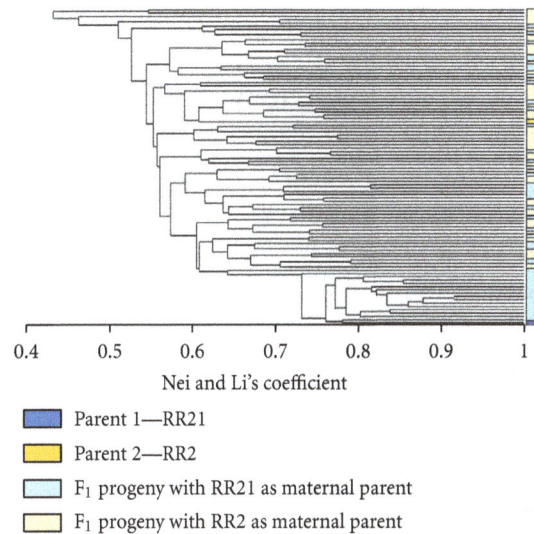

Parent 1—RR21
Parent 2—RR2
F_1 progeny with RR21 as maternal parent
F_1 progeny with RR2 as maternal parent

FIGURE 4: Unrooted dendogram clustered with UPGMA of the genetic association of the F_1 progeny from a cross between prairie cordgrass lines RR21 and RR2, from the analysis of SSR regions with Nei and Li's Coefficient of genetic similarity.

with investigations into potential apomictic prairie cordgrass plants are underway. The number of seeds observed in this research appears to be larger than what was previously described by Clayton et al. [7]. The variation between the two studies can be attributed to both environmental and genetic variations. Genetic variation in seed set in prairie cordgrass, although at this stage not quantified, is demonstrated by the range in viable seed set observed in this research.

The amplification of S. spartinae with the SSR primers developed in this research are indicative of the potential colinearity amongst the genomes of Spartina spp. and other grass species; this colinearity will allow easier identification and characterization of genes. The colinearity between prairie cordgrass and other Poaceae is currently utilized

to examine genes identified in related species in prairie cordgrass, specifically genes utilized to examine phylogeny (i.e., the waxy gene for granule bound starch synthase).

The prevalence and distribution of SSR regions across plant genomes are extremely variable. Variation in SSRs is not limited to their location, but also their motif, putative function, abundance, and repeat number [28]. The results of this analysis indicate that the two dinucleotide repeats $(CA)_n$ and $(GA)_n$ are more prevalent in prairie cordgrass (31% and 46%, resp.), than the two trinucleotide repeats $(AAG)_n$ and $(CAG)_n$ (6% and 17%, resp.). The prevalence of the dinucleotide motif in prairie cordgrass is similar to what was observed in the characterization of SSRs in other Spartina sp., where Blum et al. [16] found 82% of isolated SSRs contained dinucleotide repeats and Sloop et al. [17] found 71% containing similar motifs. In all three

TABLE 2: Summary and description of primer sequences designed for SSR regions identified in prairie cordgrass and their amplification products.

Locus	Library	Primer sequence (5'3')	SSR motif	Expected size (bp)	Size range (bp)	T_a (°C)	GenBank ID
SPSD001	$(CA)_n$	F: GCTGCTCCTCTTCTCTGTCT R: ACGGCACACTTAGTTTTCTG	$(CA)_{13}$ Perfect	180	212–386	55	GQ354531
SPSD002	$(CA)_n$	F: GCACTGTTTGGTGATGCTC R: CTGACGCAAGGTTGATGAG	$(CA)_{13}$ Perfect	249	100–655	55	GQ354532
SPSD009	$(CA)_n$	F: ATGGTTTCACAAGTCGGAAGT R: CAGGGCTGCCTACAAGATG	$(GT)_{15}$ Perfect	166	215–432	55	GQ354533
SPSD010	$(CA)_n$	F: AACCAAAAGGATAGACCCTA R: ACGAAATATGTGACCGATAC	$(CA)_{28}$ Perfect	168	127–310	53	GQ354534
SPSD011	$(CA)_n$	F: CTAACGTATGTCGTTCATGTGG R: AAGGCGATTTTAAGAGGCTAAG	$(GT)_{11}$ Perfect	135	147–251	53	GQ354535
SPSD013	$(CA)_n$	F: GGGATGCTTTGTAGATAAGAAA R: TCTTCCTCTTTACTCTGTCACC	$(CA)_{41}$ Perfect	157	85–173	55	GQ354536
SPSD016	$(CA)_n$	F: CCGACTACGAGCCACATT R: GTTCCACACATACGAAGGAGA	$(CA)_{14}$ Perfect	155	156–268	53	GQ354537
SPSD019	$(CA)_n$	F: CCTGCTTACTCTTACTCCGTC R: ACCCTTTTTTCTTTTGGTCTC	$(CA)_{13}(GA)_7$ Compound	134	220–365	53	GQ354538
SPSD020	$(CA)_n$	F: ATGAGACGATAGCAGGATGAC R: AGCAGATTACGATTCAGATGG	$(CA)_{23}$ Perfect	178	146–288	55	GQ354539
SPSD025	$(CA)_n$	F: CAGTCCATGCAACTCAGAAGTA R: AACCTGATAGAAGTGGTCATGC	$(CT)_{10}(CA)_{38}$ Compound	269	204–748	55	GQ354540
SPSD026	$(CA)_n$	F: GTGGAATCAACAACACCAGA R: GTCGCTTTAGCCCGTAAG	$(GA)_{13}(GT)_{20}$ Compound	209	196–539	55	GQ354541
SPSD003	$(GA)_n$	F: ATGGAAACTGTCTGGAACTGAC R: AGCAATAACCACAGAAGAGACC	$(CT)_{28}$ Perfect	294	231–263	55	GQ354542
SPSD027	$(GA)_n$	F: TCAAACAATGGCGGAGAAG R: CTGGCTCCACCTCTTTGG	$(CT)_{17}$ Perfect	214	188–224	55	GQ354543
SPSD004	$(GA)_n$	F: GTTGCTCGGTTCCAGTTG R: CGCCACACAAAAGTAGCC	$(GA)_{22}$ Perfect	178	133 & 164	55	GQ354544
SPSD031	$(GA)_n$	F: TCGCACTTTTGATTCTCTTTAC R: TGGATGGATTAGGTTACTGTTG	$(CT)_{20}$ Perfect	188	155–346	53	GQ354545
SPSD032	$(GA)_n$	F: CTCTCGCCCATTGCTACTTA R: CCATTGCTATGTTGTTTGAGC	$(CT)_{13}$ Perfect	196	146–193	53	GQ354546
SPSD034	$(GA)_n$	F: CAGGTCTACGGAGGTCACTAC R: TCAAAAGAAGAGCACATACACA	$(CT)_{12}$ Perfect	160	144–306	53	GQ354564
SPSD036	$(GA)_n$	F: TTCACCACACCACTTTATCC R: GGAAGCAACAAACATTGATG	$(CT)_{29}$ Perfect	271	222–260	53	GQ354547
SPSD039	$(GA)_n$	F: CTTTCAGATAGCTCCACTGATC R: AGCAATAACTGTGCATACCTCT	$(GA)_{12}$ Perfect	206	193–448	55	GQ354548
SPSD040	$(GA)_n$	F: AATCGAAGTAGCAGACACCAAC R: CATGCGTTTTTCACTCATGTAG	$(GA)_{11}$ Perfect	214	190–416	55	GQ354549
SPSD041	$(GA)_n$	F: CCCAACGATGATTTCTCTTG R: TCACGGTAACACGATTAGTCC	$(GA)_{26}$ Perfect	135	104–180	53	GQ354550
SPSD042	$(GA)_n$	F: ACCTCCCACTCGTTGCTAC R: GCCATTGCTCTGTTGTTTG	$(CT)_{19}$ Perfect	211	131–257	53	GQ354551
SPSD043	$(GA)_n$	F: GTTCAAATGCGAACAAATCAG R: ATTCGATCTCACATGCAACAC	$(GA)_{30}$ Perfect	277	231–252	53	GQ354552

TABLE 2: Continued.

Locus	Library	Primer sequence (5'3')	SSR motif	Expected size (bp)	Size range (bp)	T_a (°C)	GenBank ID
SPSD044	$(GA)_n$	F: AGCTATATGACCCGAGACTGTG R: GGAATGGTCCCATACTTAATCC	$(GA)_{20}$ Perfect	275	254–276	55	GQ354553
SPSD045	$(GA)_n$	F: AACGGAGGAAGTAATAAATAGC R: AGCACACACTAGCAAGGAC	$(CT)_{16}$ Perfect	241	214–426	53	GQ354554
SPSD046	$(GA)_n$	F: CAGGTTTATCAGTGAAGACATC R: GAGGTTCTTAAAGGAACATAGC	$(TA)_9(GA)_{12}$ Compound	278	57–426	53	GQ354555
SPSD047	$(GA)_n$	F: CCACCTTCCTTGGATACAC R: CCACAACTACCACCTC	$(CT)_{17}$ Perfect	194	156–326	53	GQ354556
SPSD005	$(AAG)_n$	F: TGAACCAACATAACCTACCTG R: CCACACTAAACCGAAACTTG	$(TTC)_{19}$ Perfect	297	192–452	55	GQ354557
SPSD048	$(AAG)_n$	F: AAGGGCATAGTTTCAACCAAG R: CTTTTGCTTGTTCATCAACATG	$(AAG)_{10}$ Perfect	197	88	55	GQ354565
SPSD007	$(CAG)_n$	F: AATCCTTCGCCTATCCTACAC R: TTCACACAGCAGACAGAACTG	$(CTG)_8(CT)_9$ Compound	168	146–519	55	GQ354558
SPSD008	$(CAG)_n$	F: GCAAGAACAGACTCAAGAGC R: CTGCTGCTGAAGTAAAAGTTG	$(CAG)_9(CAA)_4(CAG)_4(CAA)_2$ Compound	276	144–309	55	GQ354559
SPSD049	$(CAG)_n$	F: TGGATTGTTTCCTGATACTCCA R: CCATAAATTGCTGCATTATTCC	$(TTG)_5(CTG)_5$ Compound	299	303–667	53	GQ354560
SPSD050	$(CAG)_n$	F: GAAGCAGAAAACACAGTATTGC R: TTGCTGGAATTTAACCTATCTG	$(CTG)_5$ Perfect	225	225–505	53	GQ354561
SPSD053	$(CAG)_n$	F: ACGCCTTCTTCACTCCAAC R: GCCACCAGTTTTCATCACC	$(CTG)_8$ Perfect	208	166–316	53	GQ354562
SPSD056	$(CAG)_n$	F: GTTCTCCAAAGTCTCCTCCT R: ATCTTTACCTTCCTTCTGGG	$(TCC)_6TTC(TCC)_2$ Interrupted	197	79–230	53	GQ354563

studies the di- and trinucleotide repeats occurred as perfect, compound, and interrupted motifs. Based on the results found in this research, the genomic libraries enriched with the dinucleotide repeats have been extensively sequenced; the resulting sequence information will be screened to isolate additional SSR regions, primers will be designed, and the resulting markers will be used to develop a molecular map of prairie cordgrass. The molecular map will then be used to find linkage between SSR markers and traits of interest allowing future MAS to be performed.

Acknowledgments

The authors want to thank S. G. Dwire for his technical assistance. The research described in this paper was funded in part by grants from the North Central Sun Grant Initiative and from the USDA Feedstock Genomics Program to J. L. Gonzalez-Hernandez, A. Boe, and V. Owens and additional support from the South Dakota Agricultural Experimental Station. The authors acknowledge the use of the SDSU-FGCF supported in part by NSF/EPSCoR Grant no. 0091948 and by the State of South Dakota.

References

[1] USDA-NRCS, *Plants Database*, USDA-NRCS, Urbandale, Iowa, USA, 2008.

[2] A. Boe and D. K. Lee, "Genetic variation for biomass production in prairie cordgrass and switchgrass," *Crop Science*, vol. 47, no. 3, pp. 929–934, 2007.

[3] J. L. Gonzalez-Hernandez, G. Sarath, J. M. Stein, V. Owens, K. Gedye, and A. Boe, "A multiple species approach to biomass production from native herbaceous perennial feedstocks," *In Vitro Cellular and Developmental Biology-Plant*, vol. 45, no. 3, pp. 267–281, 2009.

[4] X. Fang, P. K. Subudhi, B. C. Venuto, and S. A. Harrison, "Mode of pollination, pollen germination, and seed set in smooth cordgrass (Spartina alterniflora, Poaceae)," *International Journal of Plant Sciences*, vol. 165, no. 3, pp. 395–401, 2004.

[5] X. Fang, P. K. Subudhi, B. C. Venuto, S. A. Harrison, and A. B. Ryan, "Influence of flowering phenology on seed production in smooth cordgrass (Spartina alterniflora Loisel.)," *Aquatic Botany*, vol. 80, no. 2, pp. 139–151, 2004.

[6] V. G. Kern, N. J. Guarise, and A. C. Vegetti, "Inflorescence structure in species of Spartina Schreb. (Poaceae: Chloridoideae: Cynodonteae)," *Plant Systematics and Evolution*, vol. 273, no. 1-2, pp. 51–61, 2008.

[7] W. D. Clayton, M. S. Vorontsova, K. T. Harman, and H.

Williamson, *GrassBase-The Online World Grass Flora*, 2006.

[8] K. Gedye, J. Gonzalez-Hernandez, Y. Ban et al., "Investigation of the transcriptome of Prairie Cord grass, a new cellulosic biomass crop," *The Plant Genome Journal*, vol. 3, no. 2, pp. 69–80, 2010.

[9] K. M. Moncada, N. J. Ehlke, G. J. Muehlbauer, C. C. Sheaffer, D. L. Wyse, and L. R. DeHaan, "Genetic variation in three native plant species across the State of Minnesota," *Crop Science*, vol. 47, no. 6, pp. 2379–2389, 2007.

[10] L. Cardle, L. Ramsay, D. Milbourne, M. Macaulay, D. Marshall, and R. Waugh, "Computational and experimental characterization of physically clustered simple sequence repeats in plants," *Genetics*, vol. 156, no. 2, pp. 847–854, 2000.

[11] R. K. Varshney, A. Graner, and M. E. Sorrells, "Genic microsatellite markers in plants: features and applications," *Trends in Biotechnology*, vol. 23, no. 1, pp. 48–55, 2005.

[12] A. E. Van't Hof, B. J. Zwaan, I. J. Saccheri, D. Daly, A. N. M. Bot, and P. M. Brakefield, "Characterization of 28 microsatellite loci for the butterfly Bicyclus anynana," *Molecular Ecology Notes*, vol. 5, no. 1, pp. 169–172, 2005.

[13] V. Saladin, D. Bonfils, T. Binz, and H. Richner, "Isolation and characterization of 16 microsatellite loci in the Great Tit Parus major," *Molecular Ecology Notes*, vol. 3, no. 4, pp. 520–522, 2003.

[14] T. L. King, M. S. Eackles, and B. H. Letcher, "Microsatellite DNA markers for the study of Atlantic salmon (Salmo salar) kinship, population structure, and mixed-fishery analyses," *Molecular Ecology Notes*, vol. 5, no. 1, pp. 130–132, 2005.

[15] M. M. Peacock, V. S. Kirchoff, and S. J. Merideth, "Identification and characterization of nine polymorphic microsatellite loci in the North American pika, Ochotona princeps," *Molecular Ecology Notes*, vol. 2, no. 3, pp. 360–362, 2002.

[16] M. J. Blum, C. M. Sloop, D. R. Ayres, and D. R. Strong, "Characterization of microsatellite loci in Spartina species (Poaceae)," *Molecular Ecology Notes*, vol. 4, no. 1, pp. 39–42, 2004.

[17] C. M. Sloop, H. G. McGray, M. J. Blum, and D. R. Strong, "Characterization of 24 additional microsatellite loci in Spartina species (Poaceae)," *Conservation Genetics*, vol. 6, no. 6, pp. 1049–1052, 2005.

[18] S. Landjeva, V. Korzun, and A. Börner, "Molecular markers: actual and potential contributions to wheat genome characterization and breeding," *Euphytica*, vol. 156, no. 3, pp. 271–296, 2007.

[19] F. Wilde, V. Korzun, E. Ebmeyer, H. H. Geiger, and T. Miedaner, "Comparison of phenotypic and marker-based selection for Fusarium head blight resistance and DON content in spring wheat," *Molecular Breeding*, vol. 19, no. 4, pp. 357–370, 2007.

[20] A. Karakousis and P. Langridge, "High-throughput plant DNA extraction method for marker analysis," *Plant Molecular Biology Reporter*, vol. 21, no. 1, 2003.

[21] S. Rozen and H. Skaletsky, "Primer3 on the WWW for general users and for biologist programmers.," *Methods in Molecular Biology*, vol. 132, pp. 365–386, 2000.

[22] E. S. Jones, M. P. Dupal, R. Kölliker, M. C. Drayton, and J. W. Forster, "Development and characterization of simple sequence repeat (SSR) markers for perennial ryegrass (Lolium perenne L.)," *Theoretical and Applied Genetics*, vol. 102, no. 2-3, pp. 405–415, 2001.

[23] K. Gedye, J. Gonzalez-Hernandez, R. Schuelke, V. Owens, and A. Boe, "Development of SSR markers from genomic DNA and ESTs in prairie cordgrass, a cellulosic biomass crop," in *Proceedings of the Plant & Animal Genomes XIX Conference*, Town & Country Convention Center, San Diego, Calif, USA, January 2011.

[24] M. Nei and W. H. Li, "Mathematical model for studying genetic variation in terms of restriction endonucleases," *Proceedings of the National Academy of Sciences of the United States of America*, vol. 76, no. 10, pp. 5269–5273, 1979.

[25] P. H. A. Sneath and R. R. Sokal, *Numerical Taxonomy*, W.H. Freeman and Company, San Francisco, Calif, USA, 1973.

[26] W. L. Kovach, *MVSP-A Multivariate Statistical Package For Windows*, Kovach Computing Services: Pentraeth, Walesm, UK, 1998.

[27] Z. Zhang, S. Schwartz, L. Wagner, and W. Miller, "A greedy algorithm for aligning DNA sequences," *Journal of Computational Biology*, vol. 7, no. 1-2, pp. 203–214, 2000.

[28] Y.-C. Li, A. B. Korol, T. Fahima, A. Beiles, and E. Nevo, "Microsatellites: genomic distribution, putative functions and mutational mechanisms: a review," *Molecular Ecology*, vol. 11, no. 12, pp. 2453–2465, 2002.

2

A Pair of Partially Overlapping *Arabidopsis* Genes with Antagonistic Circadian Expression

Andrea Kunova, Elena Zubko, and Peter Meyer

Centre for Plant Sciences, University of Leeds, Leeds LS2 9JT, UK

Correspondence should be addressed to Peter Meyer, p.meyer@leeds.ac.uk

Academic Editor: Pierre Sourdille

A large number of plant genes are aligned with partially overlapping genes in antisense orientation. Transcription of both genes would therefore favour the formation of double-stranded RNA, providing a substrate for the RNAi machinery, and enhanced antisense transcription should therefore reduce sense transcript levels. We have identified a gene pair that resembles a model for antisense-based gene regulation as a T-DNA insertion into the antisense gene causes a reduction in antisense transcript levels and an increase in sense transcript levels. The same effect was, however, also observed when the two genes were inserted as transgenes into different chromosomal locations, independent of the sense and antisense gene being expressed individually or jointly. Our results therefore indicate that antagonistic changes in sense/antisense transcript levels do not necessarily reflect antisense-mediated regulation. More likely, the partial overlap of the two genes may have favoured the evolution of antagonistic expression patterns preventing RNAi effects.

1. Introduction

The expression of genetic information involves a sequence of molecular processes, including transcript synthesis, processing, turnover, transport and translation. Eukaryotes have developed regulatory systems at each of these levels, which can all contribute to the expression efficiency and profile of a gene. Key components in this context are natural antisense transcripts (NATs), for which a regulatory role has been demonstrated at most levels [1]. Work on yeast and animals has identified model genes that illustrate the regulatory influence of antisense transcription on sense transcript synthesis, processing, or stability. This has led to the proposal that NATs have evolved to comprise a second tier of gene expression in eukaryotes [2].

Most of our current knowledge about the mechanistic aspects of antisense-mediated gene regulation derives from work in the fungal or animal field. In plant research, antisense-mediated gene regulation systems are poorly characterised, which is contrasted by an unexpectedly high number of NATs in plants [3–5]. Among 26,939 annotated *Arabidopsis* genes, 3,027 contain antisense transcripts that are jointly expressed with their sense transcript partner. For another 7,598 genes, sense and antisense transcripts are both detected in specific tissues [3]. The abundance of NATs very strongly suggests that they are part of regulatory systems. The few examples, for which natural plant antisense transcripts have been documented to modify sense transcripts, are based on RNA interference (RNAi) mechanisms that involve dsRNA degradation into small natural antisense transcript siRNAs (nat-siRNAs). The *P5CDH* gene, which encodes a stress-response regulator, and overlaps in antisense orientation with the salt-inducible *SRO5* gene. Expression of sense and antisense transcripts results in the production of nat-siRNAs that cleave the *P5CDH* transcript [6]. Another class of nat-siRNAs, which is induced by the bacterial pathogen *Pseudomonas syringae* carrying effector avrRpt2, represses PPRL, a putative negative regulator of the RPS2 resistance pathway [7]. The biogenesis of both nat-siRNAs requires RNA-dependent RNA Polymerase 6 (RDR6), which probably synthesises secondary dsRNA substrates.

To identify novel target loci for nat-siRNA regulation, we selected T-DNA insertions in antisense genes of NATs and tested how changes in antisense transcription affect sense transcript levels. We identified a locus where the T-DNA insertion did not only alter antisense transcript levels but also transcript levels of the associated sense gene. This resembles the effects of an RNA interference mechanism between sense and antisense transcripts, but, surprisingly, we find no indication for a regulatory influence of the antisense transcription on the sense transcript.

2. Materials and Methods

2.1. Insertion Lines and Plant Material. Lines SALK_048899 [8] and SM_3.32080 [9] T-DNA insertion lines were obtained from the Nottingham *Arabidopsis* Stock Centre [10] (http://arabidopsis.info/). Plants were grown in a growth chamber under short day conditions (8 hours light, 16 hours dark, temperature 22°C and the humidity 60%). Leaf tissues were collected after four weeks and stored at −80°C. Seedlings used for the expression analysis were grown on 1/2 MS medium (Duchefa Biochmie, Haarlem, The Netherlands) with 1% sucrose (Fisher Chemicals, Loughborough, UK) and grown in long day conditions (16 hours day and 8 hours night) at 22°C. After 7 days tissues were collected and immediately frozen in liquid nitrogen. Tissues were stored at −80°C.

Seedlings used for the circadian analysis were grown according to [11]. Seeds were sterilized and sown on MS medium supplemented with 3% sucrose. Plates were kept at 4°C for four days and were transferred into a growth chamber with a cycle of 12 hours light and 12 hours darkness, at a constant temperature of 22°C and 60% humidity. After 7 days seedlings were shifted to constant white light (24 hour day). Tissues were harvested after 24 hours in constant light every four hours over a 48-hour period.

2.2. Genotyping of Insertion Lines. Genomic DNA for genotyping was extracted from 3-4 week-old leaf tissue according to [12]. After 1 hour incubation in the extraction buffer (200 µL total), samples were cleaned up with 200 µL phenol : chloroform : IAA (12 : 12 : 1). Sequence data were obtained from TAIR [13]. The presence of the T-DNA insertion and the homozygosity of insertion lines was assessed by two PCR reactions using the GoTaq master mix (Promega, Madison, USA). A first PCR was performed with the forward and reverse gene-specific primers, and in a second PCR an appropriate gene specific primer (forward or reverse) was used together with the T-DNA insertion specific primer. For the SALK insertion line, the T-DNA primer 5′-AACCAGCGTGGACCGCTTCTG-3′, forward primer 5′-AGCGAACGGTGGACAGAAAC-3′, and reverse primer 5-AGAAGTTTCGAGATCATCGTC-3′ were used. For the SM line, the T-DNA primer 5′-TACGAATAAGAGCGTCCATTTTAGAGTGA-3′ forward primer 5′-ATCTACAACAATGGCTGGAG-3′, and reverse primer 5′-TGGTTCCGAGAGAACCCTTC-3′ were used. Line *rdr6-11* (At3g49500) is a substitution mutant in *Columbia* background with C- > T substitution at position 805 bp downstream of ATG. Genotyping was performed as described by [14].

2.3. RNA Extraction and cDNA Preparation for Expression Analysis. Total RNA for the expression analysis was extracted from seedlings or 4-week-old leaves as described by [15]. Samples used in the experiments were always collected at the same time point, as the expression of genes with circadian activity was assessed. RNA was treated with DNase (Ambion, Austin, TX, USA) and was retrotranscribed with Superscript II Reverse Transcriptase (Invitrogen, Paisley, UK) according to manufacturer's information using the oligo dT primer 5′-GGCCACGCGTCGACTAGTACTTTTTTTTTTTTTTTTT-3′. For PCR, GoTaq Master mix was used, and for qPCR analysis, a SensiMix SYBR and Fluorescein Kitwas were utilised (Quantance, London, UK). Semiquantitative PCR was performed with F and R primers located downstream of the T-DNA insertion site. *Elongation Factor 1 alpha* (At1g07940) was used for the standardization of the amount of cDNA. For qPCR analysis, qF and qR primers were used. For circadian rhythm analysis, the *CCA1* gene (At2g46830) was used as a control of experimental conditions. Sequences of all primers are listed below.

At1g07940,

F primer 5′-GCGTGTCATTGAGAGGTTCG-3′,

R primer 5′-GTCAAGAGCCTCAAGGAGAG-3′

At3g16240,

F primer 5′-AACCCAGCCGTCACTTTTGG-3′,

R primer 5′-TGGTTCCGAGAGAACCCTTC-3′

At3g16250,

F primer 5′-CCGCGAACTGATATTGAGAAG-3′,

R primer 5′-AGAAGTTTCGAGATCATCGTC

At1g07940,

qF primer 5′-CTCTCCTTGAGGCTCTTGACCAG-3′,

qR primer 5′-CCAATACCACCAATCTTGTAGACA-TCC-3′

At2g46830,

qF primer 5′-AAGGCTCGATCTTCACTGGA-3′,

qR primer 5′-TCTCCTGCTCCATCTGAACC-3′

At3g16240,

5′-TTCTCCGGTGGATCCATGAACC-3′,

qR primer 5′-CCAACCCAGTAGACCCAGTG-3′

At3g16240-transgenic,

qSF primer 5′-TCTCCGGTGGATCCCGCATACC-3′

qR primer 5′-CCAACCCAGTAGACCCAGTG-3′,

At3g16250,

qF primer 5′-ACTATGGGAAGTGTACAGTTGAG-3′

qR primer 5′-TTATGGCTCGGACGGTTTTGG-3′

At3g16250-transgenic,

qASF primer 5′-CTATGGGAAGTGTACACTGCGA-3′,

qR primer 5′-TTATGGCTCGGACGGTTTTGG-3′.

2.4. Cloning and Analysis of At3g16240-At3g16250 Transgene Constructs. Promoter and central regions of the At3g16240-At3g16250 locus were cloned by PCR and individual fragments were cloned into pGreen II 0179 [16]. The At3g16240-At3g16250 locus was cloned in three parts. At3g16240 promoter (from the position 5502835 bp to 5505344 bp, according to the TAIR database, http://www.arabidopsis.org/) was amplified with primers P1-F-XhoI 5′-GTTCTCGAGAAA-GATGCAAAGC-3′ and P1-R-EcoRV 5′-TGTGGTGGG-ATATCTTGGACCCG-3′. The central region (5505345 bp to 5508454 bp) was amplified with primers OF-EcoRV 5′-AGATATCCCACCACACCACAGAAAC-3′ and OR-XmaI 5′-ACCCCGGGTAAGAGATAAAAAGAGGCACC-3′ and At3g16250 promoter (5508455 bp to 5511469 bp) was amplified with primers P2-F-XmaI 5′-TACCCGGGGTGTGTA-TGCGCCGGTTTAG-3′ and P2-R-SacI 5′-TTTGTAACC-GAGCTCAAAGAAGTGG-3′. Individual fragments were cloned into pGreen II 0179 digested with the appropriate restriction enzymes. Primers OmodF 5′-CGC-GGATCC**CGCATA**CCAGCACGTTCCTTTGGACC-3′ and OmodR 5′-CGGTGTACA**CTGCGA**TGGTTCCGGCCT-AGTAGCTTC-3′ were used to amplify a fragment that was cloned into the *BamH*I and *Bsr*GI digested ORF-pGreen0179 construct. At two positions of the central fragment, 6 nucleotides were changed to distinguish the endogenous gene from the transgene. Constructs containing the central region linked to the At3g16240 promoter or the At3g16250 promoter, respectively, were transformed into *Arabidopsis thaliana* Col-0 and single-copy transgenic lines of each construct were identified by Southern blot analysis. For qPCR analysis of transgenic lines, primers qSF and qASF were used, which contained the 6 bp modification at their 3′ end.

3. Results

3.1. A At3g16240/At3g16250 Gene Pair with Antagonistic Changes in Transcript Levels. Gene At3g16240 encodes a vacuolar membrane protein, which functions as a water channel and NH_3 transporter, and partially convergently overlaps with a gene on the opposite strand, At3g16250, which encodes a ferredoxin-related novel subunit of the chloroplast NAD(P)H dehydrogenase complex (Figure 1(a)). A T-DNA insertion into the gene At3g16250 reduces At3g16250 transcript levels and increases transcript levels of the corresponding At3g16240 gene (Figure 1(b)), while T-DNA insertion into the gene At3g16240, which almost eliminates downstream transcript levels, also reduces transcript levels of the corresponding At3g16250gene (Figure 1(c)). The joint decrease of At3g16250 transcript levels and increase in At3g16240 transcript levels are indicative of an antisense-mediated control mechanisms that involves nat-siRNA synthesis under participation of RDR6. Our analysis of the expression of both transcripts in a *rdr6* mutant did, however, show no dependence of transcript levels on RDR6 (Figures 1(d) and 1(e)).

3.2. Antagonistic Circadian Activity of At3g16240 and At3g16250. The antagonistic expression profile of At3g16240

and At3g16250 is also apparent from their circadian activity. Both genes have a circadian expression pattern with a periodicity of 20–22 hours At3g16240 transcript maxima correlating with At3g16250 transcript minima, and vice versa (Figure 2(a)). This expression pattern might indicate antagonistic regulatory effects between these two transcripts. To assess if the circadian rhythm of the two genes was dependent on the expression profile of the partner gene, we compared the two transcript levels in wildtype and T-DNA insertion lines. As observed before, down-regulation of the At3g16250 transcript correlated with an increase in At3g16240 transcript levels. The circadian activity of the At3g16240 gene in the At3g16250 insertion line seems to be abolished, as no circadian cycling in transcript levels was detected after an initial peak in expression (Figure 2(b)). Reduction of At3g16240 transcript levels in the SM3.32080 mutant line correlated with a small reduction in At3g16250 transcript levels, and moreover, circadian rhythm of the At3g16250 transcript was unchanged (Figure 2(c)). While this analysis excludes an influence of the At3g16240 transcript on the circadian activity of the At3g16250 gene, the results are less conclusive withrespect to an influence of the At3g16250 transcript on circadian activity of the At3g16240 gene. Either the reduction of At3g16250 transcription or the insertion of the T-DNA in the At3g16250 gene could interfere with circadian expression of the At3g16240 gene in the SALK_048899 insertion line.

To differentiate between T-DNA effects and effects from antisense transcription, we focused on a transgenic approach that examined expression of the two genes in the presence and absence of the corresponding antisense transcript. We designed a sense transgene construct (S transgene) that contained the complete At3g16240 gene without the At3g16250 gene promoter, and an antisense transgene construct (AS transgene) that contained the At3g16250 gene without the At3g16240 gene promoter. A third construct contained both genes with both promoters (SAS transgene). Six point mutations were introduced into the transgene constructs to provide primer regions for a separate analysis of transgene and endogen transcripts. The analysis of transformants with single copies of the transgenes confirmed previous results from the T-DNA analysis. The sense transgenic line expressed increased sense transgene levels compared to the endogenous At3g16240 copy, and antisense transgene levels were reduced compared to the endogenous At3g16250 copy that was linked to the At3g16240 gene (Figure 3). Both transgene constructs maintained the circadian transcript profiles, indicating that interactions between At3g16240 and At3g16250 transcripts are not required for the circadian expression profile of the two genes and that the respective promoters are able to drive the circadian expression without the need of the presence of the corresponding antisense partner. Transcript profiles of the sense and antisense transgenes were also independent of transcription in opposite orientation as transgene expression levels did not differ among transgenes expressing one or both genes. This suggests that neither the antagonistic circadian activity of the two genes, nor changes in the transcript levels observed in the insertion lines, is caused by antisense regulation.

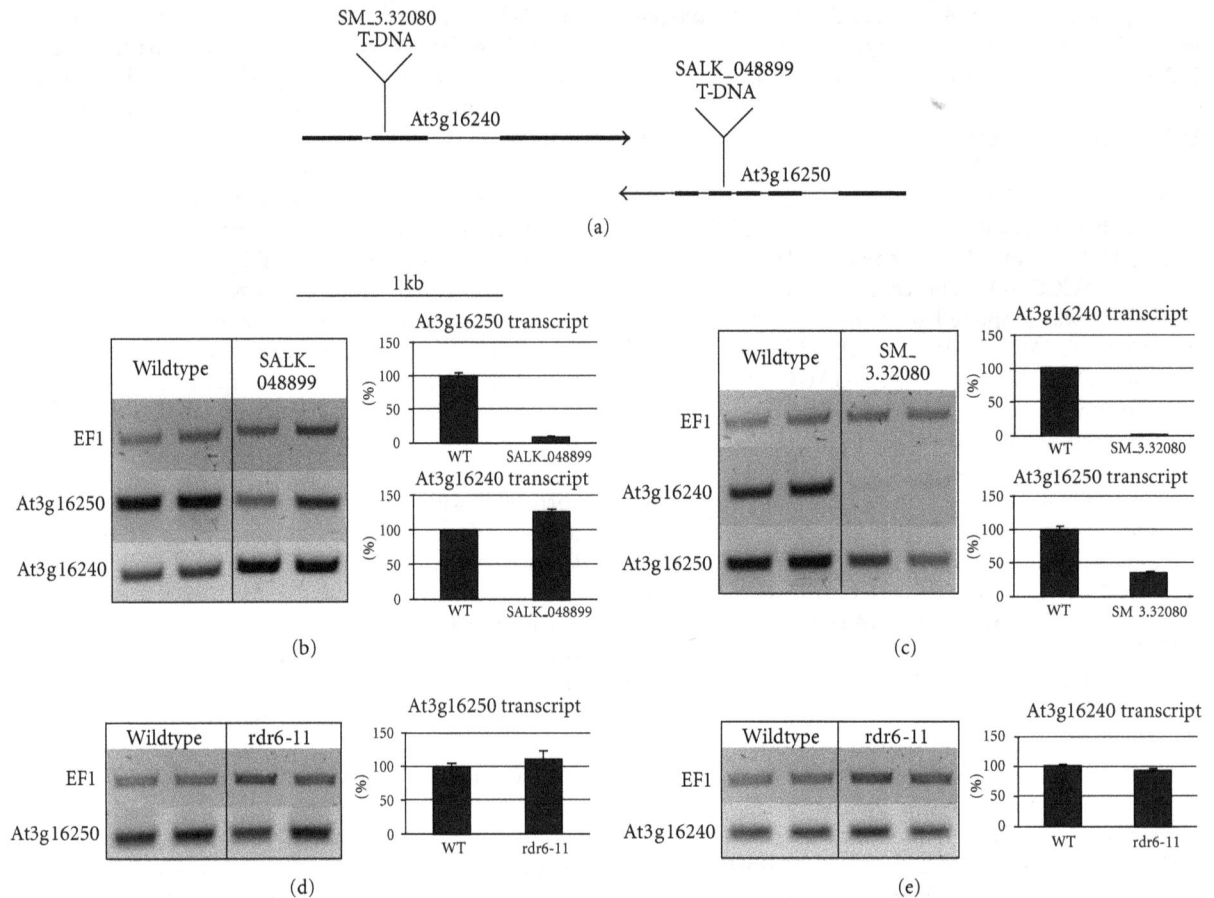

FIGURE 1: Map and transcript analysis of the locus At3g16240/At3g16250. (a) Schematic map of gene pair At3g16240/At3g16250. Thick lines indicate exons, thin lines intron regions. (b) Reduced At3g16250 transcript levels and increased At3g16240 transcript concentrations in the SALK_048899 insertion line. (c) Reduced At3g16240 and At3g16250 transcript levels in the SM3.32080 insertion line. ((d), (e)) At3g16240 and At3g16250 transcript levels are not altered in a *rdr6* mutant, compared to wildtype. The amount of cDNA of the semiquantitative PCR was calibrated to elongation factor 1 alpha (EF1a). Two replicas of semiquantitative RT-PCR data are shown (left). In the quantitative RT-PCR, wildtype levels were set to a reference level of 100% (right). The expression of both genes in wildtype and different mutants was examined at the same time point. The error bar represents standard error.

4. Discussion

The coexpression of a sense transcript and its natural antisense transcript can favor the formation of dsRNA substrates, which is degraded to nat-siRNAs that can silence sense transcripts in *cis* or transcripts from homologous loci in *trans*. Degradation of the P5CDH transcript after salt induction of its NAT SRO5 requires DCL1 and DCL2, two members of the RNase III family of nucleases that specifically cleave double-stranded RNAs, and depends on the RNA-dependent RNA polymerase RDR6, SGS3, and NRPD1A. In a two-stage process, a 24-nt siRNA is formed by a biogenesis pathway dependent on DCL2, RDR6, SGS3, and NRPD1A, which establishes a phase for the generation of 21-nt siRNAs by DCL1 and further cleavage of P5CDH transcripts [6]. Generation of a ~22 nt endogenous siRNA, nat-siRNA ATGB2, induced by *Pseudomonas syringae*, also requires RDR6, NRPD1A, and SGS3, but only one DICER enzyme, DCL1 [7]. A third example is the *Sho* gene from *Petunia hybrida*, which encodes an enzyme responsible for the synthesis of plant cytokinins. The distribution of two pools of *Sho*-specific small RNAs suggests that a partially overlapping antisense transcript can be activated in a tissue-specific response to adjust local cytokinin synthesis via degradation of *Sho* dsRNA [17].

A common feature of the two nat-siRNA producing loci in *Arabidopsis* is their sensitivity to changes in sense and antisense transcript levels, a feature that we also observe for the gene pair At3g16240/At3g16250. Reduction of At3g16250 transcript levels in a T-DNA mutant SALK_048899 correlated with an increase in At3g16240 transcript levels. The expression of the two genes was not altered in a *rdr6* mutant, which argued against an RDR6-mediated nat-siRNA pathway but it did not exclude an RDR6-independent RNA degradation mechanisms. An alternative mechanism for antagonistic effects between antisense and sense transcription would be transcriptional interference, which is based on the suppressive influence of a transcriptional process in *cis* on a second transcriptional process [18] as it was demonstrated for the *GAL10* and *GAL7* genes in *Saccharomyces cerevisiae*. When

(a)

(b)

(c)

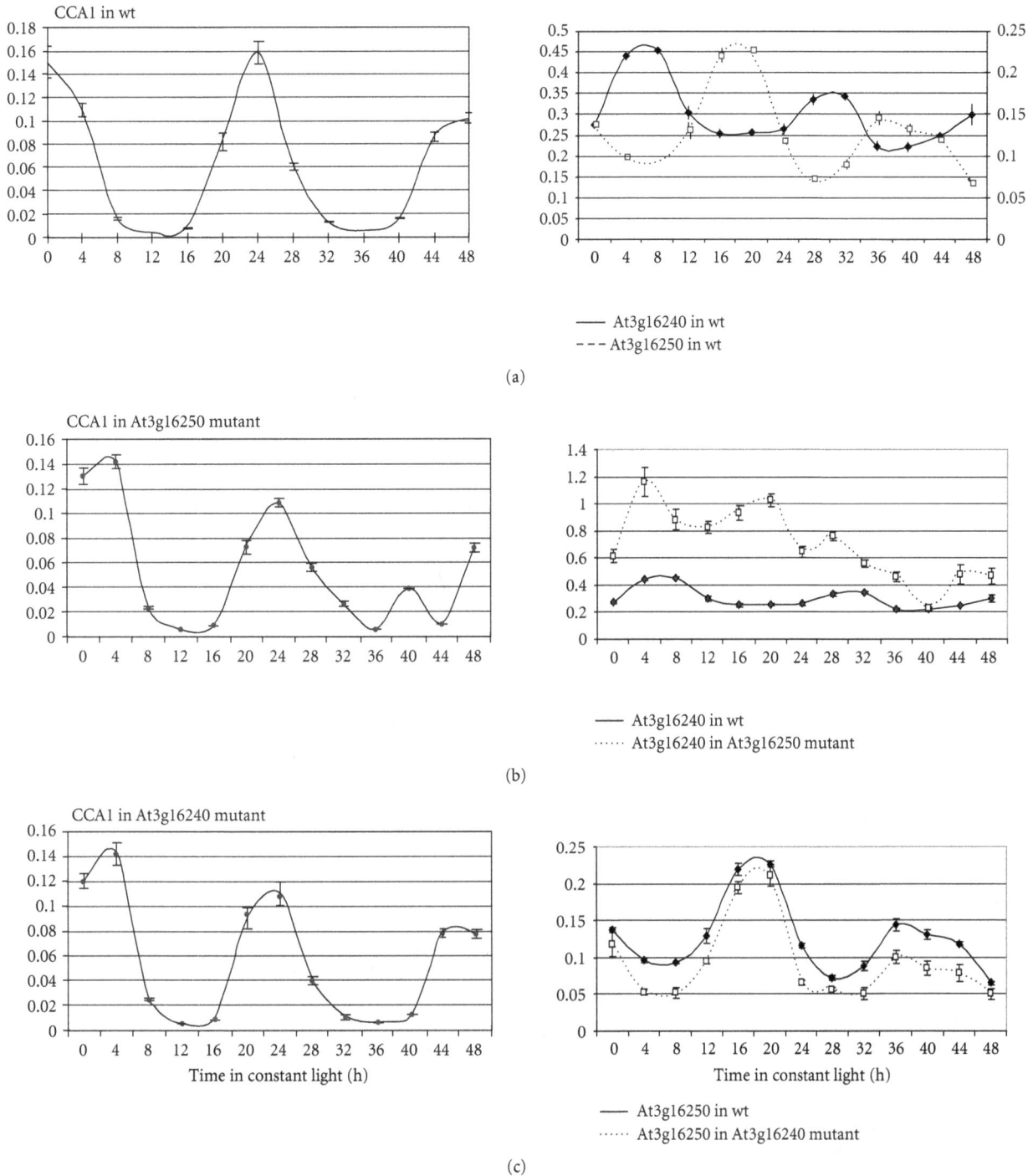

FIGURE 2: Circadian expression profile of At3g16240 and At3g16250 transcripts in wildtype and T-DNA insertion lines. *CCA1* transcript profiles were used as control of experimental conditions. The experiment was repeated two times in three replicas. The error bars represent standard error. (a) Antagonistic expression of At3g16240 and At3g16250 transcripts in wildtype. (b) Expression profile of the At3g16240 gene in wt and in the SALK_048899 insertion line. At3g16240 transcript levels are increased in the insertion line. (c) Expression profile of the At3g16250 gene in wt and in the SM3.32080 insertion line. At3g16250 transcript levels are slightly reduced in the insertion line.

both genes were arranged as convergently overlapping pairs, transcription initiation was unaffected but transcription elongation was severely inhibited. The effect was only observed in *cis*, and was proposed to reflect a collision of the two polymerase complexes [19].

The possibility of antisense-mediated gene regulation was further supported by the observation of an antagonistic circadian activity of At3g16240 and At3g16250 gene. This offered an attractive model for antisense-mediated regulation as NATs complementary to clock gene transcripts have been

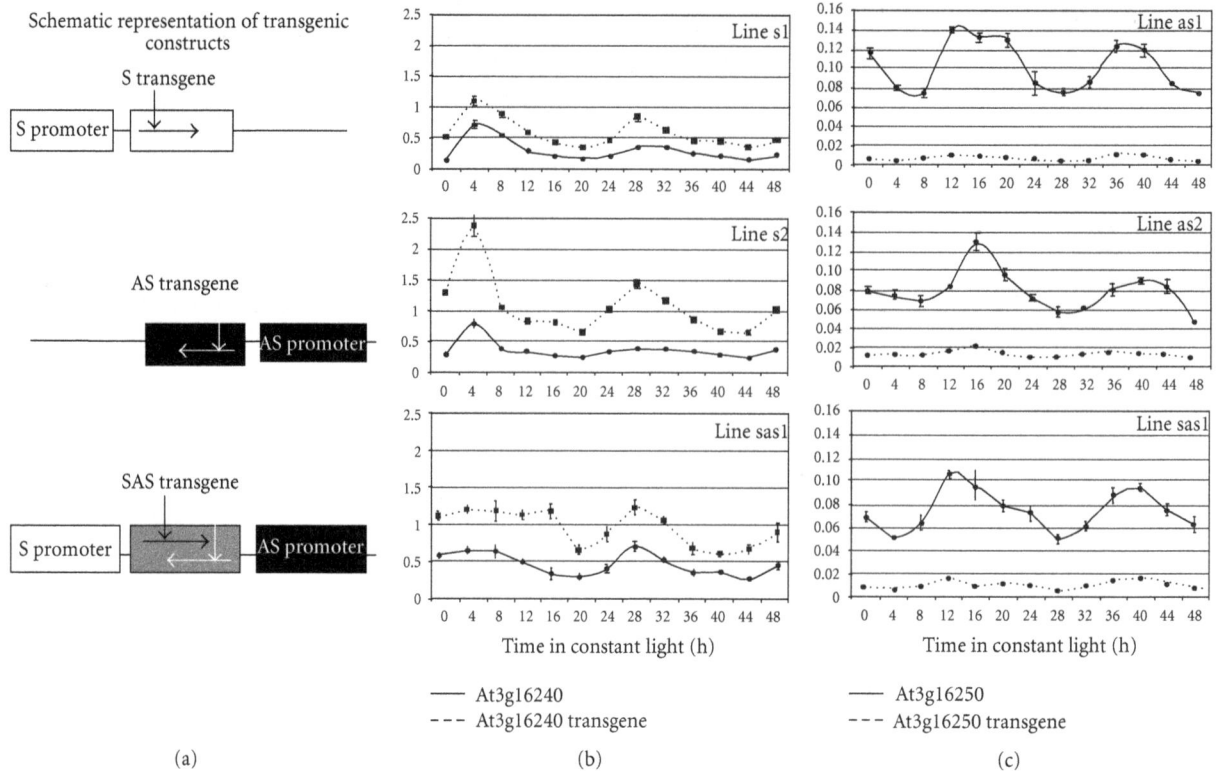

FIGURE 3: Transcript analysis of endogenous and transgenic transcripts of the At3g16240 and At3g16250 genes in transgenic lines. (a) Schematic representation of transgenic constructs used. Promoter regions and the coding region are shown. In the coding region, possible transcription of the sense or antisense transgene driven by the appropriate promoter is indicated by arrow. 6 bp substitutions in the coding region of the sense and antisense transgenes are indicated by vertical arrows. (b) In two sense transgenic lines, s1 and s2, transcript levels of the transgene are increased compared to the endogenous copy of the At3g16240 gene. The same increase occurs in a sas transgene that jointly expresses sense and antisense transgene. (c) In two antisense transgenic lines, as1 and as2, levels of the antisense transgene transcripts are decreased compared to the endogenous At3g16250 copy. The same decrease occurs in a sas transgene that jointly expresses sense and antisense transgene. The experiment was repeated twice with two replicas. The error bars represent the standard error.

reported in mammals, insects, and fungi [20]. A direct effect of antisense transcription on the circadian activity of the sense transcript has been demonstrated for the *frq* locus in *Neurospora crassa*. Antisense transcript levels of the *frq* transcript cycle in antiphase sense *frq* transcripts in the dark and are inducible by light. Inhibition of antisense transcription in mutant strains alters the time and resetting of the clock [21]. Our analysis of transgene construct does, however, not support a regulatory role of antisense transcription in cyclic expression of At3g16240 or At3g16250, as their circadian expression profile is not altered in plants that express At3g16240 or At3g16250 transgenes with or without antisense transcription.

Our results therefore suggest that joint antagonistic changes in At3g16240 and At3g16250 transcript levels must not necessarily be based on dsRNA degradation or transcriptional interference. The joint sense transcript increase and antisense transcript decrease is more likely the consequence of removing the two genes from its chromosomal environment. We speculate that transcriptional competence of the two convergently overlapping genes is influenced by the local

chromatin environment, which induces a more active conformation on the At3g16250 gene then on the At3g16240 gene. This influence may be mediated by chromatin domain or isochore structure [22], interaction with the nuclear scaffold [23], epigenetic marks [24], chromatin torsion [25], or other effects that influence transcriptional competence. Both T-DNA integration and translocation could disturb this local chromosomal impact on gene expression and cause the observed antagonistic changes that mimic RNAi effects.

Our results do not challenge the phenomenon of the existence and significance of antisense-mediated gene regulation but they highlight that antagonistic expression patterns must not necessarily be the consequence of antisense transcription. Partially overlapping sense-antisense gene pairs may have coevolved with mechanisms that prevent interference between their transcripts. One example is the evolution of alternative polyadenylation regions that shorten sense transcript length and prevent transcript overlap with antisense transcripts [26]. The presence of a partly overlapping gene in antisense orientation may favour the evolution of antagonistic expression patterns, which would avoid the generation of large amounts of double-stranded RNA as substrates for RNA degradation mechanisms.

Acknowledgment

T-DNA insertion and transcript data were obtained from TAIR (http://arabidopsis.org/).

References

[1] C. Vanhée-Brossollet and C. Vaquero, "Do natural antisense transcripts make sense in eukaryotes?" *Gene*, vol. 211, no. 1, pp. 1–9, 1998.

[2] J. S. Mattick, "Non-coding RNAs: the architects of eukaryotic complexity," *EMBO Reports*, vol. 2, no. 11, pp. 986–991, 2001.

[3] K. Yamada, J. Lim, J. H. Dale et al., "Empirical analysis of transcriptional activity in the Arabidopsis genome," *Science*, vol. 302, no. 5646, pp. 842–846, 2003.

[4] X. J. Wang, T. Gaasterland, and N. H. Chua, "Genome-wide prediction and identification of *cis*-natural antisense transcripts in *Arabidopsis thaliana*," *Genome biology*, vol. 6, no. 4, p. R30, 2005.

[5] C. H. Jen, I. Michalopoulos, D. R. Westhead, and P. Meyer, "Natural antisense transcripts with coding capacity in *Arabidopsis* may have a regulatory role that is not linked to double-stranded RNA degradation," *Genome biology*, vol. 6, no. 6, p. R51, 2005.

[6] O. Borsani, J. Zhu, P. E. Verslues, R. Sunkar, and J. K. Zhu, "Endogenous siRNAs derived from a pair of natural cis-antisense transcripts regulate salt tolerance in Arabidopsis," *Cell*, vol. 123, no. 7, pp. 1279–1291, 2005.

[7] S. Katiyar-Agarwal, R. Morgan, D. Dahlbeck et al., "A pathogen-inducible endogenous siRNA in plant immunity," *Proceedings of the National Academy of Sciences of the United States of America*, vol. 103, no. 47, pp. 18002–18007, 2006.

[8] J. M. Alonso, A. N. Stepanova, T. J. Leisse et al., "Genome-wide insertional mutagenesis of *Arabidopsis thaliana*," *Science*, vol. 301, no. 5633, pp. 653–657, 2003.

[9] A. F. Tissier, S. Marillonnet, V. Klimyuk et al., "Multiple independent defective Suppressor-mutator transposon insertions in Arabidopsis: a tool for functional genomics," *Plant Cell*, vol. 11, no. 10, pp. 1841–1852, 1999.

[10] R. L. Scholl, S. T. May, and D. H. Ware, "Seed and molecular resources for Arabidopsis," *Plant Physiology*, vol. 124, no. 4, pp. 1477–1480, 2000.

[11] K. D. Edwards, P. E. Anderson, A. Hall et al., "FLOWERING LOCUS C mediates natural variation in the high-temperature response of the *Arabidopsis* circadian clock," *Plant Cell*, vol. 18, no. 3, pp. 639–650, 2006.

[12] Z. Vejlupkova and J. E. Fowler, "Maize DNA preps for undergraduate students: a robust method for PCR genotyping," *Maize Genetics Cooperation Newsletter*, vol. 77, pp. 24–25, 2003.

[13] D. Swarbreck, C. Wilks, P. Lamesch et al., "The Arabidopsis Information Resource (TAIR): gene structure and function annotation," *Nucleic Acids Research*, vol. 36, supplement 1, pp. D1009–D1014, 2008.

[14] A. Peragine, M. Yoshikawa, G. Wu, H. L. Albrecht, and R. S. Poethig, "*SGS3* and *SGS2/SDE1/RDR6* are required for juvenile development and the production of *trans*-acting siRNAs in *Arabidopsis*," *Genes and Development*, vol. 18, no. 19, pp. 2368–2379, 2004.

[15] M. Stam, R. De Bruin, R. Van Blokland, R. A. L. Van Der Hoorn, J. N. M. Mol, and J. M. Kooter, "Distinct features of post-transcriptional gene silencing by antisense transgenes in single copy and inverted T-DNA repeat loci," *Plant Journal*, vol. 21, no. 1, pp. 27–42, 2000.

[16] R. P. Hellens, E. Anne Edwards, N. R. Leyland, S. Bean, and P. M. Mullineaux, "pGreen: a versatile and flexible binary Ti vector for Agrobacterium-mediated plant transformation," *Plant Molecular Biology*, vol. 42, no. 6, pp. 819–832, 2000.

[17] E. Zubko and P. Meyer, "A natural antisense transcript of the *Petunia hybrida* Sho gene suggests a role for an antisense mechanism in cytokinin regulation," *Plant Journal*, vol. 52, no. 6, pp. 1131–1139, 2007.

[18] K. E. Shearwin, B. P. Callen, and J. B. Egan, "Transcriptional interference—a crash course," *Trends in Genetics*, vol. 21, no. 6, pp. 339–345, 2005.

[19] E. M. Prescott and N. J. Proudfoot, "Transcriptional collision between convergent genes in budding yeast," *Proceedings of the National Academy of Sciences of the United States of America*, vol. 99, no. 13, pp. 8796–8801, 2002.

[20] S. K. Crosthwaite, "Circadian clocks and natural antisense RNA," *FEBS Letters*, vol. 567, no. 1, pp. 49–54, 2004.

[21] C. Kramer, J. J. Loros, J. C. Dunlap, and S. K. Crosthwaite, "Role for antisense RNA in regulating circadian clock function in Neurospora crassa," *Nature*, vol. 421, no. 6926, pp. 948–952, 2003.

[22] J. Dekker, K. Rippe, M. Dekker, and N. Kleckner, "Capturing chromosome conformation," *Science*, vol. 295, no. 5558, pp. 1306–1311, 2002.

[23] A. K. Linnemann, A. E. Platts, and S. A. Krawetz, "Differential nuclear scaffold/matrix attachment marks expressed genes," *Human Molecular Genetics*, vol. 18, no. 4, pp. 645–654, 2009.

[24] M. Shogren-Knaak, H. Ishii, J. M. Sun, M. J. Pazin, J. R. Davie, and C. L. Peterson, "Histone H4-K16 acetylation controls chromatin structure and protein interactions," *Science*, vol. 311, no. 5762, pp. 844–847, 2006.

[25] K. Havas, A. Flaus, M. Phelan et al., "Generation of superhelical torsion by ATP-dependent chromatin remodeling activities," *Cell*, vol. 103, no. 7, pp. 1133–1142, 2000.

[26] E. Zubko, A. Kunova, and P. Meyer, "Sense and antisense transcripts of convergent gene pairs in *Arabidopsis thaliana* can share a common polyadenylation region," *PLoS One*, vol. 6, no. 2, Article ID e16769, 2011.

3

Local Assemblies of Paired-End Reduced Representation Libraries Sequenced with the Illumina Genome Analyzer in Maize

Stéphane Deschamps,[1] Kishore Nannapaneni,[2] Yun Zhang,[2] and Kevin Hayes[2]

[1] DuPont Agricultural Biotechnology, P.O. Box 80353, Wilmington, DE 19880, USA
[2] DuPont Pioneer, P.O. Box 1004, Johnston, IA 50131, USA

Correspondence should be addressed to Stéphane Deschamps, stephane.deschamps@usa.dupont.com

Academic Editor: Yunbi Xu

The use of next-generation DNA sequencing technologies has greatly facilitated reference-guided variant detection in complex plant genomes. However, complications may arise when regions adjacent to a read of interest are used for marker assay development, or when reference sequences are incomplete, as short reads alone may not be long enough to ascertain their uniqueness. Here, the possibility of generating longer sequences in discrete regions of the large and complex genome of maize is demonstrated, using a modified version of a paired-end RAD library construction strategy. Reads are generated from DNA fragments first digested with a methylation-sensitive restriction endonuclease, sheared, enriched with biotin and a selective PCR amplification step, and then sequenced at both ends. Sequences are locally assembled into contigs by subgrouping pairs based on the identity of the read anchored by the restriction site. This strategy applied to two maize inbred lines (B14 and B73) generated 183,609 and 129,018 contigs, respectively, out of which at least 76% were >200 bps in length. A subset of putative single nucleotide polymorphisms from contigs aligning to the B73 reference genome with at least one mismatch was resequenced, and 90% of those in B14 were confirmed, indicating that this method is a potent approach for variant detection and marker development in species with complex genomes or lacking extensive reference sequences.

1. Introduction

DNA-based genetic markers are pivotal tools for applications as diverse as QTL mapping, marker assisted selection, association mapping, and fine mapping for the detection of genes linked to a particular phenotype [1]. Among the variety of genetic markers that have been developed, those derived from single nucleotide polymorphisms (SNPs) have become the marker of choice for many mapping applications because of their abundance and the availability of high-throughput and cost-effective technologies for detection and diagnostics [2–4]. One popular tool for SNP identification and detection has been the construction of reduced-representation libraries (RRL) and their sequencing with massively parallel sequencing platforms, in species as varied as cattle, worm, soybean, rice, maize, or common bean [5–10]. However, one major limitation of such platforms is the relatively short length of individual sequencing reads. While the availability of a high quality reference sequence may render short reads sufficient for alignment and subsequent SNP detection, this limitation may be further compounded in crop species due to (1) the inherent complexity of genomes (and transcriptomes) in economically important species, such as maize, soybean, or canola, due to an elevation in ploidy and/or the frequent expansion of paralogous sequences and gene families, and the need to generate very long sequencing reads for resolving highly duplicated sequences within a single genome, and (2) the potentially large number of polymorphisms between lines in those same plant species (including indels and in regions flanking a polymorphism of interest) and the need to provide long line-specific sequences for identifying variants, open reading frames, or other biologically active regions for lines whose genome sequence has a significantly altered composition in comparison to the reference assembly. Because of those limitations, the Roche 454 FLX platform [11] is often used as the instrument of choice for providing long sequencing reads and generating an appropriate sequencing scaffold. However, its relatively lower sequencing throughput when compared to other second-generation sequencing instruments often means that a large number of runs are needed to generate a high quality sequencing assembly from very large genomes. More recently, the PacBio RS platform from Pacific Biosciences [12] has garnered a lot of attention due to its ability to generate very long sequencing reads, in the kilobase range, but its low raw accuracy rate makes it, for now, unpractical to use as a variant detection platform.

In a previous study [8], we developed a methodology for rapid SNP detection in rice and soybean that can be applied to a wide range of moderately or highly complex plant genomes where sufficient genomic reference sequences are available. The methodology, based on digesting the genome with a methylation-sensitive restriction endonuclease (RE) followed by a secondary digestion with the 4-bp restriction endonuclease $DpnII$, generates short DNA fragments that are sequenced at one end using the Illumina Genome Analyzer II. The resulting 32 bp sequencing reads are immediately adjacent to the $DpnII$ site. Thus, they can be directly compared and aligned to a reference assembly for SNP detection. However, one problem inherent to this approach is the lack of sequencing information flanking the 32 bp read of interest and the need to rely on reference genomic sequences from a different line or species for primer design and SNP assay development. The relative short length of the sequence also can lead to relatively high false positive discovery rates in species lacking robust reference sequences.

Several studies already have shown that DNA fragments sequenced at both ends on a massively parallel sequencing platforms can be locally assembled into contigs hundreds or even thousands of bases long. Hiatt et al. [13] developed a methodology to generate long consensus sequences in $Pseudomonas\ aeruginosa$ from short second-generation sequencing reads. In there, ~500 bps DNA fragments are sheared randomly to produce fragments shorter than the original fragment set. Those shorter fragments then are end-sequenced at both ends with the Illumina Genome Analyzer II and the resulting read pairs are grouped together and assembled to recreate the sequence of the original ~500 bp DNA fragment. More recently, Etter et al. [14] developed a paired-end restriction-site associated DNA (RAD-PE) method, where ~350–850 bps RAD fragments from three-spined sticklebacks ($Gasterosteus\ aculeatus$), with one end anchored by a restriction endonuclease site and the other end randomly sheared, were sequenced on the Illumina Genome Analyzer II, to locally assemble staggered randomly sheared ends associated to a particular restriction site into restriction site-specific contigs several hundred bases in length. The RAD-PE protocol requires several steps, including the digestion of genomic DNA, ligation of two independent adapters, and the selective PCR amplification of RAD fragments flanked only by the two distinct adapters. A variation in protocol, namely, partial digestion of the genomic DNA rather than full digestion, creates overlapping RAD fragments, which in turn can be assembled into overlapping contigs covering sequences several thousand bases in length. A similar concept using RAD fragments was tested by Willing et al. [15] on two guppy ($Poecilia\ reticulata$) populations, generating consensus ~200–400 bps sequences

that were used for polymorphism discovery, checking for the presence or absence of restriction-site associated sequences in the two populations to identify them as polymorphic (as the absence of a sequence suggests the presence of a polymorphism in the restriction site).

In this study, a revised version of the RAD-PE method was tested and implemented in maize, in order to assess the feasibility of using such a strategy for SNP detection on the large and complex genome of an economically important crop. The revised RAD-PE protocol shown here extends the concept presented in Deschamps et al. [8], where SNP detection in complex crop genomes had been successfully attempted. Briefly, the digestion of DNA fragments with $DpnII$ is replaced by random shearing, and the sheared DNA fragments containing the methylation-sensitive RE site at one end are recovered, size-selected, enriched, and sequenced at both ends on the Illumina Genome Analyzer IIx. Two steps are used to select for DNA fragments prior to paired-end sequencing, namely, a biotin selection step performed after the initial digestion with a methylation-sensitive RE, and a second selection step when DNA fragments are selectively amplified via PCR prior to sequencing. The resulting individual read pairs then are assembled $de\ novo$ to create larger contigs that can be used both for SNP detection and SNP assay design for marker development. These contigs can be several hundreds bps long and their expected length can be modulated simply by selecting appropriate DNA fragment size ranges on gel. In addition to SNP discovery and assay design, they can be used in multiple species for applications whose purpose is facilitated by long sequencing reads, such as copy number variant detection, open reading frame discovery, or metagenomics.

2. Materials and Methods

2.1. Tissue Preparation and Genomic DNA Isolation. Maize genomic DNA samples were extracted according to the protocol described in Deschamps et al. [8].

2.2. Genomic DNA Preparation and Library Construction. B73 and B14 genomic DNA (10 μg) was digested for 4 h at 37°C with 15 U of $PstI$ (Promega) in a total volume of 20 μL of 1X $PstI$ buffer (Promega) containing 1X acetylated BSA (Promega). The enzyme was inactivated at 65°C for 20 min then the digested genomic DNA was purified with QIAquick PCR purification spin columns (Qiagen). Biotinylated $PstI$-specific adapters were created by mixing 1500 pmol each of two synthetic oligonucleotides (upper strand, /5′Bio/GGTTGACATGCTGGATTGAGACCTGCAGGTGC *A, where * is a phosphorothioate bond; lower strand, /5′PO4-/CCTGCAGGTCTCAATCCAGCATGTC) in 100 μL water, heating them up at 95°C for 2 min, and then allowing them to cool slowly to room temperature. A ~50-fold excess of $PstI$-specific adapters (37.5 pmol) relative to available $PstI$ ends was ligated to the digested DNA, in a total volume of 20 μL of 1X ligase buffer (New England Biolabs) containing 10,000 U of T_4 DNA ligase (New England Biolabs). The reaction was incubated overnight at 16°C then 10 min at 70°C, and then purified with QIAquick PCR purification spin columns (Qiagen). The digested DNA was randomly sheared via nebulization at 32 psi for 6 min then purified with QIAquick PCR purification spin columns (Qiagen).

After random shearing, a biotin selection step allowed the capture of smaller DNA fragments containing a *PstI* site at one end. For biotin selection, 100 μL of Streptavidin-Dynabeads M-280 (Invitrogen) were washed twice with 1 mL TE buffer then resuspended in 100 μL 2X B&W buffer (10 mmol L^{-1} Tris-HCl, pH 7.5; 1 mmol L^{-1} EDTA; 2 mol L^{-1} NaCl). The DNA fragments then were added to the beads and incubated for 30 min at 30°C with gentle horizontal mixing. After withdrawing the supernatant, the beads were collected and washed three times with 1 mL 1X B&W buffer then three times with 1 mL TE buffer. The washed beads were resuspended in 200 μL of 1X NEBuffer 4 (New England Biolabs) containing 20 U *SbfI* (New England Biolabs) and incubated for 90 min at 37°C with gentle horizontal mixing. An *SbfI* recognition site (5'-CCTGCA^GG-3') located immediately upstream from the *PstI* site on the biotinylated adapter allowed for the cleavage and recovery of the biotin-bound DNA fragments. The supernatant containing the released DNA fragments was then transferred to a new tube, extracted with 100 μL phenol/chloroform/isoamyl alcohol (25 : 24 : 1) and precipitated with ethanol and 3 mol L^{-1} NaOAc. DNA was resuspended in 30 μL EB buffer (Qiagen).

End-repair, "A" base addition and ligation of adapters were performed according to the protocol developed by Illumina for preparing DNA samples for paired-end sequencing. Buffer and enzymatic reagents were obtained directly from Illumina's PE Sample Prep Kit (catalog number PE-102-1002). Briefly, DNA was end repaired for 30 min at 20°C in a total volume of 100 μL of 1X T$_4$ DNA ligase buffer with 10 mM ATP containing 4 μL 10 mM dNTP mix, 5 μL T$_4$ DNA polymerase, 1 μL Klenow enzyme, and T$_4$ PNK. After incubation, DNA was purified with a QIAquick PCR Purification Kit (Qiagen) and resuspended in 32 μL EB buffer (Qiagen). "A" bases then were added to the 3' ends of blunt-ended DNA fragments. DNA was incubated at 37°C for 30 min in 1X Klenow buffer containing 10 μL 1 mM dATP and 3 μL Klenow exo (3' to 5' exo minus). After incubation, DNA was purified with a QIAquick MinElute Purification Kit (Qiagen) and resuspended in 10 μL EB buffer (Qiagen). Illumina PE adapters then were ligated to DNA fragments at 20°C for 15 min in a total volume of 50 μL of 1X DNA ligase buffer containing 1 μL PE adapter oligo mix and 5 μL DNA ligase. After incubation, DNA was purified with a QIAquick PCR Purification kit (Qiagen) and resuspended in 30 μL EB buffer (Qiagen). Ligated DNA then were loaded on an E-Gel 2% with SYBR Safe agarose gel (Invitrogen) in the presence of low molecular weight DNA ladder (New England Biolabs) for size selection. After running the gel for 26 min in an E-Gel PowerBase v4 electrophoresis unit (Invitrogen), a 200–500 bp DNA smear was cut out and purified with the QIAquick Gel Extraction kit (Qiagen). DNA was resuspended in 30 μL EB buffer (Qiagen).

After ligation of Illumina adapters and gel-based size selection of the 200–500 bps ligated DNA fragments, a PCR amplification step was performed using the Illumina Paired-End (PE) PCR primer 1.0 and a modified PE PCR primer 2.0 ending in 5'-TGCAGGTGCA-3' (matching adjacent *SbfI* and *PstI* recognition sites). The use of Illumina PE adapters and of a modified PE PCR primer 2.0 further allowed the selective amplification of DNA fragments containing a *PstI* recognition site at one end. For PCR amplification, 1 μL of size-selected DNA was incubated in 50 μL of 1X Phusion HF master mix (Finnzymes) in the presence of PCR primer PE 1.0 (Illumina) and the modified PCR primer PE 2.0 (CAAGCAGAAGACGGCATAC-GAGATCGGTCTCGGCATTCCTGCTGAACCGCTCTTCC-GATCTTGCAGGTGC*A, where * is a phosphorothioate bond and "TGCAGGTGC*A" is a signature 3' sequence containing leftovers from *SbfI* & *PstI* restriction sites). After PCR amplification (30 s at 98°C, followed by 20 rounds of 10 s at 98°C, 30 s at 65°C, 30 s at 72°C, and a final extension step of 5 min at 72°C), DNA was purified with a QIAquick PCR Purification Kit (Qiagen) then subjected to a second round of gel-based size selection using conditions similar to the ones described above. The 200–500 bps amplified DNA fragments were size-selected on gel and the presence of Illumina adapters at both ends suggests that the actual sizes of the fragments that were sequenced were between ~70 bps and ~370 bps.

2.3. Cluster Generation and Paired-End Sequencing. Cluster generation and paired-end sequencing were performed on an Illumina Cluster Station and Genome Analyzer IIx, respectively, according to protocols and recipes developed by Illumina. Sequencing was performed for 76 cycles at both ends of the clustered DNA fragments using PE sequencing primers for Read 1 and Read 2 (Illumina), generating read pairs, "reads 1" and "reads 2", mapping the ends of PCR products generated with PE PCR primer 1.0 and the modified PE PCR primer 2.0, respectively. The resulting read 1 and read 2 sequences were grouped into "read pairs" according to the X and Y coordinates of the corresponding DNA cluster on the flow cell. Sequencing reads and quality scores were generated in a real-time fashion with the Illumina Data Collection Software v2.6. After initial base calling, additional custom filtering was performed using calibrated quality scores generated by the Illumina pipeline. Reads generated from both ends of DNA fragments (reads 1 and reads 2) were trimmed by removing from the 3' ends bases with a PHRED-equivalent quality score below 10. A length threshold of 24 was applied to filtering, indicating that all bases <24 bases in length after trimming were removed from further analysis.

The use of a modified PE PCR primer 2.0 led to a vast majority of read 2 starting with the signature sequence 5'-TGCAGGTGCA-3'. However, in spite of a significant bias in base composition for the first 10 bases of reads 2, a majority of all read 2 data were high quality reads that paired well with their read 1 counterparts. Nonetheless, it must be noted that since Illumina uses images of the first few cycles to calculate phasing, such low 5' sequence variation could cause base calling errors in subsequent sequencing cycles if a different Illumina instrument or upgraded base calling software were to be used for sequencing. In that particular case, increasing the amount of phiX control DNA spiked into each lane (from ~1% to ~25%) would be recommended. The replacement of *SbfI* (and of its corresponding recognition sequences in the biotinylated *PstI*-specific adapter) by a Type IIs enzyme (such as *BseRI*), thus removing the 10 bp "TGCAGGTGCA" signature sequence from the 5′ ends of reads 2, also is being explored on separate sets of genomic DNA samples.

2.4. Alignment to Reference Genome Assembly and De Novo Assembly. The 10 bp "TGCAGGTGCA" signature sequence at the 5′ end of read 2 sequences was identified by calculating the Hamming distance with the first 10 bases on read 2. Only read 2 sequences that have a Hamming distance less than or equal to 2 were considered to have a *Pst1* site. Alignment of the reads and contigs to B73 were performed using Bowtie [16]. Local assemblies for each distinct read 2 sequences and their corresponding read 1 sequences were generated using the Velvet software [17] with k-mer values ranging from 23 to 63 with increments of 4 and coverage cutoff values of 4 and 8 for each k-mer value. Of the 22 assemblies generated with varying k-mer and coverage cutoff values, the assembly with the largest contig size was considered the final one for that distinct read 2 sequence and its paired read 1 sequences.

2.5. Sanger-Based Validation of Single-Nucleotide Polymorphisms. Sequences from 100 random B14 and B73 contigs with at least 1 polymorphism to the B73 reference genome v1.0 assembly were used to design PCR primers (length 18–24 bps, Tm 60–65°C) using a local version of the Primer3 primer design software tool. CROSSMATCH analysis was performed on contig sequences using a local maize repeat database to mask repetitive DNA sequences prior to primer design and M13 forward and reverse "tails" were added to the 5′ ends of the PCR primer sequences before ordering. B73 and B14 genomic DNA was subjected to PCR amplification (15 min at 95°C, followed by 40 rounds of 30 s at 95°C, 30 s at 60°C, 1 min at 72°C, and a final extension step of 10 min at 72°C) by using 5 pmol each of the "tailed" PCR primers in a total volume of 10 μL of 1X HotStarTaq Master Mix (Qiagen) containing 25 U HotStarTaq DNA polymerase (Qiagen). 1.5 mmol L^{-1} MgCl$_2$ and 200 μmol L^{-1} dNTPs. PCR cleanup reactions were performed by mixing 2 μL of PCR products with 0.75 μL of ExoSAP-IT (USB Corporation) in a total volume of 17 μL with sterile distilled water, and incubating at 37°C for 25 min then 80°C for 25 min. 5 μL of the cleaned-up amplified DNA then were end-sequenced using M13 forward and reverse oligonucleotides and the ABI BigDye version 3.1 Prism sequencing kit. After ethanol-based cleanup, cycle sequencing reaction products were resolved and detected on Life Technologies (Carlsbad, CA, USA) ABI3730xl automated capillary DNA sequences. Individual sequences from each genotype were combined into a single project (one project per amplified fragment) and assembled with the Phred/Phrap/Consed package (see http://www.phrap.org/phredphrapconsed.html). Confirmed SNPs identified on the Genome Analyzer were validated by comparison to single-base mismatches between the B73 and B14 genotypes located in regions that matched the original

contig sequence generated by assembling individual read pair sequences with Velvet.

3. Results and Discussion

Library construction and massively parallel sequencing were performed on two public maize inbred lines, B73 and B14, according to the method described in Figure 1. DNA samples first were digested with the methylation-sensitive RE *PstI* (5′-C*TGCA^G-3′). The *PstI* activity is blocked by 5-C methylation of the cytosine in the first position (C*) in the recognition sequence. The choice of *PstI* was guided by (1) the intention of enriching for genic regions and avoiding the capture of the repeated fraction of the genome [18, 19], and (2) the potential number of unique sites digested by this enzyme in both samples.

3.1. Illumina Sequencing. After trimming and filtering, a total of 63.9 million and 94.9 million high-quality read pairs (reads 1 and reads 2) were obtained for B73 and B14, respectively, from one run on the Illumina Genome Analyzer II (Table 1). The disparity between those numbers is due in part to the fact that 3 lanes and 4 lanes of a flow cell were used for B73 and B14, respectively. A total of 61.5 million (96.2% of total) and 92.2 million (97.1% of total) reads 2 for B73 and B14, respectively, contained the signature 5′-TGCAGGTGCA-3′ sequence at their 5′ ends.

Since reads 2 are anchored by a methylation-sensitive RE cut site, each read pairs can be grouped based on the identity of their respective read 2 sequences. In order to assign read pairs to specific regions of the genome, and perform local assemblies of such regions, the following analysis was performed: (1) assessing the number of read 2 sequences aligning uniquely to the B73 reference genome assembly, (2) determining the degree of stacking (and the number of distinct reads 2) for all uniquely aligned paired reads 2, and (3) assembling each stacks of distinct read 2 regions covered by more than 100 reads 2 with their corresponding read 1 sequences to generate contigs in targeted regions of the genome.

3.2. Alignment to the Reference Genome Assembly. Only read pairs containing the 10 bp signature 5′-TGCAGGTGCA-3′ sequence at the 5′ end of reads 2 were considered for alignment. After filtering, the 10 bp signature sequence was removed and paired reads 2 were aligned to the B73 reference genome v2.0 assembly using Bowtie, allowing up to 2 mismatches per individual reads. 31.1 million B73 paired reads 2 (50.5% of all paired reads 2 containing the 10 bp signature sequence) and 57.0 million B14 paired reads 2 (61.8%) were aligned uniquely to the B73 reference genome (Table 1). In addition to uniquely aligned reads 2, 13.3 million paired reads 2 in B73 and 22.8 million paired reads 2 in B14 were aligned to multiple regions of the genome, indicating that 72.2% and 86.5% of all paired reads 2 for B73 and B14, respectively, containing the 10 bp signature sequence, align at least once to the B73 reference genome. Finally, a relatively significant fraction of B73 (17 million) and B14 (12.3 million) reads 2 did not align to

TABLE 1: Run metrics. The numbers of paired reads 2, paired reads 2 containing the 10 bp "TGCAGGTGCA" signature sequence at their 5′ ends, and paired reads 2 aligning to the public B73RefGen_v2 reference genome sequence are indicated.

Run metrics	B73	B14
Number of paired reads 2	63,964,770	94,976,365
Number of paired reads 2 with signature sequence	61,512,151	92,262,878
Alignment against the B73 reference genome sequence[a]		
Align once	31,121,355	57,009,568
Align more than once	13,306,206	22,869,021
Do not align	17,084,590	12,384,289

[a] Best match to reference sequence of reads aligning uniquely or multiple times to the reference sequence with no more than 2 mismatches.

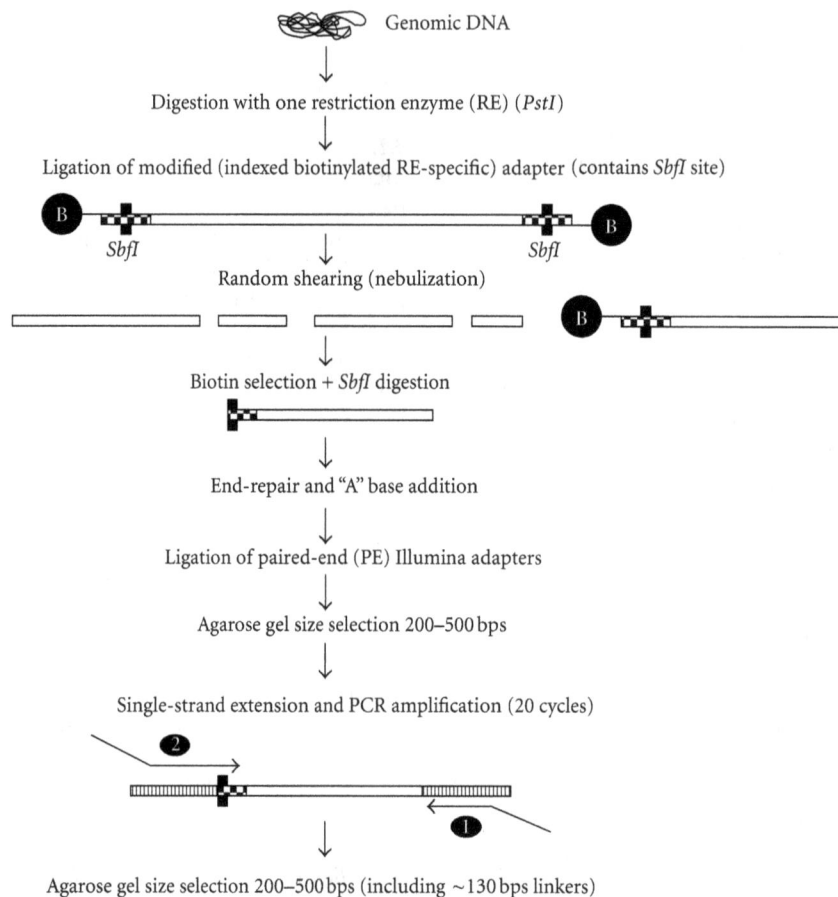

FIGURE 1: Preparation of paired-end reduced representation libraries. Genomic DNA is digested with methyl-sensitive restriction endonuclease PstI. After random shearing, DNA fragments containing the PstI end are selected via biotin selection and end-sequenced. Resulting sequences are assembled locally to create large contig sequences.

the B73 reference genome (Table 1). It is possible that those unaligned reads were excluded from the alignment due to the presence of more than 2 mismatches per individual reads that were caused by natural polymorphisms between lines, sequencing errors, or assembly errors in the B73 reference genome v2.0 assembly. Additional BLAST search of a subset of the remaining unaligned B73 and B14 read pairs indicates that at least a fraction of the unaligned sequences contain adapter sequences, suggesting the possibility of DNA fragments ligating to adapters or adapter-adapter ligations during the library construction process.

3.3. Genomic Distribution of Distinct Reads 2. Sequencing coverage for all uniquely aligned regions of the B73 genome reference v2.0 assembly was assessed by determining distinct read 2 coverage information (Figure 2). Distinct reads 2 (or "regions") were generated from all uniquely aligned reads 2, gathering all similar and uniquely aligned read 2 sequences into a unique sequence entry. In both lines, 312,924 and 401,379 regions for B73 and B14, respectively, are covered by at least 11 uniquely aligned reads 2 and a majority of uniquely aligned reads 2 (76.7% for B73 and 88.2% for B14) are contained within regions covered by

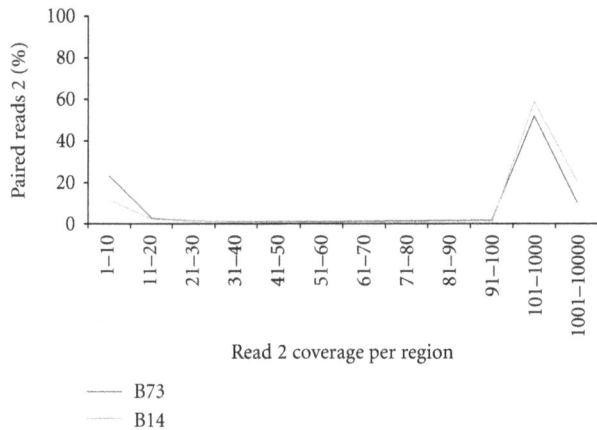

FIGURE 2: Read 2 coverage of regions by percentages of reads 2 sequences. The percentage of high quality paired reads 2 uniquely aligned to the B73 reference genome is shown in relation to their presence in regions with variable coverage. Y-axis: percentage of all high quality reads 2 uniquely aligned to B73 reference genome; X-axis: variations in sequencing coverage for regions covered by high quality paired read 2 (e.g., 1–10 = sequencing coverage varying from one read 2 to ten reads 2 for each covered region).

TABLE 2: Position overlap between B73 and B14. The numbers of distinct B73 and B14 reads overlapping at the same genomic position (as determined by the B73_RefGen v2.0 reference genome) are shown, including redundant and nonredundant positions.

	B73	B14
Redundant positions		
Not overlapping	657,961	421,056
Overlapping	2,367,323	
Nonredundant positions		
Not overlapping	731,837	407,130
Overlapping	1,616,620	

11 or more reads (and by extension read pairs) (Figure 2), potentially enabling high quality genome-wide *de novo* assembly. The relatively large percentage of distinct reads 2 covered by one to ten reads (1–10 in Figure 2) suggests the presence of sequencing singlets with unchecked sequencing errors and/or the possibility of extraneous contamination during the library construction procedure. Nonetheless, that number is low enough not to decrease the overall quality and usefulness of the analysis. The disparity in number of covered regions between the two lines can be explained at least in part by the higher number of read pairs sequenced for B14 in comparison to B73 and suggest that additional sequencing could further increase the coverage at each individual region adjacent to *Pst*I sites in their respective genomes. The number of regions covered by 11 or more reads 2 in B73 in comparison to B14 also suggests that additional sequencing mostly would increase the number of regions covered by 11 reads 2 or more, rather than generating contigs aligning to still unknown regions of interest, although that number also may vary depending on genome content and organization.

3.4. Distinct Read 2 Overlap in B73 and B14. Genomic position overlap between the B73 and B14 distinct read 2 datasets was determined by aligning all distinct reads 2 against the B73 reference genome, using Bowtie and allowing up to two mismatches. The resulting genomic coordinates then were compared between B73 and B14 to assess genomic position overlap. Redundant and nonredundant genomic positions are listed on Table 2 (where distinct reads 2 aligning to more than one region of the genome are considered as redundant). Interestingly, 68.8% and 79.8% of distinct reads 2 in B73 and B14, respectively, and sequenced at least once in one or both genotypes (i.e., sequencing coverage of 1X, or one read 2 sequence) also are sequenced in the other

genotype (Table 2). After including reads aligning to multiple regions (up to two mismatches were allowed when aligning distinct reads using Bowtie), 78.2% and 84.9% of distinct reads 2 in B73 and B14, respectively, overlap at the same genomic position. These data suggests that read 2 sequence data, or contigs resulting from assembling paired read 1 and read 2 data, can be used for direct comparison between genotypes for SNP detection and genotyping, in a manner similar to the approach described in Deschamps et al. [8].

3.5. De Novo Assembly of Individual Read Pairs. Individual read pairs were assembled *de novo* into contigs with the Velvet short read de novo assembler. Only distinct reads 2 generated from uniquely aligned reads 2 and with a sequencing coverage greater than 100 were considered and grouped with their corresponding read 1 sequences for *de novo* assembly. Each group was assembled individually. A total of 183,609 and 129,018 contigs were generated for B14 and B73, respectively. Read usages in both lines as well as the possibility of multiple contigs generated within the same regions likely explain why a higher number of contigs were generated in comparison to the number of regions with a sequencing coverage greater than 100. As suggested by Figure 3, 77% of the B73 contigs (99,431 contigs) and 76.2% of the B14 contigs (139,968 contigs) are >200 bps in length, including several contigs (176 for B73 and 310 for B14) above 500 bps in length, likely artifacts from the gel size-selection process but an indication nonetheless of the potential of the method if larger DNA fragments were to be size-selected on gel.

Contigs were aligned with Bowtie to the B73 reference genome v2.0 assembly, allowing up to two mismatches. Out of those, 30,763 for B73 and 62,524 for B14 were uniquely mapped to the assembly (Table 3). The lower number of uniquely aligned contigs compared to the total number of contigs shown above may be explained by the fact that most contigs did not extend to the stacked and unique read 2 sequences. As shown on Table 3, Bowtie alignments indicate that a significant fraction of these contigs align perfectly to the assembly (0 mismatch). As expected, a larger fraction of B73 uniquely mapped contigs (72.9%) align perfectly to the assembly, when compared to B14, where only 67.6% align with 0 mismatches. This is due in part to the different genome organization between the two inbred lines and the likely presence of various polymorphisms, including indels

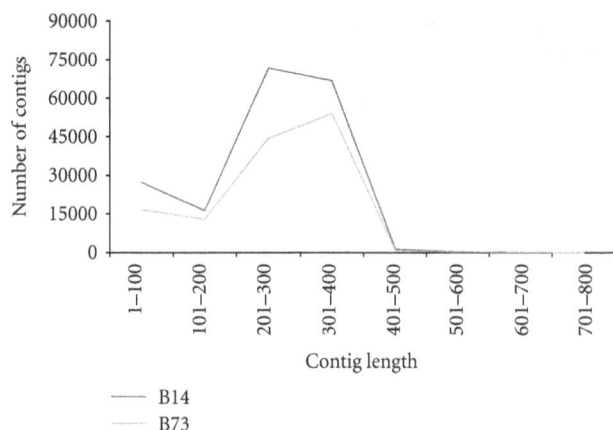

FIGURE 3: Contig length distribution. The number of contigs generated *de novo* is shown in relation to their length in bps. *Y*-axis: number of contigs generated by assembling paired read 1 and read 2 data extracted from regions with at least 100 stacked read 2 sequences uniquely aligned to the B73 reference genome; *X*-axis: contig length distribution (in bps) (e.g., 1–100 = contigs <100 bps in length).

in the B14 contigs when compared to the B73 reference assembly. Interestingly, 13.4% of the B73 contigs align to the B73 v2.0 assembly with 1 mismatch and 13.6% align with 2 mismatches, which could be an indication of sequencing or assembly errors in our assembled data or assembly errors in the public data.

Due to the palindromic nature of *PstI* recognition sites, one valuable side effect of this method is the possibility of generating contigs immediately adjacent to the 5′ end and the 3′ end of the same *PstI* site. The expected outcome would be that such adjacent contigs are in most cases separated by a variable number of bases, as determined by the reference sequence they are aligned against, assuming that contigs are extended from ends opposite the *PstI* ends in DNA fragments assembled together, creating "contig pairs" spanning larger regions of the genome useful for marker development and *de novo* assembly applications. Alignments of contig sequences to the B73 reference genome sequence suggest that 6,495 of the >100 bps uniquely mapped B73 contigs and 28,605 of the >100 bps uniquely mapped B14 contigs generate contig pairs separated by a distance less than 1,000 bps, and centered on a given *PstI* site in their respective genome. In addition, out of those, 680 B73 contigs and 3356 B14 contigs are immediately adjacent contigs centered on a unique *PstI* site. It is expected that deeper sequencing would increase the number of contig pairs per experiment by increasing sequencing coverage and contig length for each discrete genomic regions.

3.6. Sequencing Confirmation of Single Nucleotide Polymorphisms. One application of generating longer contigs from individual read pairs, anchored at a given *PstI* site is the possibility of detecting new SNP in complex genomes and using long contig sequences for primer design and SNP assay development. To determine the feasibility of using contig assemblies for SNP detection, a subset of B73 and B14 contigs

TABLE 3: Contig alignment to the B73 reference genome. The number of contigs uniquely aligned to the B73 reference genome assembly and exhibiting 0, 1, or 2 mismatches in relation to the reference are shown.

Number of contigs	B73	B14
0 mismatch	22,436	42,279
1 mismatch	4,142	11,875
2 mismatches	4,185	8,370
Total	30,763	62,524

with one or more mismatch to the B73 reference genome was selected randomly. Mismatches detected by Bowtie (putative "SNPs") then were confirmed by Sanger sequencing of PCR products generated using primers derived from their respective contig sequences. A total of 41 amplicons, containing 50 putative SNPs, were sequenced in B14. A total of 45 SNPs were confirmed via Sanger sequencing, while the remaining 5 showed sequencing peaks typical of a heterozygous call, possibly the results of assembly errors in the B73 reference genome of highly conserved regions differing by only one base, assembly errors in the B73 contigs generated by assembling read 2 data from different regions grouped by similar read 1 sequences, or duplication of the corresponding regions in the B14 genome in relation to B73. These data nonetheless confirm that the putative SNPs are present in the DNA fragments of interest, and the quality and value of contig sequences for detecting SNPs in a complex genome-like maize.

To determine the origin of mismatches in the B73 contig sequences aligned to the B73 reference genome, a similar approach was used, where 40 amplicons, containing 52 putative SNPs, were sequenced via Sanger sequencing of PCR products. Out of the 52 putative SNPs, only 20 were confirmed, while 17 were not confirmed, exhibiting a monomorphic pattern at the base position of interest, and 15 exhibit sequencing peaks characteristic of a heterozygous call. The confirmation of 20 SNPs and the existence of 15 putative SNPs exhibiting sequencing peaks characteristic of a heterozygous call suggests assembly errors in the B73 reference genome or natural variations between DNA materials. The presence of 17 unconfirmed mismatches could be explained by possible sequencing errors. Alternatively, errors in the B73 reference genome assembly could have led to artificial grouping of reads 1 data, led by the fact that uniquely aligned read 2 sequences actually correspond in reality to separate regions of the genomes. A rapid BLAST search of the B73 contig sequences carrying unconfirmed SNPs or SNPs exhibiting sequencing peaks characteristic of a heterozygous call indicates that all or parts of a majority of the contigs (63%, data not shown) align with more than one B73 BAC clone sequence (mapped to different chromosomes), confirming the potential value of filtering such contig sequences prior to SNP discovery or detection.

4. Conclusions

A revised RAD-PE protocol for generating long contig sequences mapping to discrete regions of the maize genome has been tested and implemented. Contig length essentially varies with the size of the DNA fragments selected during the library construction procedure and is expected to be limited by the maximum length of the DNA fragments sequenced on the Illumina platform (separate experiments have shown that DNA fragments up to ~1 Kbps can be sequenced effectively on the Illumina Genome Analyzer). Resulting contigs can be used for several applications, including SNP discovery and marker assay development. They also are expected to facilitate SNP discovery efforts in complex genomes lacking robust reference sequence information, not only in plants but also any other species of interest.

Acknowledgments

The authors would like to thank Hai Zhu and Chris Cornelison for providing the sequencing data, Andrew Tan for assistance with loading the samples, Amber Long and Jamie Broomall for performing the Sanger-based validation, and Victor Llaca for insightful discussions.

References

[1] S. R. Eathington, T. M. Crosbie, M. D. Edwards, R. S. Reiter, and J. K. Bull, "Molecular markers in a commercial breeding program," *Crop Science*, vol. 47, pp. S154–S163, 2007.

[2] J. Shendure and H. Ji, "Next-generation DNA sequencing," *Nature Biotechnology*, vol. 26, no. 10, pp. 1135–1145, 2008.

[3] S. Deschamps and M. A. Campbell, "Utilization of next-generation sequencing platforms in plant genomics and genetic variant discovery," *Molecular Breeding*, vol. 25, no. 4, pp. 553–570, 2010.

[4] R. K. Varshney, S. N. Nayak, G. D. May, and S. A. Jackson, "Next-generation sequencing technologies and their implications for crop genetics and breeding," *Trends in Biotechnology*, vol. 27, no. 9, pp. 522–530, 2009.

[5] L. W. Hillier, G. T. Marth, A. R. Quinlan et al., "Whole-genome sequencing and variant discovery in *C. elegans*," *Nature Methods*, vol. 5, no. 2, pp. 183–188, 2008.

[6] C. P. Van Tassell, T. P. L. Smith, L. K. Matukumalli et al., "SNP discovery and allele frequency estimation by deep sequencing of reduced representation libraries," *Nature Methods*, vol. 5, no. 3, pp. 247–252, 2008.

[7] M. A. Gore, M. H. Wright, E. S. Ersoz et al., "Large-scale discovery of gene-enriched SNPs," *The Plant Genome*, vol. 2, pp. 121–133, 2009.

[8] S. Deschamps, M. la Rota, J. P. Ratashak et al., "Rapid genome-wide single nucleotide polymorphism discovery in soybean and rice via deep resequencing of reduced representation libraries with the Illumina Genome Analyzer," *The Plant Genome*, vol. 3, pp. 53–68, 2010.

[9] D. L. Hyten, Q. Song, E. W. Fickus et al., "High-throughput SNP discovery and assay development in common bean," *BMC Genomics*, vol. 11, no. 1, article 475, 2010.

[10] X. Wu, C. Ren, T. Joshi, T. Vuong, D. Xu, and H. T. Nguyen, "SNP discovery by high-throughput sequencing in soybean," *BMC Genomics*, vol. 11, no. 1, article 469, 2010.

[11] M. Margulies, M. Egholm, W. E. Altman et al., "Genome sequencing in microfabricated high-density picolitre reactors," *Nature*, vol. 437, pp. 376–380, 2005.

[12] J. Eid, A. Fehr, J. Gray et al., "Real-time DNA sequencing from single polymerase molecules," *Science*, vol. 323, pp. 133–138, 2009.

[13] J. B. Hiatt, R. P. Patwardhan, E. H. Turner, C. Lee, and J. Shendure, "Parallel, tag-directed assembly of locally derived short sequence reads," *Nature Methods*, vol. 7, no. 2, pp. 119–122, 2010.

[14] P. D. Etter, J. L. Preston, S. Bassham, W. A. Cresko, and E. A. Johnson, "Local de novo assembly of rad paired-end contigs using short sequencing reads," *PLoS ONE*, vol. 6, no. 4, article e18561, 2011.

[15] E. M. Willing, M. Hoffmann, J. D. Klein, D. Weigel, and C. Dreyer, "Paired-end RAD-seq for de novo assembly and marker design without available reference," *Bioinformatics*, vol. 27, no. 16, pp. 2187–2193, 2011.

[16] B. Langmead, C. Trapnell, M. Pop, and S. L. Salzberg, "Ultrafast and memory-efficient alignment of short DNA sequences to the human genome," *Genome Biology*, vol. 10, no. 3, article R25, 2009.

[17] D. R. Zerbino and E. Birney, "Velvet: algorithms for de novo short read assembly using de Bruijn graphs," *Genome Research*, vol. 18, no. 5, pp. 821–829, 2008.

[18] P. D. Rabinowicz, K. Schutz, N. Dedhia et al., "Differential methylation of genes and retrotransposons facilitates shotgun sequencing of the maize genome," *Nature Genetics*, vol. 23, no. 3, pp. 305–308, 1999.

[19] P. D. Rabinowicz, R. Citek, M. A. Budiman et al., "Differential methylation of genes and repeats in land plants," *Genome Research*, vol. 15, no. 10, pp. 1431–1440, 2005.

4

Evolutionary and Molecular Aspects of Indian Tomato Leaf Curl Virus Coat Protein

Sivakumar Prasanth Kumar,[1] **Saumya K. Patel,**[1] **Ravi G. Kapopara,**[1]
Yogesh T. Jasrai,[1,2] **and Himanshu A. Pandya**[1,2]

[1] *Department of Bioinformatics, Applied Botany Center, University School of Sciences, Gujarat University, Ahmedabad 380 009, India*
[2] *Department of Botany, University School of Sciences, Gujarat University, Ahmedabad 380 009, India*

Correspondence should be addressed to Yogesh T. Jasrai, yjasrai@yahoo.com

Academic Editor: Akhilesh Kumar Tyagi

Tomato leaf curl disease (ToLCD) is manifested by yellowing of leaf lamina with upward leaf curl, leaf distortion, shrinking of the leaf surface, and stunted plant growth caused by tomato leaf curl virus (ToLCV). In the present study, using computational methods we explored the evolutionary and molecular prospects of viral coat protein derived from an isolate of Vadodara district, Gujarat (ToLCGV-[Vad]), India. We found that the amino acids in coat protein required for systemic infection, viral particle formation, and insect transmission to host cells were conserved amongst Indian strains. Phylogenetic studies on Indian ToLCV coat proteins showed evolutionary compatibility with other viral taxa. Modeling of coat protein revealed a topology similar to characteristic Geminate viral particle consisting of antiparallel β-barrel motif with N-terminus α-helix. The molecular interaction of coat protein with the viral DNA required for encapsidation and nuclear shuttling was investigated through sequence- and structure-based approaches. We further emphasized the role of loops in coat protein structure as molecular recognition interface.

1. Introduction

Tomato leaf curl virus (ToLCV) is one of the most devastating causal agents of tomato (*Solanum lycopersicum*) crop which had emerged causing damage and encroaching new areas in tropical and subtropical continents every year. Plant-infecting geminiviruses belong to the family Geminiviridae in which *Begomovirus* is one among the genera possessing both mono- and bipartite genomes that infect especially dicotyledonous plant species [1]. The disease is marked by symptoms such as yellowing of leaf lamina with upward leaf curl as well as distortion, reduction in internodes, new leaves size reduction, wrinkle facade, stunted growth, and dissemination of flower from plant before onset of fruiting. ToLCV is primarily transmitted by sweet potato whitefly (*Bemisia tabaci*) and silver leaf whitefly (also called Biotype B; *Bemisia argentifolii*). Whiteflies harboring virus can nonspecifically infect a wide spectrum of plant crops and weeds including eggplant, potato, tobacco, pepper, and common bean.

Infected plants seem healthy but develop symptoms leading to enormous economic loss [2].

In Indian subcontinent, ToLCV is a major problem for tomato-growing regions as several reports on new strains have been documented including New Delhi, Lucknow, Bangalore, Varanasi, Mirzapur, Vadodara, and so forth and posed a threat to crop productivity [6]. Indian ToLCV isolates are mostly monopartite (DNA-A) in nature with few isolates possessing bipartite (DNA-A and DNA-B) genome organization such as tomato leaf curl New Delhi virus (ToLCNDV) and tomato leaf curl Palampur virus (ToLCPalV) [7]. Both DNA-A and DNA-B are single-stranded (ss) DNA genomes of approximately 2.7 kb size and encode viral factors essential for viral replication, encapsidation, transmission, and systemic spread [8]. Jyothsna et al. 2012 reported tomato leaf curl Gujarat virus (ToLCGV) possesses monopartite genome and is infectious expressing systemic symptoms in *Nicotiana benthamiana* and tomato [9]. An increased symptom severity and shortened incubation period required for symptom expression was noticed when ToLCGV was coinoculated with betasatellite of tomato yellow leaf curl virus Thailand (TYLCTHB) resulted in yellow mottling [9]. The molecular relationship of ToLCGV-[Vad], an isolate from Vadodara district of Gujarat, with other strains revealed that it belongs to Old World *Begomoviruses* and established a closely related cluster with other North Indian strains including ToLCGV-(Varanasi)-[Var] and ToLCGV-(Mirzapur)-[Mir] based on the DNA-A sequence alignment [10].

The measure "breeding for resistance" conceptualizes the introduction of resistance genes found in wild tomato species into tomato cultivars to develop resistance against diseases. Kunik et al. 1994 demonstrated that tomato plants transformed with TYLCV coat protein were found to be virus-resistant [11]. In India, *Agrobacterium tumefaciens* mediated transformation of coat protein gene was carried out to develop ToLCV tolerant/resistant transgenic tomato plants under glass house conditions [12]. Transgenic tomato plants containing cucumber mosaic virus coat protein gene was also successfully transformed [13]. An asymmetric synergism and virulent pseudorecombinant between ToLCNDV and ToLCGV was reported by Chakraborty et al. 2008 and found enhanced pathogenicity when tested in *N. benthamiana*, *N. tabacum*, and *S. lycopersicum* [14]. An evidence for natural recombination was observed between tomato leaf curl Bangalore virus (ToLCBV), ToLCBV [Ban 5], and ToLCBV [Kolar] and examined the possibility of recombination between strains/species that coexist within the same geographical location [15]. Hence, tremendous consideration should be given to study the biological and molecular properties of this newly emerging causal agent.

In the present study, we examined the evolutionary and molecular prospects of ToLCGV-[Vad] coat protein. Sequence analysis of coat protein revealed that amino acids essential for systemic infection, viral particle formation, and insect transmission to host cells were evolutionarily compatible when compared to non-Indian isolates giving clues of evolutionary conservativeness. Further, molecular modeling of coat protein provided a topology similar to characteristic Geminate viral particle. Electronic properties of coat protein facilitated its interaction with viral DNA with the loop element acting as molecular recognition interface which is one of the major findings of the present study.

2. Materials and Methods

2.1. Protein Sequence Retrieval. The coat protein of ToLCGV-[Vad] (accession no. AAL78666.1) was retrieved from NCBI database (http://www.ncbi.nlm.nih.gov/) [16]. Coat proteins from Indian strains (Bangalore-CAA88227.1, Bangalore (Ban4)-AAD51286.1, Bangalore (Ban5)-AAK19178.1, Bangalore (Kolar)-AAL26553.1, Varanasi-AAO25668.1, Kelloo-AAM21566.1, Karnataka-AAB08929.1, New Delhi (Mild)-AAA92817.1, and Lucknow-CAA76209.1) were also obtained for multiple sequence alignment and phylogenetic analysis.

2.2. Protein Family Classification. The family of coat protein was studied using a combination of programs, namely, NCBI

CD- Search (http://www.ncbi.nlm.nih.gov/Structure/cdd/wrpsb.cgi, database searched: CDD v3.03–42251 PSSMs) [17], PSI-BLAST (http://blast.ncbi.nlm.nih.gov/Blast.cgi) [18], and Pfam (http://pfam.sanger.ac.uk/) [19]. CD-Search is a NCBI's interface to search Conserved Domain Database (CDD) which utilizes RPS-BLAST (Reverse-PSI-BLAST; a variant of PSI-BLAST) to scan a set of precalculated position specific scoring matrices (PSSMs) using a protein query. PSI-BLAST (position-specific-iterated BLAST) uses initial matches to query sequence to build scoring matrix and appends additional matches to the matrix by an iterative search method in order to detect remote homologs. Pfam designates protein family by HMM (Hidden Markov Model)-based search (default settings were chosen and Pfam-A significant matches were only considered) over known protein family classifiers.

2.3. Analysis of Nuclear Localization Signal and Its Prediction. Nuclear localization signals (NLSs) were predicted using cNLS Mapper (http://nls-mapper.iab.keio.ac.jp/) [20] as coat protein that is known to be karyophilic [21]. cNLS Mapper is a computer program that predicts NLS by activity-based profile search and an additivity-based motif scoring function in different classes of importin-α/β pathway-specific NLS. The prediction was made with a score cut-off of 5.0 and searched for both mono- and bipartite NLSs with a long linker (13–20 amino acid length) as ToLCGV possesses mono-bipartite genome organization [6]. Classic NLS typically rich in basic amino acids such as lysine and arginine, the counts of basic amino acids was performed manually in the above predicted NLSs. Comparison with literature-reported NLS specific to BR1 nuclear export family was carried out to examine the pattern of nuclear localization. This was achieved using pairwise sequence alignment using EMBOSS Stretcher (http://www.ebi.ac.uk/Tools/psa/emboss_stretcher/) [22] with a representative family protein member (BR1 nuclear shuttle protein from squash leaf curl virus(SqLCV); NCBI accession No. NP_047247.2) against the coat protein of study. Both sequences were aligned using EBLOSUM62 scoring matrix with a gap opening and extending penalty of 12 and 2.

2.4. Multiple Sequence Alignment and Phylogenetic Analysis of Coat Proteins. Coat protein sequences of Indian strains were used for the analysis of multiple sequence alignment (MSA). MSA was performed using EBI ClustalW program (http://www.ebi.ac.uk/Tools/msa/clustalw2/) [23] in which the sequences were aligned pairwise initially (gap open penalty = 10, gap extension penalty = 0.1, matrix = Gonnet) and then the best local pairs (gap open penalty = 10, gap extension penalty = 0.20, matrix = Gonnet) were clustered by Neighbour-joining (NJ) technique. Subsequently, an alignment file in ClustalW format was generated and specified as input to draw phylogenetic tree using Phylip version 3.68 package (http://evolution.genetics.washington.edu/phylip/) [24]. NJ algorithm was used to draw tree with inclusion of branch length.

2.5. Structure Modeling of Coat Protein. Sequence-based similarity searching was initially executed in NCBI nonredundant (NR) database using BLASTp program (http://blast.ncbi.nlm.nih.gov/Blast.cgi) [25] with default settings to find a close homolog with known 3D protein structure information is known. Similarly, BLAST based homolog search in RCSB Protein Data Bank (PDB; http://www.rcsb.org/pdb/home/home.do) [26] was also carried out. Both of these procedures yielded no close homologs. So, we opted to model the coat protein using homology domain modeling and remote-based homology modeling.

2.6. Disorderness Prediction. In order to characterize regions of sequences in coat protein which can be efficiently modeled, disorderness prediction was made. Disordered residue was identified using DISOPRED server (http://bioinf.cs.ucl.ac.uk/disopred/) [27] with a filter threshold of 5% and a false positive threshold of 2%. The disorderness is predicted by scanning the available sequence records in the PDB and then matches the electron density map to identify the missing coordinates. As a result, atomic coordinates of such amino acids will not be available for modeling of the protein and has the greater possibility of producing an irregular loop region in the modeled coat protein. Thus, manual search (only at the N-and C-terminals) for disordered sequence window in DISOPRED predictions was performed with the intention of excluding the corresponding region for modeling the coat protein. It was also ensured that disordered residue reported in the intervening sequence positions was left out so that the structure model did not possess any gaps.

2.7. Homology Domain Modeling. Robetta server (http://robetta.bakerlab.org/) was used for modeling the coat protein in which Ginzu, a hierarchical domain parsing and modeling protocol was adopted [28]. The input sequence (coat protein excluded with disorderness) was initially searched using BLAST, PSI-BLAST, FFAS03 (http://ffas.burnham.org/), and 3D-Jury (http://meta.bioinfo.pl/) to obtain information on homologous regions which are then modeled with their comparative modeling protocol. Unassigned (i.e., nonhomologous as identified in the first stage) regions were then parsed to model as domain linkers using a combination of approaches, namely, HMMER search (http://hmmer.janelia.org/) over Pfam-A database and an MSA (produced from initial PSI-BLAST results) based search over NCBI NR database. Subsequently, K*Sync alignment method was utilized to predict elements that are obligated to the fold to produce a single default alignment by dynamic programming. Best five models were generated after loop modeling, domain assembly, and side-chain packing.

2.8. Remote-Based Homology Modeling. Efforts were also carried out to predict structure of coat protein using remote-based homology modeling approach with the help of protein homology/analogy recognition engine (Phyre) version 2.0 (http://www.sbg.bio.ic.ac.uk/phyre2/) [29]. In the first step, profile was constructed using five iterations of PSI-BLAST against NR sequence database. The query secondary

structure was predicted using three independent prediction programs (PSI-PRED (http://bioinf.cs.ucl.ac.uk/psipred/), SSPro (http://download.igb.uci.edu/sspro4.html) and Jnet (http://www.compbio.dundee.ac.uk/Software/JNet/jnet.html)) and a consensus prediction was made consequently. This profile and secondary structure were then scanned against the fold library using a profile-profile alignment algorithm to generate 3D models. Followed by a reconstruction procedure in the last stage, side-chains are packed using rotamer library and best models (selected based on confidence and sequence coverage) were returned.

2.9. Energy Minimization and Structure Validation of Models. The modeled structures were energy minimized using a utility in Tripos Benchware 3D Explorer (academic version; Tripos: A Certara company, http://www.tripos.com/) [30] with AMBER7 force field. Modeled structures were then validated for structure correctness and stereochemistry using Ramachandran plot [31] from RAMPAGE server (http://mordred.bioc.cam.ac.uk/~rapper/rampage.php) [32]. Based on the percentage of favourness and frequency of outliers, the models were selected and used for further analysis.

2.10. Prediction of Sequence- and Structure-Based DNA Binding Properties. In vitro studies showed that coat proteins from Geminiviridae family bind nonspecifically with both ss- and ds-viral DNA [21, 33, 34]. To elucidate the role of DNA binding abilities of coat protein, sequence- and structure-based approaches were used. BindN (http://bioinfo.ggc.org/bindn/) employs support vector machines (SVMs) trained from data instances such as side chain pK_a value, hydrophobicity index, and molecular mass of an amino acid [35]. A specificity of 80% with a filter threshold of 5% was chosen to avoid overwhelmed predictions. PreDs (http://pre-s.protein.osaka-u.ac.jp/preds/) makes use of molecular surface to evaluate electrostatic potential, local, and global curvatures of the PDB queried structure to predict potential dsDNA binding sites [36]. The modeled coat protein was defined as input with validation chosen from scoring functions.

2.11. Viral DNA Structure Modeling and Docking with Coat Protein. Canonical viral DNA was modeled using 3D-DART (3DNA-Driven DNA Analysis and Rebuilding Tool; http://haddock.chem.uu.nl/dna/dna.php) web service with default introduction of parameters for bends (roll, tilt, and twist) [37]. It uses 3DNA "fiber" module to generate canonical DNA structure and "find_pair" and "analyze" modules to produce a corresponding base pair (step) parameter file. The parameter file was used to set up local and global bends in the DNA structure file which are then remodeled finally using "rebuild" component to return PDB formatted DNA structure file. The docking phase was carried out using HADDOCK (High Ambiguity Driven biomolecular DOCKing; http://haddock.chem.uu.nl/) program [38] with the modeled coat protein and ds-viral DNA as inputs. Residues encompassed in a DNA binding region predicted by PreDs

```
ToLCGV-[Vad]      18  RRRLNFDSPSvSRAAAPIVRVT-KAKAWANRPMYRKPRIYRMYRSPdvpkgcegPCKVQSFD-----AKNDIGHMGKVIC 91
CDD: pfam00844     1  RRRLNFDSRR-SRARAPTRRSTnKKRGWSKRPMRRKGRSYRMYRSP--------PLKVQSFEinqfgPDFVVSNTGKVSL 71
Pfam HMM search       Rrrlnfds.rrsraraplkrstnkkksrskrpmrrkgrvvvrvarsp........plkvqsfeinqegsefvvsntgkvsl

ToLCGV-[Vad]      92  LSDVTRGIGLTHRVGKRFCVKSLYFVGKIWMDENIKVKNHTNTVMFWIVRDRRPSG-TPSDFQQVFNVYDNEPSTATVKN 170
CDD: pfam00844    72  ISDPTRGKGETNRVRKYIKLKSLYFKGTVWIDENIKDKNMTGVFSFWLVRDRRPHGnTPGTFDELFGMYDNEPGTLTVKR 151
Pfam HMM search       isdpsrGkgetnRvrkriklkslaikgtvaidenaakknmtgvvvfwLVrDrkPtgstpgtfdeiFgaydnepstltvkr

ToLCGV-[Vad]     171  DQRDRFQVLRRFQATVTGGQ------YAAKEQAIIRRFFR-VNNYVVYNHQEAGKYENHTENALLLYMACTHASNPVYAT 243
CDD: pfam00844   152  DLRDRFVVKRKWKRVVSSEGdtnmvdFPGKNQLSVRRFFKwLGVKDEWKNSETGKYGNIKKNALLLYYCWMSAPNSGAST 231
Pfam HMM search       alkdRfvVlrkwkvvvssegdtrvvdvegkeqlsvrrFfkwlnvktewknsetGkygnikknALllyyawlsavnskast

ToLCGV-[Vad]     244  LKVRSYFYDSVTN 256
CDD: pfam00844   232  TKIRVYFKLSYNG 244
Pfam HMM search       tkirvyfkssgnn
```

FIGURE 1: Sequence alignment of ToLCGV-[Vad] coat protein with nuclear export factor of BR1 family (Pfam entry: 00844 recovered from NCBI CDD) and HMM profile of the geminivirus coat protein.

with a Parea of greater than 250Å^2 was specified as active site residues whereas passive residues were automatically defined around the active site which forms the boundary of the DNA-binding region. This specification was introduced to enhance the conformational search space for docking simulation as well as to avoid blindfold docking experiments. Definition of residues takes the form of experimental data which were converted into ambiguous interaction restraints (AIRs) in order to generate topology of the structures subsequently. The docking procedure consists of three stages: an energy minimization in a rigid-body manner, a semiflexible refinement in torsional space, and a final refinement in explicit solvent. After execution of each of these stages, the resultant structures are scored and ranked and the best fitted structures are employed in next stages. The best docked conformation can be obtained (usually clustered at the top) by inspecting the HADDOCK score which is a summation of intermolecular energies, namely, van der Waals (vdW), electrostatic (Elec), desolvation (Dsolv) and AIRS together with buried surface area (BSA): rigid-body score = $1.0 * \text{Elec} + 1.0 * \text{vdW} - 0.05 * \text{BSA} + 1.0 * \text{Dsolv} + 1.0 * \text{AIR}$; final score = $1.0 * \text{Elec} + 1.0 * \text{vdW} + 1.0 * \text{Dsolv} + 1.0 * \text{AIR}$.

2.12. Generation of Electrostatic Potential Map for Docked Structures. The influence of electrostatics for enabling DNA-protein interaction was studied using continuum Poisson-Boltzmann (PB) electrostatic approach. It was achieved by PBEQ-Solver (PBEQuation-Solver; http://www.charmm-gui.org/?doc=input/pbeqsolver) [39] for which PQR files were required as molecular inputs. Hence, the docked complexes in PDB format were converted into PQR format using PDB2PQR server (http://nbcr-222.ucsd.edu/pdb2pqr_1.8/) [40]. PQR format embodies the replacement of occupancy column in a PDB file ("*P*") with the atomic charge ("*Q*") and the temperature factor column with the atomic radius ("*R*"). The inputted PDB file was subjected to following structural manipulations: rebuilding missing heavy atoms, building and optimizing hydrogens and assignment of atomic charges and radii based on force field parameters from CHARMM22 (selected option), AMBER99 or PARSE. All the PB calculations on PBEQ-Solver were performed in a coarse grid space (before focusing = 1.5 Å and after focusing = 1.0 Å)

and utilized molecular surface (computed with a probe radius of 1.4 Å) to set up the dielectric boundary. The resultant electrostatic potential grid map in data explorer (*dx*) format was recovered and specified as input to PyMol version 2.5 program (academic version; Schrodinger LLC) [41] to view the PBEQ electrostatic map.

3. Results and Discussion

3.1. Prediction of Protein Family of Coat Protein. The protein family of ToLCV coat protein was predicted using a combination of programs. NCBI CD-Search using protein sequence revealed that it belongs to Gemini-coat protein superfamily (Pfam entry: pfam00844, accession no. Q8QYY9). Upon carefully examining the sequence alignment generated (*E*-value: 5.53e-100; bit score: 290.36) with SqLCV BR1 nuclear shuttle protein, it was studied that ToLCGV belongs to nuclear export factor BR1 family (Figure 1). BR1 is a ssDNA binding protein that shuttles between the nucleus and cytoplasm in plant cells [33].

PSI-BLAST sequence hit (PSI-BLAST threshold: 0.005 maximum iterations: 7; *E*-value: 1e-105; bit score: 383; sequence coverage in alignment: 99.64%) with a capsid protein of *Begomovirus* taxa (UniRef90 P03560; tomato golden mosaic virus) was observed. Further, sequence-based query over Pfam-A (Pfam-B not chosen as we focused on obtaining highly curated data) database produced a result similar to NCBI CD-Search. This HMM-based search provided an alignment with an *E*-value of 2.3e-87 and bit score of 292.1 (Figure 1). Manual inspection of PubMed references in the pfam00844 entry in NCBI CDD disclosed that coat proteins of *Geminiviridae* family binds ss- and ds-viral DNA *in vitro* [42]. For instance, TYLCV coat protein [43], maize streak virus (MSV) coat protein [21], SqLCV nuclear shuttle protein [44], and bean dwarf mosaic geminivirus(BDMV) movement protein [34] have the same function of binding which helps them to establish infection by nuclear shuttling of viral DNA across cell boundaries. Besides, coat protein also possesses binding function necessary for encapsidation of viral DNA. It is well known that the genomic component DNA-B in bipartite *Begomovirus* such as ToLCNDV encodes two movement proteins, namely, nuclear shuttle protein

(a) Predicted and experimental NLS

SqLCV	25	------KRs----YGAARGDD RR----Rp---	39
MSV	1	MSTSkRKRGDDSNWS--KR VTKK----KPS--	24
TYLCV	1	MS---KRpGDIIISTPVSKvRRRLNFDSPYSS	29
ToLCV	1	MS---KRpADMLIFTPASKvRRR---------	20

(b) Crucial amino acids

TYLCV-Sic	125	NIKKPNHTNQVMFFLVRDRRPYGTSPMEFGQ	155
TYLCV-SicRv	125	NIKKPNHTNHVMFFLVRDRRPYGTSPMEFGQ	155
TYLCV-Sar	125	NIKKQNHTNQVMFFLVRDRRPYGTSPMDFGQ	155

Indian wild ToLCV strains

Bangalore	125	NIKTKNHTNSVMFFLVRDRRPVDKPQDFGDV	155
Bangalore (Ban4)	125	SIKTKNHTNSVMFFLVRDRRPVDKPQDFGDV	155
Bangalore (Ban5)	125	NIKTKNHTNSVMFFLVRDRRPVDKPQDFGDV	155
Bangalore (Kolar)	125	NIKTKNHTNSVMFFLVRDRRPIDKPQDFGDV	155
Karnataka	125	NIKTKNHTNSVMFFLVRDRRPVDKPQDFGEV	155
New Delhi (Mild)	125	NIKTKNHTNSVMFFLVRDRRPTGSPQDFGEV	155
Lucknow	125	NIKTKNHTNSVMFFLVRDRRPTGAPHDFGEV	155
Gujarat (Vadodara)	125	NIKVKNHTNTVMFWIVRDRRPSGTPSDFQQV	155
Varanasi	125	NIKVKNHTNTVMFWIVRDRRPSGTPNDFQQV	155
Kelloo	125	NIKVKNHTNTVMFWIVRDRRPSGTPSDFQQV	155

FIGURE 2: (a) MSA of predicted and experimental NLS of selected geminivirus coat proteins. (b) MSA of coat proteins from Indian and non-Indian strains corresponding to sequence region of interest containing the key amino acids (highlighted in larger fonts) required for systemic infection, viral particle formation, and insect transmission.

(NSP) and cell-to-cell movement protein (MP) that direct the viral genome to the cortical cytoplasm and across the barrier of the cell wall for infection multiplication [45]. On the other hand, monopartite *Begomovirus* including ToLCGV [9] produces coat protein and other associated proteins to alleviate movement inside the host while coat protein acts as nuclear shuttler facilitating import and export of DNA [46]. Therefore, it is anticipated that coat protein of ToLCGV may also function as nuclear shuttler.

3.2. Prediction of NLS in the Coat Protein Sequence. Mutagenesis study on MSV coat protein [4] and TYLCV [5] NLS region resulted in the cytoplasmic accumulation of the mutant protein. Thus, ToLCV coat protein must possess a NLS region in its sequence in order to be translocated to nucleus. A NLS signal was predicted with a score 10.2 by the cNLS Mapper in the coat protein N-terminal with a composition of 20 amino acids (predicted bipartite NLS: MSKRPADMLIFTPASKVRRR, predicted monopartite NLS: none).

The occurrence of basic amino acids in the predicted NLS showed that lysine and arginine constituted 2 and 4 counts which proposed to have a classic NLS pattern (Table 1) and can be comparable to experimentally identified TYLCV coat protein NLS [5] (Figure 2(a)). Despite the impressive number of receptor-cargo interactions that have been studied, the prediction of NLSs in candidate proteins remains extremely difficult. So, we step forwarded our search in

scientific literatures related to BR1 nuclear export family in order to infer the predictions made. Pairwise sequence alignment of NLS region from MSV and ToLCGV-[Vad] coat proteins resulted in an identity and similarity percentage of 25% and 33.3% with a score of −6 whereas TYLCV and ToLCGV-[Vad] yielded an identity and similarity percentage of 51.7% and 58.6% with a score of 49. This pairwise alignment suggested that ToLCV coat protein is much more conserved with TYLCV rather than SqLCV [3] (identity and similarity = 15%, score = −20), a representative protein member of BR1 family (pfam00844) in Pfam database.

3.3. MSA and Phylogenetic Analysis of Indian Strains. ToLCV coat protein sequences from Indian strains were retrieved from NCBI database to construct MSA in order to study amino acids crucial in conserved domain and responsible for systemic infection, viral particle formation, and insect transmission. Norris et al. 1998 reported that a functional coat protein having amino acids in the following sequence positions, namely, Pro/Gln129, Gln/His134, and Glu/Asp152 on TYLCV isolates is essential for correct assembly of virions and transmission by the insect vector [47]. These key residues were identified by *B. tabaci* transmissibility studies in the field isolates of TYLCV-Sic (Sicily), TYLCV-Sar (Sardinia), and TYLCV-SicRv (engineered mutant of Sicily) [47]. Examination of corresponding positions in our MSA cluster revealed that Lys129, Ser/Thr134, and Asp151 (instead of 152nd position as a result of single residue deletion) were

TABLE 1: Known and predicted NLS pattern of BR1 nuclear export family.

Organism	Predicted NLS pattern	Frequency of lysine residues	Frequency of arginine residues
SqLCV coat protein	KRSYGAARGDDRRRP (Sanderfoot et al., 1996 [3])	1	5
MSV coat protein	MSTSKRKRGDDSNWSKRVTKKKPS (Liu et al., 1999 [4])	6	3
TYLCV coat protein	MSKRPGDIIISTPVSKVRRRLNFDSPYSS (Kunik et al., 1998 [5])	2	4
ToLCGV-[Vad]	MSKRPADMLIFTPASKVRRR (predicted by cNLS Mapper)	2	4

conserved amongst Indian strains in comparison to non-Indian isolates and are found to be wild-type. The comparison of chemical properties of the template with that of MSA showed that a positively charged amino acid (Lys129) was identified in the uncharged polar (Gln129) position. The second important residue (Ser/Thr134 in replacement with Gln/His134) was conserved in terms of polarity while a negatively charged residue (Glu/Asp152; Asp151) was preserved in the third crucial position (Figure 2(b)). This amino acids combination (Lys129, Ser/Thr134, and Asp121) is also conserved in coat proteins among different wild-type viruses, namely, tomato golden mosaic virus, tomato mottle virus-[Florida], sinaloatomato leaf curl virus, tomato leaf crumple virus, taino tomato mottle virus, abutilon mosaic virus-[Hawaii], bean golden mosaic virus-[Brazil], SqLCV and papaya leaf curl virus [47].

A phylogenetic tree based on NJ algorithm was constructed using Phylip version 3.68 to study the sequence conservativeness among Indian strains. Surprisingly, coat proteins characteristic from districts, namely, Vadodara, Varanasi, and Kelloo were clustered in a node with a branch length of 0.167. It should be noted that these members were representing different states in the Northern India contrasting to other members which were sufficiently diverged to each other. Besides the fact that Bangalore isolates were conserved among each other, they were distinct with one of the state member, Karnataka with a length of 0.011. Isolates from New Delhi and Lucknow were conserved as expected in terms of area nearness (Figure 3). The key amino acids required for biochemical functions were indeed conserved amongst each other with respect to the comparison using MSA made above.

3.4. Disorderness and Their Link with Predicted NLS.
Disorderness was predicted in the coat protein to identify the sequence regions that cannot be modeled efficiently with the protein modeling procedures adopted by us. This scrutiny was taken to eliminate the loop region in the sequence terminals. Disorder profile produced with 5% filter threshold showed that a window with sequence positions from 1 to 50 was scattered with peaks demonstrating residue disorderness (Figure 4). This region corresponds to NLS in the N-terminal. There exists a relationship between the predicted NLS and the disorderness as the corresponding sequence position

was predicted as loop region with a variety of secondary structure prediction programs including PSI-Pred, GORIV, and so forth. So, we decided to exclude NLS signal from the protein sequence for modeling due to the consideration of disordered profile and the increased possibility of generating loop geometry.

3.5. Structure Modeling of Coat Protein.
No close template was obtained in an attempt to find structurally known homolog of ToLCV coat protein using BLASTp and sequence based BLAST search in PDB with default settings. Therefore, we decided to use homology domain modeling using Robetta program and remote-based homology modeling using Phyre program.

After evaluation of predictions related to secondary structure features, Ginzu, a domain parser of Robetta modeled two domains with a sequence span of 1–59 and 60–235 using Pfam and HHSearch as sources. K∗Sync alignment method was employed subsequently to generate structurally good scoring decoys followed by loop modeling, domain assembly, and side-chain packing. This procedure yielded five models. Remote-based homology modeling using Phyre provided a model based on the Nucleoplasmin-like/viral protein (viral coat and capsid proteins) as folding unit (PDB ID: 2BUK chain A; structure of satellite tobacco necrosis virus after crystallographic refinement at 2.5 Å resolution) [48] derived from satellite virus with a confidence of 97.3. Fortunately, Robetta used the same structural template for models generation. Phyre provided a list of other models sorted by its confidence level were found to only range from 8.03 to 23.9.

3.6. Selection of Best Scoring Models.
A total of six coat protein models (five obtained from Robetta program and one from Phyre program) were subjected to energy minimization with AMBER7 force field and 250 as maximum number of evaluations using minimize energy module engineered in Tripos Benchware 3D Explorer. Energy minimized structures were then validated using stereochemistry check with the help of Ramachandran plot. The Robetta model (energy minimized to −2682.00 Kcal/mol) was chosen based on the residue occurrence of more than 95% estimated by summing up favorable and allowed regions (Figure 5) in φ-ψ

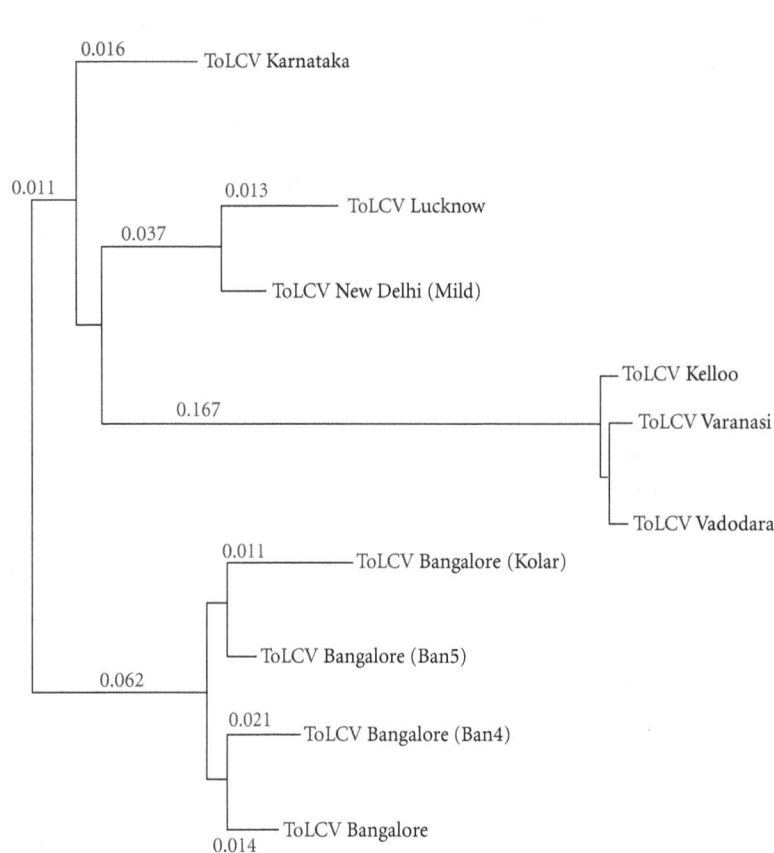

FIGURE 3: Phylogenetic tree of ToLCV coat protein representing closeness among Indian strains.

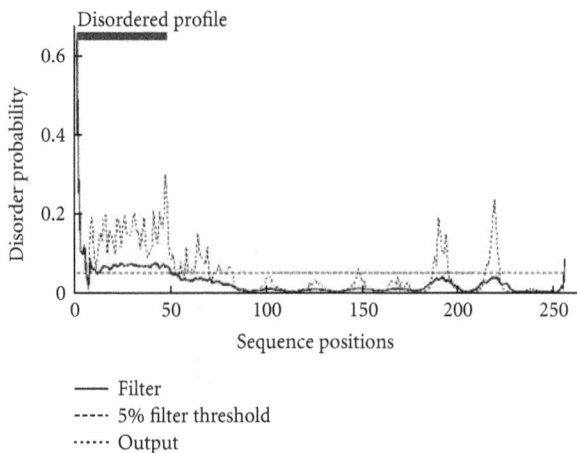

FIGURE 4: Disordered profile of ToLCV coat protein with disorderness in the sequence positions 1–50 which contains predicted NLS.

core areas and the presence of only one outlier (Glu204) whereas the best scored Phyre model was discarded due to the loose packing of loop regions and its close resemblance to Robetta model (root mean squared deviation (RMSD): 20.8578 Å over 1041 matched atoms). There is one more reason to unconsider the Phyre model as the N-terminus was constituted with loop elements, two helices in the intervening

region (two helices expected at the N-terminal) and eleven β-strands (only eight were expected instead of eleven) as it was not complied with characteristic Geminate viral particle. So, we discarded these models for further analysis.

3.7. Resemblance to Geminate Viral Particle. The modeled ToLCV coat protein was compared with the structure of the MSV Geminate particle, a member of Geminiviridae family determined using cryo-electron microscopy and three-dimensional image reconstruction methods [49]. The modeled protein possessed an N-terminal helix with an eight-stranded antiparallel β-barrel motif characteristic of Geminate particle. The β-barrel motif is a dominant structural unit in all ssDNA virus structures that have been determined to atomic resolution. Unlike the model generated by Zhang et al. 2001 [49], our Robetta model has 7 reliable and 1 short (total 8 strands) antiparallel β-strands and 2 helices at the N-terminal instead of 1 in comparison to Zhang's Geminate model. We also noticed 1 helix at the intervening region (Figure 6). We expected that this additional accumulation of secondary structural elements is beyond the evolution and may be an additional procurement in the dsDNA virus structures in disparity to ssDNA virus structures due to conservational pressures as described by Zhang et al. 2001 [49] or due to the insertional sequences or it may be due to the loop geometry as it was undistinguished by the present

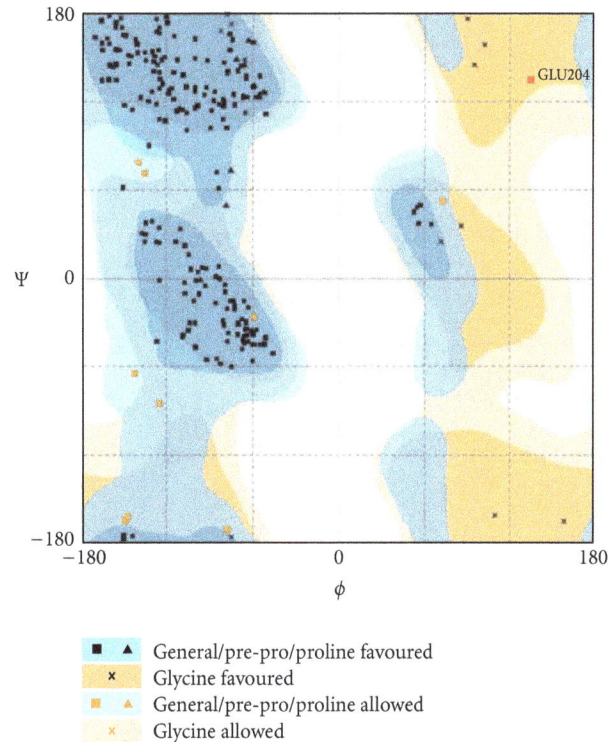

FIGURE 5: Best scored ToLCV Robetta model showing one outlier (Glu204) on Ramachandran plot.

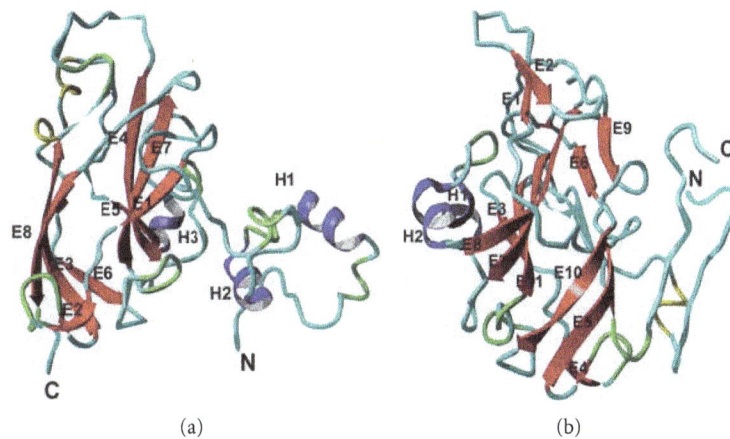

FIGURE 6: Developed ToLCV models using homology domain modeling (Robetta program; (a)) and remote-based homology modeling (Phyre program; (b)). H-Helix, E-Extended β-strands.

programs due to the nonavailability of experimental atomic information.

3.8. Sequence and Structure-Based DNA Binding Properties.

ToLCV must possess DNA binding properties for accomplishing several cellular functions such as nuclear targeting of viral DNA, encapsidation of viral DNA, systemic infection, viral particle formation, insect transmission, and correct assembly of virions as experimentally studied in other members of the family, namely, MSV, SqLCV, BDMV, and so forth. Interestingly, the coat proteins bind both ss- and ds-viral DNA in sequence independent fashion [21, 33, 34]. To reveal the crucial amino acids involved in DNA interaction, sequence- and structure-based approaches were utilized. BindN predicted spatially distributed residues as component of DNA binding interface with 74 amino acids and 28.90% contribution (Figure 7). We expected that this widespread distribution of DNA binding residues will come together during protein folding and will interact with viral DNA. Thus, we step forwarded to identify those prominent amino acids using our generated ToLCV coat model. PreDs revealed loop regions as potential DNA binding region accompanied with all those amino acids predicted by BindN. This prediction was validated by inspecting the scoring functions such as

```
Sequence:   MSKRPADMLIFTPASKVRRRLNFDSPSVSRAAAPIVRVTKAKAWANRPMYRKPRIYRMYR
Prediction: -+++-----------++-+++----+-+-++------+-++-+----+--++++-++-++
Confidence: 37884868984234683999433663575965755794776962429236994937 9257

Sequence:   SPDVPKGCEGPCKVQSFDAKNDIGHMGKVICLSDVTRGIGLTHRVGKRFCVKSLYFVGKI
Prediction: +------------+-++--------------------+----+++--++--+--+-----
Confidence: 62267255453648466592599857929864247276447748568853542 6557939

Sequence:   WMDENIKVKNHTNTVMFWIVRDRRPSGTPSDFQQVFNVYDNEPSTATVKNDQRDRFQVLR
Prediction: ---------+--+-+-------+-++-+-+-----------------+++-+-++--+-+----+
Confidence: 79886925823635967688657626273256549869474355546655425 5663586

Sequence:   RFQATVTGGQYAAKEQAIIRRFFRVNNYVVYNHQEAGKYENHTENALLLYMACTHASNPV
Prediction: +---+-+-------+-----++-+-------------++----------------+--+---
Confidence: 75346553322565627676646583437836436654422227489872764 5256225

Sequence:   YATLKVRSYFYDSVTN
Prediction: +-+-+-+++-+-+-+-
Confidence: 6493959664435742
```

FIGURE 7: Sequence-based DNA binding properties of ToLCV coat protein. DNA binding residues are shown in red text.

Pscore and Parea. Pscore is an indicator for the ratio of the predicted area possessing maximum value while Parea represents area of the predicted ds-DNA binding region on the protein surface. We achieved 0.31 as Pscore (threshold: >0.12) and 2102.26 Å^2 (threshold: >250 Å^2) as Parea, respectively (Figure 8).

The TYLCV coat protein gets imported into the plant and insect cells nuclei via using its N- terminal NLS [5]. It is also been proposed that TYLCV coat protein functions as BR1 protein facilitating DNA trafficking across cell boundaries and demonstrated that the coat protein binds DNA cooperatively [43]. It is also highlighted that TYLCV coat protein may also aid in protecting the transported coat protein-DNA complex from intracellular nucleolytic degradation as this complex was highly resistant to *in vitro* S1 nuclease activity [43]. In MSV coat protein, the DNA binding domain was mapped to the N-terminal 104 amino acids inclusive of NLS [21]. Immuno-electron microscopy revealed that DNA binding domain between residues 5 and 22 suggested that this region could be involved in transporting geminivirus coat protein towards nuclei [50]. We predicted that certain N-terminal amino acids of ToLCV coat protein scored a confidence of 6–9 indicating DNA binding properties (Figure 7). Besides, structural analysis of modeled ToLCV indicated that N-terminal residues contributing to loop secondary structure form the major element in interacting with viral DNA as described below.

3.9. Viral DNA Structure Modeling and Docking with Coat Protein.

We developed a canonical ds-viral DNA using 3D-DART with generalized geometrical constraints to explore the interaction with coat protein. Molecular docking was performed using HADDOCK program with predicted DNA binding region as active site. Best scoring clusters were sorted based on HADDOCK score. The most reliable (top) cluster having four similar docked conformations (HADDOCK score: −12.0 ± 11.0) were recovered. The DNA binding

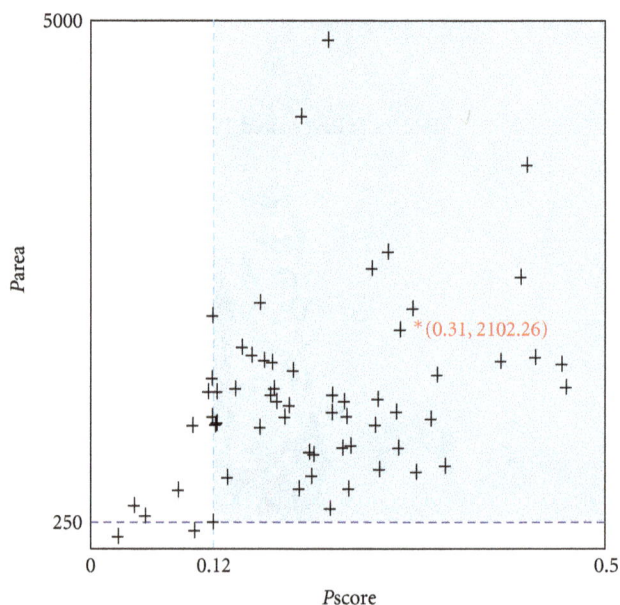

FIGURE 8: Structure based DNA binding properties of ToLCV coat protein having reliable Pscore and Parea.

interface of coat protein was found to be the loop region whereas major groove was the molecular interface of viral DNA in which the best conformers were sampled (Figure 9). The intermolecular energies (unit in KJ/mol) obtained are as follows: vdW: −66.9 ± 5.5, Elec: 827.4 ± 87.8, Dsolv: 125.7 ± 11.7, AIR: 947.0 ± 50.09, and BSA: 2099.6 ± 137.9. We also observed that RMSD of overall lowest-energy structure was 4.3 ± 3.0 with respect to structures in different clusters. The internal energy in apo form (free molecules) was 278722.00 KJ/mol whereas in bound form (DNA-protein complex) was 15494.00 KJ/mol and the binding energy was predicted as −264139.00 KJ/mol.

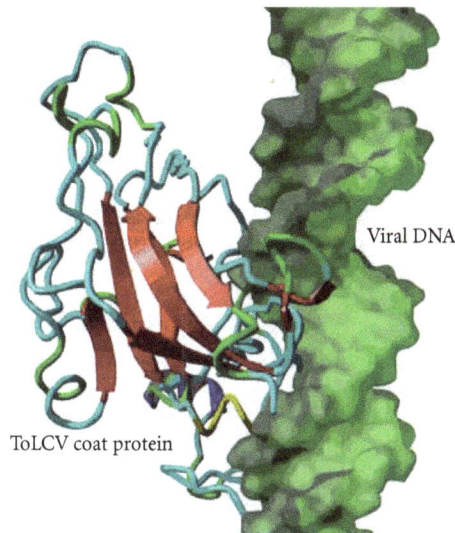

FIGURE 9: The binding mode of ToLCV coat protein with viral dsDNA where loop element of protein and major groove of DNA acting as interfaces.

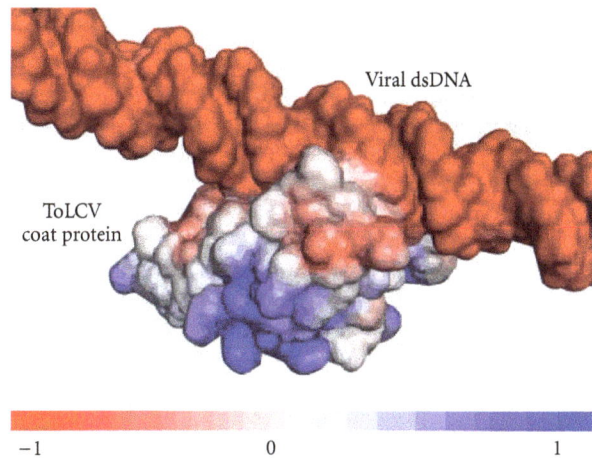

FIGURE 10: Electrostatic potential map of ToLCV coat protein-viral dsDNA complex. Neutral patch of coat protein interacts with negatively charged DNA.

3.10. Electrostatic Interaction of Coat Protein. DNA binding interface predicted by PreDs was found to be electrostatically favored as its prediction was principally based on electrostatic potential, local, and global curvatures present on the DNA surface. HADDOCK revealed coat protein's loop topology as its molecular interface unit. DNA-protein interaction is predominantly influenced by electrostatics and can be efficiently studied using adaptive Poisson-Boltzmann solver (APBS) approach. The docked conformation was charged appropriately based on physiological environment with preprocessing using PDB2PQR and all PB calculations were carried out using PBEQ-Solver in a coarse grid space. Electrostatic grid map analyzed using PyMol showed that neutral patches in coat protein were found to interact with viral DNA (Figure 10). We expected that an isosurface of positive patch will tend to interact with negatively charged viral DNA. Upon manual inspection of charged clusters in the coat protein, the positive charged clusters were cornered in surface with significantly low negative patches and a greater deal of positive regions corresponded to β-barrel motif. We ruled out the requirement of positive patch contributing to DNA interaction as geometrical flexibility of loop region in the coat protein having neutrality was much favored rather than charge-charge attraction. This view was further inferred by the best docked conformation in which electrostatic energy term was investigated and found to be the major intermolecular descriptor representing interaction.

Zrachya et al. 2007 reported that TYLCV coat protein could be targeted by small interfering ds RNAs (siRNAs) derived from intron-hairpin RNA construct to develop disease resistant transgenic tomato cultivars and showed its potential in *N. benthamiana* transient assays [51]. Similar studies targeting against antisense replicase gene (AC1) in ToLCV helped in developing trait-stable transgenics [52]. Besides, the midsized aggregation of coat protein inside

nucleus is associated with resistance whereas large aggregates leading to infection susceptibility [53]. We highlighted sequence regions in ToLCV coat protein possessing both DNA binding properties and the functional amino acids combination essential for virulence and these regions of interest can be targeted by developing siRNA. In addition, these molecular properties of ToLCV coat protein can be accounted in developing *Begomovirus* resistant-engineered plants.

4. Conclusion

ToLCV coat protein possesses DNA binding properties to function similar to BR1 nuclear shuttler. The amino acid combinations crucial for virulence were investigated through MSA and the evolutionary relationship was traced by phylogenetic analysis which indicated that the Indian strains are closely related in the context of geographical locations. The predicted NLS of ToLCV coat protein shares more similarity with experimentally known TYLCV coat protein NLS. Molecular modeling represented ToLCV coat protein as Geminate viral particle. Further, sequence- and structure-based approaches identified that ToLCV coat protein through its loop topology interacts with viral DNA as surface complementarity proven to be the major promoting factor followed by electrostatic interaction. We anticipate the conserved region of ToLCV coat protein prominent for DNA binding and functional sequence pattern can be targeted by RNA interference to develop disease-resistant transgenic tomato plants.

Acknowledgments

S. P. Kumar acknowledges generous support from DST, New Delhi as Innovation in Science Pursuit for Inspired Research (INSPIRE) Fellowship. The authors would like to thank three anonymous reviewers for their fruitful suggestions.

References

[1] B. Gronenbor, "The *Tomato yellow leaf curl virus* genome," in *Tomato Yellow Leaf Curl Virus Disease: Management, Molecular Biology, Breeding For Resistance*, H. Czosnek, Ed., vol. 8, pp. 67–84, Springer, Netherlands, 2007.

[2] M. J. Melzer, D. Y. Ogata, S. K. Fukuda et al., "First report of *Tomato yellow leaf curl* virus in Hawaii," *Plant Disease*, vol. 94, no. 5, pp. 641–643, 2010.

[3] A. A. Sanderfoot, D. J. Ingham, and S. G. Lazarowitz, "A viral movement protein as a nuclear shuttle: the geminivirus BR1 movement protein contains domains essential for interaction with BL1 and nuclear localization," *Plant Physiology*, vol. 110, no. 1, pp. 23–33, 1996.

[4] H. Liu, M. I. Boulton, C. L. Thomas, D. A. M. Prior, K. J. Oparka, and J. W. Davies, "Maize streak virus coat protein is karyophyllic and facilitates nuclear transport of viral DNA," *Molecular Plant-Microbe Interactions*, vol. 12, no. 10, pp. 894–900, 1999.

[5] T. Kunik, K. Palanichelvam, H. Czosnek, V. Citovsky, and Y. Gafni, "Nuclear import of the capsid protein of *Tomato yellow leaf curl virus* (TYLCV) in plant and insect cells," *Plant Journal*, vol. 13, no. 3, pp. 393–399, 1998.

[6] S. Chakraborty, P. K. Pandey, M. K. Banerjee, G. Kalloo, and C. M. Fauquet, "*Tomato leaf curl Gujarat virus*, a New begomovirus species causing a severe leaf curl disease of tomato in Varanasi, India," *Phytopathology*, vol. 93, no. 12, pp. 1485–1495, 2003.

[7] R. W. Briddon, J. K. Brown, E. Moriones et al., "Recommendations for the classification and nomenclature of the DNA-β satellites of *Begomoviruses*," *Archives of Virology*, vol. 153, no. 4, pp. 763–781, 2008.

[8] A. Kheyr-Pour, M. Bendahmane, V. Matzeit, G. P. Accotto, S. Crespi, and B. Gronenborn, "*Tomato yellow leaf curl virus* from Sardinia is a whitefly-transmitted monopartite geminivirus," *Nucleic Acids Research*, vol. 19, no. 24, pp. 6763–6769, 1991.

[9] P. Jyothsna, R. Rawat, and V. G. Malathi, "Predominance of *Tomato leaf curl Gujarat virus* as a monopartite *Begomovirus*: association with Tomato yellow leaf curl Thailand betasatellite," *Archives of Virology*. In press.

[10] P. Pandey, S. Mukhopadhya, A. R. Naqvi, S. K. Mukherjee, G. S. Shekhawat, and N. R. Choudhury, "Molecular characterization of two distinct monopartite *Begomoviruses* infecting tomato in india," *Virology Journal*, vol. 7, article no. 337, 2010.

[11] T. Kunik, R. Salomon, D. Zamir et al., "Transgenic tomato plants expressing the *Tomato yellow leaf curl virus* capsid protein are resistant to the virus," *Nature Biotechnology*, vol. 12, no. 5, pp. 500–504, 1994.

[12] S. K. Raj, R. Singh, S. K. Pandey, and B. P. Singh, "*Agrobacterium*-mediated tomato transformation and regeneration of transgenic lines expressing *Tomato leaf curl virus* coat protein gene for resistance against TLCV infection," *Current Science*, vol. 88, no. 10, pp. 1674–1679, 2005.

[13] D. Pratap, S. K. Raj, S. Kumar, S. K. Snehi, K. K. Gautam, and A. K. Sharma, "Coat protein-mediated transgenic resistance in tomato against a IB subgroup *Cucumber mosaic virus* strain," *Phytoparasitica*, vol. 40, no. 4, pp. 375–382, 2012.

[14] S. Chakraborty, R. Vanitharani, B. Chattopadhyay, and C. M. Fauquet, "Supervirulent pseudorecombination and asymmetric synergism between genomic components of two distinct species of *Begomovirus* associated with severe tomato leaf curl disease in India," *Journal of General Virology*, vol. 89, no. 3, pp. 818–828, 2008.

[15] N. Kirthi, S. P. Maiya, M. R. N. Murthy, and H. S. Savithri, "Evidence for recombination among the *Tomato leaf curl virus* strains/species from Bangalore, India," *Archives of Virology*, vol. 147, no. 2, pp. 255–272, 2002.

[16] D. A. Benson, I. Karsch-Mizrachi, D. J. Lipman, J. Ostell, and D. L. Wheeler, "GenBank," *Nucleic Acids Research*, vol. 34, pp. D16–D20, 2006.

[17] A. Marchler-Bauer and S. H. Bryant, "CD-Search: protein domain annotations on the fly," *Nucleic Acids Research*, vol. 32, pp. W327–W331, 2004.

[18] S. F. Altschul, T. L. Madden, A. A. Schäffer et al., "Gapped BLAST and PSI-BLAST: a new generation of protein database search programs," *Nucleic Acids Research*, vol. 25, no. 17, pp. 3389–3402, 1997.

[19] A. Bateman, E. Birney, L. Cerruti et al., "The pfam protein families database," *Nucleic Acids Research*, vol. 30, no. 1, pp. 276–280, 2002.

[20] S. Kosugi, M. Hasebe, T. Entani, S. Takayama, M. Tomita, and H. Yanagawa, "Design of peptide inhibitors for the importin α/β nuclear import pathway by activity-based profiling," *Chemistry and Biology*, vol. 15, no. 9, pp. 940–949, 2008.

[21] H. Liu, M. I. Boulton, and J. W. Davies, "Maize streak virus coat protein binds single- and double-stranded DNA *in vitro*," *Journal of General Virology*, vol. 78, no. 6, pp. 1265–1270, 1997.

[22] P. Rice, L. Longden, and A. Bleasby, "EMBOSS: the European molecular biology open software suite," *Trends in Genetics*, vol. 16, no. 6, pp. 276–277, 2000.

[23] M. A. Larkin, G. Blackshields, N. P. Brown et al., "Clustal W and Clustal X version 2.0," *Bioinformatics*, vol. 23, no. 21, pp. 2947–2948, 2007.

[24] J. Felsenstein, "PHYLIP- phylogeny inference package (version 3. 2)," *Cladistics*, vol. 5, no. 2, pp. 164–166, 1989.

[25] M. Johnson, I. Zaretskaya, Y. Raytselis, Y. Merezhuk, S. McGinnis, and T. L. Madden, "NCBI BLAST: a better web interface," *Nucleic acids research*, vol. 36, pp. W5–9W, 2008.

[26] F. C. Bernstein, T. F. Koetzle, and G. J. B. Williams, "The protein data bank: a computer based archival file for macromolecular structures," *Journal of Molecular Biology*, vol. 112, no. 3, pp. 535–542, 1977.

[27] J. J. Ward, L. J. McGuffin, K. Bryson, B. F. Buxton, and D. T. Jones, "The DISOPRED server for the prediction of protein disorder," *Bioinformatics*, vol. 20, no. 13, pp. 2138–2139, 2004.

[28] D. E. Kim, D. Chivian, and D. Baker, "Protein structure prediction and analysis using the Robetta server," *Nucleic Acids Research*, vol. 32, pp. W526–W531, 2004.

[29] L. A. Kelley and M. J. Sternberg, "Protein structure prediction on the web: a case study using the Phyre server," *Nature protocols*, vol. 4, no. 3, pp. 363–371, 2009.

[30] Tripos, A Certara Company, "Tripos benchware 3D explorer for interactive molecular visualizations and structure manipulations," St. Louis, Miss, USA, http://www.tripos.com/.

[31] G. N. Ramachandran, C. Ramakrishnan, and V. Sasisekharan, "Stereochemistry of polypeptide chain configurations," *Journal of Molecular Biology*, vol. 7, pp. 95–99, 1963.

[32] S. C. Lovell, I. W. Davis, W. B. Arendall et al., "Structure validation by Cα geometry: φ,ψ and Cβ deviation," *Proteins*, vol. 50, no. 3, pp. 437–450, 2003.

[33] E. Pascal, A. A. Sanderfoot, B. M. Ward, R. Medville, R. Turgeon, and S. G. Lazarowitz, "The geminivirus BR1 movement protein binds single-stranded DNA and localizes the cell nucleus," *Plant Cell*, vol. 6, no. 7, pp. 995–1006, 1994.

[34] M. R. Rojas, A. O. Noueiry, W. J. Lucas, and R. L. Gilbertson, "*Bean dwarf mosaic Geminivirus* movement proteins recognize

DNA in a form- and size-specific manner," *Cell*, vol. 95, no. 1, pp. 105–113, 1998.

[35] L. Wang and S. J. Brown, "BindN: a web-based tool for efficient prediction of DNA and RNA binding sites in amino acid sequences," *Nucleic Acids Research*, vol. 34, pp. W243–W248, 2006.

[36] Y. Tsuchiya, K. Kinoshita, and H. Nakamura, "PreDs: a server for predicting dsDNA-binding site on protein molecular surfaces," *Bioinformatics*, vol. 21, no. 8, pp. 1721–1723, 2005.

[37] M. van Dijk and A. M. Bonvin, "3D-DART: a DNA structure modelling server," *Nucleic Acids Research*, vol. 37, no. 2, pp. W235–W239, 2009.

[38] S. J. de Vries, M. van Dijk, and A. M. Bonvin, "The HAD-DOCK web server for data-driven biomolecular docking," *Nature protocols*, vol. 5, no. 5, pp. 883–897, 2010.

[39] S. Jo, M. Vargyas, J. Vasko-Szedlar, B. Roux, and W. Im, "PBEQ-Solver for online visualization of electrostatic potential of biomolecules," *Nucleic acids research*, vol. 36, pp. W270–W275, 2008.

[40] T. J. Dolinsky, J. E. Nielsen, J. A. McCammon, and N. A. Baker, "PDB2PQR: an automated pipeline for the setup of Poisson-Boltzmann electrostatics calculations," *Nucleic Acids Research*, vol. 32, pp. W665–W667, 2004.

[41] M. A. Lill and M. L. Danielson, "Computer-aided drug design platform using PyMOL," *Journal of Computer-Aided Molecular Design*, vol. 25, no. 1, pp. 13–19, 2011.

[42] A. O. Noueiry, W. J. Lucas, and R. L. Gilbertson, "Two proteins of a plant DNA virus coordinate nuclear and plasmodesmal transport," *Cell*, vol. 76, no. 5, pp. 925–932, 1994.

[43] K. Palanichelvam, T. Kunik, V. Citovsky, and Y. Gafni, "The capsid protein of *Tomato yellow leaf curl virus* binds cooperatively to single-stranded DNA," *Journal of General Virology*, vol. 79, no. 11, pp. 2829–2833, 1998.

[44] S. Qin, B. M. Ward, and S. G. Lazarowitz, "The bipartite geminivirus coat protein aids BR1 function in viral movement by affecting the accumulation of viral single-stranded DNA," *Journal of Virology*, vol. 72, no. 11, pp. 9247–9256, 1998.

[45] M. Hussain, S. Mansoor, S. Iram, A. N. Fatima, and Y. Zafar, "The nuclear shuttle protein of *Tomato leaf curl New Delhi virus* is a pathogenicity determinant," *Journal of Virology*, vol. 79, no. 7, pp. 4434–4439, 2005.

[46] M. R. Rojas, H. Jiang, R. Salati et al., "Functional analysis of proteins involved in movement of the monopartite *Begomovirus, Tomato yellow leaf curl virus*," *Virology*, vol. 291, no. 1, pp. 110–125, 2001.

[47] E. Noris, A. M. Vaira, P. Caciagli, V. Masenga, B. Gronenborn, and G. P. Accotto, "Amino acids in the capsid protein of *Tomato yellow leaf curl virus* that are crucial for systemic infection, particle formation, and insect transmission," *Journal of Virology*, vol. 72, no. 12, pp. 10050–10057, 1998.

[48] T. A. Jones and L. Liljas, "Structure of satellite tobacco necrosis virus after crystallographic refinement at 2.5 Å resolution," *Journal of Molecular Biology*, vol. 177, no. 4, pp. 735–767, 1984.

[49] W. Zhang, N. H. Olson, T. S. Baker et al., "Structure of the maize streak virus geminate particle," *Virology*, vol. 279, no. 2, pp. 471–477, 2001.

[50] M. S. Pinner, V. Medina, K. A. Plaskitt, and P. G. Markham, "Viral inclusions in monocotyledons infected by Maize streak and related viruses," *Plant Pathology*, vol. 42, no. 1, pp. 75–87, 1993.

[51] A. Zrachya, P. P. Kumar, U. Ramakrishnan et al., "Production of siRNA targeted against TYLCV coat protein transcripts leads to silencing of its expression and resistance to the virus," *Transgenic Research*, vol. 16, no. 3, pp. 385–398, 2007.

[52] S. V. Ramesh, A. K. Mishra, and S. Praveen, "Hairpin RNA-mediated strategies for silencing of tomato leaf curl virus AC1 and AC4 genes for effective resistance in plants," *Oligonucleotides*, vol. 17, no. 2, pp. 251–257, 2007.

[53] R. . Gorovitsa, A. Moshea, M. Kolotb, I. Sobola, and H. Czosneka, "Progressive aggregation of *Tomato yellow leaf curl virus* coat protein in systemically infected tomato plants, susceptible and resistant to the virus," *Virus Research*. In press.

5

Phylogenetic, Molecular, and Biochemical Characterization of Caffeic Acid *o*-Methyltransferase Gene Family in *Brachypodium distachyon*

Xianting Wu,[1,2] Jiajie Wu,[3] Yangfan Luo,[1] Jennifer Bragg,[1] Olin Anderson,[1] John Vogel,[1] and Yong Q. Gu[1]

[1] Western Regional Research Center, USDA-ARS, 800 Buchanan Street, Albany, CA 94710, USA
[2] Department of Plant Sciences, University of California, Davis, CA 95616, USA
[3] State Key Laboratory of Crop Biology, Shandong Agricultural University, 61 Daizong Avenue, Tai'an, Shandong 271018, China

Correspondence should be addressed to Yong Q. Gu; ygu1997@yahoo.com

Academic Editor: Pierre Sourdille

Caffeic acid *o*-methyltransferase (COMT) is one of the important enzymes controlling lignin monomer production in plant cell wall synthesis. Analysis of the genome sequence of the new grass model *Brachypodium distachyon* identified four COMT gene homologs, designated as *BdCOMT1*, *BdCOMT2*, *BdCOMT3*, and *BdCOMT4*. Phylogenetic analysis suggested that they belong to the COMT gene family, whereas syntenic analysis through comparisons with rice and sorghum revealed that *BdCOMT4* on Chromosome 3 is the orthologous copy of the COMT genes well characterized in other grass species. The other three COMT genes are unique to *Brachypodium* since orthologous copies are not found in the collinear regions of rice and sorghum genomes. Expression studies indicated that all four *Brachypodium* COMT genes are transcribed but with distinct patterns of tissue specificity. Full-length cDNAs were cloned in frame into the pQE-T7 expression vector for the purification of recombinant *Brachypodium* COMT proteins. Biochemical characterization of enzyme activity and substrate specificity showed that BdCOMT4 has significant effect on a broad range of substrates with the highest preference for caffeic acid. The other three COMTs had low or no effect on these substrates, suggesting that a diversified evolution occurred on these duplicate genes that not only impacted their pattern of expression, but also altered their biochemical properties.

1. Introduction

Temperate grains like wheat and barley, along with forage grasses, contribute greatly to the human food and animal feed supply. However, the large and complex genomes in these economically important grasses present challenges for genomics studies and map-based cloning of target genes for crop improvement. Similarly, although large perennial grasses like switchgrass and *Miscanthus* are being developed as dedicated herbaceous energy crops, our knowledge about the biological and genetic basis of important bioenergy traits remains limited [1–4]. *Brachypodium distachyon* (hereafter referred as *Brachypodium*) is an attractive experimental

system and genomics model for grass research. It has many desirable attributes (small physical stature, short generation time, easy growth requirement, etc.) and numerous freely available genomics resources (high quality genome sequence, EST collection, large-insert BAC libraries, expression/tilling microarray, T-DNA mutant population, etc.) [5]. Thus, *Brachypodium* can serve as a useful model system to address issues unique to grasses ranging from grain improvement to the development of superior bioenergy crops [5–7].

Plant cell walls are a complex composite structure composed of polysaccharides (cellulose and hemicellulose) that are embedded and crosslinked by other polymers and small

Phylogenetic, Molecular, and Biochemical Characterization of Caffeic Acid o-Methyltransferase Gene Family in Brachypodium distachyon

37

molecules (lignin, pectin, and ferulic acid). The type of hemicellulose and other components differs significantly between the grasses and dicots [8]. Briefly, primary grass cell walls contain glucuronoarabinoxylans and mixed-linkage glucans as the hemicelluloses, high levels of ferulic acid and p-coumaric acids, and the relatively little pectin and protein. Dicot primary cell walls contain xyloglucan, mannans, and glucomannans as the hemicelluloses and high levels of pectin and structural proteins. Secondary cell walls of both grasses and dicots are composed of cellulose, glucuronoarabinoxylans, and lignin. However, the structure of the glucuronoarabinoxylans differs between dicots and grasses. Given the compositional and architectural difference between grass and dicot cell walls, we have turned to *Brachypodium* as our model system to study grass cell walls.

The use of cellulosic biomass crops as a feedstock for the production of transportation fuel offers significant potential environmental and economic advantages. However, due to the difficulty in converting the sugars locked in cellulose and hemicellulose into fuel, our ability to use this renewable feedstock is limited. A deeper understanding of the genes that control cell wall biosynthesis and architecture may allow us to tailor the cell wall to be more amenable to conversion into biofuel [10–14]. Lignin in particular is an impediment to the production of biofuels and forage digestibility [15–20]. Due to crosslinking, it blocks access of cell wall degrading enzymes to cellulose and hemicellulose [21–23]. This makes it necessary to employ harsh pretreatment to disrupt the cell wall structure prior to enzyme hydrolysis. This harsh pretreatment create compounds that are inhibitory to organisms used to convert the free sugars into biofuels like ethanol and butanol. However, due to its phenolic ring structure, lignin is an energy dense compound and thus a high lignin content is desirable if the biomass is used for direct combustion.

Lignins are complex heteropolymers derived from three monolignols, p-coumaryl, coniferyl, and sinapyl alcohol. These monolignols produce p-hydroxyphenyl (H), guaiacyl (G), and syringyl (S) units, respectively, when incorporated into the lignin polymer. Although composed of only three building blocks, the composition and structure of lignin varies considerably within and among plants because the ratio of monomers varies and there is no repeating linkage structure. Rather, the lignin monomers undergo what appears to be a random cross-linking in the cell wall via several linkages. The S content and the S to G ratio are critical parameters that measure for characterizing lignin composition in the cell wall of angiosperm plants [15, 20]. Enzymes involved in the lignin biosynthesis pathway have been characterized in different plant species [24]. o-methylation modulates the chemical and physical properties of the lignin polymer. o-methylation mediated by o-methyltransferases (OMTs) transfers a methyl group from S-adenosyl-L-methionine (SAM) to the hydroxyl group of a methyl acceptor molecule. Plants contain a large family of OMT enzymes [24]. One of the essential OMTs in lignin biosynthesis is caffeoyl CoA 3-o-methyltransferase (CCoAOMT; EC 2.11.104) which is primarily responsible for the initial o-methylation

of the 3-hydroxyl group and specifically methylates the 3,4-dihydroxy substrate as a CoA-linked thioester. A second important OMT termed caffeic acid o-methyltransferase (COMT; EC 2.1.1.68) generally catalyzes the o-methylation of the 5-hydroxyl group of 3-methoxy-4,5-dihydroxy precursors. The function of both OMTs has been inferred from the effects on lignin composition through the downregulation of these enzymes in transgenic plants as well as in mutant lines affecting the expression of these genes. For instance, the COMT gene was first cloned in maize by differential screening of a root cDNA library [25]. Characterization of the maize COMT promoter indicated that it directed GUS expression to the xylem and other tissues undergoing lignifications [26]. Transgenic plants with a downregulation of the maize COMT showed a strong decrease of Klason lignin content, a decrease in syringyl units, lower p-coumaric acid contents, and the occurrence of an unusual 5-OH guaiacyl unit [27]. These features are similar to those observed in the maize bm3 (brown-midrib3) mutant lacking the functional COMT gene [27, 28]. Furthermore, it appeared that a decrease in COMT activity led to improved forage digestibility, suggesting that the downregulation of COMT might alters the overall cell wall organization in a way that walls become more accessible to bacterial enzymes [27]. The effects of caffeic acid o-methyltransferase (COMT) gene modification on reduction of S monomer level, which is linked directly to the lignin reduction, have been shown in other modified plant species [29–36].

OMT genes have been cloned and characterized in a number of plant species and provide valuable information for the study of the evolution, expression, and function of OMTs in plants [29, 32, 33, 36–40]. However, the characterization of OMT genes in *Brachypodium* has not been reported. In this study, we identified, isolated, and characterized four *Brachypodium* COMT genes. Comparative and phylogenetic analyses revealed that while related grass genomes (rice and sorghum) contain only one copy of COMT, *Brachypodium* possesses four copies located in three different chromosomes. Functional characterization indicated that only the orthologous copy has the biochemical properties similar to other COMTs in grass species although all the *Brachypodium* COMT genes are expressed during plant development. The isolation and characterization of *Brachypodium* COMTs will permit us to develop an efficient system to analyze these enzymes in the lignin biosynthesis pathway through forward and reverse genetics approaches in this tractable model species.

2. Results

2.1. Genomic Organization of COMT Genes in Brachypodium. To identify COMT genes in *Brachypodium*, the rice COMT gene (XP480185) and *Arabidopsis* COMT gene (AAB96879) were used in a BLASTP search against the *Brachypodium* genome database v1.0 annotation (http://www.brachypodium.org/). The top four hits matched annotated *Brachypodium* genes: Bradi1g14870, Bradi2g02380, Bradi2g02390, and Bradi3g16530, with e-values lower than $1e-100$ (data not shown). These four genes were renamed *BdCOMT1*, *BdCOMT2*, *BdCOMT3*, and *BdCOMT4*, respectively, in this study. From the fifth hit, the e-values dropped considerably to be higher than e^{-50} and are, therefore, considered distantly related to COMT. *BdCOMT1* and *BdCOMT4* are located on *Brachypodium* chromosomes 1 and 3, respectively, while both *BdCOMT2* and *BdCOMT3* are located on chromosome 2. Interestingly, the *BdCOMT2* and *BdCOMT3* genes are next to each other with inverted orientation (Figure 1). When the sequences of the four *BdCOMT* genes were blasted against both the rice and sorghum genome databases, they all retrieved a single strong match to the corresponding COMT gene, suggesting that only *Brachypodium* contains multiple copies of COMT gene.

To analyze the genomic organization of these *Brachypodium* genes, genomic regions surrounding the four *BdCOMT* genes were used to search for orthologous regions in the rice and sorghum genomes. The collinearity of these regions was compared to the orthologous regions in the rice and sorghum genomes (Figure 1). The rice and sorghum regions orthologous to the *BdCOMT4* region contained their single COMT genes (Figure 1(a)), indicating that *BdCOMT4* is the orthologous counterpart of the rice and sorghum COMT genes. The rice and sorghum regions orthologous to the other *Brachypodium* COMT gene regions did not contain orthologous COMT genes although the surrounding genes showed general collinearity (Figures 1(b) and 1(c)). This collinearity analysis supports the sequence blast search results indicating that multiple COMT genes exist only in *Brachypodium*.

2.2. Phylogenetic Analysis of COMTs in Brachypodium and Other Species.

To further confirm the evolutionary relatedness of the *Brachypodium* COMT genes, previously characterized COMTs from both dicot and monocot species were used to perform phylogenic analysis (Figure 2). Dicot COMTs formed one clade and monocot COMTs were grouped into another clade. BdCOMT4 is more closely related to the other grass COMTs than the other three *Brachypodium* COMTs, which formed a distinct clade separated from the orthologous COMT of different grass species. Nevertheless, the observation that BdCOMT1, BdCOMT2, and BdCOMT3 were grouped into monocot COMT clade suggests that duplicated *Brachypodium* COMT genes were generated after the divergence of dicot and monocot.

We also tested the possibility that BdCOMT1, BdCOMT2, and BdCOMT3 belong to other related o-methyltransferases, caffeoyl-CoA o-methyltransferase (CCoAOMT) groups, by including a number of characterized CCoAOMTs in our phylogenic analysis. These *Brachypodium* COMTs did not group with CCoAOMTs (see Supplemental Figure S1 available at http://dx.doi.org/10.1155/2013/423189), providing further supporting evidence that these *Brachypodium* COMT genes are likely to be paralogs of *BdCOMT4* that resulted from gene duplication.

2.3. Sequence Analyses of BdCOMTs.

ClustalW analysis was used to align the *Brachypodium* COMT genes at both nucleotide and protein sequence levels. The results showed that they shared over 70% nucleotide identity between any two sequences (data not shown). At the amino acid level, these COMTs shared at least 60% sequence similarity between any two sequences. As expected, BdCOMT2 and BdCOMT3 are the most closely related of the four BdCOMTs. The physical proximity indicated that they are derived from a tandem duplication event (Figure 1(c)). Sequence alignment with COMTs from other species indicated that BdCOMT4 protein shares 58%, 59%, 78%, and 83% sequence similarity with *Arabidopsis thaliana, Medicago, Zea mays,* and *Oryza sativa* COMT proteins, respectively.

From examination of crystal structure of the alfalfa COMT protein, amino acids responsible for catalytic sites and substrate binding sites have been inferred [41]. There are three catalytic sites predicted to be conserved within COMT genes. Two catalytic sites, E297 and E329, in MsCOMT are conserved among all four BdCOMT proteins (Figure 3). The other catalytic site H269 in MsCOMT is conserved in BdCOMT4 (H266) and BdCOMT1 (H270). However, in BdCOMT2 and BdCOMT3, it was replaced with N at this site. Thirteen substrate binding sites have been previously predicted based on MsCOMT crystal structure; twelve of these sites are conserved in BdCOMT4. The only difference in sites is a change at I316 in MsCOMT to V313 in BdCOMT4. It seems this mutation occurred in MsCOMT since OsCOMT, ZmCOMT, and AtCOMT all encode V at this position. However, this site was changed to D in BdCOMT1, and it was mutated to encode I in BdCOMT2. In comparison with BdCOMT4, more changes in these substrate binding sites were observed in BdCOMT2 (8 sites), BdCOMT3 (8 sites), and BdCOMT1 (7 sites). Thus, although BdCOMTs have an overall higher amino acid similarity among themselves when each is compared to MsCOMT, BdCOMT4 and MsCOMT showed higher conservation in the substrate binding sites. The methyl donor (SAM) binding sites were also analyzed. Among all five donor binding regions, there is only one site (W271 in MsCOMT) where BdCOMT2 and BdCOMT3 are similar to each other but different from other COMTs. Otherwise, all sites are generally conserved among BdCOMTs and COMT genes from other species (Figure 3).

2.4. Tissue-Specific Expression of BdCOMT Genes.

PCR primers specific for each *Brachypodium* COMT gene were designed to characterize the expression of these genes in various tissues including root, stem node, stem internode, leaf sheath, leaf blade and callus (Figure 4). In this semiquantitative RT-PCR analysis experiment, *BdCOMT4* was found to be expressed uniformly and constitutively in all the tissues examined as compared to the ubiquitous *Brachypodium* actin gene expression. In contrast, the other three COMT genes showed very different expression patterns. For example, *BdCOMT2* was only expressed in the internode, leaf sheath and leaf blade, but absent in the other tissues, whereas *BdCOMT3* appeared to be expressed in all the tissues with the highest expression level detected in root and leaf blade, suggesting that these tandemly duplicated genes are now regulated differently at the transcriptional level. *BdCOMT1* gene was strongly expressed in the stem node and internode but barely detected in the other tissues (Figure 4).

Phylogenetic, Molecular, and Biochemical Characterization of Caffeic Acid o-Methyltransferase Gene Family in Brachypodium distachyon

39

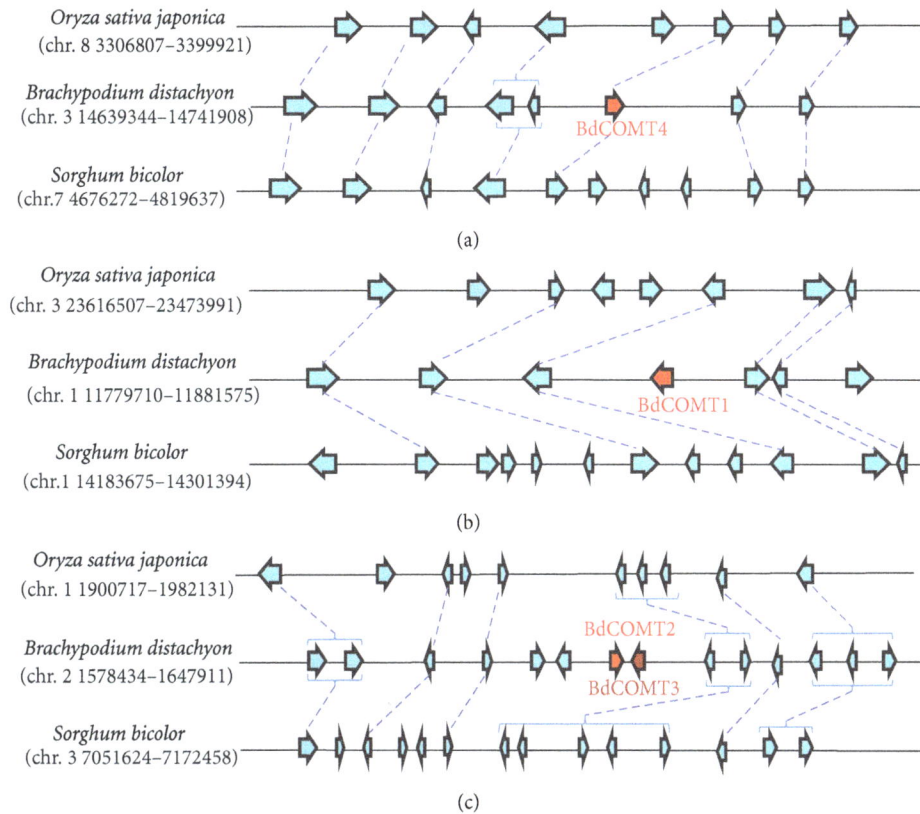

FIGURE 1: Collinearity analysis of *Brachypodium* COMT gene-containing regions with the orthologous regions of rice and sorghum genomes. Genomic regions harboring *BdCOMT4* (a), *BdCOMT1* (b), *BdCOMT2* (c), and *BdCOMT3* (c) were extracted and the annotated genes in these regions were used to identify orthologous regions from rice and sorghum genomes for collinearity analyses. Annotated genes are represented by arrows with the direction of arrow heads indicating the orientation of transcription. *Brachypodium* COMT genes are labeled in red. Orthologous genes that are shared in two or more genomes are connected with dashed lines. Duplicate genes in a genome are bracketed.

2.5. Expression and Biochemical Characterization of Recombinant BdCOMT Proteins. The gene expression analysis indicated that all the *Brachypodium* COMT genes are active genes. To biochemically characterize these *Brachypodium* COMTs, full-length cDNA for each COMT gene was isolated and cloned in frame into His-tagged pQE-T7 expression vector. The recombinant proteins were expressed in *E. coli* and purified with nickel-conjugated agarose beads (Figure 5(a)). Minor nonspecific protein contamination was detected in purified recombinant proteins despite imidazole concentrations in the wash series up to 50 mM (Figure 5(b)). Nevertheless, we estimated that the purified proteins reached at least 85% purity. The purified recombinant proteins were used for enzymatic activity assays with S-[methyl-^{14}C]-adenosyl-L-methionine (SAM) as the methyl donor and six substrates including caffeic acid, caffeoyl aldehyde, and four flavonoid compounds (Table 1). Among these substrates, BdCOMT4 showed the highest activity on caffeic acid, and this activity was set as 100% for comparison with other substrates. BdCOMT4 activity on caffeoyl aldehyde was less than half of that on caffeic acid (41.1%). BdCOMT4 also showed high catalytic activity on luteolin (85.0%). This flavonoid compound was a preferred substrate for rice COMT [38]. In addition, BdCOMT4 displayed activity on three other

flavonoid substrates at different levels, ranging from 17.0 to 31.1% (Table 1). These results suggest that BdCOMT4 prefers the most likely *in vivo* substrate, caffeic acid, but is also active on a range of substrates. Kinetic parameters of purified recombinant BdCOMT4 were determined with caffeic acid as the substrate. The apparent K_m and V_{max} values calculated based on the nonlinear Michaelis-Menten plot were 187 μM and 44 nkat/mg protein, respectively. Compared to the kinetic parameter of rice COMT on caffeic acid [38], BdCOMT4 has a higher K_m (rice COMT; K_m 69 μM) and a greater V_{max} (rice COMT; V_{max} 5.5 nkat/mg protein). In contrast to BdCOMT4, the three other purified *Brachypodium* COMTs failed to show significant enzymatic activity on the substrates examined (Table 1). Further studies will be required to determine if these proteins possess enzymatic activity with different substrates. However, based on sequence similarity and phylogenetic analysis result, currently, we still classified them as COMT genes in this study.

3. Discussion

To characterize COMT in the model grass *Brachypodium*, four COMT genes were identified at different locations

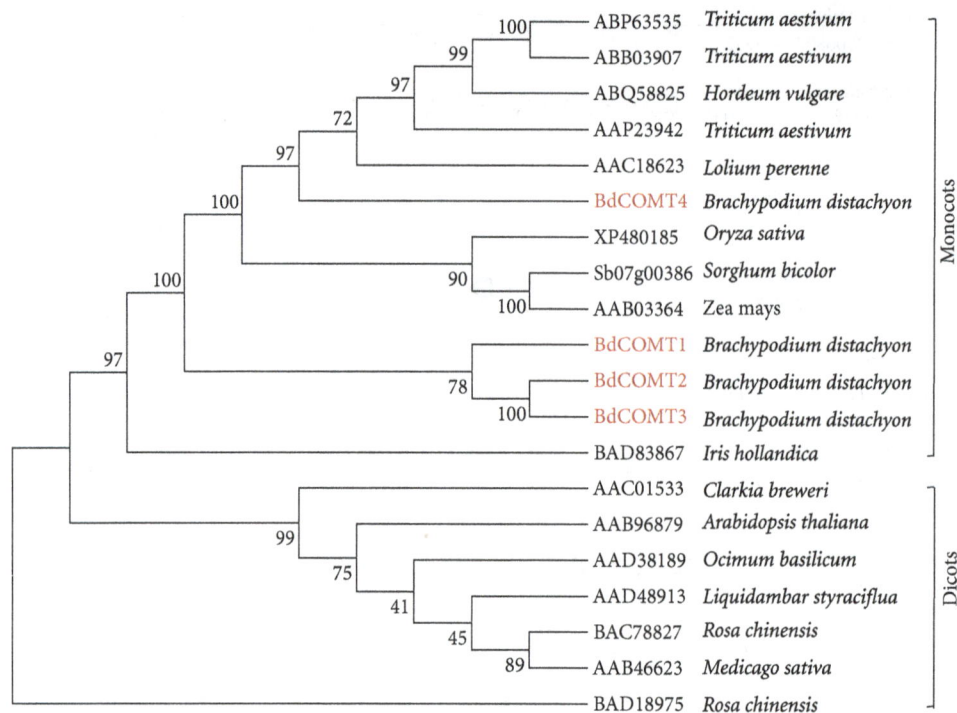

FIGURE 2: Phylogenetic analysis of COMTs from different plant species. COMT sequences represented by their GenBank or gene IDs from different plant species were extracted from NCBI database and used to construct a phylogenetic tree with phylogenic tree using neighbor-joining method.

in three chromosomes. Phylogenetic analysis suggests that they can be grouped into the COMT family. Comparative analysis indicated that BdCOMT4 is the orthologous copy of COMT genes in the rice and sorghum, while the other three *Brachypodium* COMT genes do not have orthologous copies in rice and sorghum genomes, suggesting that they are unique to *Brachypodium* (Figure 1). Phylogenetic analysis also indicated that they do not belong to CCoAOMTs, a different type of OMT closely related to COMT [24]. Therefore, BdCOMT4 is the ancestral COMT gene since its counterpart is present in rice and sorghum. The other *Brachypodium* COMT genes are likely to be paralogs of BdCOMT4, resulting from gene duplications and translocations into different locations. It has been shown that gene duplication followed by translocation often results in the presence of noncollinear genes in syntenic regions of two closely related genomes [42]. The fact that both rice and sorghum do not have these paralogs suggests that it is most likely that the gene duplications occurred after *Brachypodium* lineage diverged from the rice and sorghum lineages. Evolutionarily, *Brachypodium* is more closely related to wheat and barley since they both belong to the subfamily Pooideae. Rice belongs to the subfamily Ehrhartoideae and sorghum to the subfamily Panicoideae [5]. Collinearity analyses of *Brachypodium* with wheat and barley are very limited because their genome sequences are not available yet. Since wheat is polyploid species, we searched diploid barley EST collection for expressed COMT transcripts. Phylogenetic tree analysis indicated the coding regions of four barley EST contigs aligned with four *Brachypodium* COMT genes

(Supplemental Figure S2), suggesting both *Brachypodium* and Triticeae developed multiple COMT genes. However, a direct orthologous relationship among the four BdCOMTs and the four barley genes cannot be not determined here.

On the other hand, we also cannot exclude the possibility that COMT gene duplications occurred prior to the divergence of the *Brachypodium*, rice, and sorghum genomes, and duplicate paralogs were then deleted in the rice and sorghum lineages, resulting in their different evolutionary fates. The different fates of duplicated genes in different species can be explained by the birth-death evolution model in which new genes are created by gene duplication and some of these duplicate genes stay in the genome over time, whereas others are inactivated or deleted from the genome [43].

The birth-death model provides a plausible explanation for the evolution of the OMT gene family. Plant OMTs constitute a large family of enzymes that methylate the oxygen atom of a variety of secondary metabolites including phenylpropanoids, flavonoids, and alkaloids [24]. These enzymes play a key role in lignin biosynthesis, stress tolerance, and disease resistance in plants. Duplication of OMT genes provides raw materials for diversified evolution by mutations to create new OMT genes with different structures and functions. The four *BdCOMT* genes and those barley COMTs could represent a process of diversified evolution of duplicate genes in *Brachypodium*. These duplicate *Brachypodium* genes now display different expression patterns and tissue specificities in development compared to the ancestral *BdCOMT4* (Figure 4). In addition, sequence analysis indicated that the

Phylogenetic, Molecular, and Biochemical Characterization of Caffeic Acid o-Methyltransferase Gene Family in
Brachypodium distachyon

41

```
BdCOMT2 MA-------------EEEA-CMYALQLAVSSVLPMTLKTVIELGILETLVS-A----G 39
BdCOMT3 MA-------------EEEAGCMHALMLASSVVQPMAVRTAIELGLLEILVAGA----G 41
BdCOMT1 MGSARTEMVVPGAAGAGDEEA-CMYALQLAASSILPMTLKNAIELGMLEILVA------- 52
BdCOMT4 MGS--T---AADMAATADEEA-CMFALQLASSSILPMTLKNAIELGLLDTLVQAS----G 50
OsCOMT  MGS--T---AADMAAAADEEA-CMYALQLASSSILPMTLKNAIELGLLETLQSAAVAGGG 55
ZmCOMT  MGS--T---AGDVAAVVDEEA-CMYAMQLASSSILPMTLKNAIELGLLEVLQKEAG--GG 52
AtCOMT  MGS--TAETQLTPVQVTDDEA-ALFAMQLASASVLPMALKSALELDLLEIMAKN------ 51
MsCOMT  MGS--TGETQITPTHISDEEA-NLFAMQLASASVLPMILKSALELDLLEIIAKA-----G 52
        *:        ::**  :*  :*::*::*::::::**:..:..**:.*.::

BdCOMT2 REAPLTPEDLAAKLPAKAN--PEAASMVDRMLRVLASFNVVSCVVEEAKDGSLSRRYGPA 97
BdCOMT3 YGKTMSPEEVTAKLPT-SN--PEAASMVDRLLRVLASYSVVSCVVEEAKDGSLSRRYGPA 98
BdCOMT1 AGKTLSPSQVAERLQAKPG--PDAPAMLDRMLRLLASYNVVSCEVEEGQEGLLARRYGPA 110
BdCOMT4 KS--LTPAEVAAKLPSSSN--PAAPDMVDRMLRLLASYGVVSCAVEEGENGKLSRRYAAA 106
OsCOMT  KAALLTPAEVADKLPSKAN--PAAADMVDRMLRLLASYNVVRCEMEEGADGKLSRRYAAA 113
ZmCOMT  KAA-LAPEEVVARMPAAPSDPAAAAAMVDRMLRLLASYDVVRCQMED-RDGRYERRYSAA 110
AtCOMT  -GSPMSPTEIASKLPTKNP---EAPVMLDRILRLLTSYSVLTCSNRKLSGDGVERIYGLG 107
MsCOMT  PGAQISPIEIASQLPTTNP---DAPVMLDRMLRLLACYIILTCSVRTQQDGKVQRLYGLA 109
        ::*  ::. ::  :         *. *:**:**:*:.: :: *  .     * *. .

BdCOMT2 PVCKWLTPNEDGVSMAAFALAAQDKVHMATWPYMKDAVLEG-GDPFTKALGMSWFEYAGA 156
BdCOMT3 PVCKWLTPNEDGVSMAPFCLLAQDRVFTETWCYMKEAILEGRGGAFNKAFGTTWFEHAGV 158
BdCOMT1 PVCKWLTPNDDGVSMDPLALLIQDKVSMESWYHLKDVVLDG-GLPFNKAHGIIAFEYHGK 169
BdCOMT4 PVCKWLTPNEDGVSMAALALMNQDKVLMESWYYLKDAVLDG-GIPFNKAYGMSAFEYHGT 165
OsCOMT  PVCKWLTPNEDGVSMAALALMNQDKVLMESWYYLKDAVLDG-GIPFNKAYGMTAFEYHGT 172
ZmCOMT  PVCKWLTPNEDGVSMAALALMNQDKVLMESWYYLKDAVLDG-GIPFNKAYGMTAFEYHGT 169
AtCOMT  PVCKYLTKNEDGVSIAALCLMNQDKVLMESWYHLKDAILDG-GIPFNKAYGMSAFEYHGT 166
MsCOMT  TVAKYLVKNEDGVSISALNLMNQDKVLMESWYHLKDAVLDG-GIPFNKAYGMTAFEYHGT 168
        .*.*:*. *:****.  .: *  **:*   :*  ::*:::* * .*.** *   **: *

BdCOMT2 DTRFNRMYNEAMTHHSGIITKKFLELYTGLDGIGTLIDVGGGIGATIHAVTSKYPTIKGI 216
BdCOMT3 DTRFNNLFNEAMKQHSVIITKKLLELYKGFEGISVLVDVAGGVGATTHAITSRYPSIKGI 218
BdCOMT1 DARFDRVFNEAMKNHSTILTKKFLEFYTGFDDVKTLVDVGGGVGATIRAIISKYPHISGV 229
BdCOMT4 DPRFNRVFNEGMKNHSIIITKKLLDLYPGFEGLGTLVDVGGGVGATVGAIVARHPAIKGI 225
OsCOMT  DARFNRVFNEGMKNHSVIITKKLLDLYTGFDAASTVVDVGGGVGATVAAVVSRHPIRGI 232
ZmCOMT  DARFNRVFNEGMKNHSVIITKKLLDFYTGFEGVSTLVDVGGGVGATLHAITSRHPHISGV 229
AtCOMT  DPRFNKVFNNGMSNHSTITMKKILETYKGFEGLTSLVDVGGGIGATLKMIVSKYPNLKGI 226
MsCOMT  DPRFNKVFNKGMSDHSTITMKKILETYTGFEGLKSLVDVGGGTGAVINTIVSKYPTIKGI 228
        *.**:.::*:.*..** * **:*: * *::   :*:*.** **. : :::* . *:

BdCOMT2 NYDLPHVIADAPAYPGGRVQHVGGNMFEKVPSGADAILMKWILNCFRDEECATLLKNCYD 276
BdCOMT3 NFDLPHVVAEAPAYPGGRVQHVGGDMFEKVPSG-DAIFMKWILNCFSDKDCATLLKNCYD 277
BdCOMT1 NFDLPHVISSAPTCPG--VQHIGGDMFKKVPSG-DAILMKWILHDWTDDHCMMLLRNCYD 286

BdCOMT4 NFDLPHVISEGIPFPG--VTHVGGDMFQKVP-SGDAILMKWILHDWSDAHCATLLKNCYD 282
OsCOMT  NYDLPHVISEAPPFPG--VEHVGGDMFASVPRGGDAILMKWILHDWSDEHCARLLKNCYD 290
ZmCOMT  NFDLPHVISEAPPFPG--VRHVGGDMFASVP-AGDAILMKWILHDWSDAHCATLLKNCYD 286
AtCOMT  NFNLPHVIEDAPSHPG--IEHVGGDMFVSVPKG-DAIFMKWICHDWSDEHCVKFLKNCYE 283
MsCOMT  NFDLPHVIEDAPSYPG--VEHVGGDMFVSIPKA-DAVFMKWICHDWSDEHCLKFLKNCYE 285
        *::****:... ** : *:**:**  .:*  . **:*****  : * .*  :*:***:

BdCOMT2 ALPAHGKVINVECILPVNPDETPSARGLIQIDMSLLAYSPGGKERYLRELEKLAKGAGFA 336
BdCOMT3 ALPAHGKVINLECIMPVNPEPTHGAQGLISVDVSLLAYSPGGKERYLRELEKLAKGAGFA 337
BdCOMT1 ALPVGGKLIIIESILPVNPEATPRARMAFEDDMIMLTYTPGGKERYKREFEVLAKGARFA 346
BdCOMT4 ALPAHGKVVIVECILPVNPEATPKAQGVFHVDMIMLAHNPGGKERYEREFEELARGAGFT 342
OsCOMT  ALPEHGKVVVVECVLPESSDATAREQGVFHVDMIMLAHNPGGKERYEREFRELARAAGFT 350
ZmCOMT  ALPENGKVIVVECVLPVNTEATPKAQGVFHVDMIMLAHNPGGKERYEREFRELAKGAGFS 346
AtCOMT  SLPEDGKVILAECILPETPDSSLSTKQVVHVDCIMLAHNPGGKERTEKEFEALAKASGFK 343
MsCOMT  ALPDNGKVIVAECILPVAPDSSLATKGVVHIDVIMLAHNPGGKERTQKEFDLAKGAGFQ 345
        :** **:: *.::* .: : . *  :*::.****** :*:. **:.: *

BdCOMT2 AVKATYIYANFWAIEYTK-- 354
BdCOMT3 DVKATYIYADFWAIEYTK-- 355
BdCOMT1 SVRTTYIYANSWAIEYTK-- 364
BdCOMT4 GVKATYIYANAWAIEFTK-- 360
OsCOMT  GFKATYIYANAWAIEFTK-- 368
ZmCOMT  GFKATYIYANAWAIEFIK-- 364
AtCOMT  GIKVVCDAFGVNLIELLKKL 363
MsCOMT  GFKVHCNAFNTYIMEFLKKV 365
        .:.   .   :*  *
```

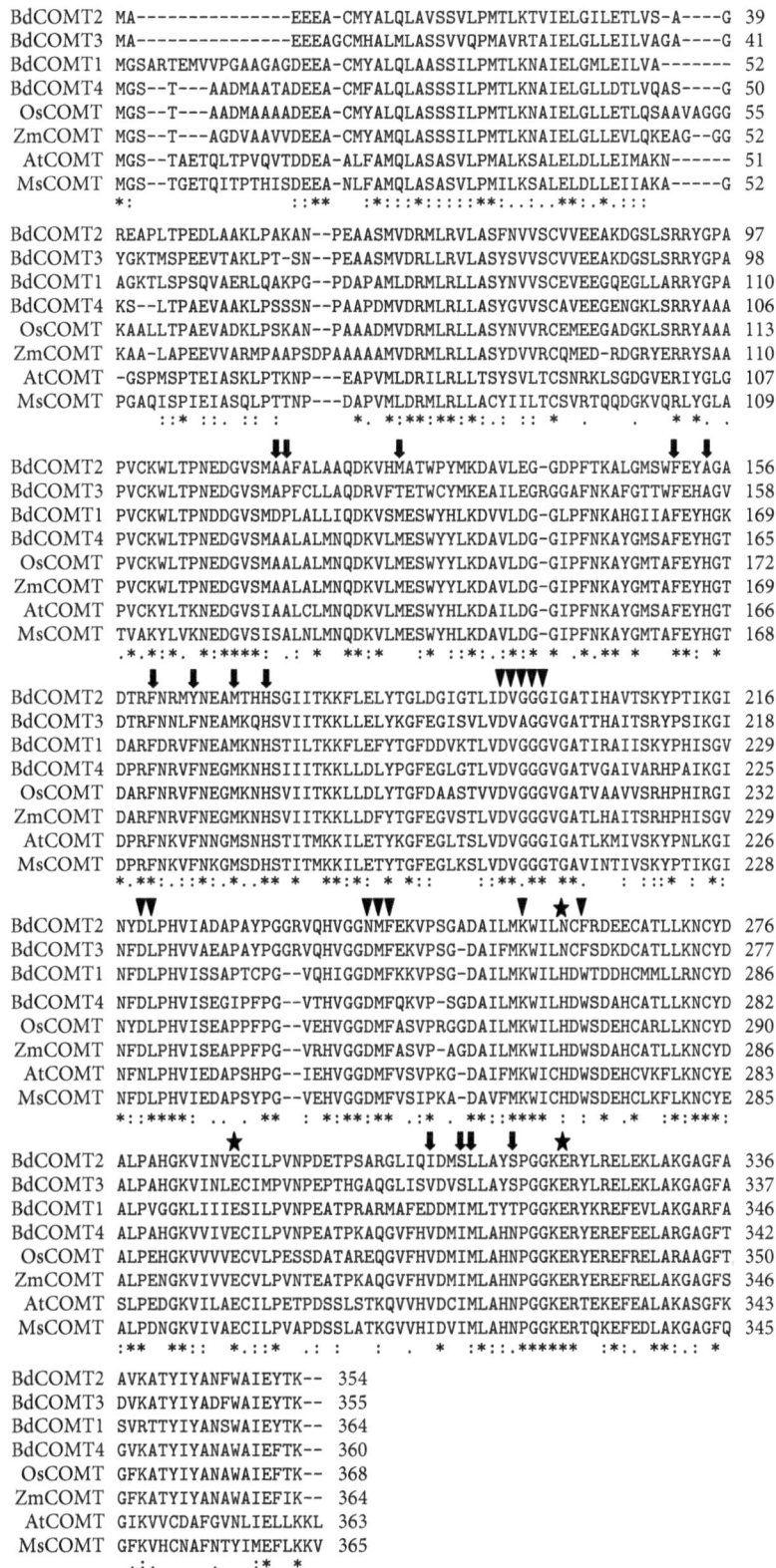

FIGURE 3: Comparison of amino acid sequences of four BdCOMTs (BdCOMT1, BdCOMT2, BdCOMT3, and BdCOMT4) with COMTs from *Oryza sativa* (OsCOMT), *Zea mays* (ZmCOMT), *Arabidopsis thaliana* (AtCOMT), and *Medicago sativa* (MsCOMT). Inferred from MsCOMT crystallographic structure, substrate binding residues are indicated by arrows, methyl donor binding sites by triangles, and catalytic residues by stars. An asterisk "∗" indicates identical residues in all sequences. A semicolon ":" indicates strongly conserved residues (score >0.5), and a dot "." indicates weaker conserved residues (score <0.5) [9].

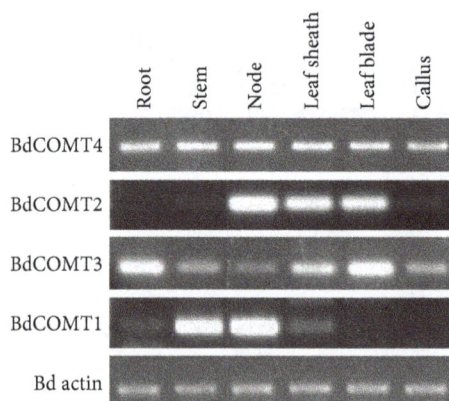

FIGURE 4: Tissue-specific expression of BdCOMT genes. Total RNAs were extracted from different *Brachypodium* tissues as indicated and used for synthesis of cDNA for RT-PCR. Gene-specific primers for each *Brachypodium* COMT genes were designed and used to amply corresponding transcripts. Primers specific for actin expression are used as the internal control to indicate that equal amount of cDNAs was used in each RT-PCR assay.

TABLE 1: Relative activity (%) of BdCOMT1 recombinant protein purified with six substrates (^{a}34 nkat mg^{-1}).

Substrate structure	Substrate name	BdCOMT4	BdCOMT1	BdCOMT2	BdCOMT3
	Caffeic Acid	100^a	1.34	0.95	0.92
	Caffeoyl Aldehyde	43.1	1.21	1.08	1.10
	Luteolin	83.5	1.13	0.95	1.03
	Pyrogallol	27.4	1.40	1.14	1.08
	Propyl Gallate	17.0	1.15	1.18	1.46
	Methyl Gallate	31.5	1.15	0.93	0.97

amino acids responsible for enzyme catalytic activity and donor binding are highly conserved in these proteins. Less conservation was observed in substrate binding sites (Figure 3), suggesting that they have evolved or are still evolving to have altered substrate binding properties. This notion is in agreement with the low or lack of activity of these proteins on the common COMT substrates (Table 1). Genes with high sequence similarity, but distinct substrate specificity, have been observed in the OMT gene family [33, 44]. Due to the changes in substrate specificity, they are often reclassified

Phylogenetic, Molecular, and Biochemical Characterization of Caffeic Acid o-Methyltransferase Gene Family in Brachypodium distachyon

43

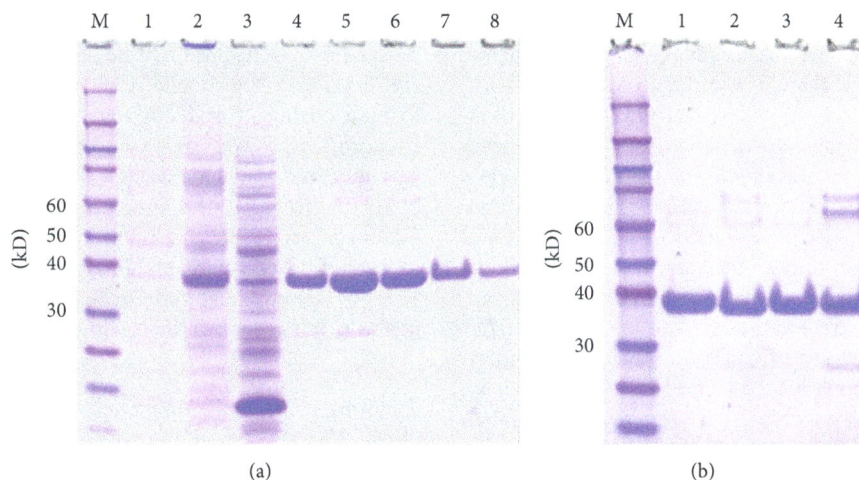

FIGURE 5: SDS-PAGE gel analysis of purification of recombinant BdCOMTs in *E. coli*. (a) Expression and purification of BdCOMT4 in *E. coli*. Lane M, protein size standards. Lane 1, total lysate of BL21(DE2) pLysS *E.coli.* containing pQE-BdCOMT1 before IPTG induction. Lane 2, total lysate of BL21(DE3)pLysS *E. coli.* containing pQE-BdCOMT1 with 1 mM IPTG induction for 4 hrs. Lane 3, flow-through lysate. Lanes 4–8, BdCOMT1 protein samples collected in different 100 μL fractions after adding the elution buffer. (b) SDS-PAGE gel analysis of purified BdCOMT proteins. Lane M, Protein size standards. Lane 1, 3 μg BdCOMT4; Lane 2, 3 μg BdCOMT2; Lane 3, 3 μg BdCOMT3; Lane 4, 3 μg BdCOMT1.

into different OMT classes [33]. Although the preferred substrates for these three *Brachypodium* COMTs were not identified in this study, the expression and conservation of amino acids responsible for enzyme activity suggest that they are likely to be active genes. The differential expression pattern in different tissues might suggest that these genes might have different roles in plant development. Further characterization of BdCOMT1, BdCOMT2, and BdCOMT3 by examining more substrates or gene silencing through transgenic approaches is needed to understand their functions in *Brachypodium*. It is also interesting to note that *Brachypodium* has the most compact genome of all the species examined and it tends to have fewer members within gene family [6]. These facts suggest that the BdCOMT1-3 have been retained because they might have adopted some other functions.

Enzymatic assays of the *Brachypodium* COMTs revealed that BdCOMT4 had strong COMT activity on caffeic acid and caffeoyl aldehyde substrates with > twofold preference for caffeic acid over caffeoyl aldehyde (Table 1). In addition, BdCOMT4 showed high activities on several flavonoid compounds. These results indicate that BdCOMT1's substrate specificity is quite similar to that of rice COMT [38], but differs from the specificity of the COMT (MsCOMT) from alfalfa, a dicot species. MsCOMT exhibited the highest activity on 5-hydroxyconiferaldehyde and caffeoyl aldehyde and the lowest activity on caffeic acid [45]. Low activity on caffeic acid substrates was also observed for COMTs from other dicots such as sweetgum and *Vanilla planifolia* [45–48]. In *Vanilla planifolia*, Van COMT showed higher preference for caffeoyl aldehyde to caffeic acid as a substrate. In addition, Van COMT has low activity on 1,2,3-trihydroxybenzene derivatives such as propyl gallate and methyl gallate, but these two substrates were preferred by Van OMT-2 and Van OMT-3, which are closely related to Van COMT at the sequence

level [33]. Interestingly, BdCOMT4 also showed considerable activity on flavonoid substrates such as propyl gallate and methyl gallate substrates, although the most preferred substrate is caffeic acid (Table 1). A similar preference for these different types of substrates was also observed in rice COMT [38]. Whether rice COMT and BdCOMT4 have roles in pathways involved in flavonoid compound biosynthesis is not known. It appears that the substrate preference for COMTs might even not be shared by the monocot species. It has been shown that several monocot COMTs have substrate preference for aldehydes over alcohols and acids such as ryegrass and sorghum [49, 50]. It is also interesting to note that the expression pattern of COMT is different among monocot species. For example, while BdCOMT4 and wheat COMT have a similar expression level in different tissues [51], COMT from maize is highly expressed only in roots [25]. COMT from ryegrass is abundantly expressed in stems [52]. Whether the differential expression of COMTs in different species suggests that they may serve additional function in some species in comparison to others is also not clear.

The difference in substrate specificity among COMTs from different plant species is intriguing. Several reports have demonstrated that a few or single amino acid changes can drastically alter substrate specificity among plant OMTs [35, 44]. Structure analysis of alfalfa COMT by crystallography suggests motifs or amino acids critical for enzyme activity on substrates [41]. Alignment of BdCOMT4 with COMTs from other species including alfalfa revealed high conservation on these amino acid sites; the change from I316 in MsCOMT to V313 in BdCOMT4 marks the only difference among the 13 active substrate binding sites (Figure 3). Interestingly, *Arabidopsis* COMT has valine at the corresponding position (Table 1) and its activity on caffeic acid is higher than that on caffeoyl aldehyde substrate [53]. In addition, this isoleucine in MsCOMT is changed to leucine with a similar amino acid structure in Van COMT in *Vanilla planifolia*. The substrate preference of Van COMT is also similar to MsCOMT with caffeoyl aldehyde as a more highly preferred substrate than caffeic acid [33]. The question of whether a single amino acid change at this site caused the difference in the substrate specificity toward caffeic acid and caffeoyl aldehyde among these COMTs merits further investigation. As we understand more of the effects of specific amino acids on enzyme activity, it may someday become possible to engineer COMT that can act on a particular substrate of interest.

Despite the importance of COMT in lignin biosynthesis, extensive characterization of COMTs has mainly been reported on dicot species such as *Arabidopsis* and alfalfa. Recent studies on the characterization of monocot COMTs using both biochemical and gene silencing approaches have greatly facilitated a better understanding of their functions in lignin biosynthesis in grass species [27, 38, 39, 49, 54]. Lignin content and composition are known to differ among plant species, contributing to the architecture difference of cell walls between dicots and monocots [8]. Several monocot species such as switchgrass and *Miscanthus* are targeted for development into an herbaceous biomass fuel crops. However, those dedicated biofuel crops are often not ideal for basic research due to their large physical stature and complex genome structure. The demonstration of many favorable characteristics of *Brachypodium* as a desirable experimental system promised to provide unique knowledge on grass biology including the lignin biosynthesis pathway. The notion that *Brachypodium* and *Miscanthus* have similar cell wall compositions further supports the use of *Brachypodium* as a model for investigating grass cell walls [55]. In this study, we showed that, by evolution, BdCOMT4 is an ortholog of COMTs in grasses including rice-and-sorghum-based gene collinearity analysis. It may also share similar biochemical and functional properties with COMTs in other grass species. Thus, knowledge gleaned from cell wall biology in *Brachypodium* could greatly benefit grass bioenergy crop research for sustainable fuel production.

4. Materials and Methods

4.1. Plant Materials. *Brachypodium* seeds (accession Bd21-3) were sown in a soilless mix (Supersoil, Rod McLellan Co., Marysville, OH, USA), placed in 4°C cold room in dark for a two-week vernalization period, and then transferred to a greenhouse with temperature range of 24°C in the day and 18°C at night with supplemental lighting to extend day length to 16 hours. Plant tissues (leaf blade, leaf sheath, stem node, stem internode, and root) were collected 3-4 weeks after growth in the greenhouse and frozen in liquid nitrogen for mRNA extraction.

4.2. Identification of Brachypodium COMT Genes and Collinearity Analysis. The rice and *Arabidopsis* COMT gene sequence were used in a BLASTN search against the *Brachypodium* genome database. Four hits matched annotated *Brachypodium* genes Bradi3g16530, Bradi2g02380,

Bradi2g02390, and Bradi1g14870. These genes were designated *BdCOMT1*, *BdCOMT2*, *BdCOMT3*, and *BdCOMT4*, respectively. Colinearity analysis was performed using CoGe's GEvo tool (http://genomevolution.org/CoGe/index.pl/). Genomic regions (~100 kb in size) surrounding *Brachypodium* COMT genes were analyzed by comparison with orthologous regions in rice and sorghum. Collinear genes present in more than two genomes were identified based on BLASTN algorithm with an *e*-value of $1e - 30$. Noncollinear gene sequences that are present in one genome, but absent in the other genomes, were also marked.

4.3. Protein Sequence Alignment. COMT protein sequences of alfalfa, *Arabidopsis*, maize, and rice were collected from the NCBI database. The four *Brachypodium* COMT protein sequences annotated in the genome database were confirmed with the full-length cDNA sequences amplified from *Brachypodium* total mRNA (see below). A total of eight COMT protein sequences were aligned with ClustalW2 (multiple sequence alignment tool) (http://www.ebi.ac.uk/Tools/msa/clustalw2/). Predicted substrate binding sites, methyl donor binding sites, and catalytic sites were inferred according to alfalfa COMT crystal structure analysis [41].

4.4. Phylogeny Analysis. The coding sequences of the four *Brachypodium* COMT genes were used to construct a phylogeny tree with available COMT sequences from different plant species by MEGA5 program with the neighbor-joining method. The evolutionary history was inferred using the neighbor-joining method [56]. The bootstrap consensus tree inferred from 1000 replicates was taken to represent the evolutionary history of the taxa analyzed [57]. Branches corresponding to partitions reproduced in less than 50% bootstrap replicates were collapsed. The percentage of replicate trees in which the associated taxa clustered together in the bootstrap test (1000 replicates) were shown next to the branches [57]. All positions containing gaps and missing data were eliminated from the dataset for analysis.

4.5. RT-PCR. Plant tissues were ground in liquid nitrogen and extracted with TRIzol Reagent (Invitrogen, Inc.). Total mRNA was quantified by ND-1000 Spectrophotometer (NanoDrop Technology, Inc.). cDNA was synthesized by the SuperScript III kit (Invitrogen, Inc.). PCR primers specific for each *Brachypodium* COMT gene were designed as below: BdCOMT1: forward, 5'-GAGGTCGAGGAGGGGCAG-GAG-3', reverse, 5'-CGCCAACATCCACGAGGGTCTT-3'. BdCOMT2: forward, 5'-TGAGCGCCGGCAGGGAAG-C-3', reverse, 5'-TGGCCTCGTTGTACATGCGGTTGA-3'; BdCOMT3: forward, 5'-CGTCTCCATGGCCCCCTT-CTG-3', reverse, 5'-CCGGCGACATCCACGAGGAC-3'; BdCOMT4: forward, 5'-CCGCCCTCGCGCTCATGAA-3' reverse, 5'-ACGTGGGGGAGGTCGAAGTTGAT-3'; The *Brachypodium* actin gene served as an expression control for RT-PCR. The primer set for amplifying its transcript is

Phylogenetic, Molecular, and Biochemical Characterization of Caffeic Acid o-Methyltransferase Gene Family in
Brachypodium distachyon

45

as follow: forward: 5$'$-AAGTACCCTATTGAGCATGG-3$'$;
reverse: 5$'$-CGTAGTCAAGAGCCACATATG-3$'$.

4.6. cDNA Isolation and Expression Vector Cloning. Full-length cDNA of each BdCOMT gene was amplified by gene-specific primers. To facilitate cloning, NdeI and XhoI restriction sites were added at the beginning of 5$'$ primers and in the end of 3$'$ primers, respectively. Amplified fragments were cloned in frame at the *NdeI* and *XhoI* sites into pQE-T7-1 vector (Qiagen, Inc.). Cloned plasmids were named as pQE-BdCOMT1, pQE-BdCOMT2, pQE-BdCOMT3, and pQE-BdCOMT4. Each plasmid was sequenced to confirm the sequence accuracy and then retransformed into *E. coli* Bd21 (DES) strains (Invitrogen, Inc.) for protein expression.

The cloning primers for each COMT gene are listed below: *BdCOMT4*: forward, 5$'$-GGAATTCCATAT-GAAACAGATGGGTTCCACGGCGGCGGA-3$'$; reverse, 5$'$-CCGCTCGAGTCTTACTACTACTTGGTGAAC-TCGATGG-3$'$; *BdCOMT2*: forward, 5$'$-GGAATTCCA-TATGAAACAGATGGCCGAGGAGGAGGCGTG-3$'$; reverse, 5$'$-CCGCTCGAGTCTTACTACTACTTGGTG-TACTCGATGGCCCAGA A-3$'$; *BdCOMT3*: forward, 5$'$-GGAATTCCATATGAAACAGATGGCGGAAGA-GGAAGCCGGGTG-3$'$; reverse, 5$'$-CCGCTCGAGTCT-TACTACTACTTGGTGTACTCGATGGCCCAGAA-3$'$; *BdCOMT1*: forward, 5$'$-GGAATTCCATATGAAACA-GATGGGCTCTGCCCGCACCG-3$'$; reverse, 5$'$-CCGCTC GAGTCTTACTACTATTTGGTGTACTCAATGGCC-CAGGAG-3$'$.

4.7. Protein Expression, Purification, and Quantification. A single *E. coli* clone carrying the expression construct was inoculated into 6 mL Luria broth (LB) medium with 50 μg/mL kanamycin and incubated in 37°C shaker overnight. The overnight culture (1 mL) was added to 500 mL (1 : 500 ratio) LB medium with 50 μg/mL kanamycin. When OD 600 of the culture reached 0.7, IPTG was added to the final concentration of 0.2 mM to induce protein expression and incubated at room temperature for 16 hours with shaking.

Cells were harvested and suspended in 30 mL lysis buffer (50 mM NaH_2PO_4, 300 mM NaCl, and 10 mM imidazole). A lysozyme was added to the final concentration of 1 mg/mL and the mixture was placed on ice for 30 min before sonication. *E. coli* soluble fraction was collected after centrifugation and incubated with 1 mL Ni-NTA agarose beads (Qiagen) by shaking at room temperature for 1 hour. Beads were washed three times in a column each with 50 mL lysis buffer containing imidazole with step increase of concentration at 20 mM, 30 mM, and 50 mM. Proteins were eluted by elution buffer (50 mM NaH_2PO_4, 300 mM NaCl, 200 mM imidazole, and 0.05% Tween 20) and stored at −20°C following addition of glycerol to a final concentration of 30%. Proteins were quantified by the BCA assay kit (Thermal Scientific, Inc.).

4.8. Enzyme Activity Assay. Assay reaction mixture contained 5 mM substrate, 3 μg COMT protein, 0.5 mM S-[me-thyl-^{14}C]-adenosyl-L-methionine (SAM) (47 mCi mmol^{-1}), 0.1 M Tris-HCl (pH = 7.5) in a final volume of 50 μL reaction system. After 20-minute incubation at 30°C, the reaction was stopped by adding 50 μL 0.2 M HCl and 200 μL hexane : ethyl acetate (1 : 1 ratio) mixture to extract methylated substrate. 20 μL of the supernatant layer containing ^{35}S-labeled methylated products from the reaction mixture was moved into 2 mL cocktail solution for liquid scintillation counting by LS6500 Scintillation System (Beckman Coulter, Inc.). Each assay was performed with three replicates and repeated three times. Control reaction was performed in the same manner without addition of COMT protein.

4.9. Determination of K_m and V_{max} Value of BdCOMT1 Protein. Steady kinetic curves were determined using different caffeic acid substrate concentrations under the same reaction conditions as described in the enzyme activity assay except 3 μL [^{14}C]SAM and 10 μL of unlabeled S-adenosyl-L-methionine (SAM) (Sigma, Inc.) were used. The substrate concentrations in this array were 0.05, 0.1, 0.25, 0.5, 1.0, and 2.0 mM. The K_m and V_{max} were calculated from the nonlinear Michaelis-Menten plot enzyme activity assay by Prism 5 for Windows (GraphPad Software, Inc.). The counts per second were calculated to nanomoles of product produced per second (nkat), based on the specific activity (SA) of the ^{14}C-labeled substrate and the disintegrations per Minute (DPMs) read by the scintillation counter.

Abbreviations

CCoAOMT: Caffeoyl CoA 3-*o*-methyltransferase
COMT: Caffeic acid *o*-methyltransferase
PCR: Polymerase chain reaction.

Authors' Contribution

X. Wu and J. Wu contributed equally to the work.

Conflict of Interests

The authors do not have a direct financial relation with the commercial identity mentioned in this work that might lead to a conflict of interests.

Acknowledgments

The authors thank Xiaohua He for the critical reading of this paper. This work was supported in part by the United States Department of Agriculture, Agriculture Research Service CRIS 5325-21000-015 and 532521000-17 and by the office of Science (DER), US Department of Energy, Interagency Agreement no. DE-AI02-07ER64452.

References

[1] N. C. Carpita and M. C. McCann, "Maize and sorghum: genetic resources for bioenergy grasses," *Trends in Plant Science*, vol. 13, no. 8, pp. 415–420, 2008.

[2] S. K. Chakrabarti, P. K. Roychoudhury, and P. K. Bajpai, "Biphasic biomethanation of wood-hydrolysate effluent," *Artificial Cells, Blood Substitutes, and Immobilization Biotechnology*, vol. 27, no. 5-6, pp. 461–467, 1999.

[3] A. M. Missaoui, A. H. Paterson, and J. H. Bouton, "Investigation of genomic organization in switchgrass (*Panicum virgatum* L.) using DNA markers," *Theoretical and Applied Genetics*, vol. 110, no. 8, pp. 1372–1383, 2005.

[4] M. N. Somleva, K. D. Snell, J. J. Beaulieu, O. P. Peoples, B. R. Garrison, and N. A. Patterson, "Production of polyhydroxybutyrate in switchgrass, a value-added cO-product in an important lignocellulosic biomass crop," *Plant Biotechnology Journal*, vol. 6, no. 7, pp. 663–678, 2008.

[5] J. Brkljacic, E. Grotewold, R. Scholl et al. et al., "Brachypodium as a model for the grasses: today and the future," *Plant Physiology*, vol. 157, pp. 3–13, 2011.

[6] J. P. Vogel, D. F. Garvin, T. C. Mockler et al., "Genome sequencing and analysis of the model grass *Brachypodium distachyon*," *Nature*, vol. 463, no. 7282, pp. 763–768, 2010.

[7] E. Wolny and R. Hasterok, "Comparative cytogenetic analysis of the genomes of the model grass *Brachypodium distachyon* and its close relatives," *Annals of Botany*, vol. 104, no. 5, pp. 873–881, 2009.

[8] J. Vogel, "Unique aspects of the grass cell wall," *Current Opinion in Plant Biology*, vol. 11, no. 3, pp. 301–307, 2008.

[9] J. D. Thompson, T. J. Gibson, F. Plewniak, F. Jeanmougin, and D. G. Higgins, "The CLUSTAL_X windows interface: flexible strategies for multiple sequence alignment aided by quality analysis tools," *Nucleic Acids Research*, vol. 25, no. 24, pp. 4876–4882, 1997.

[10] R. E. Nordon, S. J. Craig, and F. C. Foong, "Molecular engineering of the cellulosome complex for affinity and bioenergy applications," *Biotechnology Letters*, vol. 31, no. 4, pp. 465–476, 2009.

[11] J. Porter, R. Costanza, H. Sandhu, L. Sigsgaard, and S. Wratten, "The value of producing food, energy, and ecosystem services within an agrO-ecosystem," *Ambio*, vol. 38, no. 4, pp. 186–193, 2009.

[12] P. M. Schmitz and A. Kavallari, "Crop plants versus energy plants—on the international food crisis," *Bioorganic and Medicinal Chemistry*, vol. 17, no. 12, pp. 4020–4021, 2009.

[13] M. C. Suh, M. J. Kim, C. G. Hur et al., "Comparative analysis of expressed sequence tags from *Sesamum indicum* and *Arabidopsis thaliana* developing seeds," *Plant Molecular Biology*, vol. 52, no. 6, pp. 1107–1123, 2003.

[14] J. S. Yuan, K. H. Tiller, H. Al-Ahmad, N. R. Stewart, and C. N. Stewart Jr., "Plants to power: bioenergy to fuel the future," *Trends in Plant Science*, vol. 13, no. 8, pp. 421–429, 2008.

[15] F. Chen and R. A. Dixon, "Lignin modification improves fermentable sugar yields for biofuel production," *Nature Biotechnology*, vol. 25, no. 7, pp. 759–761, 2007.

[16] V. L. Chiang, "From rags to riches," *Nature Biotechnology*, vol. 20, no. 6, pp. 557–558, 2002.

[17] C. Fu, J. R. Mielenz, X. Xiao et al., "Genetic manipulation of lignin reduces recalcitrance and improves ethanol production from switchgrass," *Proceedings of the National Academy of Sciences of the United States of America*, vol. 108, no. 9, pp. 3803–3808, 2011.

[18] M. C. McCann and N. C. Carpita, "Designing the deconstruction of plant cell walls," *Current Opinion in Plant Biology*, vol. 11, no. 3, pp. 314–320, 2008.

[19] C. Weber, A. Farwick, F. Benisch et al., "Trends and challenges in the microbial production of lignocellulosic bioalcohol fuels," *Applied Microbiology and Biotechnology*, vol. 87, no. 4, pp. 1303–1315, 2010.

[20] J. K. Weng, X. Li, N. D. Bonawitz, and C. Chapple, "Emerging strategies of lignin engineering and degradation for cellulosic biofuel production," *Current Opinion in Biotechnology*, vol. 19, no. 2, pp. 166–172, 2008.

[21] M. Pauly and K. Keegstra, "Cell-wall carbohydrates and their modification as a resource for biofuels," *Plant Journal*, vol. 54, no. 4, pp. 559–568, 2008.

[22] W. S. York and M. A. O'Neill, "Biochemical control of xylan biosynthesis—which end is up?" *Current Opinion in Plant Biology*, vol. 11, no. 3, pp. 258–265, 2008.

[23] R. Vanholme, J. Ralph, T. Akiyama et al., "Engineering traditional monolignols out of lignin by concomitant up-regulation of F5H1 and down-regulation of COMT in Arabidopsis," *Plant Journal*, vol. 64, no. 6, pp. 885–897, 2010.

[24] K. C. Lam, R. K. Ibrahim, B. Behdad, and S. Dayanandan, "Structure, function, and evolution of plant O-methyltransferases," *Genome*, vol. 50, no. 11, pp. 1001–1013, 2007.

[25] P. Collazo, L. Montoliu, P. Puigdomènech, and J. Rigau, "Structure and expression of the lignin O-methyltransferase gene from *Zea mays* L.," *Plant Molecular Biology*, vol. 20, no. 5, pp. 857–867, 1992.

[26] M. Capellades, M. A. Torres, I. Bastisch et al., "The maize caffeic acid O-methyltransferase gene promoter is active in transgenic tobacco and maize plant tissues," *Plant Molecular Biology*, vol. 31, no. 2, pp. 307–322, 1996.

[27] J. Piquemal, S. Chamayou, I. Nadaud et al., "Down-regulation of caffeic acid O-methyltransferase in maize revisited using a transgenic approach," *Plant Physiology*, vol. 130, no. 4, pp. 1675–1685, 2002.

[28] F. Vignols, J. Rigau, M. A. Torres, M. Capellades, and P. Puigdomenech, "The brown midrib3 (bm3) mutation in maize occurs in the gene encoding caffeic acid O-methyltransferase," *Plant Cell*, vol. 7, no. 4, pp. 407–416, 1995.

[29] T. Goujon, R. Sibout, B. Pollet et al., "A new *Arabidopsis thaliana* mutant deficient in the expression of O-methyltransferase impacts lignins and sinapoyl esters," *Plant Molecular Biology*, vol. 51, no. 6, pp. 973–989, 2003.

[30] S. Guillaumie, D. Goffner, O. Barbier, J. P. Martinant, M. Pichon, and Y. Barrière, "Expression of cell wall related genes in basal and ear internodes of silking brown-midrib-3, caffeic acid O-methyltransferase (COMT) down-regulated, and normal maize plants," *BMC Plant Biology*, vol. 8, article 71, 2008.

[31] D. Guo, F. Chen, K. Inoue, J. W. Blount, and R. A. Dixon, "Downregulation of caffeic acid 3-O-methyltransferase and caffeoyl CoA 3-O-methyltransferase in transgenic alfalfa: impacts on lignin structure and implications for the biosynthesis of G and S lignin," *Plant Cell*, vol. 13, no. 1, pp. 73–88, 2001.

[32] P. Kota, D. Guo, C. Zubieta, J. Noel, and R. A. Dixon, "O-Methylation of benzaldehyde derivatives by "lignin specific" caffeic acid 3-O-methyltransferase," *Phytochemistry*, vol. 65, no. 7, pp. 837–846, 2004.

[33] H. M. Li, D. Rotter, T. G. Hartman, F. E. Pak, D. Havkin-Frenkel, and F. C. Belanger, "Evolution of novel O-methyltransferases from the *Vanilla planifolia* caffeic acid O-methyltransferase," *Plant Molecular Biology*, vol. 61, no. 3, pp. 537–552, 2006.

Phylogenetic, Molecular, and Biochemical Characterization of Caffeic Acid o-Methyltransferase Gene Family in
Brachypodium distachyon

47

[34] F. Lu, J. M. Marita, C. Lapierre et al., "Sequencing around 5-hydroxyconiferyl alcohol-derived units in caffeic acid *O*-methyltransferase-deficient poplar lignins," *Plant Physiology*, vol. 153, no. 2, pp. 569–579, 2010.

[35] Q. H. Ma and Y. Xu, "Characterization of a caffeic acid 3-*O*-methyltransferase from wheat and its function in lignin biosynthesis," *Biochimie*, vol. 90, no. 3, pp. 515–524, 2008.

[36] N. Yoshihara, M. Fukuchi-Mizutani, H. Okuhara, Y. Tanaka, and T. Yabuya, "Molecular cloning and characterization of *O*-methyltransferases from the flower buds of *Iris hollandica*," *Journal of Plant Physiology*, vol. 165, no. 4, pp. 415–422, 2008.

[37] S. Bout and W. Vermerris, "A candidate-gene approach to clone the sorghum Brown midrib gene encoding caffeic acid *O*-methyltransferase," *Molecular Genetics and Genomics*, vol. 269, no. 2, pp. 205–214, 2003.

[38] F. Lin, G. Yamano, M. Hasegawa, H. Anzai, S. Kawasaki, and O. Kodama, "Cloning and functional analysis of caffeic acid 3-*O*-methyltransferase from rice (*Oryza sativa*)," *Journal of Pesticide Science*, vol. 31, no. 1, pp. 47–53, 2006.

[39] Q. H. Ma, "The expression of caffeic acid 3-*O*-methyltransferase in two wheat genotypes differing in lodging resistance," *Journal of Experimental Botany*, vol. 60, no. 9, pp. 2763–2771, 2009.

[40] J. M. Zhou, Y. W. Seo, and R. K. Ibrahim, "Biochemical characterization of a putative wheat caffeic acid *O*-methyltransferase," *Plant Physiology and Biochemistry*, vol. 47, no. 4, pp. 322–326, 2009.

[41] C. Zubieta, P. Kota, J. L. Ferrer, R. A. Dixon, and J. P. Noel, "Structural basis for the modulation of lignin monomer methylation by caffeic acid/5-hydroxyferulic acid 3/5-*O*-methyltransferase," *Plant Cell*, vol. 14, no. 6, pp. 1265–1277, 2002.

[42] J. C. Schnable, M. Freeling, and E. Lyons, "Genome-wide analysis of syntenic gene deletion in the grasses," *Genome Biology and Evolution*, vol. 4, no. 3, pp. 265–277, 2012.

[43] M. T. Rutter, K. V. Cross, and P. A. Van Woert, "Birth, death and subfunctionalization in the *Arabidopsis* genome," *Trends in Plant Science*, vol. 17, no. 4, pp. 204–212, 2012.

[44] D. R. Gang, N. Lavid, C. Zubieta et al., "Characterization of phenylpropene *O*-methyltransferases from sweet basil: facile change of substrate specificity and convergent evolution within a plant *O*-methyltransferase family," *Plant Cell*, vol. 14, no. 2, pp. 505–519, 2002.

[45] K. Parvathi, F. Chen, D. Guo, J. W. Blount, and R. A. Dixon, "Substrate preferences of *O*-methyltransferases in alfalfa suggest new pathways for 3-*O*-methylation of monolignols," *Plant Journal*, vol. 25, no. 2, pp. 193–202, 2001.

[46] W. Boerjan, J. Ralph, and M. Baucher, "Lignin Biosynthesis," *Annual Review of Plant Biology*, vol. 54, pp. 519–546, 2003.

[47] L. Li, J. L. Popko, T. Umezawa, and V. L. Chiang, "5-Hydroxyconiferyl aldehyde modulates enzymatic methylation for syringyl monolignol formation, a new view of monolignol biosynthesis in angiosperms," *Journal of Biological Chemistry*, vol. 275, no. 9, pp. 6537–6545, 2000.

[48] K. Osakabe, C. C. Tsao, L. Li et al., "Coniferyl aldehyde 5-hydroxylation and methylation direct syringyl lignin biosynthesis in angiosperms," *Proceedings of the National Academy of Sciences of the United States of America*, vol. 96, no. 16, pp. 8955–8960, 1999.

[49] G. V. Louie, M. E. Bowman, Y. Tu, A. Mouradov, G. Spangenberg, and J. P. Noel, "Structure-function analyses of a caffeic acid *O*-methyltransferase from perennial ryegrass reveal the molecular basis for substrate preference," *Plant Cell*, vol. 22, no. 12, pp. 4114–4127, 2010.

[50] N. A. Palmer, S. E. Sattler, A. J. Saathoff, and G. Sarath, "A continuous, quantitative fluorescent assay for plant caffeic acid *O*-methyltransferases," *Journal of Agricultural and Food Chemistry*, vol. 58, no. 9, pp. 5220–5226, 2010.

[51] Q. H. Ma, Y. Xu, Z. B. Lin, and P. He, "Cloning of cDNA encoding COMT from wheat which is differentially expressed in lodging-sensitive and -resistant cultivars," *Journal of Experimental Botany*, vol. 53, no. 378, pp. 2281–2282, 2002.

[52] F. M. McAlister, C. L. D. Jenkins, and J. M. Watson, "Sequence and expression of a stem-abundant caffeic acid *O*-methyltransferase cDNA from perennial ryegrass (*Lolium perenne*)," *Australian Journal of Plant Physiology*, vol. 25, no. 2, pp. 225–235, 1998.

[53] S. G. A. Moinuddin, M. Jourdes, D. D. Laskar et al., "Insights into lignin primary structure and deconstruction from *Arabidopsis thaliana* COMT (caffeic acid *O*-methyl transferase) mutant *Atomt1*," *Organic and Biomolecular Chemistry*, vol. 8, no. 17, pp. 3928–3946, 2010.

[54] Y. Tu, S. Rochfort, Z. Liu et al., "Functional analyses of caffeic acid *O*-methyltransferase and Cinnamoyl-CoA-reductase genes from perennial ryegrass (*Lolium perenne*)," *Plant Cell*, vol. 22, no. 10, pp. 3357–3373, 2010.

[55] L. D. Gomez, J. K. Bristow, E. R. Statham, and S. J. McQueen-Mason, "Analysis of saccharification in *Brachypodium distachyon* stems under mild conditions of hydrolysis," *Biotechnology for Biofuels*, vol. 1, article 15, 2008.

[56] N. Saitou and M. Nei, "The neighbor-joining method: a new method for reconstructing phylogenetic trees," *Molecular Biology and Evolution*, vol. 4, no. 4, pp. 406–425, 1987.

[57] J. Felsenstein, "Confidence limits on phylogenies: an approach using the bootstrap," *Evolution*, vol. 39, pp. 783–791, 1985.

6

Mapping of Micro-Tom BAC-End Sequences to the Reference Tomato Genome Reveals Possible Genome Rearrangements and Polymorphisms

Erika Asamizu,[1] **Kenta Shirasawa,**[2] **Hideki Hirakawa,**[2] **Shusei Sato,**[2] **Satoshi Tabata,**[2]
Kentaro Yano,[3] **Tohru Ariizumi,**[1] **Daisuke Shibata,**[2] **and Hiroshi Ezura**[1]

[1] *Faculty of Life and Environmental Sciences, University of Tsukuba, 1-1-1 Tennodai, Tsukuba 305-8572, Japan*
[2] *Kazusa DNA Research Institute, 2-6-7 Kazusa-kamatari, Kisarazu 292-0818, Japan*
[3] *School of Agriculture, Meiji University, 1-1-1 Higashi-mita, Tama-ku, Kawasaki 214-8571, Japan*

Correspondence should be addressed to Erika Asamizu, asamizu@gene.tsukuba.ac.jp

Academic Editor: Pierre Sourdille

A total of 93,682 BAC-end sequences (BESs) were generated from a dwarf model tomato, cv. Micro-Tom. After removing repetitive sequences, the BESs were similarity searched against the reference tomato genome of a standard cultivar, "Heinz 1706." By referring to the "Heinz 1706" physical map and by eliminating redundant or nonsignificant hits, 28,804 "unique pair ends" and 8,263 "unique ends" were selected to construct hypothetical BAC contigs. The total physical length of the BAC contigs was 495, 833, 423 bp, covering 65.3% of the entire genome. The average coverage of euchromatin and heterochromatin was 58.9% and 67.3%, respectively. From this analysis, two possible genome rearrangements were identified: one in chromosome 2 (inversion) and the other in chromosome 3 (inversion and translocation). Polymorphisms (SNPs and Indels) between the two cultivars were identified from the BLAST alignments. As a result, 171,792 polymorphisms were mapped on 12 chromosomes. Among these, 30,930 polymorphisms were found in euchromatin (1 per 3,565 bp) and 140,862 were found in heterochromatin (1 per 2,737 bp). The average polymorphism density in the genome was 1 polymorphism per 2,886 bp. To facilitate the use of these data in Micro-Tom research, the BAC contig and polymorphism information are available in the TOMATOMICS database.

1. Introduction

Tomato (*Solanum lycopersicum*) is one of the most important vegetable crops cultivated worldwide. Tomato has a diploid (2n = 2x = 24) and relatively compact genome of approximately 950 Mb [1]. Recently, its genome has been completely sequenced by the international genome sequencing consortium [2].

Genetic linkage maps of tomato have been created by crossing cultivated tomato (*S. lycopersicum*) with several wild relatives, *S. pennellii*, *S. pimpinellifolium*, *S. cheesmaniae*, *S. neorickii*, *S. chmielewskii*, *S. habrochaites*, and *S. peruvianum* [3]. Introgression lines generated from a cross between *S. lycopersicum* and *S. pennellii* have contributed to the isolation of important loci and quantitative trait loci (QTLs) related

to fruit size by utilizing DNA markers on the Tomato-EXPEN 2000 genetic map [4–9]. Such interspecies genetic mapping is effective because the divergent genomes provide many polymorphic DNA markers. In contrast, intraspecies mapping is less popular in tomato because of the low genetic diversity within cultivated tomatoes that has resulted from the domestication process and subsequent modern breeding [10]. Recently, we developed SNP, simple sequence repeat (SSR), and intronic polymorphic markers using publicly available EST information and BAC-end sequences (BESs) derived from "Heinz 1706," a standard line for tomato genomics [11, 12], and applied these markers to create linkage maps between Micro-Tom and either Ailsa Craig, a greenhouse tomato, or M82, a processing tomato, by mapping 1,137 markers [12].

Mapping of Micro-Tom BAC-End Sequences to the Reference Tomato Genome Reveals Possible Genome Rearrangements and Polymorphisms

49

Micro-Tom, a dwarf cultivar, is regarded as a model cultivar for functional genomics of tomato because of several characteristics, including small size (20 cm plant height), short life cycle (3 months), existence of indoor cultivation protocols under normal fluorescent conditions, and high-efficiency transformation methods that have been developed for this line [13–15]. The dwarf phenotype of Micro-Tom is the result of mutations in at least two major recessive loci. *dwarf* (*d*) encodes a cytochrome P450 protein, which functions in the brassinosteroid biosynthesis pathway [16]. Another locus, *miniature* (*mnt*), is suggested to associate with gibberellin (GA) signaling without affecting GA metabolism, but the causal gene has not been identified to date [17]. In Japan, Micro-Tom genomics resources have been extensively accumulated, mainly in the framework of the National BioResource Project (NBRP) (http://tomato .nbrp.jp/indexEn.html). Large-scale ethyl methanesulfonate (EMS) and gamma-ray-mutagenized populations have been created, and visible phenotype data have been accumulated [18–20]. The availability of Micro-Tom genome sequence data will accelerate the mapping of mutant alleles.

BAC-end sequencing has been performed in the tomato standard line "Heinz 1706" genome project to order BAC clones along the chromosomes [21]. Currently, about 90,000 BESs are available at the Sol Genomics Network (SGN, http://solgenomics.net/). BAC-end sequencing has been conducted for other crop species. In the rice *indica* cultivar "Kasalath," 78,427 BESs were generated from 47,194 clones and mapped onto the "Nipponbare" reference genome. As a result, 12,170 paired BESs were mapped that covered 80% of the rice genome [22]. Recently, BAC-end sequencing has been performed in crop plants with higher genome complexity. BESs from a commercial sugarcane variety, an interspecific hybrid with complex ploidy, were generated to analyze microsynteny between sugarcane and sorghum [23]. In wheat, which has a complex hexaploid genome, the short arm of chromosome 3A was flow sorted to make a BAC library, and chromosome arm-specific BESs were generated for DNA marker development [24]. In switchgrass, more than 50,000 SSRs were identified from 330,000 BESs, and this enabled detailed analysis on the evolution of this species [25]. A low level of genetic variation has been observed for cultivated peanuts. Polymorphic SSRs were accumulated from the BESs and successfully used in the construction of a genetic map [26]. BAC-end sequencing can be useful as a resource for performing comparative genomic studies through mapping of the sequences to a reference genome and by facilitating the development of polymorphic DNA markers.

In the present study, we generated 93,682 single-pass end sequences from a Micro-Tom BAC library. To compare the structures between the reference tomato "Heinz 1706" genome, mapping of unique ends was performed, and possible genome rearrangements and polymorphisms were identified.

2. Materials and Methods

2.1. Micro-Tom BAC Library Construction. Micro-Tom (TOMJPF00001) seeds were obtained from the NBRP (MEXT, Japan) and sent to the Clemson University Genomics Institute (CUGI) for BAC library construction. The genomic DNA was partially digested, and fragments were cloned into the *Hin*d III site of pIndigoBAC536. A total of 55,296 clones in *Escherichia coli* DH10B cells were arrayed in 144 384-well plates.

2.2. End Sequencing of Micro-Tom BAC Clones. To analyze BESs, the BAC DNAs were amplified using a TempliPhi large-construction kit (GE Healthcare, UK), and the end sequences were analyzed according to the Sanger method, using a cycle sequencing kit (Big Dye-terminator kit, Applied Biosystems, USA) with a type 3730xl DNA sequencer (Applied Biosystems). The resulting sequence reads were quality checked with PHRED [27, 28], allowing the identification and removal of low-quality (QV < 20) sequences. The 93,682 reads clearing the quality criteria were submitted to DDBJ/GenBank with accession numbers FT227487-FT321168.

2.3. Mapping to the Reference Genome and Analyses. BES reads were subjected to similarity search using the BLASTN program [29, 30]. To isolate unique sequences from repetitive ones, 93,682 BESs were searched against the repeat database in ITAG2.3 (http://solgenomics.net/) using a cutoff E-value of less than 10^{-50}. The remaining sequences were searched against the published version of the "Heinz 1706" genome (SL2.40), which was accessed from the SGN database (http://solgenomics.net/). From all of the BLAST alignments, BESs were extracted according to the following criteria, suggested in a previous report [22]: (1) sequence identity > 90% and alignment coverage > 50%; (2) mapped positions of each pair of ends < 200 kb apart in the same chromosome; (3) direction of each paired end is correct; (4) BLASTN $E <$ 10^{-100}; (5) a minimum of one hit for one of the paired ends; (6) no redundant chromosomal locations. Sequence polymorphisms (SNPs and Indels) between Micro-Tom and "Heinz 1706" were predicted based on the BLASTN alignment. Since we did not allow a gap exceeding 27 bases, only Indels up to 26 bases in length were counted.

2.4. Database and Clone Distribution. Mapped data and SNP/Indel sites were made accessible through the database TOMATOMICS at http://bioinf.mind.meiji.ac.jp/tomatomics/. BAC clones are available upon request from NBRP tomato (http://tomato.nbrp.jp/indexEn.html).

3. Results

3.1. General Features of the Generated BESs. The BAC insert size distribution was deduced based on the mapping results. According to these results, 45.4% (6,396 out of 14,101) of the BACs ranged from 100 to 120 kb, with average and median sizes of 101.3 kb and 101.8 kb, respectively (Figure 1). By multiplying by the number of clones (55,296), this BAC library covers 5.9x of the 950 Mb tomato genome.

Micro-Tom BES mapping to the "Heinz 1706" genome was processed as indicated in Figure 2. By eliminating repetitive, redundant, and unmapped sequences, 28,804 "unique

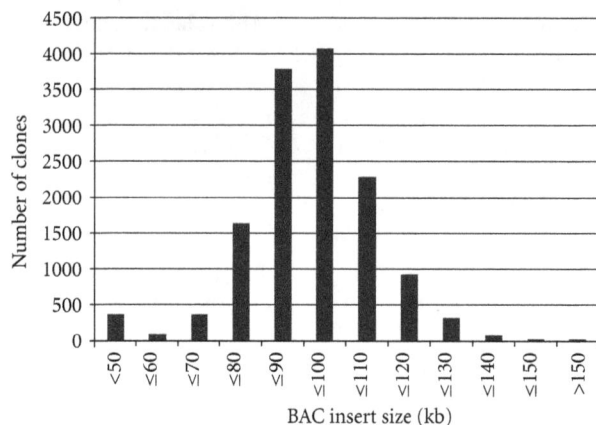

FIGURE 1: Distribution of BAC clone insert size. The insert size was deduced by mapping BESs onto the reference "Heinz 1706" genome (SL2.40).

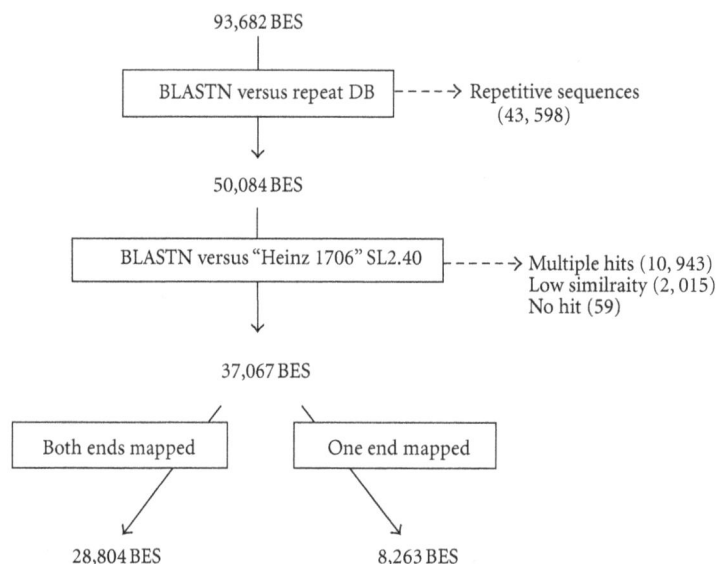

FIGURE 2: Flow of the BES analysis. To eliminate repetitive sequences, 93,682 BESs were initially searched against the repeat dataset of ITAG 2.3 with a BLASTN cutoff value of $E < 10^{-50}$. Next, the remaining sequences were mapped onto the "Heinz 1706" pseudomolecule sequences (SL2.40) under the following criteria: identity >90%, coverage >50%; $E < 10^{-100}$; the inclusion of single hits only. Mapped BESs were classified as either unique pair ends, for which both ends were mapped, or unique ends, for which only one end was mapped.

pair ends" and 8,263 "unique ends" were selected. Paired-end sequences were mapped onto the reference tomato genome sequence, and 2,248 hypothetical BAC contigs were constructed (see details at TOMATOMICS, http://bioinf.mind.meiji.ac.jp/tomatomics/). The integrity of the hypothetical contigs was confirmed by linking to the DNA markers on two genetic maps, AMF_2 and MMF_2 (see Supplementary Table 1 in Supplementary Material available online at doi:10.1155/2012/437026).

The genome coverage of the hypothetical BAC contigs was assessed by applying euchromatin/heterochromatin boundary information from the genetic map EXPEN2000 [2]. The results indicated that the euchromatin coverage ranged between 45.1% and 71.1% (average, 58.9%) among the different chromosomes, while heterochromatin coverage ranged between 57.4% and 75.3% (average, 67.3%). The total

physical length of the BAC contigs was 495,833,423 bp, covering 65.3% of the total chromosomes (Table 1).

3.2. Possible Genome Rearrangements. To assess the occurrence of genome rearrangements, Micro-Tom and the reference tomato "Heinz 1706" were compared. Possible inversions, translocations, and insertions were considered. To eliminate an artificial effect (e.g., chimeric BAC clones), only regions covered by more than two BAC clones were selected. After removing regions that had cleared the criteria for extraction (see Section 2) but were either shown to be multicopy by manual evaluation of the BLAST results or displayed similarity to transposable elements, we obtained two cases of a possible rearrangement between Micro-Tom and "Heinz 1706" (Table 2). On chromosome 2, a possible inversion was detected. The size of this inversion could be 20–220 kb

TABLE 1: Coverage of chromosomes by hypothetical Micro-Tom BAC contigs.

		SL2.40ch01	SL2.40ch02	SL2.40ch03	SL2.40ch04	SL2.40ch05	SL2.40ch06	SL2.40ch07	SL2.40ch08	SL2.40ch09	SL2.40ch10	SL2.40ch11	SL2.40ch12	Total (ch01–ch12)
Euchromatin	Chromosome length	27,903,720	24,734,122	16,423,960	13,871,288	10,836,573	17,576,248	17,480,118	15,552,430	10,522,300	9,129,273	11,175,203	12,034,427	**187,239,662**
Euchromatin	no. of Contigs	100	78	45	34	32	52	53	55	37	25	45	44	**600**
Euchromatin	no. of BACs	533	504	401	231	170	339	224	279	176	197	184	317	**3,555**
Euchromatin	Covered bases	17,310,734	14,644,412	11,678,941	6,261,540	5,377,576	10,621,719	8,336,310	9,541,847	6,047,365	5,473,701	6,855,876	8,119,058	**110,269,079**
Euchromatin	Uncovered bases	10,592,986	10,089,710	4,745,019	7,609,748	5,458,997	6,954,529	9,143,808	6,010,583	4,474,935	3,655,572	4,319,327	3,915,369	**76,970,583**
Euchromatin	% Coverage	62.0%	59.2%	71.1%	45.1%	49.6%	60.4%	47.7%	61.4%	57.5%	60.0%	61.3%	67.5%	**58.9%**
Heterochromatin	Chromosome length	62,400,524	25,184,172	48,416,754	50,193,024	54,184,865	28,465,388	47,788,503	47,480,227	57,139,791	55,705,032	42,210,822	53,451,826	**572,620,928**
Heterochromatin	no. of Contigs	175	74	147	131	169	76	135	150	149	159	128	155	**1,648**
Heterochromatin	no. of BACs	1,000	391	903	1,022	752	544	959	856	1,209	1,056	746	992	**10,430**
Heterochromatin	Covered bases	39,941,033	15,405,507	32,458,031	34,993,238	31,099,727	19,672,865	35,988,229	33,980,702	40,376,427	37,964,128	27,534,109	36,150,348	**385,564,344**
Heterochromatin	Uncovered bases	22,459,491	9,778,665	15,958,723	15,199,786	23,085,138	8,792,523	11,800,274	13,499,525	16,763,364	17,740,904	14,676,713	17,301,478	**187,056,584**
Heterochromatin	% Coverage	64.0%	61.2%	67.0%	69.7%	57.4%	69.1%	75.3%	71.6%	70.7%	68.2%	65.2%	67.6%	**67.3%**
Total	Chromosome length	90,304,244	49,918,294	64,840,714	64,064,312	65,021,438	46,041,636	65,268,621	63,032,657	67,662,091	64,834,305	53,386,025	65,486,253	**759,860,590**
Total	no. of Contigs	275	152	192	165	201	128	188	205	186	184	173	199	**2,248**
Total	no. of BACs	1,533	895	1,304	1,253	922	883	1,183	1,135	1,385	1,253	930	1,309	**13,985**
Total	Covered bases	57,251,767	30,049,919	44,136,972	41,254,778	36,477,303	30,294,584	44,324,539	43,522,549	46,423,792	43,437,829	34,389,985	44,269,406	**495,833,423**
Total	Uncovered bases	33,052,477	19,868,375	20,703,742	22,809,534	28,544,135	15,747,052	20,944,082	19,510,108	21,238,299	21,396,476	18,996,040	21,216,847	**264,027,167**
Total	% Coverage	63.4%	60.2%	68.1%	64.4%	56.1%	65.8%	67.9%	69.1%	68.6%	67.0%	64.4%	67.6%	**65.3%**

TABLE 2: Possible genome rearrangement events observed in the Micro-Tom and "Heinz 1706" genome.

No.	BAC	End1	Acc	Chr	Direction	From	To	End2	Acc	Chr	Direction	From	To	Possible event
1	MTBAC102D20	T7	FT290741	SL2.40ch02	—	29,374,874	29,375,640	SP6	FT290742	SL2.40ch02	—	29,494,209	29,494,781	*Inversion*
1	MTBAC084K15	T7	FT278701	SL2.40ch02	—	29,375,421	29,376,188	SP6	FT278702	SL2.40ch02	—	29,462,866	29,463,675	
2	MTBAC041L05	T7	FT251747	SL2.40ch03	—	6,601,537	6,602,368	SP6	FT251748	SL2.40ch03	—	55,664,754	55,665,559	*Translocation and Inversion*
2	MTBAC077O14	SP6	FT274148	SL2.40ch03	—	6,602,568	6,603,163	T7	FT274147	SL2.40ch03	—	55,665,296	55,666,020	

depending on which end of the BAC clone is inversed. Translocation and inversion were observed on chromosome 3. For each of two BAC clones (MTBAC041L05 and MTBAC077O14), one of the ends was mapped to 6,601 kb of chromosome 3, while the other end was mapped to 55,665 kb, more than 49 megabases apart. In addition, both ends were mapped on the minus strand.

3.3. Polymorphisms between Micro-Tom and the Reference Tomato. SNPs and Indels between Micro-Tom and "Heinz 1706" were identified. Among the SNPs and Indels found, 171,792 were mapped on 12 chromosomes, and 2,635 were mapped on pseudomolecules with no chromosomal information (SL2.40ch00 of the tomato whole-genome shotgun chromosomes) (Table 3 and Supplementary Table 2, see details at TOMATOMICS). According to these results, among the mapped SNPs and Indels, a total of 30,930 polymorphisms were found in the euchromatin (1 out of 3,565 bp), and 140,862 were found in the heterochromatin (1 out of 2,737 bp). The average polymorphism density in the genome was 1 polymorphism per 2,886 bp. Transversion-type SNPs were observed in 83,262 cases, while 60,631 were transition-type SNPs. Among the 30,534 Indels, single-base insertions (on the SL2.40 version of the tomato whole-genome shotgun chromosomes) were observed in 10,740 cases, and single-base deletions were seen in 17,064 cases. The remainder were larger Indels, ranging from 2 to 26 bp (Supplementary Table 2). Classification of polymorphisms regarding genic or intergenic regions is shown in Table 4.

4. Discussion

By selecting unique end sequences from 93,682 reads, 28,804 paired ends (14,402 pairs) and 8,263 unpaired ends were obtained. The majority of the nonselected sequences (43,598) were derived from repetitive regions. For the rest, 10,943 had redundant hits to the "Heinz 1706" genome, possibly including repetitive sequences that were not represented in the repeat database in ITAG2.3 (http://solgenomics.net/), 2,015 showed weak similarity, and 59 showed no similarity (Figure 2). Considering that the genome has been previously estimated to be composed of 25% gene-rich euchromatin [31, 32], BES selection in this study (39.6%, (28,804 + 8,263)/93,682)) could have eliminated repetitive regions to a moderate degree. We identified 59 reads showing no significant similarity to the "Heinz-1706" genome. Micro-Tom was bred by crossing the home-gardening cultivars, Florida Basket and Ohio 4013-3. The pedigree of Ohio 4013-3 suggested that a wild relative species was used in the breeding history [18, 33]. Such introgressed segments may lead to the introduction of genomic regions not harbored by "Heinz 1706." The Micro-Tom genome is now being sequenced (draft sequence data available at DDBJ with the accession number DRA000311), and mapping of orphan BESs to the *de novo* assembly of Micro-Tom genome data will help to clarify this question.

The total physical length of Micro-Tom BAC contigs was 495,833,423 bp, which covers approximately 65.3% of the

DNA from all 12 chromosomes. In the Kasalath rice BES analysis, chromosomal coverage in relation to the reference Nipponbare pseudomolecule was about 80%, despite the lower number (78,427) of analyzed BESs [22]. Because we used the same criteria for repetitive sequence selection ($E < 10^{-50}$), the discrepancy between the two studies might be due to the larger genome size of tomato (950 Mb) compared with rice (430 Mb) [34]. Our Micro-Tom BAC coverage is reasonable, taking into account the scale of the BAC library used.

Micro-Tom has been considered as a model cultivar to promote functional genomics studies of tomato by taking advantage of its characteristics. Currently, many tools and platforms have been developed, and some of these are already available to the research community. The present study characterized the overall polymorphisms found between Micro-Tom BESs and the reference tomato "Heinz 1706" genome. In addition, two possible genome rearrangement events, on chromosome 2 and chromosome 3, were observed (Table 2). In the case of translocation and inversion on chromosome 3, a gene annotated as reverse transcriptase was found in the flanking region (Solyc03g104840.1). We speculate that this region was translocated by the activity of a retrotransposon, as it was in the case of *SUN*. Enhanced expression of *SUN* caused by a gene duplication event mediated by the retrotransposon *Rider* led to an elongated fruit shape [35]. In the future, we plan to sequence the entire BAC and expect that this will help us to characterize these events in more detail. In the case of the other rearrangement possibility, on chromosome 2, we could not find any trace of a retrotransposon. Since these rearrangements took place in euchromatin, which is rich in genes, these regions could represent an interesting target to investigate their possible effects on phenotypic variation between Micro-Tom and the reference tomato.

We mapped the polymorphisms and depicted them, alongside maps showing covered regions and gaps, in Figure 3. On chromosomes 2, 5, and 11, polymorphisms seemed to be concentrated in the heterochromatic regions; however, this tendency was not clearly observed in the other chromosomes. For the other regions, the polymorphism discovery rate seemed to be somehow correlated with the BAC coverage. Although our analysis indicated little possibility of large-scale genome rearrangement between Micro-Tom and "Heinz 1706" (Table 2), this uneven polymorphism distribution suggests the existence of highly divergent chromosomal regions. The gaps in the hypothetical Micro-Tom BAC contigs could have resulted from low coverage of the BAC library, but the occurrence of chromosomal segments specific to either Micro-Tom or "Heinz 1706" is also possible. The ongoing Micro-Tom genome sequencing and *de novo* assembly of the Micro-Tom genome will clarify the genome structure in detail, enabling a more solid assessment of the differences between Micro-Tom and "Heinz 1706."

We had previously developed SNP markers among several cultivated tomatoes [12]. By selecting SNPs through *in silico* analysis using public EST information and previously developed SSR markers, 1,137 markers were obtained and successfully mapped on linkage groups between Micro-Tom

Mapping of Micro-Tom BAC-End Sequences to the Reference Tomato Genome Reveals Possible Genome Rearrangements and Polymorphisms

53

TABLE 3: Number of polymorphisms found in each chromosome.

		SL2.40ch01	SL2.40ch02	SL2.40ch03	SL2.40ch04	SL2.40ch05	SL2.40ch06	SL2.40ch07	SL2.40ch08	SL2.40ch09	SL2.40ch10	SL2.40ch11	SL2.40ch12	Total
Euchromatin	no. of polymorphisms	4,152	4,123	3,700	2,863	969	2,417	3,504	2,113	1,932	1,302	1,694	2,161	30,930
Euchromatin	Covered bases	17,310,734	14,644,412	11,678,941	6,261,540	5,377,576	10,621,719	8,336,310	9,541,847	6,047,365	5,473,701	6,855,876	8,119,058	110,269,079
Euchromatin	kb/polymorphism	4,169	3,552	3,156	2,187	5,550	4,395	2,379	4,516	3,130	4,204	4,047	3,757	3,565
Heterochromatin	no. of polymorphisms	12,319	10,694	10,408	9,995	30,951	5,134	8,347	8,562	10,231	9,209	14,937	10,075	140,862
Heterochromatin	Covered bases	39,941,033	15,405,507	32,458,031	34,993,238	31,099,727	19,672,865	35,988,229	33,980,702	40,376,427	37,964,128	27,534,109	36,150,348	385,564,344
Heterochromatin	kb/polymorphism	3,242	1,441	3,119	3,501	1,005	3,832	4,312	3,969	3,946	4,123	1,843	3,588	2,737
Total	no. of polymorphisms	16,471	14,817	14,108	12,858	31,920	7,551	11,851	10,675	12,163	10,511	16,631	12,236	171,792
Total	Covered bases	57,251,767	30,049,919	44,136,972	41,254,778	36,477,303	30,294,584	44,324,539	43,522,549	46,423,792	43,437,829	34,389,985	44,269,406	495,833,423
Total	kb/polymorphism	3,476	2,028	3,129	3,208	1,143	4,012	3,740	4,077	3,817	4,133	2,068	3,618	2,886

TABLE 4: Number of polymorphisms found in genic and intergenic regions in each chromosome.

			SL2.40ch01	SL2.40ch02	SL2.40ch03	SL2.40ch04	SL2.40ch05	SL2.40ch06	SL2.40ch07	SL2.40ch08	SL2.40ch09	SL2.40ch10	SL2.40ch11	SL2.40ch12	Total
Genic	Exon	3′ UTR	115	100	127	108	62	58	59	87	47	15	0	0	778
		5′ UTR	157	58	48	23	26	34	28	14	38	2	0	0	428
		CDS	1,152	967	1,031	758	820	603	757	605	669	570	662	615	9,209
	Intron	Intron	2,035	1,938	2,023	1,971	1,547	1,122	1,778	1,154	1,425	994	1,947	1,172	19,106
		Splice junction	19	15	12	13	14	15	13	8	8	7	12	12	148
Intergenic			12,993	11,739	10,867	9,985	29,451	5,719	9,216	8,807	9,976	8,923	14,010	10,437	142,123
Total			16,471	14,817	14,108	12,858	31,920	7,551	11,851	10,675	12,163	10,511	16,631	12,236	171,792

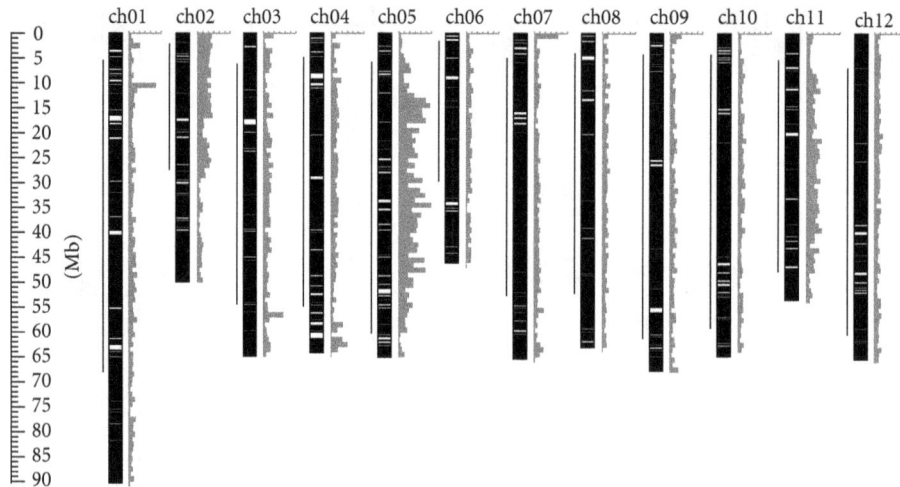

FIGURE 3: Micro-Tom BAC coverage with respect to the "Heinz 1706" chromosomes and detected polymorphisms. Black boxes indicate covered regions, and white boxes indicate gaps. Bars represent heterochromatic regions. The scale bars for polymorphisms indicate the number of SNPs or Indels per megabase (200 polymorphisms/scale).

and either Ailsa Craig or M82. In the present study, we identified 171,792 SNPs and Indels and mapped them on 12 chromosomes. The average density was 1 SNP per 3,565 bp in euchromatin and 1 SNP per 2,886 bp in the genome in general (including both euchromatin and heterochromatin). Previously, large-scale Micro-Tom full-length cDNA analysis and comparison of exon regions with those on the "Heinz 1706" genome revealed a mean sequence mismatch of 0.061% (1/1,640 bp) [36]. One possible explanation for the difference is the quality of the reference "Heinz 1706" genome sequence used in the two studies. We used the published version of the "Heinz 1706" genome sequence, which has higher coverage, giving rise to greater accuracy, although our selection may still contain sequence errors because BESs are single-pass sequences.

The information provided in this study will be useful in the development of DNA markers between Micro-Tom and cultivated tomatoes, which will facilitate a better understanding of the physiological and metabolic differences between them. It would also be useful in the genetic mapping of Micro-Tom mutants through the generation of F_2 segregating populations.

Authors' Contribution

E. Asamizu and K. Shirasawa equally contributed to this work.

Acknowledgments

The authors are grateful to Shinobu Nakayama and Akiko Watanabe for technical assistances. Construction of the Micro-Tom BAC library was supported by Japan Solanaceae Consortium (JSOL). This work was supported by the Genome Information Upgrading Program "Micro-Tom BAC-end sequencing" of the National BioResource Project in 2009.

References

[1] K. Arumuganathan and E. D. Earle, "Nuclear DNA content of some important plant species," Plant Molecular Biology Reporter, vol. 9, no. 3, pp. 208–218, 1991.

[2] Tomato Genome Consortium, "The tomato genome sequence provides insights into fleshy fruit evolution," Nature, vol. 485, no. 7400, pp. 635–641, 2012.

[3] M. R. Foolad, "Genome mapping and molecular breeding of tomato," International Journal of Plant Genomics, vol. 2007, Article ID 64358, 52 pages, 2007.

[4] I. Paran, I. Goldman, S. D. Tanksley, and D. Zamir, "Recombinant inbred lines for genetic mapping in tomato," Theoretical and Applied Genetics, vol. 90, no. 3-4, pp. 542–548, 1995.

[5] T. M. Fulton, J. C. Nelson, and S. D. Tanksley, "Introgression and DNA marker analysis of Lycopersicon peruvianum, a wild relative of the cultivated tomato, into Lycopersicon esculentum, followed through three successive backcross generations," Theoretical and Applied Genetics, vol. 95, no. 5-6, pp. 895–902, 1997.

[6] A. Frary, T. C. Nesbitt, A. Frary et al., "fw2.2: a quantitative trait locus key to the evolution of tomato fruit size," Science, vol. 289, no. 5476, pp. 85–88, 2000.

[7] J. Liu, J. Van Eck, B. Cong, and S. D. Tanksley, "A new class of regulatory genes underlying the cause of pear-shaped tomato fruit," Proceedings of the National Academy of Sciences of the United States of America, vol. 99, no. 20, pp. 13302–13306, 2002.

[8] B. Cong, L. S. Barrero, and S. D. Tanksley, "Regulatory change in YABBY-like transcription factor led to evolution of extreme fruit size during tomato domestication," Nature Genetics, vol. 40, no. 6, pp. 800–804, 2008.

[9] H. Xiao, N. Jiang, E. Schaffner, E. J. Stockinger, and E. Van Der Knaap, "A retrotransposon-mediated gene duplication underlies morphological variation of tomato fruit," Science, vol. 319, no. 5869, pp. 1527–1530, 2008.

[10] C. M. Rick, E. Kesicki, J. F. Fobes, and M. Holle, "Genetic and biosystematic studies on two new sibling species of Lycopersicon from interandean Peru," Theoretical and Applied Genetics, vol. 47, no. 2, pp. 55–68, 1976.

Mapping of Micro-Tom BAC-End Sequences to the Reference Tomato Genome Reveals Possible Genome
Rearrangements and Polymorphisms

55

[11] K. Shirasawa, E. Asamizu, H. Fukuoka et al., "An interspecific linkage map of SSR and intronic polymorphism markers in tomato," *Theoretical and Applied Genetics*, vol. 121, no. 4, pp. 731–739, 2010.

[12] K. Shirasawa, S. Isobe, H. Hirakawa et al., "SNP discovery and linkage map construction in cultivated tomato," *DNA Research*, vol. 17, no. 6, pp. 381–391, 2010.

[13] R. Meissner, Y. Jacobson, S. Melamed et al., "A new model system for tomato genetics," *Plant Journal*, vol. 12, no. 6, pp. 1465–1472, 1997.

[14] E. Emmanuel and A. A. Levy, "Tomato mutants as tools for functional genomics," *Current Opinion in Plant Biology*, vol. 5, no. 2, pp. 112–117, 2002.

[15] H. J. Sun, S. Uchii, S. Watanabe, and H. Ezura, "A highly efficient transformation protocol for Micro-Tom, a model cultivar for tomato functional genomics," *Plant and Cell Physiology*, vol. 47, no. 3, pp. 426–431, 2006.

[16] G. J. Bishop, K. Harrison, and J. D. G. Jones, "The tomato Dwarf gene isolated by heterologous transposon tagging encodes the first member of a new cytochrome P450 family," *Plant Cell*, vol. 8, no. 6, pp. 959–969, 1996.

[17] E. Martí, C. Gisbert, G. J. Bishop, M. S. Dixon, and J. L. García-Martínez, "Genetic and physiological characterization of tomato cv. Micro-Tom," *Journal of Experimental Botany*, vol. 57, no. 9, pp. 2037–2047, 2006.

[18] S. Watanabe, T. Mizoguchi, K. Aoki et al., "Ethylmethanesulfonate (EMS) mutagenesis of *Solanum lycopersicum* cv. Micro-Tom for large-scale mutant screens," *Plant Biotechnology*, vol. 24, no. 1, pp. 33–38, 2007.

[19] C. Matsukura, I. Yamaguchi, M. Inamura et al., "Generation of gamma irradiation-induced mutant lines of the miniature tomato (*Solanum lycopersicum* L.) cultivar 'Micro-Tom'," *Plant Biotechnology*, vol. 24, no. 1, pp. 39–44, 2007.

[20] T. Saito, T. Ariizumi, Y. Okabe et al., "TOMATOMA: a novel tomato mutant database distributing micro-tom mutant collections," *Plant and Cell Physiology*, vol. 52, no. 2, pp. 283–296, 2011.

[21] L. A. Mueller, S. D. Tanskley, J. J. Giovannoni et al., "The tomato sequencing project, the first cornerstone of the International Solanaceae Project (SOL)," *Comparative and Functional Genomics*, vol. 6, no. 3, pp. 153–158, 2005.

[22] S. Katagiri, J. Wu, Y. Ito et al., "End sequencing and chromosomal in silico mapping of BAC clones derived from an indica rice cultivar, Kasalath," *Breeding Science*, vol. 54, no. 3, pp. 273–279, 2004.

[23] T. R. Silva Figueira, V. Okura, F. R. da Silva et al., "A BAC library of the SP80-3280 sugarcane variety (*saccharum* sp.) and its inferred microsynteny with the sorghum genome," *BMC Research Notes*, vol. 5, p. 185, 2012.

[24] S. K. Sehgal, W. Li, P. D. Rabinowicz et al., "Chromosome arm-specific BAC end sequences permit comparative analysis of homoeologous chromosomes and genomes of polyploid wheat," *BMC Plant Biology*, vol. 12, p. 64, 2012.

[25] M. K. Sharma, R. Sharma, P. Cao et al., "A genome-wide survey of switchgrass genome structure and organization," *PLoS One*, vol. 7, no. 4, Article ID e33892, 2012.

[26] H. Wang, R. V. Penmetsa, M. Yuan et al., "Development and characterization of BAC-end sequence derived SSRs, and their incorporation into a new higher density genetic map for cultivated peanut (*Arachis hypogaea* L.)," *BMC Plant Biology*, vol. 12, p. 10, 2012.

[27] B. Ewing, L. Hillier, M. C. Wendl, and P. Green, "Base-calling of automated sequencer traces using phred. I. Accuracy assessment," *Genome Research*, vol. 8, no. 3, pp. 175–185, 1998.

[28] B. Ewing and P. Green, "Base-calling of automated sequencer traces using phred. II. Error probabilities," *Genome Research*, vol. 8, no. 3, pp. 186–194, 1998.

[29] S. F. Altschul, W. Gish, W. Miller, E. W. Myers, and D. J. Lipman, "Basic local alignment search tool," *Journal of Molecular Biology*, vol. 215, no. 3, pp. 403–410, 1990.

[30] S. F. Altschul, T. L. Madden, A. A. Schäffer et al., "Gapped BLAST and PSI-BLAST: a new generation of protein database search programs," *Nucleic Acids Research*, vol. 25, no. 17, pp. 3389–3402, 1997.

[31] X. B. Zhong, P. F. Fransz, J. W. V. Eden et al., "FISH studies reveal the molecular and chromosomal organization of individual telomere domains in tomato," *Plant Journal*, vol. 13, no. 4, pp. 507–517, 1998.

[32] Y. Wang, X. Tang, Z. Cheng, L. Mueller, J. Giovannoni, and S. D. Tanksley, "Euchromatin and pericentromeric heterochromatin: comparative composition in the tomato genome," *Genetics*, vol. 172, no. 4, pp. 2529–2540, 2006.

[33] J. W. Scott and B. K. Harbaugh, *MICRO-TOM: A Miniature Dwarf Tomato (Circular)*, Agricultural Experiment Station, Institute of Food and Agricultural Sciences, University of Florida, 1989.

[34] J. Yu, S. Hu, J. Wang et al., "A draft sequence of the rice genome (*Oryza sativa* L. ssp. indica)," *Science*, vol. 296, no. 5565, pp. 79–92, 2002.

[35] H. Xiao, N. Jiang, E. Schaffner, E. J. Stockinger, and E. Van Der Knaap, "A retrotransposon-mediated gene duplication underlies morphological variation of tomato fruit," *Science*, vol. 319, no. 5869, pp. 1527–1530, 2008.

[36] K. Aoki, K. Yano, A. Suzuki et al., "Large-scale analysis of full-length cDNAs from the tomato (*Solanum lycopersicum*) cultivar Micro-Tom, a reference system for the Solanaceae genomics," *BMC Genomics*, vol. 11, no. 1, article no. 210, 2010.

7

Gel-Based and Gel-Free Quantitative Proteomics Approaches at a Glance

Cosette Abdallah,[1,2] Eliane Dumas-Gaudot,[2] Jenny Renaut,[1] and Kjell Sergeant[1]

[1] Environment and Agro-Biotechnologies Department, Centre de Recherche Public-Gabriel Lippmann, 41 rue du Brill, 4422 Belvaux, Luxembourg
[2] UMR Agroécologie INRA 1347/Agrosup/Université de Bourgogne, Pôle Interactions Plantes Microorganismes ERL 6300 CNRS, Boite Postal 86510, 21065 Dijon Cedex, France

Correspondence should be addressed to Kjell Sergeant, Sergeant@lippmann.lu

Academic Editor: Hikmet Budak

Two-dimensional gel electrophoresis (2-DE) is widely applied and remains the method of choice in proteomics; however, pervasive 2-DE-related concerns undermine its prospects as a dominant separation technique in proteome research. Consequently, the state-of-the-art shotgun techniques are slowly taking over and utilising the rapid expansion and advancement of mass spectrometry (MS) to provide a new toolbox of gel-free quantitative techniques. When coupled to MS, the shotgun proteomic pipeline can fuel new routes in sensitive and high-throughput profiling of proteins, leading to a high accuracy in quantification. Although label-based approaches, either chemical or metabolic, gained popularity in quantitative proteomics because of the multiplexing capacity, these approaches are not without drawbacks. The burgeoning label-free methods are tag independent and suitable for all kinds of samples. The challenges in quantitative proteomics are more prominent in plants due to difficulties in protein extraction, some protein abundance in green tissue, and the absence of well-annotated and completed genome sequences. The goal of this perspective assay is to present the balance between the strengths and weaknesses of the available gel-based and -free methods and their application to plants. The latest trends in peptide fractionation amenable to MS analysis are as well discussed.

1. Introduction

"In the wonderland of complete sequences, there is much that genomics cannot do, and so the future belongs to proteomics, the analysis of complete complements of proteins" [1]

Originally coined by Wilkins et al. in 1996, proteomics by name is now over 15 years old. The term "proteome" refers to the entire PROTEin complement expressed by a genOME [2]. Proteomics is thus the large-scale analysis of proteins in a cell, tissue, or whole organism at a given time under defined conditions. The cutting-edge proteomics techniques offer several advantages over genome-based technologies as they directly deal with the functional molecules rather than genetic code or mRNA abundance. Even though there is only one definitive genome of an organism, it codes for multiple proteomes since the accumulation of a protein changes in relation to the environment and is the result of a combination of transcription, translation, protein turnover, and posttranslational modifications.

The field of proteomics has grown at an astonishing rate, mainly due to tremendous improvements in the accuracy, sensitivity, speed and throughput of mass spectrometry (MS), and the development of powerful analytical software. It appears to be gaining momentum as proteomic techniques become increasingly widespread and applied to an expanding smorgasbord of biological assays. Recently, proteomics has expanded from mere protein profiling to accurate and high-throughput protein quantification between two or multiple biological samples.

Most of the early developments in quantitative proteomics were driven by research on yeast and mammalian cell lines [3]. The incidence of proteomic studies on plants has increased over the past years but still lags behind human and animal proteomics, moreover model organisms and cash crops (e.g., *Arabidopsis* and rice) continue to be dominant in the plant proteomic literature. Most quantitative proteomic techniques used for human, animal, or other eukaryotic organisms can essentially also be employed for plant systems but plants, possessing distinct properties with regard to their genome, physiology, and culture, can impose high demands on proteomic sample handling. However, these advanced strategies have helped and facilitated the study of plant proteins and many new reports on differential expression, as well as global and organellar proteomic elucidation, have been put forth.

Quantitative proteomic approaches can be classified as either gel-based or gel-free methods as well as "label-free" or "label-based," of which the latter can be further subdivided into the various types of labelling approaches such as chemical and metabolic labelling. In the present work, the thorough description and current status of commonly used gel-based and -free proteomic methodologies is provided. An overview of their suitability, potential, and bottleneck applications in plant proteomics is discussed.

2. Gel-Based Proteomics

"Electrophoresis today and tomorrow: helping biologist's dreams come true" [4]

2.1. Two-Dimensional Gel Electrophoresis (2-DE): The Workhorse of Proteomics. Since it was first introduced in 1975 [5], 2-DE has evolved at different levels and became the workhorse of protein separation and the method of choice for differential protein expression analysis. Proteins first undergo isoelectric focusing (IEF) based on their net charge at different pH values and in the orthogonal second dimension further separation is performed based on the molecular weight (MW). This technique has an excellent resolving power, and today, it is possible to visualize over 10,000 spots corresponding to over 1,000 proteins, multiple spots containing different molecular forms of the same protein, on a single 2-DE gel [3]. Due to the pivotal problem of protein solubility, the overwhelming majority of electrophoretic protein separations is made under denaturing conditions. Two types of reagents are used in 2-DE buffers to ensure protein solubility and denaturation. The first type, chaotropes (e.g., urea and thiourea) used at multimolar concentrations, is able to unfold proteins by weakening noncovalent bonds (hydrophobic interactions, hydrogen bonds) between proteins [6]. The second one is ionic detergents, in which SDS (sodium dodecyl sulfate) is the archetype. It is made of a long and flexible hydrocarbon chain linked to an ionic polar head. The detergent molecules will bind through their hydrophobic hydrocarbon tail to hydrophobic amino acids. This binding favours amino acid-detergent interactions over amino acid-amino acid interactions, thereby promoting denaturation. Moreover, nonionic or zwitterionic detergents such as Triton X-100 are also used for protein solubilisation, since IEF requires low ion concentration in the sample [7]. The detection method postgel migration is achieved either

by the use of visible stains such as silver and Coomassie or fluorescent stains such as Sypro Ruby, Lava and Deep Purple.

Nevertheless, 2-DE has lately come under assault due to its known limitations and in part to the development of alternative MS-based approaches. Some of the reasons behind this trend include issues related to reproducibility [8], poor representation of low abundant proteins [9], highly acidic/basic proteins, or proteins with extreme size or hydrophobicity [10], and difficulties in automation of the gel-based techniques [11]. Moreover, the comigration of multiple proteins in a single spot renders comparative quantification rather inaccurate.

Although no technique has a better resolving power than classical 2-DE, many endeavours were made to step forward and make it suitable to study membrane proteins [7], and to overcome the protein ratio errors due to low gel-to-gel reproducibility by the inclusion of difference gel electrophoresis (2D-DIGE) [12]. This technique enables protein detection at subpicomolar levels and relies on preelectrophoretic labelling of samples with one of three spectrally resolvable fluorescent CyDyes (Cy2, Cy3, and Cy5). These dyes have an NHS-ester reactive group that covalently attaches to the ε-amino group of protein lysines via an amide linkage. The ratio of dye to protein is specifically designed to ensure that the dyes are limiting in the reaction and approximately cover 1-2% of the available proteins where only a single lysine per protein is labelled. Intergel comparability is achieved by the use of an internal standard (mixture of all samples in the experiment) labelled with Cy2 and coresolved on the gels that each contains individual samples labelled with Cy3 or Cy5. Since every sample is multiplexed with an equal aliquot of the same Cy2 standard mixture, each resolved feature can be directly related to the Cy2-labelled internal standard, and ratios can be normalized to all other ratios from other samples and across different gels. This can be done with extremely low technical variability and high statistical power [13–15].

For quantitative analysis, imaging software is required to align gel spots and measure their intensities. To this end, gels need to be digitalised either by using a scanner recording light transmitted through or reflected from the stained gel or fluorescent scanner. The images are subsequently imported into dedicated commercially available 2-DE image analysis softwares such as DeCyder (GE Healthcare), Proteomweaver (Bio-Rad), PDQuest (Bio-Rad), and Progenesis Same Spots (Nonlinear Dynamics). Most of these analysis software tools are user-friendly and allow (i) image alignment and spot matching across the gels, (ii) normalization, background adjustment and noise removal, (iii) spot detection, and (iv) quantification by calculation of the spot volumes and statistical analysis to highlighting differentially present proteins. Background cleaning allows the enhancement of the protein signal and distinguishes the noise from a spot. The global background correction consists of subtraction of all pixels below a set threshold of the maximum intensity. For matching, typically a reference gel is chosen and all gels are then automatically matched to the master one. Matching represents the most laborious step since frequent mistakes are made due to gel-to-gel and spot migration variability. Therefore, user intervention is needed to manually correct the software and improve the accuracy in spot matching. The quantification is performed through a summation of the pixel intensities localized within the defined spot area. The softwares use multivariate statistical packages, such as ANOVA (analysis of variance) based on spot size and intensity, spots are then assigned to P values, fold changes between groups. Most packages furthermore apply FDR (false discovery rates) or q-values to avoid the wrongful assignment of significant changes. PCA (Principle Component Analysis) is also often carried out. These available statistical tests make the 2-DE analyses and quantification more straightforward. However, the challenges associated with computational 2-DE analysis are technical problems such as experimental variation between gels and a high probability of piling several proteins under one spot.

Gel-based proteomics has so far been the main approach used in plant proteomics. 2D-DIGE has been successfully applied to investigate symbiosis- and pathogenesis-related protein in *Medicago truncatula* [16, 17] and to study the impact of abiotic stresses such as drought in oak [18], frost in *Arabidopsis* [19], ozone, and heavy metals in poplar [20–22].

2.2. Electrophoretic Separations of Native Proteins.

In their endeavour to study the protein complexes of the respiratory chain of mitochondria, Schägger and von Jagow developed a gel-based system able to separate protein complexes involved in oxidative phosphorylation in their native state [23]. This technique enables the separation of protein complexes under native conditions followed by the separation of individual proteins under denaturing conditions, thereby providing insight into the stoichiometry of the complexes. A charge-shifting agent, the dye Coomassie Brilliant Blue G-250, is added to the cathode buffer in order to stick to proteins conferring a uniform electric charge without unfolding the protein structure. Thus, intact protein complexes can be separated on a nondenaturing gradient gel roughly according to their MW, but the size and shape of each complex also influence how far that complex migrates into the gel. The gel lane is then cut out and separated on a second gel and orientated perpendicularly to the first axis of separation. This second dimension, a classic SDS-PAGE, is performed to separate the component proteins of each complex according to their MW. Blue Native-PAGE (BN-PAGE) studies were mainly focused on the analysis of electron transfer chain complexes in plastids and mitochondria; the potential application of this technique in plant proteomics was previously discussed and reviewed [24]. More recently, this strategy was used efficiently to analyze the proteome of wheat chloroplast protein complexes [25]. BN-PAGE was highly linked to membrane proteomics showing a deep interest to improve the hydrophobic proteome coverage of gel-based approaches [26].

BN-PAGE appears to be unsuitable to resolve small protein complexes (<100 kDa) due to the small separation distance of the first gel step, nevertheless a protocol for bacteria and eukaryotic cells allowing the identification of complexes in the range of 20–1,300 kDa was recently reported [27]. However, distinct complexes of similar molecular masses may comigrate and the constitutive proteins appear then to be present in the same complex. Despite the trick of the use of a charge-shifting agent, BN-PAGE is difficult to optimize and it is quite common to observe some trailing of the bands, which indicates insufficient protein solubilisation. To improve the resolution, three-dimensional electrophoresis can be performed, combining 2 variants of native electrophoresis in the first and second dimension, and SDS-PAGE in the third dimension [7].

2.3. One-Dimensional Gel Electrophoresis (1-DE): The Birth of Proteomics.

Soon after its inception, one-dimensional gel electrophoresis (1-DE) became the most popular method for at least two purposes: fast determination of protein MWand assessing the protein purity. Today, this widespread technique is used for many applications: comparison of protein composition of different samples, analysis of the number and size of polypeptide subunits, western blotting coupled to immunodetection, and, of course, as a second dimension in 2-DE maps.

Taking advantage of both gel-based protein and gel-free peptide separation properties 1-DE is, nowadays, coupled to subsequent analysis in liquid chromatography (LC) prior to MS. After protein separation on SDS gel, the entire gel lane is excised and divided into slices prior to the proteolytic digestion. Afterwards, peptide fractions are subjected to a second separation in LC prior to MS/MS analysis. The main advantages of this technique are the harsh ionic detergent use of the SDS that ensures protein solubility during the size-separation step and the reduced sample complexity prior to LC which renders the chance of identifying low abundant proteins higher. Recently comparisons of 1-DE-LC approach to other fractionation methods (e.g., cation exchange, isoelectric focusing, etc.), at both protein and peptide level, demonstrated its superior performance and higher proteome coverage [28–30]. Thus, by increasing the solubility (the major bottleneck in protein separations) and dwindling the complexity of the system by cutting the protein gel lane, 1-DE coupled to LC/MS analysis represents an attractive technique in proteomics studies. In plants, 1-DE-LC-MS/MS approach has been broadly applied, as an example the study on *M. truncatula* plasma membrane changes in response to arbuscular mycorrhizal symbiosis [31] and on *Arabidopsis thaliana* chloroplast envelope [32]. Lately, this approach has also been used for the compilation of a protein expression map of the *Arabidopsis* root providing the identity and cell type-specific localization of nearly 2,000 proteins [33].

3. Proteomics: From Gel-Based to Gel-Free Techniques

> *"A la carte proteomics with an emphasis on gel-free techniques"* [34]

Two-dimensional gel electrophoresis is a now a mature and well-established technique, however it suffers from some ongoing concerns regarding quantitative reproducibility and limitations on the ability to study certain classes of proteins.

Therefore in recent years, most developmental endeavours have been focused on alternative approaches, such as promising gel-free proteomics. With the appearance of MS-based proteomics, an entirely new toolbox has become available for quantitative analysis. In shotgun proteomics (bottom-up strategy) complex peptide fractions, generated after protein proteolytic digestion, can be resolved using different fractionation strategies, which offer high-throughput analyses of the proteome of an organelle or a cell type and provide a snapshot of the major protein constituents.

Although these novel approaches were initially pitched as replacements for gel-based methods, they should probably be regarded as complements to rather than replacements of 2-DE. There are many points of comparison and contrast between the standard 2-DE and shotgun analyses, such as sample consumption, depth of proteome coverage, analyses of isoforms and quantitative statistical power. Both platforms have the ability to resolve hundreds to thousands of features, so the choice between the different platforms is often determined by the biological question addressed. Currently there is no single method, which can provide qualitative and quantitative information of all protein components of a complex mixture. Ultimately, these approaches are both of great value to a proteomic study and often provide complementary information for an overall richer analysis.

4. Peptide Fractionation Procedures

"The introduction of multidimensional peptide resolving techniques is of unquestionable value for the characterization of complex proteomes" [35]

Since there is no method or instrument that is capable of identifying and quantifying the components of a complex sample in a single-step operation, there is ample evidence that high dimensional fractionation is required for deep exploration of complex proteomes and low abundant proteins. The basic principle of multidimensional fractionation is to separate peptides according to various orthogonal physicochemical properties and/or affinity interactions, resulting in much less complex fractions. There are numerous methodologies of separation available that can be used in tandem to perform a reduction in sample complexity. Each method has its own merits and drawbacks; therefore, the downstream needs of the workflow determine the optimal method for sample analysis.

4.1. Ion-Exchange Chromatography (IEC).
This type of chromatography involves peptide separation according to electric charge. In cation-exchange chromatography (CX), negative functional groups attract positively charged peptides at acidic pH, while in anion-exchange chromatography (AX), positive functional groups have affinity for negatively charged peptides at basic pH. Strong cation-exchange chromatography (SCX) encompasses a strong exchanger group that can be ionised over a broad pH range. For peptide separation using SCX columns, the peptide mixture is loaded under acidic conditions so that the positively charged peptides bind to the column. By increasing the salt concentration, peptides are displaced according to their charge, while by applying a pH gradient, peptides are resolved according to their isoelectric point (pI). Thus, positively charged peptides bind to the SCX column when the actual buffer pH is lower than their pI.

4.2. Reversed-Phase Chromatography (RP).
This most widespread LC-method applied in proteomics allows neutral peptide separation according to their hydrophobicity. The separation is based on the analyte partition coefficient between the polar mobile phase and the hydrophobic (nonpolar) stationary phase. The trapped peptides are then eluted using an organic phase gradient, usually acetonitrile. The ion-pair chromatography relies upon the addition of ionic compounds to the mobile phase to promote the formation of ion pairs with charged analytes. These reagents are comprised of an alkyl chain with an ionisable terminus. The introduction of ion-pair reagents increased the retention of charged analytes and improved peak shapes. Trifluoroacetic acid (TFA) and formic acid (FA) have been extensively used as ion-pairing reagents [35].

4.3. Two-Dimensional Liquid Chromatography (2D-LC).
Multidimensional analytical methods, having orthogonal separation power, are required to reduce sample complexity and increase the proteome coverage. The separation of peptide mixtures by 2D-LC has been performed using several orthogonal combinations such as AX coupled to RP (AX/RP), size exclusion chromatography coupled to RP (SEC/RP), and affinity chromatography coupled to RP (AFC/RP). In most shotgun proteomic analyses, the second dimension is performed by RP because the mobile phase is compatible with MS [36].

It has been shown that SCX is an excellent match to RP for multidimensional proteomic separations. In offline mode, the eluted fractions of the first dimension (SCX) are collected and then subjected to the second dimension (RP). Online approaches are faster with less sample loss due to the direct coupling of the two dimensions. In multidimensional protein identification technology (MudPIT) the SCX and RP stationary phase are packed together in the same microcapillary column. It was developed in the Yates laboratory and the results showed a high number of protein identifications, including low abundant ones [37]. This technology shows good separation power and presents a prime example of the enhanced proteome coverage in bottom-up proteomic approaches [38]. Several studies employed MudPIT in plant proteomics and its usage in this field was been previously reviewed [39, 40].

4.4. OFFGEL Electrophoresis (OGE).
The recently developed OFFGEL fractionator allows liquid-phase peptide IEF. The separation is carried out in a two-phase system with an upper liquid phase, containing carrier ampholytes and buffer-free solution, divided into 12 or 24 compartments and a lower phase, which is the IPG strip [41]. After sample loading into the wells and application of a voltage gradient, peptides migrate through the IPG strip until they reach their pI at

a given compartment. After IEF, peptides can be easily recovered in solution for further analysis. OGE has high loading capacity and resolution power [41]. Unlike LC fractionation, OGE provides additional physiochemical information such as peptide pI, which is a highly valuable tool to corroborate MS results, sort false positive rates, and increase the reliability of the identification procedure. While a study comparing MudPIT to OGE fractionation for the high-resolution separation of peptides revealed comparable results using both platforms [42], others showed that the IPG as a first dimension separation strategy is superior to SCX with a salt gradient [43] or pH gradient [44] for the analysis of complex mixtures. In contrast, Yang and coworkers reported that RP-LC offered better resolution and yielded more unique peptide and protein identifications in comparison to OGE in proteomic analysis of differentially expressed proteins in long-term cold storage of potato tubers [45]. During the last few years the use of OGE in plant proteomics has increased. Its application allowed the recovering of wheat soluble proteins extracted from leaves [46]. OGE was furthermore compared to classical IEF on microsomal fractions of 5 plant species. OGE performed slightly better in the identification of proteins with transmembrane domains and significantly increased the number of proteins in the alkaline range [47]. Finally, this technique has also been used on microsomal proteins extracted from *M. truncatula* roots to investigate the iTRAQ labelling effect on peptide isoelectric point and thus their focusing behaviour in OGE [48].

The long running time of OGE (which varies from few hours to 2-3 days) in comparison with other offline technique was the main disadvantage associated to this novel technique.

5. MS-Based Quantitation

"Mass spectrometry-based proteomics turns quantitative" [49]

In the last decade, MS has known a tremendous progress in proteomics and has increasingly established itself as a key tool for the analysis of complex protein samples notably after the availability of protein sequence databases and the development of more sensitive and user-friendly MS equipment [50]. A new toolbox of label-based and label-free quantitative proteomic methods is currently available. "To label or not to label," to answer this question and select the appropriate quantitative approach some considerations should be taken into account. Different proteomic approaches vary in their sensitivity, and the variability of each method should be defined *a priori* together with the workflow and sample-specific characteristics [51]. The number of biological and technical replicates is also critical, the greater the number of replicates, the more representative the results will be for the general population. Several studies have focused on the comparison of label-based and label-free methods for quantitative proteomics and the results showed that there is no superiority and that the accuracy of the acquired results depends on the experimental set-up [52].

6. Overview of Label-Based Proteomic Approaches

"Stable isotope methods for high-precision proteomics" [53]

The labelling methods for relative quantification studies can be classified into two main groups: chemical isotope tags and metabolic labelling. These approaches are based on the fact that both labelled and unlabelled peptides exhibit the same chromatographic and ionisation properties but can be distinguished from each other by a mass-shift signature. In metabolic labelling, the label is introduced to the whole cell organism through the growth medium, while in chemical labelling, proteins or peptides are tagged through a chemical reaction [3].

6.1. Chemical Labelling

6.1.1. Proteolytic Labelling. ^{18}O stable-isotope labelling is a simple, fast, and reliable method that takes place during proteolytic digestion in presence of heavy water ($H_2^{18}O$) [54]. Samples undergo enzymatic digestion either in presence of $H_2^{16}O$ (unlabelled sample) or $H_2^{18}O$ (labelled sample). The natural catalytic activity of serine proteases (e.g., trypsin, Lys-C, and Arg-C) can exchange both C-terminal oxygen atoms with a "heavy" ^{18}O from water in the surrounding solution. The first ^{18}O atom is introduced upon the cleavage of the peptidic amide bond, while the second ^{18}O atom is introduced when the cleaved peptide is bound to the enzyme as a reaction-mechanism intermediate (Table 1). The resulting peptides, 2 or 4 Da heavier than their unlabelled counterparts, are pooled with the unlabelled peptide mixture and peak intensities of the isotopic envelopes are compared, which can be resolved in medium-high resolution mass spectrometers [55]. Trypsin-catalyzed ^{18}O isotopic labelling has not often been used in plant proteomics and only one application was found (Table 2). Nelson and coauthors has used ^{18}O isotopic labelling for relative quantification of the degree of enrichment of *Arabidopsis* plasma membrane proteins [56]. The main drawback of this technique, despite optimization by Staes et al. [57], is that the exchange reaction is rarely complete for all peptides, resulting in a complex isotopic pattern due to the overlap of the unlabeled and singly and doubly labelled peptides.

6.1.2. Isotope-Coded Affinity Tags (ICAT). One of the first labels used for differential isotope labelling consists of three functional elements: a specific chemical reactive group that binds to sulfhydryl groups of cysteinyl residues, an isotopically coded linker with light or heavy isotopes, and a biotin tag for affinity purification (Table 1) [58]. The proteins containing cysteine residues are labelled either with light or heavy isotopes, where the latter form has eight ^{13}C atoms. Afterwards, light- and heavy-labelled samples are pooled and proteolytically cleaved. Subsequently, the complexity of the sample is reduced prior to MS analysis through the purification of tagged cysteine-containing peptides by affinity chromatography using biotin-avidin affinity columns. Peptide

TABLE 1: The various available isotopic labels, their sites of labeling, and structures.

Label	Modified site	Structure
^{18}O one atom of ^{18}O is introduced by the reaction cleaving the peptide bond A second ^{18}O is introduced by repeated binding/hydrolysis cycles with the proteolytic peptide fragment as a pseudosubstrate [52]	^{18}O incorporation at lysine and arginine via trypsin during the proteolytic digestion	
ICAT isotope-coded affinity tags [56]	Thiol group	Biotin affinity group — Labelled linker — Thiol-specific reactive group Heavy reagent: X = deuterium Light reagent: X = hydrogen
ICPL isotope-coded protein labelling [65]	Free amino group	NH$_2$ reactive group Heavy reagent: X = deuterium Light reagent: X = hydrogen
iTRAQ isobaric tags for relative and absolute quantification [70]	Peptide N-termini and ε-amino group of lysine	Reporter group mass: 114–117 — Balance group mass: 31–28 — Peptide reactive group (NHS)
TMT tandem mass tag [85]	Free amino group	Mass reporter group — Cleavage linker — Mass normalizer — Protein reactive group
SILAC stable isotopic labelling with amino acids in cell culture [88]	Metabolic incorporation of lysine or arginine	Lysine Heavy lysine: 6X^{13}C

TABLE 2: An overview of the latest MS-based quantitative proteomic studies on plant systems. The table shows the implemented quantitative approaches, plant species, biological questions, and reference of the corresponding paper.

Quantitative approach	Plant	Biological study	Authors
^{18}O labelling	Arabidopsis thaliana	Quantification of the degree of plasma membrane protein enrichment	[56]
ICAT	Arabidopsis thaliana	Localization of integral membrane proteins by using the localization of organelle proteins by isotope tagging (LOPIT)	[60]
ICAT	Hordeum vulgare	Identification of specific disulfide targets of barley thioredoxin in proteins released from barley aleurone layers	[63]
ICAT	Arabidopsis thaliana	Understanding of AtMPK6 role in transducing ozone-derived signals	[64]
ICAT	Arabidopsis thaliana	Functional information about S-nitrosylation sites in plants	[65]
ICAT	Triticum aestivum	Identification of wheat seed proteins and their related expression to chromosome deletion	[66]
ICAT, 2-DE, label-free	Zea mays	Quantitative comparative proteome analysis of purified mesophyll and bundle sheath chloroplast stroma in maize	[67]
ICAT	Oryza sativa	Protein profiling of uninucleate stage rice anther and identification of the CMS-HL-related proteins	[90]
ICAT	Arabidopsis thaliana	ProCoDeS (proteomic complex detection using sedimentation) for profiling the sedimentation of a large number of proteins	[91]
ICPL, iTRAQ	Ricinus communis	Quantitative proteomic comparison of ICPL versus iTRAQ on ricinus communis seeds	[71]
iTRAQ	Solanum tuberosum	Comparative proteomic approach of potato tubers after 0 and 5 months of storage at 5°C	[45]
iTRAQ	Arabidopsis thaliana	Quantitative study of the secreted proteins from Arabidopsis cells in response to Pseudomonas syringae	[76]
iTRAQ	Vitis vinifera	Comparative proteomic study of dynamic changes in control and infected Vitis vinifera	[77]
iTRAQ	Vitis vinifera	Comparative analysis of differentially expressed proteins in Erysiphe necator infected grape	[78]
iTRAQ	Citrus sinensis	Comparative proteomic approach of the pathogenic process of HLB in affected sweet orange leaves	[79]
iTRAQ	Zea mays	Proteomic approach of two maize inbreds in the early infection by Fusarium graminearum	[80]
iTRAQ	Arabidopsis thaliana	Changes tack of the Arabidopsis phosphoproteome during the defence response to Pseudomonas syringae	[81]
iTRAQ	Arabidopsis thaliana	Investigation of the proteomic changes in the chloroplasts of clpr2-1	[82]
iTRAQ	Hordeum vulgare	Comparative proteomic study of boron-tolerant and -intolerant barley	[83]
iTRAQ	Hordeum vulgare	Quantitative proteomic approach to unravel the contribution of vacuolar transporters to Cd^{2+} detoxification	[84]
iTRAQ, 2D-DIGE	Brassica juncea	Quantitative proteomic approaches to understand the effect of cadmium on Brassica juncea roots	[85]
iTRAQ, label-free	Oryza sativa	Quantitative proteomic response of rice seedling to 48, 72, and 96 h of cold stress	[86]
iTRAQ, BN-PAGE, label-free	Zea mays	Comparative analysis of protein abundance in chloroplast thylakoid and envelope membrane proteomes in maize	[87]
Cys-TMT	Solanum lycopersicum	Study of the redox proteomic analysis of the Pseudomonas syringae tomato DC3000 treated tomato leaves	[89]
SILAC	Chlamydomonas reinhardtii	Dynamic changes of proteome turnover under salt stress	[92]
SILAC	Ostreococcus tauri	Quantitative proteomics on synthesis and degradation rate constants of individual proteins in autotrophic organisms	[93]
SILAC	Chlamydomonas reinhardtii	Comparative proteomics on the iron deficiency impact in Chlamydomonas reinhardtii	[94]

TABLE 2: Continued.

Quantitative approach	Plant	Biological study	Authors
^{14}N/^{15}N labelling	Solanum tuberosum	Effectiveness of fully label a plant with ^{15}N isotopes	[95]
^{14}N/^{15}N labelling	Arabidopsis thaliana	Demonstration of plant ^{15}N labelling as a powerful comparative quantitative proteomic approach	[96]
^{14}N/^{15}N labelling	Arabidopsis thaliana	Comparative analysis of Arabidopsis cells following a cadmium exposure	[97]
^{14}N/^{15}N labelling	Nicotiana tabacum	Quantitative proteomic approach of the detergent-resistant membranes of tobacco cells in response to cryptogenin	[98]
^{14}N/^{15}N labelling	Arabidopsis thaliana	Quantitative approach of phosphorylated sites in signaling and protein response in flg22 or xylanase Arabidopsis-treated cells	[99]
HILEP	Arabidopsis thaliana	Demonstration of HILEP suitability for relative plant quantitative proteomic subjected to oxidative stress	[100]
SILIP	Solanum lycopersicum	SILIP development for homogeneously ^{15}N incorporation within the whole plant proteome	[101]
^{14}N/^{15}N labelling	Arabidopsis thaliana	Investigation of both partial and full ^{15}N labelling effect on quantitative analysis in a complex mixture	[102]
Spectral counting	Arabidopsis thaliana	Proteome map of Arabidopsis thaliana	[103]
Spectral counting	Arabidopsis thaliana	Comprehensive Arabidopsis chloroplast proteome analysis	[104]
Spectral counting	Glycine max	Evaluation of the suitability of spectral counting to quantitative soybean proteome study	[105]
Spectral counting	Oryza sativa	Differential proteomic response of rice leaves exposed to high- and low-temperature stress	[106]
Peak ion intensity	Arabidopsis thaliana	Sucrose-induced phosphorylation changes of plasma membrane proteins in Arabidopsis	[107]
Peak ion intensity	Solanum lycopersicum	Quantitative proteomics of phosphoproteins in tomato hypersensitive response	[108]
Peak ion intensity, 2-DE	Glycine max	Investigation of the soybean plasma membrane function in response to flooding stress	[109]
Spectral counting + peak ion intensity	Medicago truncatula	Comparison of two label-free quantitative approaches on nodule protein extracts from Medicago truncatula	[110]
Spectral counting + peak ion intensity	Arachis hypogaea	Investigation of major allergens in transgenic peanut lines	[111]
MS^E	Apium graveolens	Analysis of the Apium graveolens protein response to salicylic acid	[112]
MS^E	Zea mays	Proteomic approach assessment of the transition from dark to light in maize seedlings	[113]
MS^E	Arabidopsis thaliana	Proteomic changes in the cell wall proteome in response to salicylic acid	[114]
MS^E	Hordeum vulgare	Study of the UV-B irradiation effect on the barley proteome	[115]

pairs with 8 Da mass-shifts are detected in MS scans and their ion intensities are compared for relative quantitation. ICAT labelling takes place at the protein level allowing samples to be pooled prior to protease treatment, thus eliminating vial-to-vial variations. However, cysteine is not very abundant and approximately one in seven proteins do not contain this amino acid, greatly reducing the completeness of the study [59].

In plants, Dunkley and coworkers have studied the localization of organelle proteins by isotope tagging (LOPIT) to discriminate endoplasmic reticulum, Golgi, plasma membrane, and mitochondria or plastid proteins in Arabidopsis. This technique involves partial separation of the organelles by density gradient centrifugation followed by the analysis of protein distributions in the gradient by ICAT and MS [60]. Taking advantage of the ICAT labelling specificity to cysteinyl groups, this approach was used to study the redox-status of proteins allowing a quantitative analysis of the redox proteome and ozone stress in plants [61–64]. To increase the functional information about S-nitrosylation sites in plants, Fares and colleagues combined both "biotin-switch" method (BSM) and ICAT labelling and succeeded in identifying 53 endogenous nitrosocysteines in Arabidopsis cells [65]. ICAT was also used to identify wheat seed proteins and to understand their interactions and expression in relation to chromosome deletion, which were reported to be difficult by 2-DE due to co-synthesis of proteins by genes from three genomes, A, B and D [66]. A cross-comparison of gel-based and -free quantitative methods (2-DE, ICAT, and label-free) was performed by analysing the differential accumulation of maize chloroplast proteins in bundle sheath versus mesophyll cells. Among the 125 chloroplast proteins quantified in the 3 methods, only 20 proteins were quantified in common, demonstrating the complementary nature of these quantitative approaches [67]. More applications of ICAT quantitative approach in plant proteomics are listed in Table 2.

6.1.3. Isotope-Coded Protein Labelling (ICPL). This approach termed ICPL is based on isotopic labelling of all free amino groups in proteins. Two protein mixtures are reduced and alkylated to ensure easier access to free amino groups that are subsequently derivatised with the deuterium-free (light) or 4 deuterium containing (heavy) form, respectively (Table 1). Light- and heavy-labelled samples are then mixed, fractionated, and digested prior to high throughput MS analysis. Since peptides of identical sequence derived from the two differentially labelled protein samples differ in mass (4 Da), they appear as doublets in the acquired MS spectra. From the ratios of the ion intensities of these sister peptide pairs, the relative abundance of their parent proteins in the original samples can be determined [68]. Recently, a detailed experimental protocol called postdigest ICPL was published highlighting a better protein identification and quantification [69, 70] and when compared to iTRAQ, both techniques have shown comparable number of identified and quantified proteins in the endosperm of castor bean seeds at three developmental stages [71]. So far, the latter study is the unique reported quantitative proteomic investigation on plants using ICPL (Table 2). The main drawback of this method is the isotopic effect of deuterated tags that interferes with retention time of the labelled peptides during LC [72].

6.1.4. Isobaric Tags for Relative and Absolute Quantification (iTRAQ). Unlike ICAT and ICPL, iTRAQ tags are isobarics and primarily designed for the labelling of peptides rather than proteins. The overall molecule mass is kept constant at 145 Da and 304 Da for iTRAQ-4plex and -8plex, respectively. The structure of the iTRAQ-8plex balancer group has not been published while the iTRAQ-4plex molecule consists of a reporter group (based on N-methylpiperazine), a mass balance group (carbonyl), and a peptide reactive group (NHS ester) (Table 1) [73]. The iTRAQ reagents label peptide N-termini and ε-amino groups of lysine side chains and allow comparison of up to eight samples in the same experiment. Another difference from the pre-cited methods is that the quantification occurs in MS/MS scans after peptide fragmentation. In fact, iTRAQ-labelled peptides appear as a single unresolved precursor at the same m/z in the MS spectrum. Upon peptide fragmentation, the iTRAQ labels fragment to produce reporter ions in a "silent region," usually unpopulated, at low m/z range (e.g., 114–121). Measurements of the reporter ion intensities enable relative quantification of the peptide in each sample.

This method has quickly gained popularity in proteomics and benefits from increased MS sensitivity compared to for instance ICAT due to the contribution of all samples to the precursor ion signal. The iTRAQ reagent was furthermore reported to increase the number of lysine-terminated tryptic peptides identified by database searches to equivalence with arginine-terminated peptides [73]. Ow and coauthors evaluated iTRAQ relevance, accuracy, and precision for biological interpretation and entitled their verdict "the good, the bad and the ugly" of iTRAQ quantitation [74]. "The good" is the potential of iTRAQ to provide accurate quantification spanning two orders of magnitude. However, that potential is limited by two factors: isotopic impurities "the bad", and peptide cofragmentation (inadvertently selecting two or more closely spaced peptides for MS/MS instead of one) "the ugly" [75]. In the same study, a putative contamination of the reporter ion region with the second isotope of the phenylalanine immonium ion on the 121 m/z peak, which can interfere with peptide quantification was mentioned [74].

The iTRAQ has shown a high utility in large-scale quantitative proteomics (Table 2) to study plant responses to pathogens: *Pseudomonas syringae* in *Arabidopsis* [76], *Lobesia botrana* and *Erysiphe necator* in grape [77, 78], *Huanglongbing* in sweet orange [79], *Fusarium graminearum* in maize [80]. Quantitative shotgun proteomic approaches using iTRAQ were furthermore used for characterizing the differential phosphorylation of Arabidopsis in response to microbial elicitation [81] and the study of protein degradation in chloroplasts [82]. The potency of iTRAQ was used for better understanding mechanisms of plant tolerance to boron in barley [83], cadmium in barley [84] and *Brassica juncea* [85], and cold in potato and rice [45, 86]. An example of iTRAQ application in plant membrane proteomics is the study of differentiated state of bundle sheath and mesophyll chloroplast thylakoid and envelope membrane proteomes in maize [87].

6.1.5. Tandem Mass Tag (TMT). A novel MS/MS-based quantitative method using isotopomer labels, similar to iTRAQ, and referred to as "tandem mass tags" (TMT) was recently developed (Table 1) [88]. Both techniques share several common features. (i) These reagents employ N-hydroxysuccinimide (NHS) chemistry that permits specific tagging of primary amino groups. (ii) They were designed to allow multiplexing of several samples by chemical derivatization with different forms of the same isobaric tag that appear as a single peak in full MS scans. (iii) The release of "daughter ions" in MS/MS analysis (between 126 and 131 Da for TMT) that can be used for relative quantification. The cysteine-reactive TMT (cysTMT) reagents enable selective labelling and relative quantitation of cysteine-containing peptides from up to six biological samples. This technique has been used for the redox proteomic analysis of the tomato leaves in response to the pathogen *P. syringae* pv. tomato strain DC3000 (Table 2) [89]. Aside from this study, TMT labelling approach has so far not been fully exploited for the analysis of plant proteomes.

A study comparing TMT and iTRAQ showed that the performance of both techniques was similar in terms of quantitative precision and accuracy, however the number of identified peptides and proteins was higher with iTRAQ 4-plex compared to TMT 6-plex [116].

6.2. Metabolic Labelling. Although chemical labelling presents a wide range of approaches for quantitative proteomics, this group of techniques suffers from sample variability and induces a technical bias since the labelling occurs after the protein extraction or even after proteolytic digestion. In addition, the high cost of these reagents can be a limiting factor for large-scale experiments. Therefore metabolic labelling, which allows protein labelling at the time of protein synthesis, presents a valuable alternative strategy for quantitative proteomics.

6.2.1. Stable Isotopic Labelling with Amino Acids in Cell Culture (SILAC). *In vivo* metabolic labelling, in which two populations of cells are cultured either in a medium containing a "light" (unlabelled) amino acid or encompassing a "heavy" (labelled), one typically arginine or lysine labelled with ^{13}C and/or ^{15}N are used [117]. The mass shift induced by the incorporation of the heavy amino acid into a peptide, is known and allows comparison between a peptide in both samples (e.g., 6 Da in the case of $^{13}C_6$-Lys, Table 1). Samples are then combined prior to protein extraction, which minimizes technical variation arising during sample processing. In MS spectra, each peptide appears as a pair and the ratio of peak intensities yields the protein abundance in the sample since the light and heavy amino acids are chemically identical and only isotopically distinguished.

Although probably the most general and global labelling strategy, SILAC appears less suited for quantitative proteomic studies in plants. Being autotrophic organisms, plants are metabolic specialists capable of synthesising all amino acids from inorganic nitrogen and, therefore, have lower incorporation efficiency of the exogenously supplied labelled amino acids. The labelling efficiency achieved using exogenous amino acid feeding of *Arabidopsis* cell cultures has been found to average only 70–80% [118]. Considering these limitations and the high cost of isotopically labelled amino acids, SILAC appears likely to be inadequate for quantitative proteomics studies in plants; albeit it seems less restricted to study algae such as *Chlamydomonas reinhardtii* and *Ostreococcus tauri* (Table 2) [92–94].

6.2.2. ^{14}N/^{15}N Labelling. In this method, the label is introduced to the whole cell or organism through the growth medium. Samples can easily be labelled metabolically via growth media containing ^{15}N-labelled inorganic salts, typically $K^{15}NO_3$ [95]. The quantification process is based on the intensity of extracted ion chromatograms of survey scans containing the pair of labelled (^{15}N, heavy) and unlabelled (^{14}N, light) peptide isoforms.

Unlike SILAC, this approach achieved more than 98% incorporation in both plants [95] and cell cultures [96], and is more efficient at allowing large-scale quantitative analysis. The tradeoff is that all amino acids will incorporate the label, thus the mass shift will be peptide-sequence dependent. Metabolic ^{15}N-labelling is becoming the method of choice

for quantitative proteomics in plant studies (Table 2). It was used to study plant membrane proteome changes in response to cadmium, and cryptogenin elicitor in *Arabidopsis* and tobacco cells, respectively [97, 98]. Such a quantitative proteomic strategy was applied in quantitative phosphoproteomics to study differentiated proteins in response to fungal or microbial elicitors in *Arabidopsis* cells [99]. Moreover, other metabolic labelling strategies have been developed such as hydroponic isotope labelling of entire plants (HILEP) which has proven to be very efficient and robust method to completely label the whole mature plants. Nearly 100% of ^{15}N-labelling efficiency was achieved in *Arabidopsis* plants by growing them in hydroponic media containing 2.5 mM ^{15}N potassium nitrate and 0.5 mM ^{15}N ammonium nitrate [100, 119]. A similar quantitative proteomic method, SILIP (Stable Isotope Labelling InPlanta), was developed for labelling tomato plants growing in sand in a greenhouse environment [101]. An alternative strategy for quantitative proteomics that relies upon the subtle changes in isotopic envelope shape resulting from partial metabolic labelling to compare relative abundances of labelled and unlabelled peptides has been developed in *Arabidopsis*. Both partial and full labelling have been proven to be comparable with respect to dynamic range, accuracy, and reproducibility, and both are suitable for quantitative proteomics characterization [102].

7. Label-Free Quantitative Proteomics

> *"Comparative LC-MS: a landscape of peaks and valleys"* [120]

> *"Less label, more free"* [121]

Quantitative proteomics based on stable isotope-coding strategies often require expensive labelling reagents, high amount of starting samples, multiple sample preparation steps resulting in considerable sample loss and reduced detection sensitivity. Label-free LC/MS methods represent attractive alternatives [122] since they are amenable to all type of biological samples, are simple, reproducible, cost effective, and less prone to errors and side reactions related to the labelling process.

Given the fact that, theoretically, the peak intensity of any ion should be proportional to its abundance the ion signals in MS have been used, for decades, as a quantification technique for small molecules in analytical chemistry. However, technical variation, at both LC and ionization levels, might render comparisons of peak intensities between experiments unreliable. The recent advances in LC/MS approaches allowed circumvention of the looming replicate biases and recently the observation of a correlation between protein abundance and peak areas [123, 124] or number of MS/MS spectra [125] has widened the choice of analytical procedure in the field of quantitative proteomics. The general framework of label-free quantification can be summarised as follows: for the two samples that need to be compared quantitatively, the LC-MS/MS experiment is first performed for both samples separately, and precursor ion m/z and retention time (*Rt*) file is generated for all MS/MS spectra of each identified protein, creating a 2D map (m/z, Rt) allowing peptide match in several samples.

Depending on the MS acquisition mode, two analytical methods can be distinguished: the data-dependent analysis (DDA) and the data-independent analysis (DIA). DDA involves acquisition of a MS survey scan followed, for an allotted period of time, by precursor ion selection based on its intensity for subsequent fragmentation [126]. In this approach, quantification can be achieved using DDA-based spectral counting or spectral peak intensities. Venable and coauthors described DIA in which no parent ion is preselected; the instrument constantly operates in MS/MS mode and data acquisition of all charge states of eluted peptides is performed by rapid switching of the collision energy between low and high-energy states [127].

7.1. Spectral Counting.

Spectral counting or peptide identification frequency is becoming popular in label-free quantification due to its simple procedure that does not require chromatographic peak integration or retention time alignment. It is based on the rationale that peptides from more abundant proteins will be more selected for fragmentation and will thus produce a higher number of MS/MS spectra. Thus, the number of MS/MS scans is tabulated and the protein abundance is inferred from the total number of MS/MS spectra that match peptides from the protein [125]. The ability to accurately quantify proteins by spectral counting largely depends on the number of spectra obtained and the coverage of sampling. The relative difference in protein abundance is estimated by calculating the protein abundance index (PAI), which corresponds to the number of observed peptides in the experiment divided by the number of theoretical tryptic peptides for each protein within a given mass range of the employed mass spectrometer [128]. The exponential form of PAI minus one ($10^{PAI}-1$), exponentially modified protein abundance index (emPAI) [129], takes into account the fact that generally more peptides are detected for larger proteins and is directly proportional to the protein content in the sample. The absolute protein expression (APEX) index, a very similar approach to emPAI, is a derived measurement of protein abundance in a given sample based on the analytical features in mass spectrometric analysis [130]. It has been used to generate a protein abundance map of the Arabidopsis proteome [103] and to determine the abundance of stromal proteins in A. thaliana chloroplast [104]. Spectral counting based quantitative proteomics has been widely used in the field of plant proteomics (Table 2). The accuracy and reliability of label-free spectral counting in the relative quantitative analysis of soybean leaf proteome was evaluated by comparing nine technical replicates [105]. Gammulla and coauthors quantified and identified temperature stress responsive proteins in rice leaves by calculating the NSAF (Normalized Spectral Abundance Factor), which is given by the total number of MS/MS spectra (SpC) identifying a protein, divided by the protein's length (L), divided by the sum of SpC/L for all proteins in the experiment [106, 131]. Spectrum counting has been used to study drought stress response in root nodules of M. truncatula [132] and

in large-scale plant proteomics in response to pathogen infection in bean (Phaseolus vulgaris) [133].

7.2. Spectral Peak Intensities.

Other label-free methods use the signal intensities of individual peptides rather than the spectral counts to compare the relative abundance of proteins between samples [134]. It is based on the principle that the relative abundance of the same peptide in different samples can be estimated by the precursor ion signal intensity across consecutive LC/MS runs, given that the measurements are performed under identical conditions. In contrast to differential labelling, every biological specimen needs to be measured separately in a label-free experiment. Typically, peptide signals are detected at the MS level, their patterns are then tracked across the retention time dimension and used to reconstruct a chromatographic elution profile of the monoisotopic peptide mass. The total ion current of the peptide signal is then integrated and the measurement of the chromatographic peak areas is used as a quantitative measurement for the original peptide concentration. Profiling methods based on ion intensity were applied to define the sucrose-induced phosphorylation changes in Arabidopsis plasma membrane proteins [107]. It has been furthermore used to detect twelve phosphopeptides from 50 identified phosphoproteins in different amounts during the hypersensitive response in tomato plants [108]. Moreover, the ion intensity method was used as strategic track to study soybean plasma membrane proteins following 24 h flooding and 48 h osmotic stress (Table 2) [109, 135].

Spectral counting and spectral peak intensities were compared and results obtained from both methods are generally in good accordance [110, 136] with spectral counting covering a slightly higher dynamic range and measurements of ion abundance being more accurate for the identification of protein ratios [136]. Both techniques have also been used to investigate the major allergens in transgenic peanut lines [111].

Unlike labelling methods, in which quantitative analyses are limited to the tagged peptides, label-free approaches offer the quantitative comparison of all peptide constituents of the sample. However, they are more susceptible to errors due to parallel sample processing and thus suffer from increased analytical variability. Therefore, label-free methods are very replicate dependent. To be statistically significant, chromatographic separation reproducibility must be very high. The high-resolution power of MS, high scanning rates, high accurate mass measurements, and exact chromatogram alignment are prerequisite for the success of this quantitative technique [134, 137]. The extensive workflow ranging from peptide detection, alignment, normalization, identification, quantitative comparisons, and statistical analysis has triggered the development of several sophisticated software algorithms.

7.3. Data-Independent Analysis (DIA).

LC/MS^E, a quantitative comparison of ions emanating from identically prepared control and experimental samples, was developed by using a reproducible chromatographic separation system along with the high mass resolution and mass accuracy of an orthogonal time-of-flight mass spectrometer [134]. In this method, the instrument alternates between low and high collision energies in MS analysis. While the low collision energy scan mode leads to the determination of accurate precursor ion masses, the high-energy scan mode (MS^E) generates accurate peptide fragmentation data [26]. The use of multiplex parallel fragmentation of LC/MS^E yields uniformly product ion information of all peptides across their entire chromatographic peaks [134], which provides continuous MS data throughout the entire acquisition. Product ions are time-aligned and correlated to precursor ions to generate a list of exact mass retention time (EMRT) signatures [134]. The integrated peak areas of EMRT are compared across different biological replicates to determine the differences in protein abundances.

The LC/MS^E approach is well suited for relative and absolute quantification [112] and it was shown to increase the signal-to-noise ratio by a factor 3–5 and could identify peptides undetected in a parent ion scan [138]. This recent achievement in MS-based proteomics has provided a basis to qualitatively and quantitatively assess the transition from dark to light of maize seedlings [113] and to study the salicylic acid-induced changes in the *Arabidopsis* and *Apium graveolens* secretome [114, 139]. MS^E has also been implemented to study the changes in barley protein expression in response to UV-B treatment (Table 2) [115].

8. Conclusion

Proteomics, the promising new "omics," has become an important complementary tool to genomics providing novel information and greater insight into plant biology. The application of gel-based and -free proteomics methods to study plant physiology has strongly increased in recent years. Here, a broad perspective is offered on the available techniques.

So far, most quantitative plant proteomics was performed on *Arabidopsis thaliana*, the model plant due to various traits including its small (and annotated) genome size (125 MBp), short generation time, high transformation efficiency, and the large panel of available mutants. The completion of more plant genome sequencing projects such as rice, barley, tomato and *M. truncatula* and will permit the proteome probing of these plant systems. In the meantime, extensive EST databases for numerous important crop plants represent alternative sources of sequence information to the full genome sequences. Moreover, with the technical maturity attained in MS and protein/peptide fractionation tools, comparative plant proteomics will move out of the beginner realm and emerge as high valuable discipline to enhance the comprehension of plant systems, their subcellular membranes and organelles. It is worth noting that combining multiple quantitative proteomic techniques is highly beneficial, as these approaches yield complementary datasets which improve the understanding of biological issues and provide in-depth characterization of proteins with respect to their abundance. These technical advancements coupled to well-designed experiments will significantly reveal the protein function in plant growth and development and provide a wealth of information on plant proteome changes occurring in response to external stimuli, biotic, and abiotic stresses.

Acknowledgments

This work was supported by grants from the "Fonds National de la Recherche du Luxembourg" AFR TR-PHD BFR 08-078 and Conseil Régional de Bourgogne (PARI 20100112095254682-1). Ghislaine Recorbet and Daniel Wipf are thanked for their assistance and support.

References

[1] S. Fields, "Proteomics: proteomics in genomeland," *Science*, vol. 291, no. 5507, pp. 1221–1224, 2001.

[2] M. R. Wilkins, J. C. Sanchez, A. A. Gooley et al., "Progress with proteome projects: why all proteins expressed by a genome should be identified and how to do it," *Biotechnology and Genetic Engineering Reviews*, vol. 13, pp. 19–50, 1996.

[3] W. X. Schulze and B. Usadel, "Quantitation in mass-spectrometry-based proteomics," *Annual Review of Plant Biology*, vol. 61, pp. 491–516, 2010.

[4] K. Kleparnik and P. Bocek, "Electrophoresis today and tomorrow: helping biologists' dreams come true," *BioEssays*, vol. 32, no. 3, pp. 218–226, 2010.

[5] P. H. O'Farrell, "High resolution two dimensional electrophoresis of proteins," *Journal of Biological Chemistry*, vol. 250, no. 10, pp. 4007–4021, 1975.

[6] J. A. Gordon and W. P. Jencks, "The relationship of structure to the effectiveness of denaturing agents for proteins," *Biochemistry*, vol. 2, no. 1, pp. 47–57, 1963.

[7] A. Vertommen, B. Panis, R. Swennen, and S. C. Carpentier, "Challenges and solutions for the identification of membrane proteins in non-model plants," *Journal of Proteomics*, vol. 74, no. 8, pp. 1165–1181, 2011.

[8] K. S. Lilley, A. Razzaq, and P. Dupree, "Two-dimensional gel electrophoresis: recent advances in sample preparation, detection and quantitation," *Current Opinion in Chemical Biology*, vol. 6, no. 1, pp. 46–50, 2002.

[9] S. P. Gygi, G. L. Corthals, Y. Zhang, Y. Rochon, and R. Aebersold, "Evaluation of two-dimensional gel electrophoresis-based proteome analysis technology," *Proceedings of the National Academy of Sciences of the United States of America*, vol. 97, no. 17, pp. 9390–9395, 2000.

[10] S. E. Ong and A. Pandey, "An evaluation of the use of two-dimensional gel electrophoresis in proteomics," *Biomolecular Engineering*, vol. 18, no. 5, pp. 195–205, 2001.

[11] R. P. Tonge, J. Shaw, B. Middleton et al., "Validation and development of fluorescence two-dimensional differential gel electrophoresis proteomics technology," *Proteomics*, vol. 1, no. 3, pp. 377–396, 2001.

[12] M. Ünlü, M. E. Morgan, and J. S. Minden, "Difference gel electrophoresis: a single gel method for detecting changes in protein extracts," *Electrophoresis*, vol. 18, no. 11, pp. 2071–2077, 1997.

[13] A. Alban, S. O. David, L. Bjorkesten et al., "A novel experimental design for comparative two-dimensional gel analysis: two-dimensional difference gel electrophoresis incorporating a pooled internal standard," *Proteomics*, vol. 3, no. 1, pp. 36–44, 2003.

[14] D. B. Friedman, S. Hill, J. W. Keller et al., "Proteome analysis of human colon cancer by two-dimensional difference gel

electrophoresis and mass spectrometry," *Proteomics*, vol. 4, no. 3, pp. 793–811, 2004.

[15] N. A. Karp and K. S. Lilley, "Investigating sample pooling strategies for DIGE experiments to address biological variability," *Proteomics*, vol. 9, no. 2, pp. 388–397, 2009.

[16] G. E. Van Noorden, T. Kerim, N. Goffard et al., "Overlap of proteome changes in Medicago truncatula in response to auxin and Sinorhizobium meliloti," *Plant Physiology*, vol. 144, no. 2, pp. 1115–1131, 2007.

[17] L. Schenkluhn, N. Hohnjec, K. Niehaus, U. Schmitz, and F. Colditz, "Differential gel electrophoresis (DIGE) to quantitatively monitor early symbiosis- and pathogenesis-induced changes of the Medicago truncatula root proteome," *Journal of Proteomics*, vol. 73, no. 4, pp. 753–768, 2010.

[18] K. Sergeant, N. Spieß, J. Renaut, E. Wilhelm, and J. F. Hausman, "One dry summer: a leaf proteome study on the response of oak to drought exposure," *Journal of Proteomics*, vol. 74, no. 8, pp. 1385–1395, 2011.

[19] T. Li, S. L. Xu, J. A. Oses-Prieto et al., "Proteomics analysis reveals post-translational mechanisms for cold-induced metabolic changes in *Arabidopsis*," *Molecular Plant*, vol. 4, no. 2, pp. 361–374, 2011.

[20] S. Bohler, K. Sergeant, L. Hoffmann et al., "A difference gel electrophoresis study on thylakoids isolated from poplar leaves reveals a negative impact of ozone exposure on membrane proteins," *Journal of Proteome Research*, vol. 10, no. 7, pp. 3003–3011, 2011.

[21] T. C. Durand, K. Sergeant, S. Planchon et al., "Acute metal stress in Populus tremula x P. alba (717-1B4 genotype): leaf and cambial proteome changes induced by cadmium^{2+}," *Proteomics*, vol. 10, no. 3, pp. 349–368, 2010.

[22] P. Kieffer, J. Dommes, L. Hoffmann, J. F. Hausman, and J. Renaut, "Quantitative changes in protein expression of cadmium-exposed poplar plants," *Proteomics*, vol. 8, no. 12, pp. 2514–2530, 2008.

[23] H. Schagger and G. Von Jagow, "Blue native electrophoresis for isolation of membrane protein complexes in enzymatically active form," *Analytical Biochemistry*, vol. 199, no. 2, pp. 223–231, 1991.

[24] H. Eubel, H. P. Braun, and A. H. Millar, "Blue-native PAGE in plants: a tool in analysis of protein-protein interactions," *Plant Methods*, vol. 1, no. 1, p. 11, 2005.

[25] Q. Meng, L. Rao, X. Xiang, C. Zhou, X. Zhang, and Y. Pan, "A systematic strategy for proteomic analysis of chloroplast protein complexes in wheat," *Bioscience, Biotechnology, and Biochemistry*, vol. 75, no. 11, pp. 2194–2199, 2011.

[26] U. Kota and M. B. Goshe, "Advances in qualitative and quantitative plant membrane proteomics," *Phytochemistry*, vol. 72, no. 10, pp. 1040–1060, 2011.

[27] J. P. Lasserre and A. Menard, "Two-dimensional blue native/SDS gel electrophoresis of multiprotein complexes," *Methods in Molecular Biology*, vol. 869, pp. 317–337, 2012.

[28] H. Hahne, S. Wolff, M. Hecker, and D. Becher, "From complementarity to comprehensiveness—targeting the membrane proteome of growing *Bacillus subtilis* by divergent approaches," *Proteomics*, vol. 8, no. 19, pp. 4123–4136, 2008.

[29] Y. Fang, D. P. Robinson, and L. J. Foster, "Quantitative analysis of proteome coverage and recovery rates for upstream fractionation methods in proteomics," *Journal of Proteome Research*, vol. 9, no. 4, pp. 1902–1912, 2010.

[30] S. R. Piersma, U. Fiedler, S. Span et al., "Workflow comparison for label-free, quantitative secretome proteomics for cancer biomarker discovery: method evaluation, differential

analysis, and verification in serum," *Journal of Proteome Research*, vol. 9, no. 4, pp. 1913–1922, 2010.

[31] B. Valot, L. Negroni, M. Zivy, S. Gianinazzi, and E. Dumas-Gaudot, "A mass spectrometric approach to identify arbuscular mycorrhiza-related proteins in root plasma membrane fractions," *Proteomics*, vol. 6, supplement 1, pp. S145–S155, 2006.

[32] J. E. Froehlich, C. G. Wilkerson, W. K. Ray et al., "Proteomic study of the *Arabidopsis thaliana* chloroplastic envelope membrane utilizing alternatives to traditional two-dimensional electrophoresis," *Journal of Proteome Research*, vol. 2, no. 4, pp. 413–425, 2003.

[33] J. J. Petricka, M. A. Schauer, M. Megraw et al., "The protein expression landscape of the Arabidopsis root," *Proceedings of the National Academy of Sciences of the United States of America*, vol. 109, no. 18, pp. 6811–6818, 2012.

[34] K. Gevaert, P. Van Damme, B. Ghesquière et al., "A la carte proteomics with an emphasis on gel-free techniques," *Proteomics*, vol. 7, no. 16, pp. 2698–2718, 2007.

[35] B. Manadas, V. M. Mendes, J. English, and M. J. Dunn, "Peptide fractionation in proteomics approaches," *Expert Review of Proteomics*, vol. 7, no. 5, pp. 655–663, 2010.

[36] M. L. Fournier, J. M. Gilmore, S. A. Martin-Brown, and M. P. Washburn, "Multidimensional separations-based shotgun proteomics," *Chemical Reviews*, vol. 107, no. 8, pp. 3654–3686, 2007.

[37] M. P. Washburn, D. Wolters, and J. R. Yates 3rd, "Large-scale analysis of the yeast proteome by multidimensional protein identification technology," *Nature Biotechnology*, vol. 19, no. 3, pp. 242–247, 2001.

[38] G. Mathy and F. E. Sluse, "Mitochondrial comparative proteomics: strengths and pitfalls," *Biochimica et Biophysica Acta*, vol. 1777, no. 7-8, pp. 1072–1077, 2008.

[39] J. V. Jorrin, A. M. Maldonado, and M. A. Castillejo, "Plant proteome analysis: a 2006 update," *Proteomics*, vol. 7, no. 16, pp. 2947–2962, 2007.

[40] O. K. Park, "Proteomic studies in plants," *Journal of Biochemistry and Molecular Biology*, vol. 37, no. 1, pp. 133–138, 2004.

[41] P. Hörth, C. A. Miller, T. Preckel, and C. Wenz, "Efficient fractionation and improved protein identification by peptide OFFGEL electrophoresis," *Molecular and Cellular Proteomics*, vol. 5, no. 10, pp. 1968–1974, 2006.

[42] S. Elschenbroich, V. Ignatchenko, P. Sharma, G. Schmitt-Ulms, A. O. Gramolini, and T. Kislinger, "Peptide separations by on-line MudPIT compared to isoelectric focusing in an off-gel format: application to a membrane-enriched fraction from C2C12 mouse skeletal muscle cells," *Journal of Proteome Research*, vol. 8, no. 10, pp. 4860–4869, 2009.

[43] A. S. Essader, B. J. Cargile, J. L. Bundy, and J. L. Stephenson, "A comparison of immobilized pH gradient isoelectric focusing and strong-cation-exchange chromatography as a first dimension in shotgun proteomics," *Proteomics*, vol. 5, no. 1, pp. 24–34, 2005.

[44] B. Manadas, J. A. English, K. J. Wynne, D. R. Cotter, and M. J. Dunn, "Comparative analysis of OFFGel, strong cation exchange with pH gradient, and RP at high pH for first-dimensional separation of peptides from a membrane-enriched protein fraction," *Proteomics*, vol. 9, no. 22, pp. 5194–5198, 2009.

[45] Y. Yang, X. Qiang, K. Owsiany, S. Zhang, T. W. Thannhauser, and L. Li, "Evaluation of different multidimensional LC-MS/MS pipelines for isobaric tags for relative and absolute quantitation (iTRAQ)-based proteomic analysis of potato

tubers in response to cold storage," *Journal of Proteome Research*, vol. 10, no. 10, pp. 4647–4660, 2011.

[46] D. Vincent and S. P. Solomon, "Development of an in-house protocol for the OFFGEL fractionation of plant proteins," *Journal of Integrated OMICS*, vol. 1, no. 2, 2011.

[47] C. N. Meisrimler and S. Luthje, "IPG-strips versus off-gel fractionation: advantages and limits of two-dimensional PAGE in separation of microsomal fractions of frequently used plant species and tissues," *Journal of Proteomics*, vol. 75, no. 9, pp. 2550–2562, 2012.

[48] C. Abdallah, K. Sergeant, C. Guillier, E. Dumas-Gaudot, C. Leclercq, and J. Renaut, "Optimization of iTRAQ labelling coupled to OFFGEL fractionation as a proteomic workflow to the analysis of microsomal proteins of Medicago truncatula roots," *Proteome Science*, vol. 10, no. 1, p. 37, 2012.

[49] S. E. Ong and M. Mann, "Mass spectrometry-based proteomics turns quantitative," *Nature chemical biology*, vol. 1, no. 5, pp. 252–262, 2005.

[50] R. Aebersold and M. Mann, "Mass spectrometry-based proteomics," *Nature*, vol. 422, no. 6928, pp. 198–207, 2003.

[51] D. A. Cairns, "Statistical issues in quality control of proteomic analyses: good experimental design and planning," *Proteomics*, vol. 11, no. 6, pp. 1037–1048, 2011.

[52] M. D. Filiou, D. Martins-de-Souza, P. C. Guest, S. Bahn, and C. W. Turck, "To label or not to label: applications of quantitative proteomics in neuroscience research," *Proteomics*, vol. 12, no. 4-5, pp. 736–747, 2012.

[53] L. V. Schneider and M. P. Hall, "Stable isotope methods for high-precision proteomics," *Drug Discovery Today*, vol. 10, no. 5, pp. 353–363, 2005.

[54] X. Yao, A. Freas, J. Ramirez, P. A. Demirev, and C. Fenselau, "Proteolytic 18O labeling for comparative proteomics: model studies with two serotypes of adenovirus," *Analytical Chemistry*, vol. 73, no. 13, pp. 2836–2842, 2001.

[55] I. I. Stewart, T. Thomson, and D. Figeys, "O labeling: a tool for proteomics," *Rapid Communications in Mass Spectrometry*, vol. 15, no. 24, pp. 2456–2465, 2001.

[56] C. J. Nelson, A. D. Hegeman, A. C. Harms, and M. R. Sussman, "A quantitative analysis of *Arabidopsis* plasma membrane using trypsin-catalyzed ^{18}O labeling," *Molecular and Cellular Proteomics*, vol. 5, no. 8, pp. 1382–1395, 2006.

[57] A. Staes, H. Demol, J. Van Damme, L. Martens, J. Vandekerckhove, and K. Gevaert, "Global differential non-gel proteomics by quantitative and stable labeling of tryptic peptides with oxygen-18," *Journal of Proteome Research*, vol. 3, no. 4, pp. 786–791, 2004.

[58] S. P. Gygi, B. Rist, S. A. Gerber, F. Turecek, M. H. Gelb, and R. Aebersold, "Quantitative analysis of complex protein mixtures using isotope-coded affinity tags," *Nature Biotechnology*, vol. 17, no. 10, pp. 994–999, 1999.

[59] J. J. Thelen and S. C. Peck, "Quantitative proteomics in plants: choices in abundance," *Plant Cell*, vol. 19, no. 11, pp. 3339–3346, 2007.

[60] T. P. J. Dunkley, R. Watson, J. L. Griffin, P. Dupree, and K. S. Lilley, "Localization of organelle proteins by isotope tagging (LOPIT)," *Molecular and Cellular Proteomics*, vol. 3, no. 11, pp. 1128–1134, 2004.

[61] E. Ströher and K. J. Dietz, "Concepts and approaches towards understanding the cellular redox proteome," *Plant Biology*, vol. 8, no. 4, pp. 407–418, 2006.

[62] P. Hägglund, J. Bunkenborg, K. Maeda, and B. Svensson, "Identification of thioredoxin disulfide targets using a quantitative proteomics approach based on isotope-coded affinity tags," *Journal of Proteome Research*, vol. 7, no. 12, pp. 5270–5276, 2008.

[63] P. Hägglund, J. Bunkenborg, F. Yang, L. M. Harder, C. Finnie, and B. Svensson, "Identification of thioredoxin target disulfides in proteins released from barley aleurone layers," *Journal of Proteomics*, vol. 73, no. 6, pp. 1133–1136, 2010.

[64] G. P. Miles, M. A. Samuel, J. A. Ranish, S. M. Donohoe, G. M. Sperrazzo, and B. E. Ellis, "Quantitative proteomics identifies oxidant-induced, AtMPK6-dependent changes in *Arabidopsis thaliana* protein profiles," *Plant Signaling and Behavior*, vol. 4, no. 6, pp. 497–505, 2009.

[65] A. Fares, M. Rossignol, and J. B. Peltier, "Proteomics investigation of endogenous S-nitrosylation in *Arabidopsis*," *Biochemical and Biophysical Research Communications*, vol. 416, pp. 331–336, 2011.

[66] N. Islam, H. Tsujimoto, and H. Hirano, "Wheat proteomics: relationship between fine chromosome deletion and protein expression," *Proteomics*, vol. 3, no. 3, pp. 307–316, 2003.

[67] W. Majeran, Y. Cai, Q. Sun, and K. J. Van Wijk, "Functional differentiation of bundle sheath and mesophyll maize chloroplasts determined by comparative proteomics," *Plant Cell*, vol. 17, no. 11, pp. 3111–3140, 2005.

[68] A. Schmidt, J. Kellermann, and F. Lottspeich, "A novel strategy for quantitative proteomics using isotope-coded protein labels," *Proteomics*, vol. 5, no. 1, pp. 4–15, 2005.

[69] B. Leroy, C. Rosier, V. Erculisse, N. Leys, M. Mergeay, and R. Wattiez, "Differential proteomic analysis using isotope-coded protein-labeling strategies: comparison, improvements and application to simulated microgravity effect on *Cupriavidus metallidurans* CH34," *Proteomics*, vol. 10, no. 12, pp. 2281–2291, 2010.

[70] M. Fleron, Y. Greffe, D. Musmeci et al., "Novel post-digest isotope coded protein labeling method for phospho- and glycoproteome analysis," *Journal of Proteomics*, vol. 73, no. 10, pp. 1986–2005, 2010.

[71] F. C. Nogueira, G. Palmisano, V. Schwammle et al., "Performance of isobaric and isotopic labeling in quantitative plant proteomics," *Journal of Proteome Research*, vol. 11, no. 5, pp. 3046–3052, 2012.

[72] A. Brunner, E. M. Keidel, D. Dosch, J. Kellermann, and F. Lottspeich, "ICPLQuant—a software for non-isobaric isotopic labeling proteomics," *Proteomics*, vol. 10, no. 2, pp. 315–326, 2010.

[73] P. L. Ross, Y. N. Huang, J. N. Marchese et al., "Multiplexed protein quantitation in *Saccharomyces cerevisiae* using amine-reactive isobaric tagging reagents," *Molecular and Cellular Proteomics*, vol. 3, no. 12, pp. 1154–1169, 2004.

[74] S. Y. Ow, M. Salim, J. Noirel, C. Evans, I. Rehman, and P. C. Wright, "iTRAQ underestimation in simple and complex mixtures: 'the good, the bad and the ugly'," *Journal of Proteome Research*, vol. 8, no. 11, pp. 5347–5355, 2009.

[75] J. M. Perkel, "iTRAQ gets put to the test," *Journal of Proteome Research*, vol. 8, no. 11, p. 4885, 2009.

[76] F. A. R. Kaffarnik, A. M. E. Jones, J. P. Rathjen, and S. C. Peck, "Effector proteins of the bacterial pathogen *Pseudomonas syringae* alter the extracellular proteome of the host plant, *Arabidopsis thaliana*," *Molecular and Cellular Proteomics*, vol. 8, no. 1, pp. 145–156, 2009.

[77] M. N. Melo-Braga, T. Verano-Braga, I. R. Leon et al., "Modulation of protein phosphorylation, glycosylation and acetylation in grape (*Vitis vinifera*) mesocarp and exocarp due to *Lobesia botrana* infection," *Molecular & Cellular Proteomics*, vol. 11, no. 10, pp. 945–956, 2012.

[78] E. Marsh, S. Alvarez, L. M. Hicks et al., "Changes in protein abundance during powdery mildew infection of leaf tissues of Cabernet Sauvignon grapevine (*Vitis vinifera* L.)," *Proteomics*, vol. 10, no. 10, pp. 2057–2064, 2010.

[79] J. Fan, C. Chen, Q. Yu, R. H. Brlansky, Z. G. Li, and F. G. Gmitter Jr., "Comparative iTRAQ proteome and transcriptome analyses of sweet orange infected by, 'Candidatus Liberibacter asiaticus'," *Physiologia Plantarum*, vol. 143, no. 3, pp. 235–245, 2011.

[80] M. Mohammadi, V. Anoop, S. Gleddie, and L. J. Harris, "Proteomic profiling of two maize inbreds during early gibberella ear rot infection," *Proteomics*, vol. 11, no. 18, pp. 3675–3684, 2011.

[81] A. M. Jones, M. H. Bennett, J. W. Mansfield, and M. Grant, "Analysis of the defence phosphoproteome of *Arabidopsis thaliana* using differential mass tagging," *Proteomics*, vol. 6, no. 14, pp. 4155–4165, 2006.

[82] A. Rudella, G. Friso, J. M. Alonso, J. R. Ecker, and K. J. Van Wijk, "Downregulation of ClpR2 leads to reduced accumulation of the ClpPRS protease complex and defects in chloroplast biogenesis in *Arabidopsis*," *Plant Cell*, vol. 18, no. 7, pp. 1704–1721, 2006.

[83] J. Patterson, K. Ford, A. Cassin, S. Natera, and A. Bacic, "Increased abundance of proteins involved in phytosiderophore production in boron-tolerant barley," *Plant Physiology*, vol. 144, no. 3, pp. 1612–1631, 2007.

[84] T. Schneider, M. Schellenberg, S. Meyer et al., "Quantitative detection of changes in the leaf-mesophyll tonoplast proteome in dependency of a cadmium exposure of barley (*Hordeum vulgare* L.) plants," *Proteomics*, vol. 9, no. 10, pp. 2668–2677, 2009.

[85] S. Alvarez, B. M. Berla, J. Sheffield, R. E. Cahoon, J. M. Jez, and L. M. Hicks, "Comprehensive analysis of the Brassica juncea root proteome in response to cadmium exposure by complementary proteomic approaches," *Proteomics*, vol. 9, no. 9, pp. 2419–2431, 2009.

[86] K. A. Neilson, M. Mariani, and P. A. Haynes, "Quantitative proteomic analysis of cold-responsive proteins in rice," *Proteomics*, vol. 11, no. 9, pp. 1696–1706, 2011.

[87] W. Majeran, B. Zybailov, A. J. Ytterberg, J. Dunsmore, Q. Sun, and K. J. van Wijk, "Consequences of C4 differentiation for chloroplast membrane proteomes in maize mesophyll and bundle sheath cells," *Molecular and Cellular Proteomics*, vol. 7, no. 9, pp. 1609–1638, 2008.

[88] A. Thompson, J. Schäfer, K. Kuhn et al., "Tandem mass tags: a novel quantification strategy for comparative analysis of complex protein mixtures by MS/MS," *Analytical Chemistry*, vol. 75, no. 8, pp. 1895–1904, 2003.

[89] J. Parker, N. Zhu, M. Zhu, and S. Chen, "Profiling thiol redox proteome using isotope tagging mass spectrometry," *Journal of Visualized Experiments*, vol. 61, Article ID e3766, 2012.

[90] Q. Sun, C. Hu, J. Hu, S. Li, and Y. Zhu, "Quantitative proteomic analysis of CMS-related changes in honglian CMS rice anther," *Protein Journal*, vol. 28, no. 7-8, pp. 341–348, 2009.

[91] N. T. Hartman, F. Sicilia, K. S. Lilley, and P. Dupree, "Proteomic complex detection using sedimentation," *Analytical Chemistry*, vol. 79, no. 5, pp. 2078–2083, 2007.

[92] G. Mastrobuoni, S. Irgang, M. Pietzke et al., "Proteome dynamics and early salt stress response of the photosynthetic organism Chlamydomonas reinhardtii," *BMC Genomics*, vol. 13, no. 1, p. 215, 2012.

[93] S. F. Martin, V. S. Munagapati, E. Salvo-Chirnside, L. E. Kerr, and T. Le Bihan, "Proteome turnover in the green alga Ostreococcus tauri by time course 15N metabolic labeling

mass spectrometry," *Journal of Proteome Research*, vol. 11, no. 1, pp. 476–486, 2012.

[94] B. Naumann, A. Busch, J. Allmer et al., "Comparative quantitative proteomics to investigate the remodeling of bioenergetic pathways under iron deficiency in *Chlamydomonas reinhardtii*," *Proteomics*, vol. 7, no. 21, pp. 3964–3979, 2007.

[95] J. H. Ippel, L. Pouvreau, T. Kroef et al., "In vivo uniform 15N-isotope labelling of plants: using the greenhouse for structural proteomics," *Proteomics*, vol. 4, no. 1, pp. 226–234, 2004.

[96] W. R. Engelsberger, A. Erban, J. Kopka, and W. X. Schulze, "Metabolic labeling of plant cell cultures with $K^{(15)}NO_3$ as a tool for quantitative analysis of proteins and metabolites," *Plant Methods*, vol. 2, p. 14, 2006.

[97] V. Lanquar, L. Kuhn, F. Lelièvre et al., "^{15}N-Metabolic labeling for comparative plasma membrane proteomics in *Arabidopsis* cells," *Proteomics*, vol. 7, no. 5, pp. 750–754, 2007.

[98] T. Stanislas, D. Bouyssie, M. Rossignol et al., "Quantitative proteomics reveals a dynamic association of proteins to detergent-resistant membranes upon elicitor signaling in tobacco," *Molecular and Cellular Proteomics*, vol. 8, no. 9, pp. 2186–2198, 2009.

[99] J. J. Benschop, S. Mohammed, M. O'Flaherty, A. J. R. Heck, M. Slijper, and F. L. H. Menke, "Quantitative phosphoproteomics of early elicitor signaling in *Arabidopsis*," *Molecular and Cellular Proteomics*, vol. 6, no. 7, pp. 1198–1214, 2007.

[100] L. V. Bindschedler, M. Palmblad, and R. Cramer, "Hydroponic isotope labelling of entire plants (HILEP) for quantitative plant proteomics; an oxidative stress case study," *Phytochemistry*, vol. 69, no. 10, pp. 1962–1972, 2008.

[101] J. E. Schaff, F. Mbeunkui, K. Blackburn, D. M. Bird, and M. B. Goshe, "SILIP: a novel stable isotope labeling method for in planta quantitative proteomic analysis," *Plant Journal*, vol. 56, no. 5, pp. 840–854, 2008.

[102] E. L. Huttlin, A. D. Hegeman, A. C. Harms, and M. R. Sussman, "Comparison of full versus partial metabolic labelling for quantitative proteomics analysis in *Arabidopsis thaliana*," *Molecular and Cellular Proteomics*, vol. 6, no. 5, pp. 860–881, 2007.

[103] K. Baerenfaller, J. Grossmann, M. A. Grobei et al., "Genome-scale proteomics reveals *Arabidopsis thaliana* gene models and proteome dynamics," *Science*, vol. 320, no. 5878, pp. 938–941, 2008.

[104] B. Zybailov, H. Rutschow, G. Friso et al., "Sorting signals, N-terminal modifications and abundance of the chloroplast proteome," *PLoS ONE*, vol. 3, no. 4, Article ID e1994, 2008.

[105] B. Cooper, J. Feng, and W. M. Garrett, "Relative, label-free protein quantitation: spectral counting error statistics from nine replicate MudPIT samples," *Journal of the American Society for Mass Spectrometry*, vol. 21, no. 9, pp. 1534–1546, 2010.

[106] C. G. Gammulla, D. Pascovici, B. J. Atwell, and P. A. Haynes, "Differential metabolic response of cultured rice (*Oryza sativa*) cells exposed to high- and low-temperature stress," *Proteomics*, vol. 10, no. 16, pp. 3001–3019, 2010.

[107] T. Niittylä, A. T. Fuglsang, M. G. Palmgren, W. B. Frommer, and W. X. Schulze, "Temporal analysis of sucrose-induced phosphorylation changes in plasma membrane proteins of *Arabidopsis*," *Molecular and Cellular Proteomics*, vol. 6, no. 10, pp. 1711–1726, 2007.

[108] I. J. E. Stulemeijer, M. H. A. J. Joosten, and O. N. Jensen, "Quantitative phosphoproteomics of tomato mounting a hypersensitive response reveals a swift suppression of photosynthetic activity and a differential role for Hsp90 isoforms,"

Journal of Proteome Research, vol. 8, no. 3, pp. 1168–1182, 2009.

[109] S. Komatsu, T. Wada, Y. Abaléa et al., "Analysis of plasma membrane proteome in soybean and application to flooding stress response," *Journal of Proteome Research*, vol. 8, no. 10, pp. 4487–4499, 2009.

[110] S. Wienkoop, E. Larrainzar, M. Niemann, E. M. Gonzalez, U. Lehmann, and W. Weckwerth, "Stable isotope-free quantitative shotgun proteomics comined with sample pattern recognition for rapid diagnostics," *Journal of Separation Science*, vol. 29, no. 18, pp. 2793–2801, 2006.

[111] S. E. Stevenson, Y. Chu, P. Ozias-Akins, and J. J. Thelen, "Validation of gel-free, label-free quantitative proteomics approaches: applications for seed allergen profiling," *Journal of Proteomics*, vol. 72, no. 3, pp. 555–566, 2009.

[112] K. Blackburn, F. Mbeunkui, S. K. Mitra, T. Mentzel, and M. B. Goshe, "Improving protein and proteome coverage through data-independent multiplexed peptide fragmentation," *Journal of Proteome Research*, vol. 9, no. 7, pp. 3621–3637, 2010.

[113] Z. Shen, P. Li, R. J. Ni et al., "Label-free quantitative proteomics analysis of etiolated maize seedling leaves during greening," *Molecular and Cellular Proteomics*, vol. 8, no. 11, pp. 2443–2460, 2009.

[114] F. Y. Cheng, K. Blackburn, Y. M. Lin, M. B. Goshe, and J. D. Williamson, "Absolute protein quantification by LC/MSe for global analysis of salicylic acid-induced plant protein secretion responses," *Journal of Proteome Research*, vol. 8, no. 1, pp. 82–93, 2009.

[115] S. Kaspar, A. Matros, and H. P. Mock, "Proteome and flavonoid analysis reveals distinct responses of epidermal tissue and whole leaves upon UV-B radiation of barley (*Hordeum vulgare* L.) seedlings," *Journal of Proteome Research*, vol. 9, no. 5, pp. 2402–2411, 2010.

[116] P. Pichler, T. Köcher, J. Holzmann et al., "Peptide labeling with isobaric tags yields higher identification rates using iTRAQ 4-plex compared to TMT 6-plex and iTRAQ 8-plex on LTQ orbitrap," *Analytical Chemistry*, vol. 82, no. 15, pp. 6549–6558, 2010.

[117] M. Mann, "Functional and quantitative proteomics using SILAC," *Nature Reviews Molecular Cell Biology*, vol. 7, no. 12, pp. 952–958, 2006.

[118] A. Gruhler, W. X. Schulze, R. Matthiesen, M. Mann, and O. N. Jensen, "Stable isotope labeling of *Arabidopsis thaliana* cells and quantitative proteomics by mass spectrometry," *Molecular and Cellular Proteomics*, vol. 4, no. 11, pp. 1697–1709, 2005.

[119] M. Palmblad, L. V. Bindschedler, and R. Cramer, "Quantitative proteomics using uniform 15N-labeling, MASCOT, and the trans-proteomic pipeline," *Proteomics*, vol. 7, no. 19, pp. 3462–3469, 2007.

[120] A. H. P. America and J. H. G. Cordewener, "Comparative LC-MS: a landscape of peaks and valleys," *Proteomics*, vol. 8, no. 4, pp. 731–749, 2008.

[121] K. A. Neilson, N. A. Ali, S. Muralidharan et al., "Less label, more free: approaches in label-free quantitative mass spectrometry," *Proteomics*, vol. 11, no. 4, pp. 535–553, 2011.

[122] D. H. Lundgren, S. I. Hwang, L. Wu, and D. K. Han, "Role of spectral counting in quantitative proteomics," *Expert Review of Proteomics*, vol. 7, no. 1, pp. 39–53, 2010.

[123] P. V. Bondarenko, D. Chelius, and T. A. Shaler, "Identification and relative quantitation of protein mixtures by enzymatic digestion followed by capillary reversed-phase liquid chromatography—tandem mass spectrometry," *Analytical Chemistry*, vol. 74, no. 18, pp. 4741–4749, 2002.

[124] D. Chelius and P. V. Bondarenko, "Quantitative profiling of proteins in complex mixtures using liquid chromatography and mass spectrometry," *Journal of Proteome Research*, vol. 1, no. 4, pp. 317–323, 2002.

[125] H. Liu, R. G. Sadygov, and J. R. Yates, "A model for random sampling and estimation of relative protein abundance in shotgun proteomics," *Analytical Chemistry*, vol. 76, no. 14, pp. 4193–4201, 2004.

[126] S. J. Geromanos, J. P. C. Vissers, J. C. Silva et al., "The detection, correlation, and comparison of peptide precursor and product ions from data independent LC-MS with data dependant LC-MS/MS," *Proteomics*, vol. 9, no. 6, pp. 1683–1695, 2009.

[127] J. D. Venable, M. Q. Dong, J. Wohlschlegel, A. Dillin, and J. R. Yates, "Automated approach for quantitative analysis of complex peptide mixtures from tandem mass spectra," *Nat Methods*, vol. 1, no. 1, pp. 39–45, 2004.

[128] J. Rappsilber, U. Ryder, A. I. Lamond, and M. Mann, "Large-scale proteomic analysis of the human spliceosome," *Genome Research*, vol. 12, no. 8, pp. 1231–1245, 2002.

[129] Y. Ishihama, Y. Oda, T. Tabata et al., "Exponentially modified protein abundance index (emPAI) for estimation of absolute protein amount in proteomics by the number of sequenced peptides per protein," *Molecular and Cellular Proteomics*, vol. 4, no. 9, pp. 1265–1272, 2005.

[130] P. Lu, C. Vogel, R. Wang, X. Yao, and E. M. Marcotte, "Absolute protein expression profiling estimates the relative contributions of transcriptional and translational regulation," *Nature Biotechnology*, vol. 25, no. 1, pp. 117–124, 2007.

[131] C. G. Gammulla, D. Pascovici, B. J. Atwell, and P. A. Haynes, "Differential proteomic response of rice (*Oryza sativa*) leaves exposed to high- and low-temperature stress," *Proteomics*, vol. 11, no. 14, pp. 2839–2850, 2011.

[132] E. Larrainzar, S. Wienkoop, W. Weckwerth, R. Ladrera, C. Arrese-Igor, and E. M. González, "Medicago truncatula root nodule proteome analysis reveals differential plant and bacteroid responses to drought stress," *Plant Physiology*, vol. 144, no. 3, pp. 1495–1507, 2007.

[133] J. Lee, J. Feng, K. B. Campbell et al., "Quantitative proteomic analysis of bean plants infected by a virulent and avirulent obligate rust fungus," *Molecular and Cellular Proteomics*, vol. 8, no. 1, pp. 19–31, 2009.

[134] J. C. Silva, R. Denny, C. A. Dorschel et al., "Quantitative proteomic analysis by accurate mass retention time pairs," *Analytical Chemistry*, vol. 77, no. 7, pp. 2187–2200, 2005.

[135] M. Z. Nouri and S. Komatsu, "Comparative analysis of soybean plasma membrane proteins under osmotic stress using gel-based and LC MS/MS-based proteomics approaches," *Proteomics*, vol. 10, no. 10, pp. 1930–1945, 2010.

[136] W. M. Old, K. Meyer-Arendt, L. Aveline-Wolf et al., "Comparison of label-free methods for quantifying human proteins by shotgun proteomics," *Molecular and Cellular Proteomics*, vol. 4, no. 10, pp. 1487–1502, 2005.

[137] M. Palmblad, D. J. Mills, L. V. Bindschedler, and R. Cramer, "Chromatographic alignment of LC-MS and LC-MS/MS datasets by genetic algorithm feature extraction," *Journal of the American Society for Mass Spectrometry*, vol. 18, no. 10, pp. 1835–1843, 2007.

[138] P. C. Carvalho, X. Han, T. Xu et al., "XDIA: improving on the label-free data-independent analysis," *Bioinformatics*, vol. 26, no. 6, Article ID btq031, pp. 847–848, 2010.

[139] K. Blackburn, F. Y. Cheng, J. D. Williamson, and M. B. Goshe, "Data-independent liquid chromatography/mass spectrometry (LC/MSE) detection and quantification of the secreted *Apium graveolens* pathogen defense protein mannitol dehydrogenase," *Rapid Communications in Mass Spectrometry*, vol. 24, no. 7, pp. 1009–1016, 2010.

8

Poor Homologous Synapsis 1 Interacts with Chromatin but Does Not Colocalise with ASYnapsis 1 during Early Meiosis in Bread Wheat

Kelvin H. P. Khoo, Amanda J. Able, and Jason A. Able

School of Agriculture, Food & Wine, Waite Research Institute, The University of Adelaide, Waite Campus, PMB1, Glen Osmond, SA, 5064, Australia

Correspondence should be addressed to Jason A. Able, jason.able@adelaide.edu.au

Academic Editor: Sylvie Cloutier

Chromosome pairing, synapsis, and DNA recombination are three key processes that occur during early meiosis. A previous study of *Poor Homologous Synapsis 1* (*PHS1*) in maize suggested that PHS1 has a role in coordinating these three processes. Here we report the isolation of wheat (*Triticum aestivum*) *PHS1* (*TaPHS1*), and its expression profile during and after meiosis. While the *Ta*PHS1 protein has sequence similarity to other plant PHS1/PHS1-like proteins, it also possesses a unique region of oligopeptide repeat units. We show that *Ta*PHS1 interacts with both single- and double-stranded DNA *in vitro* and provide evidence of the protein region that imparts the DNA-binding ability. Immunolocalisation data from assays conducted using antisera raised against *Ta*PHS1 show that *Ta*PHS1 associates with chromatin during early meiosis, with the signal persisting beyond chromosome synapsis. Furthermore, *Ta*PHS1 does not appear to colocalise with the asynapsis protein (*Ta*ASY1) suggesting that these proteins are probably independently coordinated. Significantly, the data from the DNA-binding assays and 3-dimensional immunolocalisation of *Ta*PHS1 during early meiosis indicates that *Ta*PHS1 interacts with DNA, a function not previously observed in either the Arabidopsis or maize PHS1 homologues. As such, these results provide new insight into the function of PHS1 during early meiosis in bread wheat.

1. Introduction

For the majority of sexually reproducing organisms, meiosis is a cellular process required for gamete formation and is composed of one round of DNA replication, followed by two rounds of chromosome division. During meiosis I, a reductional division event leads to the segregation of homologous chromosome pairs, while an equational division during meiosis II leads to the segregation of the sister chromatids.

For the successful juxtaposition of homologous chromosomes, three key processes occur during prophase I, namely, pairing, synapsis, and recombination. Previous studies investigating the molecular mechanisms of homologous chromosome pairing have revealed complex interplay between these three tightly linked processes [1–5].

In allopolyploid organisms such as bread wheat (*Triticum aestivum*), correct alignment and pairing of homologous chromosomes are complicated by the presence of genetically similar genomes, known as homoeologous genomes. Although bread wheat possesses three homoeologous genomes (termed A, B, and D), meiosis proceeds as if the organism is a diploid, in that pairing only occurs between homologous chromosomes from the same genome ([6–9] and references within). This strict pairing interaction between homologous chromosomes has previously been shown to be controlled by *pairing homoeologous* (*Ph*) loci [10, 11]. The most extensively studied of these loci is the *Ph1* locus located on the long arm of chromosome 5B. While the molecular mechanism by which the *Ph1* locus operates is still subject to intensive research, *Ph1* appears to indirectly promote homologous chromosome pairing by suppressing homoeologous chromosome interactions through regulation of the specificity of chromosome interactions at centromeric and telomeric regions [12, 13].

In *ph1* deletion mutants, the chromatin is prematurely and asynchronously remodelled, leading to abnormal chromosome conformation that results in increased interactions between homoeologous chromosomes [13, 14]. These mutants also display other meiotic defects such as the arrest of synapsis during zygotene and the presence of uncorrected multiple axial element associations, which in wild type are normally corrected prior to entry into pachytene [15, 16]. While the deletion region in the *ph1b* mutant is extensive, the *Ph1* locus has recently been refined to an area that contains, among other genes, seven *Cyclin-dependent kinase*-like (*Cdk*-like) genes [17, 18].

Our current knowledge of other meiotic genes mostly comes from research on model species such as yeast and Arabidopsis. However, putative homologues of many of these genes have also been identified in the cereals [19]. Some of the early meiotic genes characterised in various plant species include *ASY1* (*ASYnapsis 1*) [20–24], *RAD51* (*RADiation sensitive 51*) [25, 26], and *PHS1* (*Poor Homologous Synapsis 1*) [27, 28]. In wheat, *ASY1* (*TaASY1*) is involved in chromosome synapsis and promotes homologous chromosome pairing during meiosis I [20, 21]. Interestingly, *Taasy1* knockdown mutants have been reported to display defective chromosome characteristics similar to that of the *ph1b* mutant [20]. In the absence of *Ph1*, the expression of *TaASY1* is approximately 20-fold higher compared to wild type while the localisation of its protein product is also affected. This indicates that *TaASY1* is intimately involved in the *Ph1*-dependent control of chromosome pairing in wheat [9, 20].

PHS1 was first identified in a *Mutator* transposon-mutagenised maize population, with no known homologues in yeast or other nonplant species [27]. While phenotypic analysis of the *phs1-0* mutant by Pawlowski et al. [27] revealed no vegetative defects, meiosis was disrupted resulting in male and female sterility. Transmission electron microscopy of meiotic spreads from *Zmphs1-0* meiocytes revealed significantly reduced levels of synapsis during zygotene and improper alignment of the chromosomes in the synapsed regions. Although most of the chromosomes were fully synapsed by late pachytene, the chromosomes were synapsed with multiple partners. Coupled together with results of their fluorescent *in situ* hybridisation (FISH), the data indicated that nonhomologous chromosome synapsis was present in the *Zmphs1-0* mutant.

FISH results from recent work by Ronceret et al. [28] on the Arabidopsis homologue of PHS1 (*At*PHS1) showed that PHS1 appears to function in a similar manner in different species independent of genome size and complexity. Chromosome axis formation and installation of the synaptonemal complex components in both wild type and *phs1* mutant cells of Arabidopsis and maize appeared similar albeit with ZYP1, a transverse filament protein, loading being delayed in some instances [28]. Immunolocalisation in Arabidopsis and maize revealed that PHS1 is located within the cytoplasm during the early stages of meiosis with some foci clustered along the nuclear envelope during zygotene. In some instances, a few foci are present in the nucleus during pachytene. With a reduced number of RAD50 foci observed in the nuclei of the *phs1* mutants, Ronceret and colleagues

[28] concluded that PHS1 regulates meiotic recombination and chromosome pairing by regulating the transport of RAD50, a protein which is required during double-strand break processing, into the nucleus.

Here, we report the extensive analysis of the PHS1 protein in bread wheat, providing evidence that shows *Ta*PHS1 possesses DNA-binding capabilities even though no known DNA-binding domains were identified *in silico*. Our data also show that *PHS1* is upregulated in the *ph1b* bread wheat mutant when compared to wild type, and that *Ta*PHS1 is associated with chromatin and is present on the nucleolar periphery during the early stages of meiosis as indicated through immunolocalisation analysis using an antibody that was raised against the full-length wheat PHS1 protein.

2. Materials and Methods

2.1. Plant Material. Hexaploid wheat (*Triticum aestivum* L.) cv. Chinese Spring plants and a Chinese Spring mutant lacking the *Ph1* locus (*ph1b*) were grown in a glasshouse with a 16/8 h photoperiod at 23°C. Harvesting and staging of meiotic anthers from both wild type and mutant plants, for quantitative real-time PCR (Q-PCR) and fluorescence immunolocalisation, were conducted as per Boden et al. [20]. Whole meiotic spike tissue was collected for the isolation and amplification of the gDNA and cDNA *Triticum aestivum PHS1* (*TaPHS1*) sequences.

2.2. RNA Isolation and cDNA Synthesis. Collected tissues for RNA isolation were initially ground in liquid nitrogen. Total RNA was extracted using Trizol reagent (Gibco-BRL, Carlsbad, CA, USA) according to the manufacturer's instructions. RNA concentration was determined using a Nanodrop (ND-1000) (Nanodrop, Wilmington, DE, USA). cDNA was synthesised from 2 μg of total RNA using the iScript cDNA synthesis kit (Bio-Rad, Hercules, CA, USA) according to the manufacturer's instructions.

2.3. cDNA Amplification and Sequencing of the PHS1 Coding Sequence. Primers (see Table S1 of the supplementary material available online at doi:10.1155/2011/514398) for isolating the *TaPHS1* ORF were designed using the *OsPHS1* sequence (LOC_Os06g27860, MSU Rice Genome Annotation database-http://rice.plantbiology.msu.edu) identified through a TIGR rice expressed sequence tag (EST) BLAST search (accessed 21st October 2008).

Each PCR contained 100 ng cDNA, 0.2 mM dNTPs, 0.2 μM primers, and 1 U FastStart high-fidelity *Taq* polymerase (Roche Applied Science, Mannheim, Germany) in 25 μL of 1 × high-fidelity buffer supplemented with 1 × GC-RICH solution (Roche). PCR products were cloned into pCR8/GW/TOPO (Invitrogen) for DNA sequencing (15 × coverage). Sequencing PCR and capillary separation were conducted using the same methods as described earlier except that GW1 and GW2 primers were used (see Table S1). Secondary sets of primers were designed on the sequenced products to specifically amplify the *TaPHS1* ORF. Amplification and sequencing were repeated as above. PCR cycling parameters were denatured at 95°C for 5 min, followed by 35 cycles of 96°C for 30 s, T_m°C for 30 s, and 72°C for 75 s, with a final elongation step at 72°C for 10 min (see Table S1 for T_m of primer sets). The assigned NCBI accession number for *TaPHS1* is GQ851928.

2.4. Bioinformatics Analysis.
DNA sequence alignments and comparisons were conducted with AlignX and Contig Express (Informax, VNTI Advance, Version 11, Frederick, MC, USA) software programs. VNTI software was also used to predict the molecular weight and pI of the protein. To predict the cellular localisation of *TaPHS1*, SignalP 3.0 (http://www.cbs.dtu.dk/services/SignalP/) [29] and WoLF PSORT (http://wolfpsort.org/) [30] were used, while detection for conserved domains was performed using the NCBI Conserved Domain Search Tool (http://www.ncbi.nlm.nih.gov/Structure/cdd/wrpsb.cgi), InterPro Scan (http://www.ebi.ac.uk/Tools/InterProScan/), and Pfam 23.0 (http://pfam.janelia.org/).

Amino acid alignments and comparisons of full-length PHS1 sequences (obtained from various BLAST searches using the NCBI, TIGR, and PredictProtein (http://www.predictprotein.org/; [31]) databases) and subsequent construction of the phylogenetic tree (neighbour-joining method) [32] were completed using Molecular Evolutionary Genetics Analysis (MEGA) software (version 4.0) [33]. Default parameters were used except for the following: the pair-wise deletion option was used, the internal branch test bootstrap value was set at 10,000 resamplings, and the model setting was amino acid, Poisson correction with predicted gamma parameters set at 2.0. Accession numbers of the sequences used were, *Ta* [GenBank: GQ851928], *Sb* [TIGR EST assemblies: TA33290_4558], *Zm* [GenBank: NP_001141750]; *Os* [MSU Rice Genome Annotation: LOC_Os06g27860]; *At* [GenBank: NP_172541], *Pt* [UniProtKB: B9HTU7_POPTR]; *Vv* [UniProtKB: A7QY03_VITVI], and *Rc* [UniProtKB: B9SPJ5_RICCO]. To determine whether the level of divergence between *TaPHS1* and *OsPHS1* was significant, a Tajima's Relative Rate Test [34] with the *AtPHS1* sequence as an outgroup was conducted (with one degree of freedom and a significance value of $P < 0.05$).

2.5. Southern Blot Hybridisation.
A 371 bp fragment of the *TaPHS1* gene was used as the template for the synthesis of an α-^{32}P dCTP labelled probe that was hybridised to a Chinese Spring nullisomic-tetrasomic membrane as per Lloyd et al. [35]. Autoradiography films were developed using an AGFA CP1000 Developer (AGFA, Nunawading, VIC, Australia). For *in silico* mapping experimental procedures refer to supplementary material.

2.6. Q-PCR.
Q-PCR was conducted in triplicate according to Crismani et al. [36]. Amplification of products was completed using gene-specific Q-PCR primers (see Table S1). The optimal acquisition temperature for *TaPHS1* was 80°C. Normalisation of the expression data was performed against three control genes (*actin*, *GAPdH*, and *cyclophilin*) as per Crismani et al. [36].

2.7. Protein Analysis.
The *TaPHS1* insert within the pCR8/GW/TOPO vector was cloned into a pDEST17 expression plasmid (Invitrogen) according to the manufacturer's LR clonase protocol. BL-21 A1 *E. coli* were transformed with the pDEST17-*TaPHS1* ORF vector, and protein production was induced with 0.4% L-(+)-arabinose (w/v) (Sigma-Aldrich, St Louis, MO, USA). Production of four partial *TaPHS1* peptides corresponding to the four conserved regions identified in this study were also performed as described above using DNA inserts encoding these regions. Protein isolation and purification were performed using nickel-nitrilotriacetic acid (Ni-NTA) beads (Qiagen, Clifton, VIC, Australia) according to the manufacturer's extraction protocols. Sodium dodecyl sulfate polyacrylamide gel electrophoresis (SDS-PAGE) was performed using NuPAGE Novex 4–12% Bis-Tris 7 cm mini-gels (Invitrogen) according to the manufacturer's protocol. Staining and destaining of gels were performed as previously reported [37].

The identity of the recombinant *TaPHS1* protein was confirmed by ion trap liquid chromatography-electrospray ionisation tandem mass spectrometry (LC-MS/MS). Gel slices containing the recombinant *TaPHS1* protein were washed with 100 mM ammonium bicarbonate, dried, rehydrated with 100 mM ammonium carbonate, and subjected to in-gel tryptic digestion. LC-MS/MS of the digested peptides was then conducted as reported by March et al. [38].

2.8. Polyclonal Antibody Production.
Full-length recombinant *TaPHS1* protein was dissolved in $1 \times$ PBS ($10\,\mu g\,\mu L^{-1}$), added with an equivalent amount of Freund's complete adjuvant (Sigma-Aldrich), and used for primary immunisation of two rats via subcutaneous injection. Three subsequent immunisations were administered in three-week intervals, with Freund's incomplete adjuvant (Sigma-Aldrich) added to the dissolved antigen in $1 \times$ PBS. All immunisation doses contained $200\,\mu g$ of *TaPHS1* antigen. Immune sera was collected 10.5 weeks after the first injection. Specificity of the *TaPHS1* antisera was confirmed using western analysis (see supplementary material; Figures S1 and S2).

2.9. Competitive DNA-Binding Assay.
Recombinant full-length *TaPHS1* and the four partial peptides extracted under native conditions were quantified using the Bradford assay [39]. Competitive DNA-binding assays were conducted as described by Pezza et al. [40] with modifications as per Khoo et al. [41]. The DNA-binding abilities of *TaPHS1* and its partial peptides were tested with $\Phi X174$ circular single-stranded DNA (ssDNA) (virion) ($30\,\mu M$ per nucleotide) (New England Biolabs, Beverly, MA, USA) and $\Phi X174$ linear double-stranded DNA (dsDNA) (RFI form *Pst1*-digested) ($15\,\mu M$ per base pair) (New England Biolabs).

2.10. Fluorescence Immunolocalisation.
Fluorescence immunolocalisation of *TaASY1* and *TaPHS1* was performed as per

Franklin et al. [42] and Boden et al. [20] with the following changes: anthers were fixed with 2% paraformaldehyde and cells permeabilised for 3 h. For detecting the localisation pattern of *TaPHS1*, a rat anti-*TaPHS1* antisera (1 : 100) and an AlexaFluor 488 conjugated donkey anti-rat antibody (1 : 50; Molecular Probes, Invitrogen) were used. Optical sections (90–120 per nucleus) of meiocytes were collected using a Leica TCS SP5 Spectral Scanning Confocal Microscope (Leica Microsystems, http://www.leica-microsystems.com/) equipped with an oil immersion HCX Plan Apochromat 63 × /1.4 lens, a 405 nm pulsed laser, and an Argon laser using an excitation wavelength of 468 nm. Images were processed using Leica Application Suite Advanced Fluorescence (LAS-AF; version 1.8.2, build 1465, Leica Microsystems) software to generate maximum intensity projections of each nucleus.

3. Results

3.1. PHS1 Is Highly Similar across Plant Species, and in Wheat It Encodes a Predicted Protein Product with a Unique Oligopeptide Repeat Sequence. PCR amplification from whole meiotic spike cDNA using the primers listed in Table S1 (Additional Information File 1) resulted in the isolation of *TaPHS1* which has a 1071 bp ORF. This encodes a 357 amino acid protein with a predicted molecular weight of 38.958 kDa and a pI of 5.23. Despite no nuclear localisation signal (NLS) peptides being detected within the *TaPHS1* amino acid sequence using SignalP 3.0, WoLF PSORT analysis predicted that *TaPHS1* is most likely to be located within the cell nucleus. Using AlignX, comparative amino acid sequence analysis of full-length annotated PHS1, or PHS1-like proteins obtained through database searches (refer to Methods) showed that *TaPHS1* shared relatively high levels of sequence identity with its homologues in other species (*Sorghum bicolor* [*SbPHS1*]—53.6%, *Zea mays* [*ZmPHS1*]—51.2%, *Oryza sativa* [*OsPHS1*]—41.4%, and *Arabidopsis thaliana* [*AtPHS1*-like]—21.5%). We propose that there are four prominent regions within the PHS1 amino acid sequence that are well conserved (Figure 1(a)). While a portion of Region 2 (corresponding to amino acid positions 99–145 of *TaPHS1*) was previously identified to contain two conserved domains [27], interspecies comparisons made in the current study suggest that this conserved region can now be extended by 11 amino acid residues toward the N-termini of PHS1 proteins in monocot species (Figure 1(a), dashed line). In addition, there is a short region of oligopeptide repeats from position 242 to 265 [YSGFPEGYSGFPEGYSGFPEGYSG] unique to *TaPHS1* (Figure 1(a), boxed feature).

To assess the phylogenetic relationships between the five homologues shown in Figure 1(a), a neighbour-joining tree was constructed using the full-length amino acid sequences (Figure 1(b)). As expected, Arabidopsis is the most divergent, while sorghum and maize share a higher degree of similarity with one another. Although wheat and rice fall within the same cluster, the internal branch length difference between the two species suggests that a significant level of sequence divergence has occurred. This significant level of sequence

divergence was confirmed by a Tajima's Relative Rate Test ($P = 0.00015$ with one degree of freedom). To determine whether other PHS1 sequences could be identified in the public databases, the more sensitive Hidden Markov Model (HMM) and MaxHom functions of the Predict Protein program were used. Three additional sequences were identified that were similar to *TaPHS1* and all from dicot species, namely, poplar (*Populus trichocarpa*) (*E*-value: $7E^{-97}$), grape (*Vitis vinifera*) (*E*-value: $7E^{-91}$), and castor oil (*Ricinus communis*) (*E*-value: $2E^{-97}$). The addition of these three sequences to the phylogenetic analysis shows that they cluster with Arabidopsis, the only other dicot species (Figure 1(c)).

Southern blot analysis showed that *TaPHS1* is located on chromosome group 7, with a copy on each of the three genomes (Figure 2). *In silico,* mapping revealed that *TaPHS1*, is likely to reside on the short arm of this chromosome group (Bin 7AS8-0.45-0.59, data not shown). To determine this, rice genetic markers that are located close to *OsPHS1* (on rice chromosome 6) were used to screen wheat deletion bins. One marker previously bin-mapped to wheat chromosome group 7 [GenBank: BE404111.1] was identified to be approximately 35 kb from *OsPHS1*.

3.2. TaPHS1 Interacts with DNA and Is Expressed during Meiosis. Previously reported homology searches using the maize PHS1 protein revealed low levels of sequence similarity with two families of fungal helicases, possibly indicating that PHS1 may interact with DNA [27]. Indeed, *in silico,* amino acid analysis of *TaPHS1* revealed that Region 1 contains two S/TPXX motifs (TPPP: amino acid positions 46 to 49 and SPAA: amino acid positions 71 to 74), which have previously been reported to bind DNA [45]. To determine whether *TaPHS1* interacts with DNA, a competitive DNA-binding assay using recombinant *TaPHS1* extracted under native conditions was conducted (Figure 3(a)). Interactions occurred with both single-stranded DNA (ssDNA) and double-stranded DNA (dsDNA). Interestingly, *TaPHS1* appears to have a higher affinity for ssDNA with retardation of the ssDNA species within the gel matrix being more obvious even when equivalent amounts of ss- and dsDNA are present. Although *TaPHS1* also interacts with dsDNA, the retardation of the dsDNA species only occurs at higher concentrations of *TaPHS1*. Competitive DNA binding assays using partial *TaPHS1* peptides corresponding to the four prominent conserved regions identified in this study revealed that only Region 1 possesses DNA-binding capabilities (Figures 3(b)–3(e)) and appears to have a higher affinity for ssDNA compared to dsDNA.

Quantitative real-time PCR (Q-PCR) profiling of *TaPHS1* in wild type Chinese Spring shows that it has low transcript abundance during meiosis (Figure 4). Although *TaPHS1* is expressed in wheat anther tissue throughout all stages of meiosis examined and beyond, statistical analysis suggests that expression is higher during premeiotic interphase and immature pollen. Between the pooled stages of leptotene-pachytene and diplotene-anaphase I, there is no statistically significant difference in *TaPHS1* expression. Given that Boden et al. [20] demonstrated that the *TaASY1*

```
        ++ +   +        +           + ++ +++++++++   +++ ++ ++++++ +
Ta    1 MAGAGGRSRERLTSRAEEAAGGK----RRRQRWEVEFARYFAKPRRAPSTPPPPGLRYIS   56
Sb    1 MADASDNSMALVHARLAVSAVAPPRMLRQRQKWEVEYARYFGTPRRDPSAPPPPGLRHII   60
Zm    1 MADAADSSMALVHSSLADSVLTSPRTLRQGQKWEVEYARYFGTPRRDPTAAPPSGLRYIM   60
Os    1 MADADVRSGALLPARPTPQ-------RRPQKWAVEFARFFRTPRRDPSKPPPPGLRLVA    52
At    1 MAGSLTASNRRRNAEDSSE---------IYRWTIGFARFVHYP---SSPSPHPVLKPLG    47

        ++   +  +++++   +++ ++ + +     +  +++++++++ +++ ++++++++++
Ta   57 RGKQLH--QGTWLLAASPAALCISRPTHSFAARVLTVSIGDVVYEEHYVSILNFSWPQVA  114
Sb   61 RGVHRH--QGTWIPASCPASLCVSHPSLPSAVPVLTISIGDVVFEEHFVSILNFSWPQIT  118
Zm   61 RGVHRH--QGTWIPASCPASLCVCHPSLPSAVPVLTISIGDVVFEEHFVSILNFSWPQVT  118
Os   53 RGKLRH--HGTWLPAASPAALSISCPSQSFAVPVLTVSIGDVVFVRTDPAPLPR---GAF  107
At   48 KREQYHSPHGTWLSASSSTVSLHIVDELNRSDVILSVKLGQKVLEEHYISKLNFTWPQMS  107

        ++  ++ +++++++++++  +++++++++    +++ ++ +  +        ++++ ++
Ta  115 CVTECPVRGSRVVFVSFCDRSKQIQKFAVRFPRLSDAESFLNSVIVKELSS-NTMDIMPS  173
Sb  119 CVMQCPIRGSRVVFVSFCDKSKQIQKFAVRFPQLCDAELFLSCVKECSC---ETMDMIPS  175
Zm  119 CVTQCPIRGSRVVFVSFCDKFKQIQKFAVRFPQPCDAESFLSCV-ECSCGSSGTMDIIPF  177
Os  108 CVHSQFFVASGYMCDTMPNKWEQSG---------VSLNAFLYGLFTKECST-ETMDIRPS  157
At  108 CVSGFPSRGSRAIFVTYMDSANQIQKFALRFSTCDAALEFVEALKEKIKGL-KEASTONQ  166

        ++++ ++  +++ ++++ ++++     +    ++  ++    +  ++ +
Ta  174 GSDYMCELEDSSSSEYIPSNGLQYRPDE-----------AVSFEEPTSD--HRTDAPAVG  220
Sb  176 GSDYVCE--DSSTSEYIAYNGLHHRPDD-----------ASGFEEQSSD--HTIDEPTMS  220
Zm  178 GSDYVCE--DSSASEYIVSNGLHHRLDD-----------ASNLEEQCFD--HTIDEPPMN  222
Os  158 GSDYLCE--DSSASEYIASSGIHQSFEEPDQFVHRTETPALGYHAEPDEPIHRTEAPALS  215
At  167 KNKTRCD--VSFQSDYNPSDAIIPRATQK----------EPNMVRPLNSYVPEMLPRIVY  214

           ++ + +++++  +   +                       + +
Ta  221 --------YHMEPDQPVLQSPIATNINSI[YSGFPEGYSGFPEGYSGFPEGYSG]BVKIERD  272
Sb  221 --------YHEEQDQPVLEPLSASNTSNSYSAFPPSFNQMLT---------NCSIDYDQE  263
Zm  223 --------YHEETDQHVLEPLSASNTSNN-SAFPPSFNQMLK---------SCSIDYDQE  264
Os  216 QPETPSLRHHEAPEEPLLQPLLATNIDTVFSGFPPSFTDMLT-------QFSCKTEKDAE  268
At  215 ------EAQYQKSETRSEVSFQSDYNPSIEIFPRATEEEPNMVRFFDSSVPEVLPRPEY  267

        +  +  ++    + +++  + +      +   ++  ++     + + ++++ +++++ ++
Ta  273 GGPFPATITDHAPEKAYILDTRIDAAGGNSVADKGKGAGKEIDVSDVTRDILAGIETYGG  332
Sb  264 E-PCPLAASNHALQEVYALDTSHDVANEETTAGKGLDAGEGVDTSILTYDIMARIKTYMA  322
Zm  265 E-PCPLAASNHVLQEVYVLDTSHDVANEERTAGKGMDAAEGVDASILTYDLMARIKTYMA  323
Os  269 E-PYPVTATDHAPQEVSMLDTSHNVAISTTSAN-------EIDVNRETSDIMTRIKTYIS  320
At  268 EAGQALYPSQSTLNQIPSLPPSFTTLLSGCFPDSTLDAG--QTTVKQNPDLKSQILKYME  325

        + ++ +++ ++++ +++++ ++
Ta  333 DDSFHDMLSKLDKAIDELGGDLSL------------------------------------  356
Sb  323 DESFNDMLFKLDKVIDELGGDMSL------------------------------------  346
Zm  324 DESFNDMLLKLDKAIDELGGDMSL------------------------------------  347
Os  321 DGAFHDMLFKLERVIDELGGDLSL------------------------------------  344
At  326 DSSFQDMLQKVERIIDEIGDRCVGSTIGRSSKIYIPLCFHNNCIHPFKDDCWCCLAAGTK  385

Ta      ------------------------
Sb      ------------------------
Zm      ------------------------
Os      ------------------------
At  386 KDWCWLEKDFPDAKELCMKTCTRKI  410
```

(a)

(b)

(c)

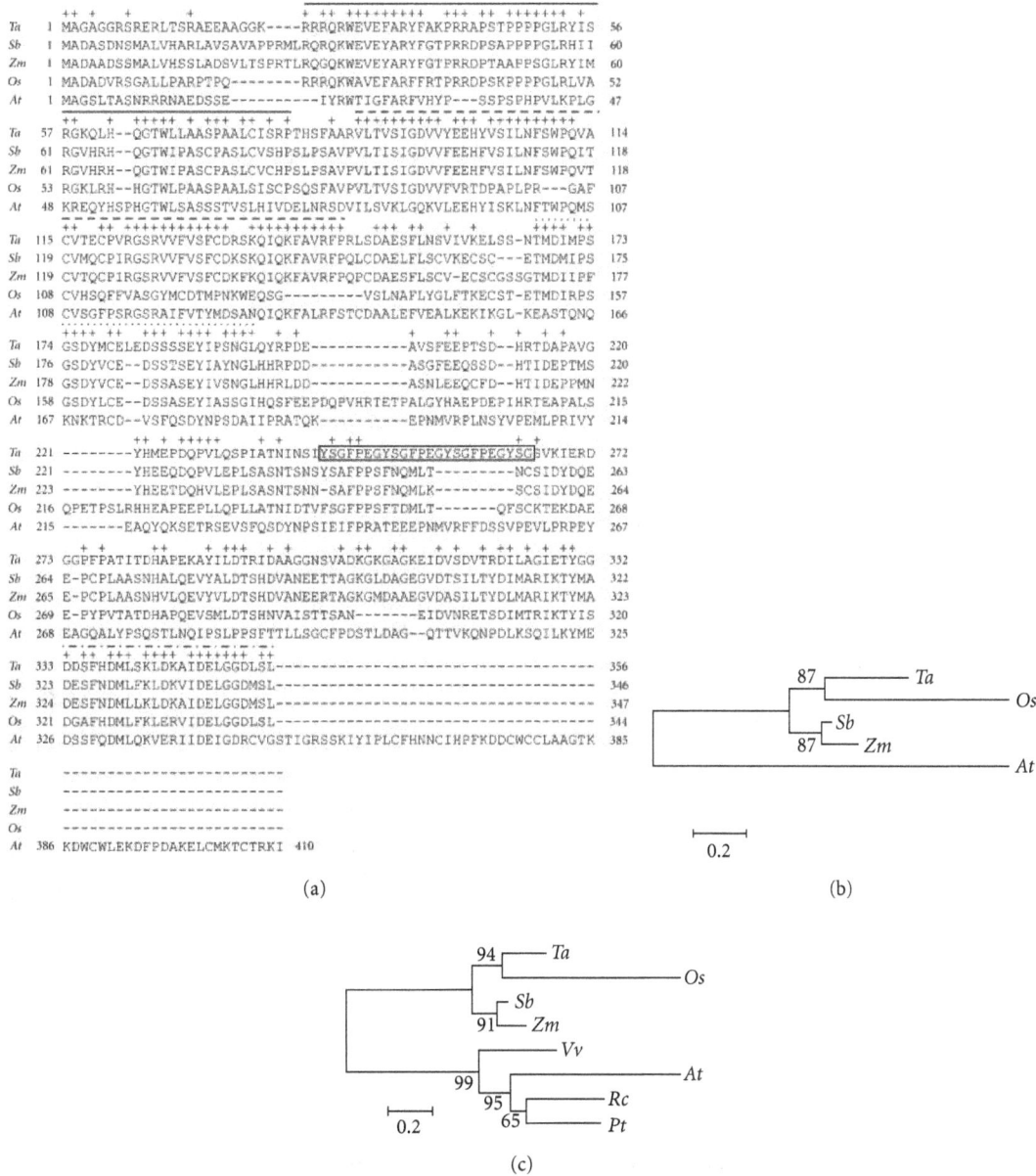

FIGURE 1: The PHS1 amino acid sequence is well conserved across plant species. (a) Alignment of *Ta*PHS1 with four homologues shows high levels of sequence conservation. Amino acid positions that are conserved across at least three species, using the *Ta*PHS1 amino acid sequence as the reference, are denoted by a "+" above. Four conserved regions were identified within the PHS1 protein and may represent functional domains: Region 1 (unbroken line), Region 2 (dashed line), Region 3 (dotted line), and Region 4 (dashed-dotted line). Oligopeptide repeat units (denoted by box) unique to *Ta*PHS1 are also highlighted. *Ta*: *Triticum aestivum* PHS1, *Sb*: *Sorghum bicolor* PHS1, *Zm*: *Zea mays* PHS1, *Os*: *Oryza sativa* PHS1, and *At*: *Arabidopsis thaliana* PHS1-like (see Methods for accession numbers). (b) The evolutionary history of *Ta*PHS1 was inferred using the Neighbor-Joining method [32]. (c) Three additional PHS1/PHS1-like sequences were obtained through Hidden Markov Model and MaxHom searches and were also assessed using phylogenetics. *Vv*: *Vitis vinifera* PHS1, *Rc*: *Ricinus communis* PHS1, and *Pt*: *Populus trichocarpa* PHS1. The reliability of the internal branches of the trees (b, c) was assessed with 10,000 bootstrap resamplings [43], with the confidence probabilities shown next to the branches. The trees are drawn to scale, with branch lengths in the same units as those of the evolutionary distances used to infer the phylogenetic tree. The evolutionary distances were computed using the Poisson correction method [44] and are in the units of the number of amino acid substitutions per site. There were a total of 442 positions in the final datasets. Phylogenetic analyses were conducted in MEGA4 [33].

transcript was significantly upregulated in a *ph1b* background when compared to wild type (approximately 20-fold), we also investigated transcription levels of *TaPHS1* in the *ph1b* mutant. While not as dramatic as that reported for

TaASY1 in Boden et al. [20], *TaPHS1* was also upregulated in the *ph1b* mutant when compared to wild type but by between 1.5-fold (premeiosis) and 2-fold (leptotene-pachytene)(Figure 4).

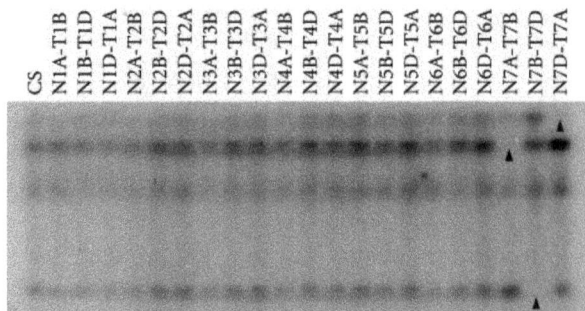

FIGURE 2: *TaPHS1* is located on chromosome group 7 of wheat. Membranes prepared with DNA from nullisomic (N), tetrasomic (T), and wheat lines of Chinese Spring (CS) were hybridised with a *TaPHS1*-specific probe showing that there is a copy on the A, B, and D genomes of chromosome group 7. Each black arrowhead indicates the absence of a band which represents the presence of a copy of *TaPHS1* in the particular genome for which the plant is nullisomic. The more intense band seen in each lane indicates the presence of the copies of *TaPHS1* in the genome for which the plant is tetrasomically compensated.

3.3. TaPHS1 Is Localised in the Nucleus and Associates with Chromatin during Early Meiosis. 3D immunolocalisation of *TaPHS1* in wild type wheat meiocytes shows that it associates with chromatin during early meiosis (Figures 5(a)–5(f)). While the signals of both *TaPHS1* and *TaASY1* were located within close proximity of each other, the two proteins do not appear to colocalise (e.g., merged panel of Figure 5(f)). In addition to its association with chromatin, the *TaPHS1* signal was also observed at the nucleolus (Figures 5(b)–5(e)). This labelling of the nucleolus appears to be on the surface, with a greater signal intensity seen at the nucleolar periphery. This signal appeared to be most intense during early-to-late zygotene/pachytene transition (Figures 5(c)–5(e)). In general, the *TaPHS1* signal appeared either as diffuse tracts or punctated foci that follow is the chromatin, unlike the distinct continuous tracts of *TaASY1*. The *TaPHS1* signal was observed from the telomere bouquet stage and persisted on the chromatin until late pachytene where it faded. Although *TaPHS1* was not detected on the chromatin in diplotene cells, detection of a weak signal was still observed in the cytoplasm in what appeared to be randomly distributed foci (Figure 5(g)).

4. Discussion

This study has reported the isolation and characterisation of *PHS1* from hexaploid wheat, with the amino acid sequences of *TaPHS1* being relatively well-conserved when compared with homologues in other plant species. *In silico*, analysis of the *TaPHS1* amino acid sequence suggests that it does not contain any known nuclear localisation signal (NLS) peptide motif. However, predictions using WoLF PSORT show that *TaPHS1* fits the profile of a nuclear protein. In addition, the immunolocalisation results (Figure 5) show that *TaPHS1* localises to the nuclei of wheat meiocytes *in vivo*. These results together indicate that *TaPHS1* and its homologues might possess an uncharacterised NLS motif. Alternatively,

the PHS1 protein may be transported into the nucleus by a yet unknown process and/or protein. Given that Region 4, referred to as the CR2 region in Ronceret et al. [28], has been identified as a putative SUMOylation site, there may be no requirement for an NLS motif. Previously, posttranslational modifications such as SUMOylation have been shown to enable transport of proteins from the cytoplasm into the nucleus [46].

Sequence alignments of *TaPHS1* with PHS1 and PHS1-like proteins of four other species obtained from BLAST searches revealed that there is a closer relationship between wheat PHS1 and PHS1 homologues in other cereals than between wheat and Arabidopsis PHS1 (Figures 1(a) and 1(b)). This was not unexpected as cereals are monocots, while Arabidopsis is a dicot. In addition, the Arabidopsis PHS1-like sequence contained an additional 61 amino acid residues on the C-terminal end that were not present in the four monocot species. With the addition of three more dicot PHS1/PHS1-like sequences, individual monocot and dicot branches were still maintained. However, a high level of divergence between the wheat and rice PHS1 sequences was evident (Figure 1(c)) and may indicate that the PHS1 proteins in these two species have evolved to function differently, as is suggested by the presence of the *TaPHS1*-specific oligopeptide repeat units from position 242 to 265 (Figure 1(a), boxed feature). Another possible explanation for this sequence divergence is that the rice PHS1 sequence, which is putatively annotated as a PHS1 protein, is not a true PHS1 homologue but instead a PHS1-like protein. However, this seems improbable for a number of reasons. Firstly, the rice genome has been sequenced and exhaustive BLAST searches identified *OsPHS1* as the most significantly similar match to both *TaPHS1* and *ZmPHS1* at the nucleic and amino acid level. Secondly, the *in silico* mapping identified a rice marker on rice chromosome 6 approximately 35 kb away from *OsPHS1* that is syntenic to a marker that has been bin-mapped to the short arm of wheat chromosome 7. Finally, a significant variance between the amino acid sequences of protein equivalents involved in meiosis has been documented across different organisms, even though their function may be conserved (see [6] and references within).

The importance of the aforementioned oligopeptide repeat units unique to *TaPHS1* remains to be determined. These repeats could be either three tandem hepta-peptide units [YSGFPEG] that span positions 242 to 262 or a series of alternating tripeptide [YSG] and tetrapeptide [FPEG] units that span positions 242 to 265. Comprehensive *in silico* database searches using amino acids 242 to 265 resulted in no significantly similar matches with any repeats reported to date. As single oligopeptide units, the tripeptides, tetrapeptides, and heptapeptides are relatively short and may therefore not form any independent structural units. However, when arranged successively, these oligopeptide units may form a regular repeating structure within *TaPHS1*, as has previously been reported in other proteins [47]. Based on work previously conducted by Yoder et al. [48], the glutamic acid (E) and glycine (G) residues on the end of the hepta-peptide unit [YSGFPEG] could represent turn residues that link the heptapeptides together, allowing the units to stack on

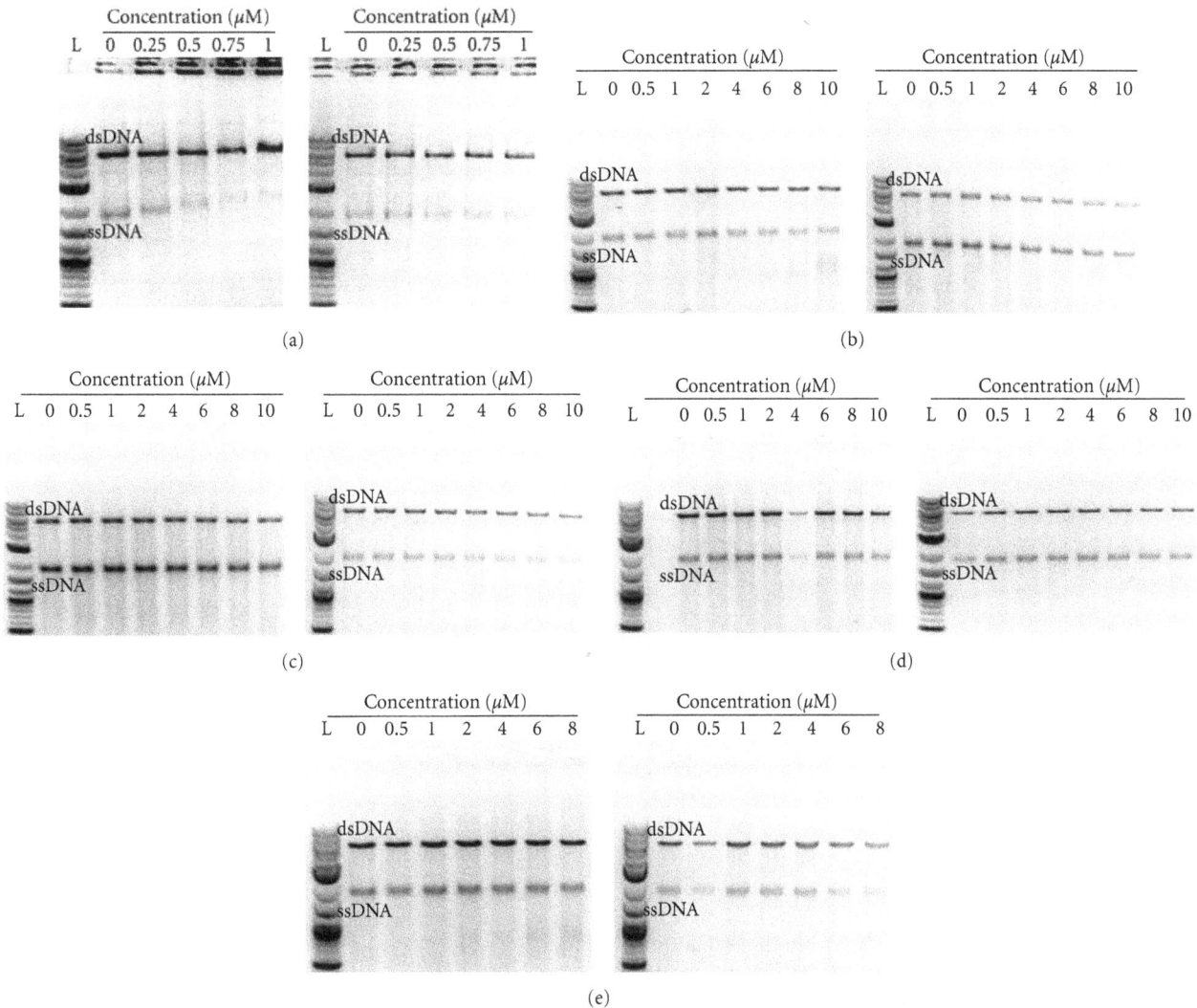

FIGURE 3: TaPHS1 interacts with DNA *in vitro*. An *E. coli* BL21 cell line containing the pDEST17-*TaPHS1* ORF construct was induced with IPTG (1 mM total concentration) to heterologously produce His-tagged *TaPHS1* protein. Total cellular protein was extracted under native conditions, and the His-tagged *TaPHS1* was isolated and purified using nickel affinity chromatography. Total cellular protein from the same cell line which was not induced was also extracted and treated identically to be used as the negative control. DNA-binding ability was only observed in assays conducted using the full-length *TaPHS1* and the Region 1 peptide, indicating that Region 1 possesses a novel/uncharacterised DNA-binding domain where two S/TPXX putative DNA-binding motifs are located. Using competitive DNA-binding assays with equivalent amounts of single- and double-stranded DNA, *TaPHS1* preferentially binds single-stranded DNA (ssDNA). This is evidenced by the increased retardation of the ssDNA species through the gel matrix with increasing concentrations of *TaPHS1* that caused more ssDNA to be bounded by *TaPHS1*. At higher concentrations, *TaPHS1* also interacts with double-stranded DNA (dsDNA). (a) Full-length *TaPHS1*, (b) Region 1 peptide, (c) Region 2 peptide, (d) Region 3 peptide, and (e) Region 4 peptide. Competitive DNA-binding assays performed with the induced samples containing the purified *TaPHS1* protein are on the left and noninduced controls are on the right. μM: protein concentration, L: Ladder.

top of each other. This series of oligopeptides may therefore impart a slightly different structure and possibly function for *TaPHS1* when compared to the rest of the PHS1/PHS1-like proteins that lack these oligopeptide repeat units. Alternatively, these heptapeptides could have a role similar to the hepta-peptide repeats in the C-terminal domain of the largest subunit of RNA polymerase II, which act as a binding scaffold for protein partners (reviewed by [49]).

In contrast to the two conserved regions (termed CR1 and CR2 by Ronceret et al. in [28]) previously described for PHS1 [27], we have shown the presence of four conserved

regions (termed Region 1 to 4) within the PHS1 amino acid sequences investigated in this study; with Regions 2 and 4 corresponding with the previously identified CR1 and CR2. Discrepancies in the lengths of the conserved regions identified in the two studies are most likely artefacts of the different alignment algorithms used, in addition to the fact that only full-length annotated transcripts of PHS1 were used in this study. The results of the DNA-binding assays in this study (Figures 3(a) and 3(b)) and the immunolocalisation of *TaPHS1* to chromatin (Figure 5) show that *TaPHS1* does have DNA-binding capabilities. In the presence of equivalent

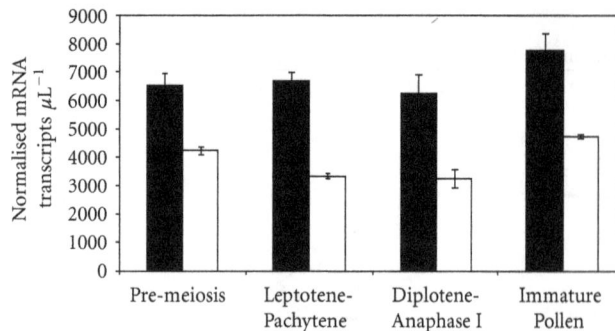

FIGURE 4: Q-PCR profiling of *TaPHS1* shows that it is expressed during meiosis. While the amount of *TaPHS1* mRNA transcript is low, it has higher levels of expression during premeiosis when compared to the other stages of meiosis examined in Chinese Spring wild type (open bars). In the *ph1b* mutant (black bars), *TaPHS1* is upregulated between 1.5- and 2-fold across the time points analysed. Normalisation of the Q-PCR data was performed against three control genes (*actin*, *GAPdH*, and *cyclophilin*) as per Crismani et al. [36]. Data represent the means ± SE of three replicates. Units on the *y*-axis represent normalised mRNA transcript μL^{-1}.

amounts of ssDNA and dsDNA, *TaPHS1* appears to preferentially bind ssDNA *in vitro* but will also bind dsDNA when the protein is present at higher concentrations. Furthermore, we have shown that Region 1 appears to be responsible for the DNA-binding ability (Figure 3(b)). Although the retardation of the DNA is less significant, this can be attributed to the reduced size of the Region 1 partial peptide in comparison to that of the full-length *TaPHS1* protein (5.854 kD versus 38.958 kD). Other regions within *TaPHS1* may also be required to further enhance the DNA-binding capabilities of Region 1. Bioinformatics analysis of the *TaPHS1* protein revealed two S/TPXX DNA-binding motifs previously identified by Suzuki [45] are located within Region 1 (TPPP: amino acid positions 46 to 49; SPAA: amino acid positions 71 to 74), possibly indicating that Region 1 may be a novel DNA interaction domain. Furthermore, our immunolocalisation data suggests that *TaPHS1* is closely associated with chromatin (and therefore dsDNA) *in vivo* during early meiosis.

Although the Q-PCR profiling suggests *TaPHS1* is a low-abundance transcript in the cell, some significant differences were detected during the stages examined, as well as between the wild type Chinese Spring and *ph1b* mutant. While the same general trend of expression is observed in both wild type and the *ph1b* mutant, *TaPHS1* is upregulated in the mutant background by approximately 1.5-fold in premeiosis, 2-fold in both leptotene-pachytene and diplotene-anaphase I, and 1.5-fold in immature pollen. This 2-fold increase in expression during early meiosis may suggest that the *Ph1* locus could directly or indirectly affect *TaPHS1*.

The immunolocalisation results showing *TaPHS1* forms diffuse tracts with punctate foci that associate with wheat chromatin are in contrast to the localisation patterns of PHS1 in both maize and Arabidopsis where *Zm*PHS1 and *At*PHS1 form granules within the cytoplasm and not in the nucleus during leptotene and zygotene [28]. In maize, PHS1

was observed to cluster around the nuclear membrane in a small proportion of maize meiocytes during the peak of *Zm*PHS1 accumulation at midzygotene while a few *Zm*PHS1 foci were present in the nucleus during late zygotene. Unlike maize, no such observations were seen with the Arabidopsis PHS1 homologue. This difference in the localisation profile of *TaPHS1* compared to its homologues in Arabidopsis and maize may suggest that *TaPHS1* instead has a direct role in chromatin interactions.

The presence of substantial *TaPHS1* signal within the nucleoli of early-stage meiocytes may indicate that *TaPHS1* is sequestered to the nucleolus either for degradation or storage until required as has been shown for other proteins (reviewed by [50, 51]). Hypothetically, should PHS1 act as a direct physical shuttling protein (as indicated but not favoured by Ronceret and colleagues [28]) that transports specific meiotic proteins into the nucleus; it is likely that it may then be sequestered to the nucleolus for degradation or to undergo further posttranslational modifications to mark it for return to the cytoplasm so that PHS1 molecules can be reused. Although our immunolocalisation data shows that both *TaPHS1* and *TaASY1* are loaded and associated with the chromatin at the same time during early meiosis in bread wheat, they do no colocalise with one another. The association of these two proteins with the chromatin appears to be particularly pronounced during late-zygotene to pachytene (Figures 5(e) and 5(f)). Could it be that *TaPHS1* is involved in a pachytene check-point mechanism to ensure that only homologous chromosomes have paired and recombined?

Another intriguing result is that the *TaPHS1* signal profile appears as faint tracts with punctate foci along regions of the chromatin. Do these punctate foci denote possible recombination sites where *TaPHS1* may be loading the recombination machinery? This is plausible as previous reports of recombination proteins including RAD51 [42], RAD50 [28], and MLH3 [52] localise to chromatin as foci as well as the fact that *TaPHS1* itself interacts with chromatin. The diffuse tracts of *TaPHS1* also suggest a direct role for *TaPHS1* in homology searching (in wheat at least) as previously suggested by both Pawlowski and Cande [53] and Ronceret et al. [28]. In the *Zmphs1-0* mutant, RAD50 is not localised to the nucleus preventing the assembly of the MRE11-RAD50-NBS1 complex (collectively known as the MRN complex) thus preventing resection of the DBS, which results in failed recruitment of the RAD51 recombinase protein [27, 28]. RAD51 has previously been shown to be capable of homology searching and promotes homologous chromosome pairing over regions of DNA several kilobases in length [54, 55]. While the reach of the RAD51/DMC1 homology searching nucleoprotein filaments is limited to only a few kilobases, *TaPHS1* molecules that form the diffuse tracts may somehow play a direct role in the homology searching process over longer distances of the chromatin. Additional data substantiating this hypothesis was uncovered by Osman and colleagues [56] when they reported that both *At*ZYP1 and *Zm*PHS1 possess regions of peptide sequences that resemble bacterial topoisomerase IV domains. The bacterial topoisomerase IV proteins are members of the topoisomerase type IIA family previously hypothesised to have

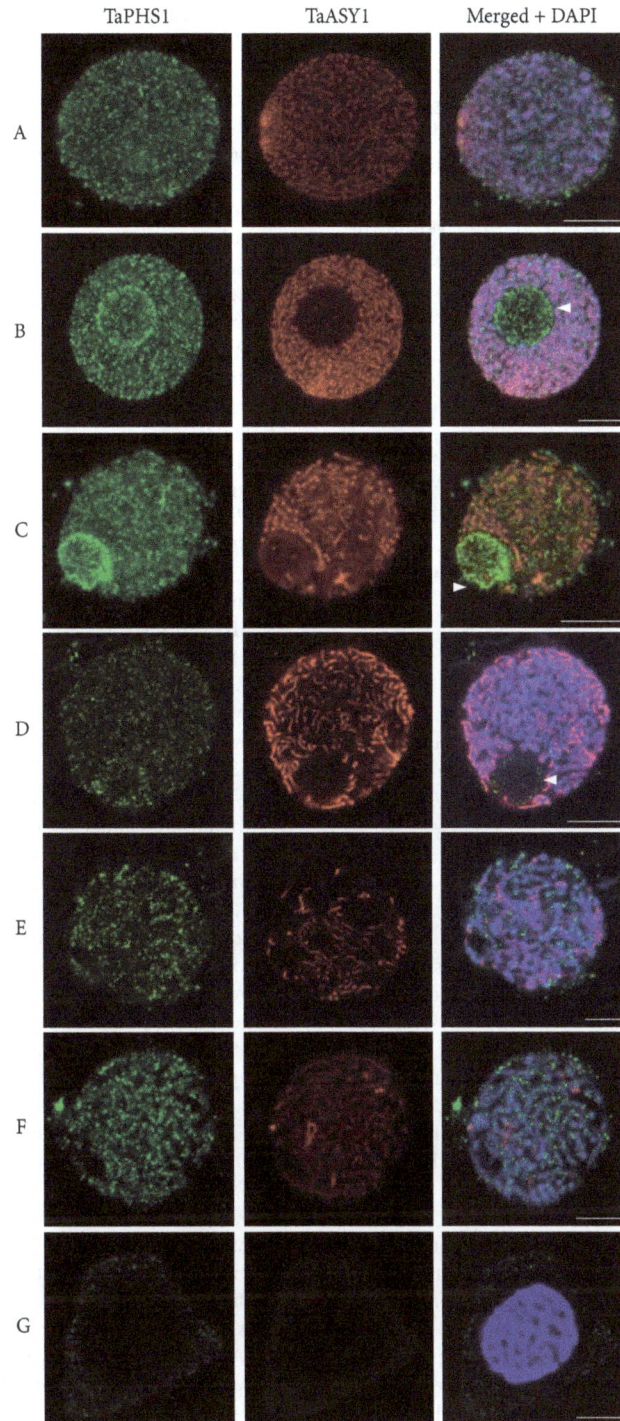

FIGURE 5: TaPHS1 localisation during early meiosis in wild type bread wheat. (a) Telomere bouquet stage, (b) leptotene, (c) early zygotene, (d) zygotene, (e) late zygotene/pachytene transition, (f) pachytene, (g) diplotene. TaPHS1 (green; left panel) localises to 4′-6-diamidino-2-phenylindole- (DAPI-) stained chromatin (blue) as diffuse tracts and/or punctated foci (as seen by viewing the merged + DAPI image). Middle panels show the TaASY1 signal (red), while the panels on the right show merged TaPHS1, TaASY1, and DAPI. Arrowheads (white) represent the nucleolus. Scale bars: 7.5 μm.

potential roles in interhomologue chromosome resolution [57]. This fits well into the hypothesis that TaPHS1 may act as a component of the homology searching mechanism in wheat.

5. Conclusions

In conclusion, the data presented demonstrates that TaPHS1 has an important and possible novel role during the early stages of wheat meiosis. Data from the DNA-binding assays as well as 3-dimensional immunolocalisation of *Ta*PHS1 during early meiosis in wild type cells indicate that *Ta*PHS1 interacts with DNA, a function not previously observed in the Arabidopsis and maize PHS1 homologues. The localisation signal profile of *Ta*PHS1 may indicate that it is a direct transporter of other meiotic proteins into the nucleus, and that it could have a role in homology searching. While the role(s) of this protein are yet to be fully understood, we are currently in the process of generating *Taphs1* knockdown and *TaPHS1* overexpression mutants to further elucidate the meiotic function in bread wheat.

Authors' Contribution

Kelvin H. P. Khoo performed all experimental procedures and drafted the manuscript. K. H. P. Khoo, J. A. Able and A. J. Able participated in the design of the study. J. A. Able and A. J. Able drafted the paper. All authors read and approved the final paper.

Acknowledgments

This research was supported in part by the Australian government under the Australia-India Strategic Research Fund (AISRF) and the School of Agriculture, Food & Wine, The University of Adelaide.

References

[1] S. J. Armstrong, F. C. H. Franklin, and G. H. Jones, "Nucleolus-associated telomere clustering and pairing precede meiotic chromosome synapsis in *Arabidopsis thaliana*," *Journal of Cell Science*, vol. 114, no. 23, pp. 4207–4217, 2001.

[2] Y. K. Chen, C. H. Leng, H. Olivares et al., "Heterodimeric complexes of Hop2 and Mnd1 function with Dmc1 to promote meiotic homolog juxtaposition and strand assimilation," *Proceedings of the National Academy of Sciences of the United States of America*, vol. 101, no. 29, pp. 10572–10577, 2004.

[3] J. D. Higgins, E. Sanchez-Moran, S. J. Armstrong, G. H. Jones, and F. C. H. Franklin, "The Arabidopsis synaptonemal complex protein ZYP1 is required for chromosome synapsis and normal fidelity of crossing over," *Genes and Development*, vol. 19, no. 20, pp. 2488–2500, 2005.

[4] C. Kerzendorfer, J. Vignard, A. Pedrosa-Harand et al., "The *Arabidopsis thaliana* MND1 homologue plays a key role in meiotic homologous pairing, synapsis and recombination," *Journal of Cell Science*, vol. 119, no. 12, pp. 2486–2496, 2006.

[5] E. Martínez-Pérez, P. Shaw, L. Aragon-Alcaide, and G. Moore, "Chromosomes form into seven groups in hexaploid and tetraploid wheat as a prelude to meiosis," *Plant Journal*, vol. 36, no. 1, pp. 21–29, 2003.

[6] J. A. Able, W. Crismani, and S. A. Boden, "Understanding meiosis and the implications for crop improvement," *Functional Plant Biology*, vol. 36, pp. 575–588, 2009.

[7] J. A. Able and P. Langridge, "Wild sex in the grasses," *Trends in Plant Science*, vol. 11, no. 6, pp. 261–263, 2006.

[8] J. A. Able, P. Langridge, and A. S. Milligan, "Capturing diversity in the cereals: many options but little promiscuity," *Trends in Plant Science*, vol. 12, no. 2, pp. 71–79, 2007.

[9] G. Moore and P. Shaw, "Improving the chances of finding the right partner," *Current Opinion in Genetics and Development*, vol. 19, no. 2, pp. 99–104, 2009.

[10] R. Riley and V. Chapman, "Genetic control of the cytologically diploid behaviour of hexaploid wheat," *Nature*, vol. 182, no. 4637, pp. 713–715, 1958.

[11] E. R. Sears, "Induced mutant with homoeologous pairing in common wheat," *Canadian Journal of Genetics and Cytology*, vol. 19, pp. 585–593, 1977.

[12] E. Martínez-Pérez, P. Shaw, and G. Moore, "The *Ph1* locus is needed to ensure specific somatic and meiotic centromere association," *Nature*, vol. 411, no. 6834, pp. 204–207, 2001.

[13] P. Prieto, P. Shaw, and G. Moore, "Homologue recognition during meiosis is associated with a change in chromatin conformation," *Nature Cell Biology*, vol. 6, no. 9, pp. 906–908, 2004.

[14] I. Colas, P. Shaw, P. Prieto et al., "Effective chromosome pairing requires chromatin remodeling at the onset of meiosis," *Proceedings of the National Academy of Sciences of the United States of America*, vol. 105, no. 16, pp. 6075–6080, 2008.

[15] P. B. Holm, "Chromosome pairing and synaptonemal complex formation in hexaploid wheat, nullisomic for chromosome 5B," *Carlsberg Research Communications*, vol. 53, no. 2, pp. 91–110, 1988.

[16] P. B. Holm and X. Z. Wang, "The effect of chromosome 5B on synapsis and chiasma formation in wheat, *Triticum aestivum* cv. Chinese Spring," *Carlsberg Research Communications*, vol. 53, no. 2, pp. 191–208, 1988.

[17] N. Al-Kaff, E. Knight, I. Bertin et al., "Detailed dissection of the chromosomal region containing the *Ph1* locus in wheat *Triticum aestivum*: with deletion mutants and expression profiling," *Annals of Botany*, vol. 101, no. 6, pp. 863–872, 2008.

[18] S. Griffiths, R. Sharp, T. N. Foote et al., "Molecular characterization of *Ph1* as a major chromosome pairing locus in polyploid wheat," *Nature*, vol. 439, no. 7077, pp. 749–752, 2006.

[19] W. D. Bovill, D. Priyanka, K. Sanjay, and J. A. Able, "Whole genome approaches to identify early meiotic gene candidates in cereals," *Functional and Integrative Genomics*, vol. 9, no. 2, pp. 219–229, 2009.

[20] S. A. Boden, P. Langridge, G. Spangenberg, and J. A. Able, "TaASY1 promotes homologous chromosome interactions and is affected by deletion of *Ph1*," *Plant Journal*, vol. 57, no. 3, pp. 487–497, 2009.

[21] S. A. Boden, N. Shadiac, E. J. Tucker, P. Langridge, and J. A. Able, "Expression and functional analysis of *Ta*ASY1 during meiosis of bread wheat (*Triticum aestivum*)," *BMC Molecular Biology*, vol. 8, article 65, 2007.

[22] A. P. Caryl, S. J. Armstrong, G. H. Jones, and F. C. H. Franklin, "A homologue of the yeast *HOP1* gene is inactivated in the Arabidopsis meiotic mutant *asy1*," *Chromosoma*, vol. 109, no. 1-2, pp. 62–71, 2000.

[23] K. I. Nonomura, M. Nakano, K. Murata et al., "An insertional mutation in the rice *PAIR2* gene, the ortholog of Arabidopsis *ASY1*, results in a defect in homologous chromosome pairing during meiosis," *Molecular Genetics and Genomics*, vol. 271, no. 2, pp. 121–129, 2004.

[24] K. J. Ross, P. Fransz, S. J. Armstrong et al., "Cytological characterization of four meiotic mutants of Arabidopsis isolated from T-DNA-transformed lines," *Chromosome Research*, vol. 5, no. 8, pp. 551–559, 1997.

[25] J. Y. Bleuyard, M. E. Gallego, F. Savigny, and C. I. White, "Differing requirements for the Arabidopsis Rad51 paralogs in meiosis and DNA repair," *Plant Journal*, vol. 41, no. 4, pp. 533–545, 2005.

[26] J. Li, L. C. Harper, I. Golubovskaya et al., "Functional analysis of maize RAD51 in meiosis and double-strand break repair," *Genetics*, vol. 176, no. 3, pp. 1469–1482, 2007.

[27] W. P. Pawlowski, I. N. Golubovskaya, L. Timofejeva, R. B. Meeley, W. F. Sheridan, and W. Z. Cande, "Coordination of Meiotic Recombination, Pairing, and Synapsis by PHS1," *Science*, vol. 303, no. 5654, pp. 89–92, 2004.

[28] A. Ronceret, M. P. Doutriaux, I. N. Golubovskaya, and W. P. Pawlowski, "PHS1 regulates meiotic recombination and homologous chromosome pairing by controlling the transport of RAD50 to the nucleus," *Proceedings of the National Academy of Sciences of the United States of America*, vol. 106, no. 47, pp. 20121–20126, 2009.

[29] J. D. Bendtsen, H. Nielsen, G. von Heijne, and S. Brunak, "Improved prediction of signal peptides: signalP 3.0," *Journal of Molecular Biology*, vol. 340, no. 4, pp. 783–795, 2004.

[30] P. Horton, K.-J. Park, T. Obayashi et al., "WoLF PSORT: protein localization predictor," *Nucleic Acids Research*, vol. 35, pp. W585–W587, 2007.

[31] B. Rost, G. Yachdav, and J. F. Liu, "The PredictProtein server," *Nucleic Acids Research*, vol. 32, pp. W321–W326, 2004.

[32] N. Saitou and M. Nei, "The neighbor-joining method—a new method for reconstructing phylogenetic trees," *Molecular biology and evolution*, vol. 4, no. 4, pp. 406–425, 1987.

[33] K. Tamura, J. Dudley, M. Nei, and S. Kumar, "MEGA4: Molecular Evolutionary Genetics Analysis (MEGA) software version 4.0," *Molecular Biology and Evolution*, vol. 24, no. 8, pp. 1596–1599, 2007.

[34] F. Tajima, "Simple methods for testing the molecular evolutionary clock hypothesis," *Genetics*, vol. 135, no. 2, pp. 599–607, 1993.

[35] A. H. Lloyd, A. S. Milligan, P. Langridge, and J. A. Able, "*TaMSH7*: a cereal mismatch repair gene that affects fertility in transgenic barley (*Hordeum vulgare L.*)," *BMC Plant Biology*, vol. 7, article 67, 2007.

[36] W. Crismani, U. Baumann, T. Sutton et al., "Microarray expression analysis of meiosis and microsporogenesis in hexaploid bread wheat," *BMC Genomics*, vol. 7, article 267, 2006.

[37] X. X. Wang, X. F. Li, and Y. X. Li, "A modified Coomassie Brilliant Blue staining method at nanogram sensitivity compatible with proteomic analysis," *Biotechnology Letters*, vol. 29, no. 10, pp. 1599–1603, 2007.

[38] T. J. March, J. A. Able, C. J. Schultz, and A. J. Able, "A novel late embryogenesis abundant protein and peroxidase associated with black point in barley grains," *Proteomics*, vol. 7, no. 20, pp. 3800–3808, 2007.

[39] M. M. Bradford, "A rapid and sensitive method for the quantitation of microgram quantities of protein utilizing the principle of protein dye binding," *Analytical Biochemistry*, vol. 72, no. 1-2, pp. 248–254, 1976.

[40] R. J. Pezza, G. V. Petukhova, R. Ghirlando, and R. D. Camerini-Otero, "Molecular activities of meiosis-specific proteins Hop2, Mnd1, and the Hop2-Mnd1 complex," *Journal of Biological Chemistry*, vol. 281, no. 27, pp. 18426–18434, 2006.

[41] K. H. P. Khoo, H. R. Jolly, and J. A. Able, "The *RAD51* gene family in bread wheat is highly conserved across eukaryotes, with *RAD51A* upregulated during early meiosis," *Functional Plant Biology*, vol. 35, no. 12, pp. 1267–1277, 2008.

[42] A. E. Franklin, J. McElver, I. Sunjevaric, R. Rothstein, B. Bowen, and W. Zacheus Cande, "Three-dimensional microscopy of the Rad51 recombination protein during meiotic prophase," *Plant Cell*, vol. 11, no. 5, pp. 809–824, 1999.

[43] J. Felsenstein, "Confidence-limits on phylogenies—an approach using the bootstrap," *Evolution*, vol. 39, pp. 783–791, 1985.

[44] E. Zuckerkandl and L. Pauling, "Evolutionary divergence and convergence in proteins," in *Evolving Genes and Proteins*, V. Bryson and H. J. Vogel, Eds., pp. 97–166, Academic Press, New York, NY, USA, 1965.

[45] M. Suzuki, "SPXX, a frequent sequence motif in gene regulatory proteins," *Journal of Molecular Biology*, vol. 207, no. 1, pp. 61–84, 1989.

[46] C. E. de Carvalho and M. P. Colaiácovo, "SUMO-mediated regulation of synaptonemal complex formation during meiosis," *Genes and Development*, vol. 20, no. 15, pp. 1986–1992, 2006.

[47] M. V. Katti, R. Sami-Subbu, P. K. Ranjekar, and V. S. Gupta, "Amino acid repeat patterns in protein sequences: their diversity and structural-functional implications," *Protein Science*, vol. 9, no. 6, pp. 1203–1209, 2000.

[48] M. D. Yoder, S. E. Lietzke, and F. Jurnak, "Unusual structural features in the parallel β-helix in pectate lyases," *Structure*, vol. 1, no. 4, pp. 241–251, 1993.

[49] H. P. Phatnani and A. L. Greenleaf, "Phosphorylation and functions of the RNA polymerase II CTD," *Genes and Development*, vol. 20, no. 21, pp. 2922–2936, 2006.

[50] M. Carmo-Fonseca, L. Mendes-Soares, and I. Campos, "To be or not to be in the nucleolus," *Nature Cell Biology*, vol. 2, no. 6, pp. E107–E112, 2000.

[51] M. O. J. Olson, M. Dundr, and A. Szebeni, "The nucleolus: an old factory with unexpected capabilities," *Trends in Cell Biology*, vol. 10, no. 5, pp. 189–196, 2000.

[52] N. Jackson, E. Sanchez-Moran, E. Buckling, S. J. Armstrong, G. H. Jones, and F. C. H. Franklin, "Reduced meiotic crossovers and delayed prophase I progression in *At*MLH3-deficient Arabidopsis," *EMBO Journal*, vol. 25, no. 6, pp. 1315–1323, 2006.

[53] W. P. Pawlowski and W. Z. Cande, "Coordinating the events of the meiotic prophase," *Trends in Cell Biology*, vol. 15, no. 12, pp. 674–681, 2005.

[54] A. L. Eggler, R. B. Inman, and M. M. Cox, "The Rad51-dependent pairing of long DNA substrates is stabilized by replication protein A," *Journal of Biological Chemistry*, vol. 277, no. 42, pp. 39280–39288, 2002.

[55] T. Nishinaka, A. Shinohara, Y. Ito, S. Yokoyama, and T. Shibata, "Base pair switching by interconversion of sugar puckers in DNA extended by proteins of RecA-family: a model for homology search in homologous genetic recombination," *Proceedings of the National Academy of Sciences of the United States of America*, vol. 95, no. 19, pp. 11071–11076, 1998.

[56] K. Osman, E. Sanchez-Moran, J. D. Higgins, G. H. Jones, and F. C. H. Franklin, "Chromosome synapsis in Arabidopsis: analysis of the transverse filament protein ZYP1 reveals novel functions for the synaptonemal complex," *Chromosoma*, vol. 115, no. 3, pp. 212–219, 2006.

[57] D. von Wettstein, "The synaptonemal complex and genetic segregation," *Symposia of the Society for Experimental Biology*, vol. 38, pp. 195–231, 1984.

9

Plant Domestication and Resistance to Herbivory

Bhupendra Chaudhary

School of Biotechnology, Gautam Buddha University, Greater Noida 201 308, India

Correspondence should be addressed to Bhupendra Chaudhary; bhupendrach@gmail.com

Academic Editor: Peter Langridge

Transformation of wild species into elite cultivars through "domestication" entails evolutionary responses in which plant populations adapt to selection. Domestication is a process characterized by the occurrence of key mutations in morphological, phenological, or utility genes, which leads to the increased adaptation and use of the plant; however, this process followed by modern plant breeding practices has presumably narrowed the genetic diversity in crop plants. The reduction of genetic diversity could result in "broad susceptibility" to newly emerging herbivores and pathogens, thereby threatening long-term crop retention. Different QTLs influencing herbivore resistance have also been identified, which overlap with other genes of small effect regulating resistance indicating the presence of pleiotropism or linkage between such genes. However, this reduction in genetic variability could be remunerated by introgression of novel traits from wild perhaps with antifeedant and antinutritional toxic components. Thus it is strongly believed that transgenic technologies may provide a radical and promising solution to combat herbivory as these avoid linkage drag and also the antifeedant angle. Here, important questions related to the temporal dynamics of resistance to herbivory and intricate genetic phenomenon with their impact on crop evolution are addressed and at times hypothesized for future validation.

1. Introduction

During speciation in crop plants, many morphological changes evolved in response to continuous selection pressure. Such characters are largely governed by genetic and epigenetic changes or are modulated according to ecological adaptations. The transition of wild progenitor species into modern elite cultivars through "domestication" entails evolutionary responses in which plant populations adapt to human selection. In response to this selection most plant species exhibit marked changes in a variety of phenotypes, most noticeably in traits consciously under selection (e.g., fruit size, yield, and evenness of maturation) [1]. As Darwin [2] profoundly recognized long ago, the study of the phenotypic variation between wild and domesticated plants presents an opportunity to generate insight into general principles of evolution, using the morphologically variable antecedent and descendant taxa.

An example of how this concept has transformed our understanding is the realization that natural selection pressure, as well as adaptation under human selection, often led to unexpected and unexplained departures from predicted phenotypes. This mainly includes traits such as enhanced yield, enhanced apical dominance, reduced seed dormancy, perennial to annual habit, and relative susceptibility to pathogens, disease, and insect pests [3, 4]. However, the latter received the least attention during the process of "agricultural evolution." The term "agricultural evolution" here, in fact, summarizes all of the changes accumulated in any wild plant form under natural selection, human-mediated artificial selection (=domestication), and modern breeding practices (Figure 1). From an evolutionary standpoint, these phenomena may be viewed as novel generators of variation in the tertiary gene pool comprised of domesticated and wild germplasms (Figure 1). Such variations occurred mostly at genetic level and provide the ability for a given species to evolve in response to the changing environmental conditions and stress factors [5, 6]. Notwithstanding the striking discoveries of the genetic basis of evolved morphological traits in crop plants [7–11], relatively little is understood about

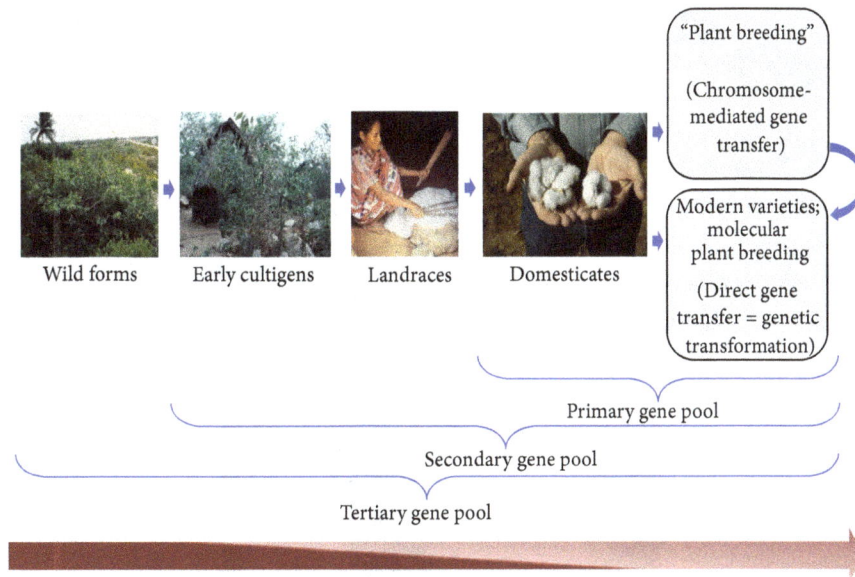

FIGURE 1: An example of cotton (*Gossypium*) evolution under human selection and contemporary breeding programs. The modern "crop" plants are the outcome of recurrent selection on wild form undergoing through early cultigens and landraces. In conventional and molecular breeding programs, it is possible to distinguish between primary, secondary, and tertiary gene pools and exchange of hereditary material. Each primary gene pool comprises one domesticated species together with those species with which it readily cross-breeds. The secondary gene pool includes species that can be cross-bred only with difficulty. The tertiary gene pool comprises those species which can be cross-bred only by using advanced techniques such as embryo rescue. (Courtesy Jonathan F. Wendel, ISU). The horizontal bar shows the reduction in genetic diversity along with domestication steps with the help of dark to lighter shades.

the manner in which gene networks and biological processes are associated with the more susceptible phenotypes of the elite forms.

Regardless of these rapidly accumulating insights, important questions remain about every stage of agricultural development. How did individual crop plants evolve from wild species and acquire agriculturally important traits? How do contemporary plant forms achieve diverse evolutionary trajectories separate from those of their progenitor(s)? How do recently formed elite accessions of crop plants become compromised of different resistance traits? To what degree has crop evolution via the process of domestication and concurrent breeding practices provided a stimulus for sustainable agriculture? Despite the domestication events followed by breeding practices across plant taxa, we do not know why agricultural evolution is so prevalent or conversely, why crop evolution is not universal if it confers some adaptive advantages and promotes species diversification. Nor do we understand the dynamics that underlie the transformation of wild forms to domesticated forms in cryptic crops such as cotton and corn.

It has been assumed that several agriculturally important traits such as resistance to abiotic and biotic stress conditions decreased significantly during evolution. For example, in domesticated accessions of the genus *Cajanus*, reduced levels of resistance have been reported against herbivores [12, 13], bacterial blight [14], and fungal diseases [15]. Among stress conditions, a reduction in drought tolerance, resistance to herbivory and pathogens, is the major threat to crop plants. It is difficult to understand what are the precise genetic

underpinnings are that make a plant species vulnerable to drought, herbivores, and pathogens after passing through the evolutionary important mechanism of crop domestication? Surprisingly, what is the extent of reduction in resistance traits across crop plants, if the reduction in any particular resistance trait is proportional to another resistance trait? The answer to these questions may not be consistent across plant taxa but may only hold true for a particular plant lineage. Generally, domestication promotes heterozygosity leading to the more successful variants under selection pressure either through fixed hybridity or by polysomic inheritance. Could it be assumed that domesticated accessions are *in general* more "successful" than their wild progenitors? This is an exceedingly difficult question to answer, in large part because "success" is an ill-defined term that can refer to anything from short-term proliferation of individuals to long-term effects on lineage diversification. The susceptible nature of modern crop plant varieties in comparison to their wild progenitors could be one of the most apparent consequences of such a megaevent. Or accelerated mutational activity in coresident genomes (in case of polyploid crops) in early generations led to a downgrade in the pathogenic and herbivore resistance of domesticated plant species. Answering these and other questions will require comparisons of wild and domesticated forms by researchers from diverse disciplines such as ecology, population biology, and physiology. Unfortunately, these important areas of biology have received far less attention than the genetics and genomics of selection [7, 16–18]. Nevertheless, even in these better-studied areas much remains to be learned, and it is only by moving beyond the idiosyncrasies of

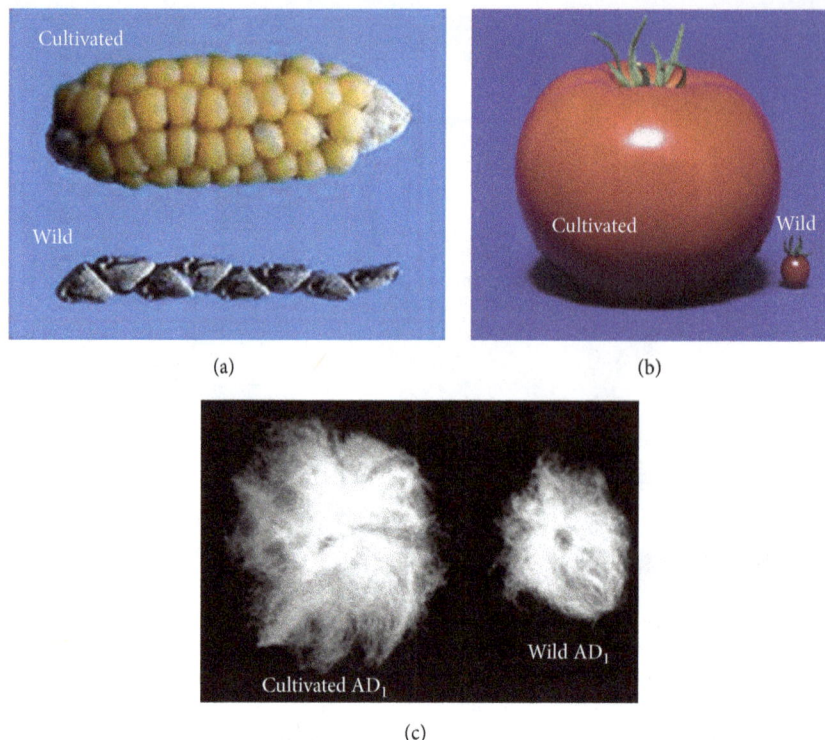

FIGURE 2: Difference in wild forms and their respective domesticated forms showing the impact of natural selection during evolution and domestication on phenotypic traits. (a) Evolution of maize; the domesticated maize (up) and wild teosinte (down). Teosinte has many lateral branches, while today's maize is unbranched. (b) Evolution of tomato; the much larger fruit (right) is from the domesticated *Solanum lycopersicum,* and the small tomato fruit (right) is from wild species *Solanum pimpinellifolium*; and (c) evolution of cotton (*Gossypium hirsutum*); the long fiber phenotype is from domesticated cotton *G. hirsutum* (AD$_1$) (left) and small fuzzy phenotype is from the wild species (right).

a handful of model crop systems that emergent properties of selection can be detected. Though crop plants are threatened by many stress conditions, most of the previous research has been focused on the study of genetic end of resistance to herbivory at the gene expression level, during the process of selection, and at the phenotypic level. The aim of this paper is to provide a broad, updated entry to the literature as well as to highlight the major unanswered questions in the field of crop evolutionary genetics.

2. Evolution of Crop Plants

A primary concern of agricultural evolution biology is to investigate where, when, and how crop plants originated. Vavilov's center of domestication [19] has been a valuable hypothesis as to where crops originated and where our sessile, agrarian cultures began. Since then, we have made great strides in pinpointing where contemporary domesticated forms have arisen and from which wild species they are derived (Figure 2). It has been believed that modern elite plant varieties with useful characters (mostly hybrids) for sustainable agriculture have been developed through (i) domestication and (ii) following research and breeding activities that were implemented by scientists and breeders worldwide (Table 1) (reviewed by [20, 21]). However, it is

essential to understand which has been the foremost drive largely in the phenomenon of "agricultural evolution."

The term "domestication" is often used to describe the process by which wild becomes stabilized. From the standpoint of morphological transformation, domesticated forms are by definition wild species with certain traits highlighted under human selection (Figure 1), showing character modifications including novel trait formation and subsequent segregation, for example, a reduction in grain shattering and seed dormancy in rice [10, 22, 23]; an increase in seed and pod size in *Vigna* [24]; an increase in fiber length and quality in cotton [16, 25]; changes in fruit size and shape in tomato [26]; increased apical dominance in maize [7, 8]; and more. As Darwin recognized [2], the study of the phenotypic variation between wild and domesticated plants presents an opportunity to generate insight into general principles of evolution, using morphologically variable antecedent and descendant in a comparative framework. This approach provides an intriguing perspective on the molecular genetics of human-mediated artificial selection. It is thus assumed that strong artificial selection coupled with introgression (=crossing with the respective wild relative) could drive the fixation of the most beneficial genes and their expression regulation in the process of crop domestication.

The domestication process and introgression under modern breeding programs must have served as effective means

TABLE 1: Some centers of origin of crop domestication and the trait under selection.

Crop	Area of origin	Traits influenced under domestication	Source
Cereals			
Rice	China	Reduction in grain shattering and seed dormancy; synchronization of seed maturation; reduction in tiller number; increase in tiller erectness; increase in panicle branches; Number of spikelets per panicle; reduction in hull and pericarp coloration and awn length	[10, 22, 23]
Barley	Fertile crescent, and Israel-Jordan area	Reduction in grain shattering; separation of seeds from hulls	[9]
Wheat	Southwest Asia (fertile crescent)	Reduction in shattering of grains (nonbrittle rachis); free-threshing trait	[27–29]
Maize	Mesoamerica	Increased apical dominance; production of seeds in relatively large numbers	[7, 8]
Brassicas			
Cabbage	—	Large number of leaves surrounding the terminal bud	[30]
Cauliflower	—	Formation of inflorescence meristems	[31]
Legumes			
Lentil	Mesoamerica	Seed dormancy	[32]
Vigna	Southeast Asia	Increase in seed and pod size, nontwining growth habit, loss of seed dormancy, and seed dispersal ability	[24]
Pea	Southwest Asia (fertile crescent)	Indehiscent pods; lack of dormancy dwarfness; less basal branches; large seeds; good seed quality day neutral flowering	[33]
Fibers			
Cotton	Mexico and Peru	Fiber length and quality	[16, 25]
Vegetables			
Tomato	Mesoamerica	Fruits' size, shape, and structure	[4, 26]
Potato	Andes and Amazonia	Shorter stolons, larger tubers, (often) colored and variously shaped tubers, and reduction of bitter tuber glycoalkaloids	[34]
Squash	Mesoamerica	increased seed length and peduncle diameter, change in fruit shape and color	[35]

to increase the genetic diversity of elite cultivars especially following the initial domestication bottleneck, and to produce cultivars adaptive to climatic conditions [36]. But is it true for all domesticated crop plants? Can it be assumed that cereals with divergent genomic backgrounds experienced one or more domestication events in their evolutionary history? To test this hypothesis, metagenome studies of population genetic structures of cereals, other important modern taxa and their wild progenitors of known origin, are required. Instances of multiple independent domestications in cereals do not provide evidence for their clear ancestry (e.g., *Oryza, Hordeum*), and even if their respective wild progenitors were identified, the multiple origins of domesticated forms and their formation would mostly remain unknown. It has also been assumed that almost all crop plants have experienced repeated polyploidization of genomes in their evolutionary history with multiple-fold duplication of ancestral angiosperm (flowering plant) genes [37]. Polyploidy has influenced flowering plant diversification and provides raw material for the evolution of novelty by relaxing purifying selection on duplicated genes. However, following polyploidization, is the formation of multiple and independent domesticated forms a synchronized event across taxa? It still remains unclear whether genetic polyploids such as wheat, unlike rice and barley, must necessarily undergo mutation in orthologous loci simultaneously [38, 39]. In fact, many domesticated plant forms appear to undergo one or more mutations in a single gene, and thus the probability of parallel and independent selection of orthologous chromosomal regions responsible for the key domestication transition is expected to be weak.

Domestication could, therefore, include all responses to plant evolution, including genetic and epigenetic effects, as reviewed recently [1, 26]. What is the postdomestication impact on the genomic architecture of a plant? The domesticated forms are expected to undergo many of the genome-level megachanges both at structural and functional levels, such as sequence loss, structural rearrangements, and changes in the regulatory sequences, respectively [40, 41]. Recent large-scale microarray studies on the comparison of wild and domesticated forms of selective plant species confirmed that global gene expression had been radically altered by domestication [16, 25]. Such changes have certainly played an important role at the evolutionary scale as domesticated plant forms are not always achieved immediately, even in relatively simpler genomes, as reviewed by [17, 36]; and inheritable changes can only occur often through mutations at genic and/or regulatory levels. For example, a QTL (*sh4*) is responsible for the reduction of grain shattering in the wild rice [10] and a loss of function mutation of Vrs1 in six-rowed barley [9]; and a QTL is responsible for free-thrashing character in wheat [27]. Accordingly, what is the preference for such target genes to undergo mutations during selection process? It is now evident that the genes involved in important domestication transitions are preferably regulatory sequences whose mutations can generate substantial phenotypic modifications serving as suitable targets for strong artificial selection in the key steps of crop evolution [42, 43].

Regarding the magnitude of the changes that occurred throughout crop formation, it appears that breeding activities of crop plants are relatively of less radical change than the conversion of wild forms to the domesticated forms. If so, will the modern elite varieties derived through hybridization of two similar domesticated varieties/forms have less evolutionary resolution than those derived through hybridization with their wild progenitor species? This is definitely an affirmative assertion and has also been validated across plant taxa with reduced genetic diversity in the breeding programs. Thus, what may be the potential risks involved with such radical loss in the genetic diversity among elite cultivars? It may also develop the weedy competitors of crop plants as well as their susceptibility to the diseases, pathogens, and herbivores leading to severe crop losses. But is there any correlation between operative stress condition and the niche of any plant population? Are there sufficient pieces of evidence for relatively less epidemic of any particular pathogen or pest population? The bewildering possibility of such prevalence may destabilize the crop productivity as well as subsequent evolution. For example, despite low crop diversity among cereals initially, successful introgression of resistance to abiotic stress conditions, pathogens, and herbivores was deployed to maintain yield. In this scenario, what is the deadline for such incorporation of resistance traits in crops for sustainable agriculture? The answer is still unclear but it may only be possible if (i) domestication, (ii) change from traditional landraces to modern breeding varieties, and (iii) their over and above decade field adaptation can work indefinitely; for example, maize hybrids in the United States now have a useful lifetime of about 4 years, half of what it was 30 years ago [44].

The advent of genomics has brought a bonafide improvement to the study of such regulatory regions and generation of molecular and expression data, knowledge, and tools which could be applied in modern breeding programs for exploitation of genes from tertiary gene pool (Figure 1). To further understand the genetic basis of domestication, tremendous variations have been revealed using molecular markers. For example, in tomato, the genetic variation present in wild species has been investigated intensively for specific traits and is being exploited for tomato breeding [45–48]. Using DNA technologies, the diversity of domesticated tomato is estimated to comprise <5% of the genetic variation as compared to the rich reservoir in wild relatives [49]. So far, even using

sensitive molecular markers, very low polymorphism within the domesticated tomato gene pool has been identified, [50, 51]. A loss of genetic diversity revealed through sequence repeat markers was also observed from wild *Triticum tauschii* to the landraces and subsequently to the elite germplasm [40]. Though the successful application of breeding programs has produced high-yielding crop varieties, ironically the plant breeding processes have threatened the genetic basis upon which the breeding depends. If assumed so, is the redundancy in the loss of diversity across taxa deleterious, or neutral? What is the spectrum of consequences of having genes and mutations underlying domestication transitions (colloquially called "domestication syndrome")? Are the answers to these questions consistent among plant lineages or between parallel domestication events? Nevertheless, intensive artificial selection does not inevitably lead to a loss of genetic diversity, and diversity can be compensated by the introgression of novel germplasm. One example of this is the development of novel rice variety [10], with the introgression of major QTL, *sh4*, responsible for the reduction of grain shattering.

3. Resistance to Herbivory

3.1. Resistance Variations Arising by Domestication. Herbivores (insect pests) are the major factor responsible for destabilizing crop productivity in agricultural ecosystems. Herbivores have been recognized as a major constraint to crop production causing significant yield loss and quality degradation. In response to herbivory, plants have acquired inherent resistance against such pests; however, the intensity of this resistance varies enormously between wild and domesticated forms. So, what could be the rationale for such large variations in the resistance levels? It is believed that the alterations in herbivory resistance are a prominent outcome of genetic reduction within and between crop species. Several studies have shown that massive expression changes accompany crop evolution (under domestication and breeding practices). The magnitude of expression changes varied greatly between species influencing genetic diversity and resistance levels tremendously. However, do enough data exist at present to reveal general trends of reduced resistance to herbivory among domesticated forms across taxa? It is evident in three wild relatives of chickpeas that are, *Cicer microphyllum, C. canariense,* and *C. macracanthus* exhibiting significant reduction in leaf feeding, larval survival, and larval weight of neonate larvae of *Helicoverpa armigera* in comparison to domesticated perennial accessions. The extent of suppression of damage is impressive in wild accessions as compared to the domesticated chickpeas.

Also, a comparison of global gene expression profiles in the wild and domesticated allotetraploid cotton *Gossypium barbadense* has specifically addressed the transcriptional effects of domestication during development. Several important genes and their functional categories related to resistance genes have been identified as downregulated in the domesticated form than its counterpart wild form supporting the assumption of reduced resistance properties of a plant species [16]. Furthermore, a comparison of wild and domesticated accessions of allopolyploid *G. barbadense* with their diploid progenitors (*G. arboreum* and *G. raimondii*; B. Chaudhary & J. F. Wendel, unpublished) revealed significant transcriptional downregulation of resistance-related genes in the expression phenotype of the domesticated form. This is a clear indication of reduction in herbivore resistance traits during the domestication of an allopolyploid crop. Since cotton allopolyploid species carry "A" and "D" genomes, derived from their diploid progenitors [52], it may also be argued that such a reduction in herbivory resistance may be the cumulative reduction occurred during both polyploid formation and domestication. However, following polyploid formation, many duplicated genes undergo transcriptional biases and do not behave as simple additive combinations of the parental genomes [53], but instead are maintained at the ancestral resistance levels at least in the wild forms. Thus, the possibility of reduction in resistance levels during polyploid formation appears to be minimal and major reductions in resistance levels may have occurred during the domestication process.

This lack of additivity in gene expression levels in most polyploid crop plants raises several fundamental questions on the consequences of the evolutionary dynamics of resistance gene expression following domestication. From a mechanistic standpoint, what is responsible for nonadditivity in gene expression, and why does this vary so much among resistance genes and between different genomic combinations? Why do such important genes from two diploid genomes demonstrate such a large disparity in the degree of suppression of the gene expression phenotype? Also from an evolutionary point of view, how does genomic evolution impact resistance gene expression variation during agricultural evolution (domestication followed by the breeding practices), and what are the potential phenotypic effects of each of these sources of variation?

3.2. Genetics of Resistance to Herbivory. One of the most recent and spectacular revelations in crop plants is the identification of a number of molecular markers, and how these markers could be applied to identify and track target genes in a marker-assisted breeding program [54–56]. Molecular markers have also been applied to increase understanding of the mechanistic and biochemical basis of herbivore resistance, as shown thoroughly in maize [57], mungbean [58], potato [59], and in soybean [60]. What are the putative regulatory genomic components and pathways providing resistance against herbivores, and at incredibly varied levels? Five QTLs have been identified in *Arabidopsis* known to regulate the glucosinolate-myrosinase system controlling the generalist herbivore *Trichoplusia ni* than specialist feeding insect *Plutella xylostella* [61] identified five QTLs in *Arabidopsis* known to regulate the glucosinolate-myrosinase system controlling the generalist herbivore *Trichoplusia ni* than specialist feeding insect *Plutella xylostella*. This demonstration of the higher levels of genetic variation for resistance to the generalist and specialist herbivores has been further verified and expanded in several subsequent studies, including one in which several QTLs from consistent resistance sources

for leaf feeding insects SWCB and FAW were mapped on and observed to be located on chromosomes 6, 7, and 9 in corn. Given the resistance to both of these insects, candidate genes were identified as *mir* cysteine proteinase gene family [62] and the *Glossly15* gene controlling adult to juvenile transition [63]. Also in soybean, 81 QTLs related to herbivory resistance were identified through meta-analysis, and the locations of true QTLs were deduced with a confidence interval of 95% [64, 65]. Thus, could it be determined whether a genetic variant having a particular QTL or haplotype of a polymorphism is associated with the resistant traits? To understand this contention, a number of herbivory-resistance QTLs have been tested for nonrandom associations in the populations derived from a cross between resistant and susceptible parents determining their proportional contribution to the phenotype in wide array of crop species [59, 66]. RFLP-based identification of herbivory-resistance QTLs in maize revealed their strong association with antixenosis (=a resistance mechanism employed by a plant to deter or prevent pest colonisation) and antibiosis (=an association of two organisms in which one is harmed or killed by the other) resistance to corn earworm [57, 67], and also in soybean [60]. The herbivory resistance QTLs discovered by Rector et al. [67] accounted for most of the genotypic variance for corn earworm resistance in the susceptible *x* resistant hybrids; however, with some exceptions those could probably be addressed later with the help of soybean insect resistance QTL database. Will the contemporary catalogues of genome-sequencing projects across plant systems support the identification of important herbivory resistance loci? In result, it may definitely be assumed that future analyses based on whole-genome sequencing data will emphasize insect resistant (IR) QTLs/IR genes identified earlier through marker-assisted selection. This will substantially reduce the time utilized for their adaptive inheritance through classical or precision breeding. Until relatively recently, family-based QTL mapping and association mapping were the primary means of searching genes involved in crop evolution [68]. Considering these studies, different chromosomal regions were identified harboring corresponding QTLs involved in the herbivory resistance phenotype. Because there is a complex correlation among different cellular traits considered important for a resistant phenotype, it becomes enormously difficult to identify such specific biochemical constituents. This observation suggests that variation in resistance traits is controlled by intricate genetic mechanisms, a suggestion further bolstered by demonstrations of resistance variation in maize synthetic hybrids, whose genomes have not undergone any subsequent selection [7, 17, 69]. The mode of resistance is of great evolutionary interest, as it may often sporadically disappear under domestication and following breeding practices [7, 17, 69]. So, what are such vital target genes, their chromosomal positions, and putative structural changes those cumulatively have influenced the loss of resistance potential? Will any type of stochastic mutations in the coding or noncoding regions lead to the differential loss of resistance potential among elite cultivars? From an evolutionary perspective, it could be hypothesized that divergence in resistance potential at the genomic level may also preserve an extensive polymorphism, thus retaining additional raw material for subsequent evolutionary tinkering if exploited under breeding programmes.

4. Susceptibility to Herbivory and Acquired Resistance

4.1. Resistance Genes: Are There Patterns? The domestication process increased a number of important traits required for agricultural innovations, though with few subsides. For example, the nascent "crop-form" (representative from independent domestication events) is more susceptible to herbivores as a result of having few important resistance loci, of which most were identified from their wild progenitor species. Is such phenotypic transformation under domestication universally advantageous or has accompanied with the loss of an "additional" benefit? What is the spectrum of consequences of having a set of important genomic loci selected under human selection? Are the answers to these questions consistent among plant lineages and/or between independent and parallel domestication events within a single species?

Traditional views maintain that domestication followed by breeding promoted the fixation of resistance loci, referred to as fixed heterozygosity [70–72]. It was thus suggested that inherited heterozygosity at resistance loci is beneficial. Modern views also support that introgression of resistance loci can be advantageous and provide a primary source of genes/alleles with new functions [73–75]. However, the identification of novel germplasm from the tertiary gene-pool is an enormously difficult task and it takes time for characterization. However, an inventive alternative is to carry out comprehensive genomic exploration of improved cultivars, primitive domesticated forms, and their wild progenitor species for the identification of candidate genes underlying resistance traits that show evidence of selection during domestication. With the latter approach, could the relationship between identified candidate genes with their phenotypic effects be envisaged? What are the confidence limits to forbid the possibility of the identified genes as false positives? The use of multiple statistical tests can certainly reduce such misreadings [76].

Analysis of a large number of loci underlying resistance to herbivory in soybean showed that resistance is an outcome of a mixture of major and minor gene effects and is not random. Some loci responsible for acquiring resistance to herbivory were underlying within regions having loci for the resistance against cyst nematode [77]. Classification of different soybean genotypes showing broad resistance has suggested important loci contributing to active synthesis and accumulation of products to stop (=antixenosis), deter (=antibiosis), and/or administer (for which mechanism is not readily established) herbivory [55, 67]. Three QTLs influencing resistance to corn borer species in maize have also been identified as overlapping with other genes of small effect in regulating the resistance phenotype, indicating the presence of pleiotropism or linkage between genes affecting resistance and other agronomic traits [78]. Recently, one

major and three minor QTLs in rice have been identified as showing resistance against green rice leafhopper along with defined microsatellites for marker-assisted selection [79]. However, at present, relatively little is understood about the temporal dynamics of resistance to herbivory in different crop plants, and this requires the study of multiple genomes with the empirical reality of long-term resistance to herbivory.

4.2. Modern Cultivars Are More Susceptible Than Their Wild Progenitor Species. Under classical models, plant resistance to herbivory and pathogens is proposed to be a primary phenotype mostly available in the wild ancestors. For instance, plant introductions (PIs) in soybean with low agronomic quality have been demonstrated to be resistant against number of defoliating insects [12, 13]. Such models agree with the theory predicting that domestication occurred by human-mediated exertion through artificial selection on a wild species, both positive and negative, over hundreds of generations resulting in the development of cultivable species. In general, wild plant forms resist attack by herbivores and pathogens mainly through constitutive and inducible defense mechanisms [80]. The evolution and maintenance of the latter are now firmly accepted as an integral component of the plant defense mechanism against herbivores. However, the question remains when, where, and how induced resistance is deployed? Based on differences in the signaling pathways and spectra of effectiveness, the induced resistance could be categorized into (i) systemic acquired resistance (SAR) occurring in the distal plant parts following localized infections and (ii) induced systemic resistance (ISR) stimulated by non-pathogenic organisms and is regulated by jasmonic acid and ethylene [81]. During crop evolution (domestication followed by breeding practices), besides all evolutionarily relevant "internal" costs (genetic or allocation) of induced resistance, what are the other costs that may also be influencing the resistance phenotypes? There has been rapid progress in the detection of other important components, such as ecological costs, which are the result of a plant's interaction with its environment. Therefore, the conceptual separation of genetic and environmental contributions throughout crop formation would help in our understanding of induced resistance [82, 83].

If artificial selection prevails all through generations at the genomic level, is the operation of such selection global or localized? It may be argued that the evolutionary event such as domestication can affect the sequence variation at wide-reaching loci within a crop plant. In that scenario, what could be the limiting factors responsible for such proposed "genetic erosion" in an elite germplasm? One explanation could be that selection in modern breeding programs instead acts on selected important loci controlling a variety of traits, concluding that selection in either case would significantly reduce species-wide polymorphism and make it more vulnerable to the stressful conditions [4]. Besides intensive selection in modern breeding programs, the narrow genetic base is often cited as a contributing factor to low diversity, at least in soybean. Analysis of 111 fragments from 102 genes in four soybean populations showed evidence of a reduction in genetic diversity within and around the selected loci creating genetic bottlenecks [3]. The reduction of genetic diversity at different loci could result in "broad susceptibility" to newly emerging diseases and herbivores, thereby threatening long-term food and feed security [4]. One or multiple domestication events in the evolutionary history of soybean provide a discernible degree of diversity compensation that is up to a 50% reduction, eliminating almost 81% rare alleles present in the wild soybean (*G. soja*), and even appear to undergo significant change in the allele frequency [3]. Could it be assumed that the most significant loss of diversity in modern cultivars occurred during domestication, or due to an unusually low level of initial genetic variability in the wild progenitor, or both? It seems that the major loss of diversity occurred during domestication leading to the bottleneck where there was a loss of rare alleles present in the wild form and landraces. Contrary to classical predictions that loci under selection pressure may be relatively free to acquire heritable changes, it has been shown among subspecies of maize and rice that single nucleotide polymorphisms encode radical changes in the regulatory regions preferentially preserved in the domesticated forms and evolve conservatively [7, 17, 57]. Any such change can lead to divergence in subgenomic expression components, for example, in allopolyploid crop plants [53]. Those in result may influence the quantitatively inherited traits such as resistance to herbivory. Since wild plant forms have large genetic diversity that could be exploited for introgression of important traits in the modern varieties, it may be assumed that some sources of resistance have been left behind during plant domestication. However, it would be relatively difficult or even unsuccessful to introgress such traits into modern cultivars because this may increase the potential of inferior yield through linkage drag as also shown earlier [84, 85]. Moreover, it will also be difficult because crops with antifeedants from wild may have toxicity and anti-nutritional angle. Here the author has a strong opinion that transgenic technologies may provide a radical solution to the herbivory as these avoid linkage drag and also the antifeedant angle.

5. Transgenic Resistance Mediated by the Expression of Foreign Proteins

5.1. Transgenic Crops and Resistance to Herbivory. Introduction of novel foreign genes into crop plants helps breeders to extend their germplasm with novel phenotypic traits. Such required traits are often related to the control of abiotic and biotic stresses, increasing the crop yield and improving the product quality, which were hitherto difficult or not possible to breed using a conventional approach [86, 87]. Given the toxicity of chemical pesticides, for the past two decades, a major emphasis has been on the control of herbivory through more rational strategies such as Integrated Pest Management (IPM). A component of IPM is the use of naturally available pesticides such as plant secondary metabolites and expression of heterologous proteins. Transgenic crops with a "modified" single gene developed for herbivore resistance are immensely

beneficial in economic, environmental, and health concerns, as recently reviewed [88, 89], and understood to be "second generation" resistant crops (detailed in next section). A number of genes have been discovered which are toxic or antifeedant to herbivores and could be of plant or bacterial origin. A major contribution of herbivory-proof crops is in the reduction in application of harmful insecticides sprays and subsequent increase in crop yield [90, 91]. For example, transgenic maize event MON863 developed using a wild type gene from bacterium *Bacillus thuringiensis* (*Bt*), resistant to corn rootworm, was first commercialized in the USA in 2003 [92] and successfully grown until recently. In such a scenario, how do we provide a global perspective of the status of biotech crops? The easiest way is to calculate the global adoption rates as a percentage of the global areas of principal crops (i.e., soybean, cotton, maize, and canola) in which biotechnology is utilized. Though, during last one decade, herbicide tolerance has consistently been the dominant trait, deployment of multiple genes for other traits such as resistance to herbivores is becoming increasingly important and most prevalent for sustainable agriculture. The best example of the dynamics of this very rapid adoption is the contemporary biotech maize for stacked traits [93]. However, at the field level evaluation, the best-studied resistance to the herbivory phenomenon in the crops is the *Bt* cotton [94], which has a documented reduction in pesticide application of more than 70% in the developing countries, utmost domesticated in India [91], though results have been very variable. In 2009, 5.6 million small and marginal resource farmers planted and benefited from ~8.4 mHa of *Bt* cotton, equivalent to 8.7% of the total area under cotton cultivation in India. The corresponding adoption rate of biotech cotton has also been increased globally from 15.5 mHa to 16.1 mHa in the year 2009 [93]. For soybean, the global hectarage of herbicide tolerant transgenics was 69.2 mHa, (up by 3.4 mHa in 2009), which leads this crop to be the largest GM crop grown worldwide [93]. These examples illustrate two major contributions of the biotech approach to plant breeding: (i) enlarging the gene pool by including novel genes that breeders could not access by crossing techniques and (ii) modifying the genes by recombinant DNA technologies to fine-tune transgene expression [95]. The latter needs more emphasis and attention to achieve success in transgenic technologies for improved traits in the crop plants.

What controls the level of a foreign gene expression in genetically modified plants? Any or all of the molecular mechanisms associated with cellular gene expression machinery could be involved. Such a definition encompasses an array of molecular mechanisms at the transcriptional level including DNA methylation, mRNA decay, and small RNA-mediated gene silencing or at translational level having protein misfolding, degradation or other modification, and nuclear/chromosomal context with respect to genomic location of transgene often referred to as "position effect." It is generally assumed that because genes of prokaryotic origin are expressed poorly in higher organisms, such as plants [96, 97], certain modifications in such genes are required in order to achieve optimal expression. This may include modification in the GC content (as plants are comparatively

GC rich than bacteria particularly *Bacillus* spp.), and in codon usage [97], after the introduction of regulatory sequences and polyadenylation signals. This fact is well supported with a significant increase in the expression of a codon-modified *Bt cry1Ac* gene in comparison to the wild-type gene from bacterial origin. Subsequently, by modifying codon usage, removing polyadenylation sequences [96], and other modifications in *Bt* genes, a number of plants have been transformed against their target pests [98, 99]. But is this true only for bacterial genes? Can it be assumed that plant-derived genes with insecticidal potential [100, 101] such as protease-inhibitors, alpha-amylase inhibitor, lectins, and hemilectins do not require any modification for their optimal efficiency prior to the delivery into other crop systems?

Under field conditions, reverse effects of transgenics were also recorded with (i) harmful effect on nontarget insects and (ii) target insect to develop resistance against insecticidal genes used. The first risk could be addressed by designing the synthetic genes to target the hypervariable regions of target insect genes, thereby avoiding their lethal effects on nontarget insect population. However, in the latter case, the potential hazard could only be equilibrated through achieving very high transgene expression in the transgenics through the system-specific codon usage modification of a gene, use of high strength constitutive promoters, "position effect-" based screening of large transgenic population, and following refuge strategy to delay the acquired resistance in the pest population. If transgenic technologies are so promising and successful over conventional breeding, what are the major constraints that delayed the worldwide cultivation of genetically modified (GM) crops? Here, one of the likely explanations may be the inappropriate resistance management strategies deployed so far for the commercially important crop plants. For rice, assessment of agricultural fields for productivity and health effects in China emphasized such issues and highlighted key concerns on policy implementation and resolution of trade barriers [102], which may also be true globally. Clearly, much remains to be learned about such issues, as how should GM crops for herbivory resistance be synchronized for sustainable agriculture?

5.2. Next Generation Herbivory-Resistant Crop Plants. Transgenic technologies are undoubtedly important for plant defense against stresses, and it has also been argued that they are useful for the incorporation of novel phenotypes into crop plants. Existing single gene biotech crop studies may be extrapolated to a hypothetical case where full coverage of all target herbivores and plant diseases would be available in a genetic stock. As mentioned previously, biotech maize in USA is the best example of the deployment of stacked multiple traits including *Bt* genes (one to control the European corn borer complex and the other to control rootworm) and herbicide tolerance (first commercialized in 2005) and continued to grow in 2009 [93]. However, hints regarding gene pyramiding, exclusively for herbivory resistance, have also been emerging rapidly in last decade. In rice (*Oryza indica*), incorporation of two *Bt* genes and one lectin gene showed the control of three major herbivores: rice leaf folder

(*Cnaphalocrocis medinalis*), yellow stemborer (*Scirpophaga incertulas*), and the brown planthopper (*Nilaparvata lugens*), respectively [103]. This indicates that, in rice, the long-term effect of multiple gene expression is an apparent enhancement of the resistance phenotype established by the synergistic effects of transgenes. Also, two *Bt* gene-transgenic cotton showed enhanced protection against *Helicoverpa zea* in comparison to the single gene transgenics with any of the two genes studied [104]. Evidence from our research laboratory on cotton indicates that transgenic stock with two *Bt* genes targeting two different lepidopteran insects shows significant improvement in the resistance trait against individual insects than do the respective single gene transgenics (B. Chaudhary and D. Pental, unpublished data). This indicates that the gene combinations may have played a strong role in reserving a durable resistance phenotype. Even without knowing the comprehensive specific mechanism(s) involved, there is clearly some association between the observation of high levels of protection against herbivores and the presence of multiple genes.

However, for any particular trait such as resistance to herbivores, can only genes from similar origin be tagged for synergistic activity? Can it be assumed that genes from distant origin be grouped together for increased resistance phenotype? The answer lies with the gut anatomy of phytophagous herbivore, which has different binding sites for toxic proteins that determine the spectrum of different lethal protein(s) activity and severity and provide clues for the use of diverse toxic genes to introgress herbivory resistance phenotype. The use of lectins, for example, *Galanthus nivalis* agglutinin (GNA), with *Bt* genes is very promising for increased insect resistance, with the ability of GNA to serve as carrier protein for the delivery of insecticidal proteins (in most cases *Bt* toxins)[105].

What should be the major selection criteria for insecticidal proteins when used for gene pyramiding? This may entail (i) toxic activity against wide spectrum of herbivores and (ii) nonhomology between or among concurrent toxins used. A well-studied example in second-generation insect-resistant transgenic plants is the use of novel vegetative insecticidal proteins (VIPs), which are produced by *Bacillus thuringiensis* during its vegetative growth along with *Bt* crystal proteins. Unlike *Bt* crystal proteins known as δ-endotoxins, VIPs are not parasporal and are secreted from the bacterial cell during vegetative growth. The full-length toxin gets activated proteolytically to a core toxin by proteases in the lepidopteran gut juice [106, 107]. Since the mode of action, structure, and binding sites of VIPs are different from *Bt* toxins in the insect gut epithelium, their use as potential insecticidal proteins for gene staking is very promising. Transgenic cotton expressing two insecticidal proteins Vip3A and Cry1Ab is estimated to be highly effective against two cotton herbivores, *Helicoverpa armigera* and *Heliothis virescens* [108]. Hence, it is assumed that enhanced resistance could be achieved by using two or more effective analogous insecticidal proteins. Gene combination *Bt cry1Ac* and snowdrop lectin GNA were also tested in cotton and showed resistance against insect pests *Heliothis armigera* and *Aphis gossypii* [109]. Also, the results from a comparison between single *Bt* toxin tobacco transgenics and two gene transgenics having the *Bt* gene and the cowpea trypsin inhibitor (CpTi) gene showed the dominance of two gene products with enhanced insecticidal efficacy against cotton bollworm (*Helicoverpa armigera*) [110]. Further, an evaluation of *Bt*-CpTi fusion protein was performed in *Brassica oleracea* to study the insecticidal effect of this fusion protein on cabbage worm, which showed high activity of trypsin inhibitor, and the overall strong resistance to the common cabbage worm [111]. The synergistic activity between unrelated genes seems very promising. Having distinct binding sites of different *Bt* genes in the mid-gut of herbivores, novel combinations of such genes have been considered good to be deployed in delaying the evolution of resistant herbivores [112]. Since most of the activated Bt toxins share a common three-domain structure with a similar mode of action [113, 114], it is possible to develop a hybrid toxin through domain swapping [115, 116]. Transgenic cotton and tobacco with a hybrid Cry1EC toxin developed against polyphagus insect *Spodoptera litura* resulted in extreme toxicity to all developmental stages of larval development [116]. Naimov et al. [117] constructed a hybrid *Bt* gene using truncated *cry1Ba* gene as a scaffold and inserted part of the second domain of *cryIIa* gene conferring resistance to both coleopteran and lepidopteran pests. These studies support the assumption that such a novel strategy may provide new avenues for resistance management studies involving multiple transgenes in crop plants [88].

The pyramiding technology has been noted to provide excellent control of a broad range of herbivores and reinforces the argument that developed resistance in selected herbivores against one toxin will still be fully susceptible to other toxin molecules present in the plants (reviewed by [118]). If so, is gene pyramiding an enduring strategy for sustainable resistance? Given the insecticidal activity, it appears very clear that single gene transgenics in the field cannot be sustained without an integrated approach [119]. Such approach will definitely delay or, with other components of IPM strategies, preclude the possible adaptation of herbivore-populations to resistant transgenic plants.

5.3. Endogenous Resistance to Herbivores is Prolonged by Transgene Stacking. As mentioned above, one strategy to delay the evolution of resistant herbivores is the stacking of multiple *Bt* genes [120, 121]. However, major concerns with this approach are (i) limited insecticidal properties of *Bt* genes to the target herbivores due to high specificity of Bt toxins, (ii) the potential cross-resistance leading to the evolution of resistant herbivores and (iii) the restricted use of possible novel combinations of *Bt* genes. A possible solution to the aforementioned problems may be either pyramiding of *Bt* genes with another transgene having a different mode of action, as also discussed earlier [108–110], or by pyramiding the native herbivory resistance genes with selected *Bt* transgenes (reviewed in [64]). In reference to the latter, Walker et al. [55] identified that when the SSR-based IR-QTL conditioning corn earworm resistance in soybean was combined with *Btcry1Ac* gene, it resulted in detrimental effects on the larval weights, and with the least foliage consumption. With

these results, can it be assumed that the combined effects of endogenous resistance and transgene-based resistance are additive while having independent mode of action(s)? Genetic modification of cotton with the *Btcry1Ab* transgene with high terpenoid levels showed more resistance to tobacco budworm than transgenics with low terpenoid levels [122]. However, transgenics developed in susceptible potato with *Btcry3A* gene against Colorado potato beetle larvae exhibit higher or at least similar mortality in the target herbivores as in the resistant potato line with leptine glycoalkaloids [123]. In the latter scenario, a better strategy may be the identification of IR-QTLs with the help of molecular markers, rather than with specific traits or compounds known to be associated with the resistance phenotype, as was earlier exampled in soybean, cotton, and potato [55, 122, 123]. The native plant resistance is suggested to be advantageous also in the controlling of the resistant herbivore populations, as demonstrated in the case of tobacco budworm, providing the "lethal dose" required for resistance management strategy [124]. Thus, staking multiple *Bt* genes along with the exploitation of native resistance through marker-assisted breeding will comprise complementary additive effects ameliorating the deployment of resistance management practices in the field.

6. Conclusions

What are the key evolutionary attributes that make the conversion of wild germplasm to "crop" such a prevalent phenomenon? Are the most important evolutionary properties of modern crop plants due to contemporary breeding efforts *per se*, or is domestication, artificial selection, just as important? In the opinion of a crop breeder, the end product is warranted by both. Plant breeding of independently selected domesticated forms began almost ten thousand years ago. Gene mutations occurred during selection, polyploidy, or artificial or natural hybridization and brought remarkable genetic variations. For management of stress conditions in crops, do parallel breeding efforts share similar genomic modifications or is it system/event specific, given the likelihood that different features of human-mediated selection may predominate among various lineages?

Very important successes during domestication in terms of crop yield and quality with other agronomic aspects have been achieved, but with compromised resistance to the herbivores and diseases. All modern crop plants are protected against herbivores by using synthetic toxic chemicals, as has been the case for many years, and this often leads to the development of resistance in herbivores against such frequently used chemicals. To circumvent this problem, introgression of important traits from the wild gene pool has been performed through classical breeding, but has met limited success due to incompatibility and numerous genetic and genomic differences among plant forms. Such principal distinctions may be critical, in that the interactions established by the initial conditions propagate from the time of initial origin through periods of stabilization and long-term evolutionary outcome.

Based on the limited number of available examples, when gene transfer occurred in the nascent interspecific crosses, genetic engineering techniques had proven useful to overcome such problems. Two potential benefits of genetic modification are precise transfer of foreign gene sequences and the evolution of adaptive transgressive traits. Two potential benefits of genetic modification are precise transfer of foreign gene sequences and the evolution of adaptive transgressive traits. Classic plant breeding programs are reinstated for heterozygosity and therefore may be more likely to experience locus-specific linkage with its evolutionary consequences. In contrast, a transgenic crop plant possesses a greater insertion of multiple alien genic sequences with immediate phenotypic effect and genetic novelty. Indeed, experimental analysis of genetically modified crops for multiple traits suggests that pyramiding of favorable genes for individual trait may yield a "super-crop" with high returns. Thus, the breadth of recurrently selected traits in the domesticated plants and the genetic transformation system together has a major effect on the creation and retention of evolutionary novelty in the crop system.

Acknowledgments

The author is thankful to Professor Jonathan F. Wendel and Kara Grupp, Iowa State University, USA, for their suggestions in the preparation of this paper. The author also thanks the Department of Science and Technology (DST), Council of Scientific and Industrial Research (CSIR), and the Department of Biotechnology (DBT), Government of India, for the financial support to carry out cotton research work in the laboratory.

References

[1] J. C. Burger, M. A. Chapman, and J. M. Burke, "Molecular insights into the evolution of crop plants," *American Journal of Botany*, vol. 95, pp. 113–122, 2008.

[2] C. Darwin, *On the Origin of Species by Means of Natural Selection, or the Preservation of Favoured Races in the Struggle for Life*, W. Clowes and Sons and Charing Cross, London, UK, 1859.

[3] D. L. Hyten, Q. Song, Y. Zhu et al., "Impacts of genetic bottlenecks on soybean genome diversity," *Proceedings of the National Academy of Sciences of the United States of America*, vol. 103, no. 45, pp. 16666–16671, 2006.

[4] S. D. Tanksley and S. R. McCouch, "Seed banks and molecular maps: unlocking genetic potential from the wild," *Science*, vol. 277, no. 5329, pp. 1063–1066, 1997.

[5] J. J. Bull and H. A. Wichmann, "Applied evolution," *Annual Review of Ecology, Evolution, and Systematics*, vol. 32, pp. 183–217, 2001.

[6] T. D. Clarkson, "Stress physiology in crop plants," *Plant, Cell & Environment*, vol. 4, p. 184, 1981.

[7] J. Doebley, "The genetics of maize evolution," *Annual Review of Genetics*, vol. 38, pp. 37–59, 2004.

[8] J. Doebley, A. Stec, and L. Hubbard, "The evolution of apical dominance in maize," *Nature*, vol. 386, no. 6624, pp. 485–488, 1997.

[9] T. Komatsuda, M. Pourkheirandish, C. He et al., "Six-rowed barley originated from a mutation in a homeodomain-leucine zipper I-class homeobox gene," *Proceedings of the National Academy of Sciences of the United States of America*, vol. 104, no. 4, pp. 1424–1429, 2007.

[10] C. Li, A. Zhou, and T. Sang, "Rice domestication by reducing shattering," *Science*, vol. 311, no. 5769, pp. 1936–1939, 2006.

[11] M. T. Sweeney, M. J. Thomson, B. E. Pfeil, and S. McCouch, "Caught red-handed: *Rc* encodes a basic helix-loop-helix protein conditioning red pericarp in rice," *Plant Cell*, vol. 18, no. 2, pp. 283–294, 2006.

[12] J. N. All, H. R. Boerma, and J. W. Todd, "Screening soybean genotypes in the greenhouse for resistance to insects," *Crop Science*, vol. 29, pp. 1156–1159, 1989.

[13] V. D. Luedders and W. A. Dickerson, "Resistance of selected soybean genotypes and segregating populations to cabbage looper feeding," *Crop Science*, vol. 17, pp. 395–396, 1977.

[14] T. Ashfield, J. R. Danzer, D. Held et al., "Rpg1, a soybean gene effective against races of bacterial blight, maps to a cluster of previously identified disease resistance genes," *Theoretical and Applied Genetics*, vol. 96, no. 8, pp. 1013–1021, 1998.

[15] A. Garcia, É. S. Calvo, R. A. De Souza Kiihl, A. Harada, D. M. Hiromoto, and L. G. E. Vieira, "Molecular mapping of soybean rust (*Phakopsora pachyrhizi*) resistance genes: discovery of a novel locus and alleles," *Theoretical and Applied Genetics*, vol. 117, no. 4, pp. 545–553, 2008.

[16] B. Chaudhary, R. Hovav, R. Rapp, N. Verma, J. A. Udall, and J. F. Wendel, "Global analysis of gene expression in cotton fibers from wild and domesticated *Gossypium barbadense*," *Evolution and Development*, vol. 10, no. 5, pp. 567–582, 2008.

[17] J. Doebley, "Unfallen grains: how ancient farmers turned weeds into crops," *Science*, vol. 312, no. 5778, pp. 1318–1319, 2006.

[18] M. Sweeney and S. McCouch, "The complex history of the domestication of rice," *Annals Botany*, vol. 100, pp. 951–957, 2007.

[19] N. I. Vavilov, "The origin, variation, immunity and breeding of cultivated plants," *Chronica Botanica*, vol. 13, no. 1–6, pp. 1–366, 1951.

[20] P. Gepts, "A comparison between crop domestication, classical plant breeding, and genetic engineering," *Crop Science*, vol. 42, no. 6, pp. 1780–1790, 2002.

[21] H. Ulukan, "The evolution of cultivated plant species: classical plant breeding versus genetic engineering," *Plant Systematics and Evolution*, vol. 280, no. 3-4, pp. 133–142, 2009.

[22] H. W. Cai and H. Morishima, "QTL clusters reflect character associations in wild and cultivated rice," *Theoretical and Applied Genetics*, vol. 104, no. 8, pp. 1217–1228, 2002.

[23] L. Z. Xiong, K. D. Liu, X. K. Dai, C. G. Xu, and Q. Zhang, "Identification of genetic factors controlling domestication-related traits of rice using an F2 population of a cross between *Oryza sativa* and *O. rufipogon*," *Theoretical and Applied Genetics*, vol. 98, no. 2, pp. 243–251, 1999.

[24] T. Isemura, A. Kaga, S. Konishi et al., "Genome dissection of traits related to domestication in azuki bean (*Vigna angularis*) and comparison with other warm-season legumes," *Annals of Botany*, vol. 100, no. 5, pp. 1053–1071, 2007.

[25] R. A. Rapp, C. H. Haigler, L. Flagel, R. H. Hovav, J. A. Udall, and J. F. Wendel, "Gene expression in developing fibres of Upland cotton (*Gossypium hirsutum* L.) was massively altered by domestication," *BMC Biology*, vol. 8, article 139, 2010.

[26] Y. Bai and P. Lindhout, "Domestication and breeding of tomatoes: what have we gained and what can we gain in the future?" *Annals of Botany*, vol. 100, no. 5, pp. 1085–1094, 2007.

[27] M. C. Simonetti, M. P. Bellomo, G. Laghetti, P. Perrino, R. Simeone, and A. Blanco, "Quantitative trait loci influencing free-threshing habit in tetraploid wheats," *Genetic Resources and Crop Evolution*, vol. 46, no. 3, pp. 267–271, 1999.

[28] V. J. Nalam, M. I. Vales, C. J. W. Watson, E. B. Johnson, and O. Riera-Lizarazu, "Map-based analysis of genetic loci on chromosome 2D that affect glume tenacity and threshability, components of the free-threshing habit in common wheat (*Triticum aestivum* L.)," *Theoretical and Applied Genetics*, vol. 116, no. 1, pp. 135–145, 2007.

[29] N. Watanabe, Y. Fujii, N. Kato, T. Ban, and P. Martinek, "Microsatellite mapping of the genes for brittle rachis on homoeologous group 3 chromosomes in tetraploid and hexaploid wheats," *Journal of Applied Genetics*, vol. 47, no. 2, pp. 93–98, 2006.

[30] D. Zohary and M. Hopf, *Domestication of Plants in the Old World*, Oxford University Press, Oxford, UK, 3rd edition, 2000.

[31] S. A. Kempin, B. Savidge, and M. F. Yanofsky, "Molecular basis of the cauliflower phenotype in arabidopsis," *Science*, vol. 267, no. 5197, pp. 522–525, 1995.

[32] G. Ladizinsky, "Lentil domestication: on the quality of evidence and arguments," *Economic Botany*, vol. 47, no. 1, pp. 60–64, 1993.

[33] N. F. Weeden, "Genetic changes accompanying the domestication of *Pisum sativum*: is there a common genetic basis to the "domestication syndrome" for legumes?" *Annals of Botany*, vol. 100, no. 5, pp. 1017–1025, 2007.

[34] D. Ugent, "The potato," *Science*, vol. 170, no. 3963, pp. 1161–1166, 1970.

[35] B. D. Smith, "The initial domestication of *Cucurbita pepo* in the Americas 10,000 years ago," *Science*, vol. 276, no. 5314, pp. 932–934, 1997.

[36] T. Sang, "Genes and mutations underlying domestication transitions in grasses," *Plant Physiology*, vol. 149, no. 1, pp. 63–70, 2009.

[37] A. H. Paterson, J. F. Wendel, H. Gundlach et al., "Repeated polyploidization of *Gossypium* genomes and the evolution of spinnable cotton fibres," *Nature*, vol. 492, pp. 423–427, 2012.

[38] J. Dubcovsky and J. Dvorak, "Genome plasticity a key factor in the success of polyploid wheat under domestication," *Science*, vol. 316, no. 5833, pp. 1862–1866, 2007.

[39] W. Li and B. S. Gill, "Multiple genetic pathways for seed shattering in the grasses," *Functional and Integrative Genomics*, vol. 6, no. 4, pp. 300–309, 2006.

[40] J. C. Reif, P. Zhang, S. Dreisigacker et al., "Wheat genetic diversity trends during domestication and breeding," *Theoretical and Applied Genetics*, vol. 110, no. 5, pp. 859–864, 2005.

[41] S. I. Wright, I. V. Bi, S. C. Schroeder et al., "Evolution: the effects of artificial selection on the maize genome," *Science*, vol. 308, no. 5726, pp. 1310–1314, 2005.

[42] J. Doebley and L. Lukens, "Transcriptional regulators and the evolution of plant form," *Plant Cell*, vol. 10, no. 7, pp. 1075–1082, 1998.

[43] J. F. Doebley, B. S. Gaut, and B. D. Smith, "The molecular genetics of crop domestication," *Cell*, vol. 127, no. 7, pp. 1309–1321, 2006.

[44] S. R. Palumbi, "Humans as the world's greatest evolutionary force," *Science*, vol. 293, pp. 1786–1790, 2001.

[45] M. P. Bretó, M. J. Asins, and E. A. Carbonell, "Genetic variability in *Lycopersicon* species and their genetic relationships," *Theoretical and Applied Genetics*, vol. 86, no. 1, pp. 113–120, 1993.

[46] H. Egashira, H. Ishihara, T. Takashina, and S. Imanishi, "Genetic diversity of the "peruvianum-complex" (*Lycopersicon peruvianum* (L.) Mill. and *L. chilense* Dun.) revealed by RAPD analysis," *Euphytica*, vol. 116, no. 1, pp. 23–31, 2000.

[47] J. Villand, P. W. Skroch, T. Lai, P. Hanson, C. G. Kuo, and J. Nienhuis, "Genetic variation among tomato accessions from primary and secondary centers of diversity," *Crop Science*, vol. 38, no. 5, pp. 1339–1347, 1998.

[48] C. M. Rick and R. T. Chetelat, "Utilization of related wild species for tomato improvement," *Acta Horticulturae*, vol. 412, pp. 21–38, 1995.

[49] J. C. Miller and S. D. Tanksley, "RFLP analysis of phylogenetic relationships and genetic variation in the genus *Lycopersicon*," *Theoretical and Applied Genetics*, vol. 80, no. 4, pp. 437–448, 1990.

[50] S. García-Martínez, L. Andreani, M. Garcia-Gusano, F. Geuna, and J. J. Ruiz, "Evaluation of amplified fragment length polymorphism and simple sequence repeats for tomato germplasm fingerprinting: utility for grouping closely related traditional cultivars," *Genome*, vol. 49, no. 6, pp. 648–656, 2006.

[51] S. M. Tam, C. Mhiri, A. Vogelaar, M. Kerkveld, S. R. Pearce, and M. A. Grandbastien, "Comparative analyses of genetic diversities within tomato and pepper collections detected by retrotransposon-based SSAP, AFLP and SSR," *Theoretical and Applied Genetics*, vol. 110, no. 5, pp. 819–831, 2005.

[52] J. F. Wendel and R. C. Cronn, "Polyploidy and the evolutionary history of cotton," *Advances in Agronomy*, vol. 78, pp. 139–186, 2003.

[53] B. Chaudhary, L. Flagel, R. M. Stupar et al., "Reciprocal silencing, transcriptional bias and functional divergence of homeologs in polyploid cotton (*Gossypium*)," *Genetics*, vol. 182, no. 2, pp. 503–517, 2009.

[54] S. J. O'Brien, *Genetic Maps*, Cold Spring Harbor Laboratory Press, Cold Spring Harbor, NY, USA, 6th edition, 1993.

[55] D. R. Walker, J. M. Narvel, H. R. Boerma, J. N. All, and W. A. Parrott, "A QTL that enhances and broadens *Bt* insect resistance in soybean," *Theoretical and Applied Genetics*, vol. 109, no. 5, pp. 1051–1057, 2004.

[56] J. G. K. Williams, A. R. Kubelik, K. J. Livak, J. A. Rafalski, and S. V. Tingey, "DNA polymorphisms amplified by arbitrary primers are useful as genetic markers," *Nucleic Acids Research*, vol. 18, no. 22, pp. 6531–6535, 1990.

[57] P. F. Byrne, M. D. McMullen, B. R. Wiseman et al., "Identification of maize chromosome regions associated with antibiosis to corn earworm (Lepidoptera: Noctuidae) larvae," *Journal of Economic Entomology*, vol. 90, no. 4, pp. 1039–1045, 1997.

[58] N. D. Young, L. Kumar, D. Menancio-Hautea et al., "RFLP mapping of a major bruchid resistance gene in mungbean (*Vigna radiata*, L. Wilczek)," *Theoretical and Applied Genetics*, vol. 84, no. 7-8, pp. 839–844, 1992.

[59] G. C. Yencho, M. W. Bonierbale, W. M. Tingey, R. L. Plaisted, and S. D. Tanksley, "Molecular markers locate genes for resistance to the Colorado potato beetle, *Leptinotarsa decemlineata*, in hybrid *Solanum tuberosum* × *S. berthaultii* potato progenies," *Entomologia Experimentalis et Applicata*, vol. 81, no. 2, pp. 141–154, 1996.

[60] J. M. Narvel, D. R. Walker, B. G. Rector, J. N. All, W. A. Parrott, and H. R. Boerma, "A retrospective DNA marker assessment of

the development of insect resistant soybean," *Crop Science*, vol. 41, no. 6, pp. 1931–1939, 2001.

[61] D. Kliebenstein, D. Pedersen, B. Barker, and T. Mitchell-Olds, "Comparative analysis of quantitative trait loci controlling glucosinolates, myrosinase and insect resistance in *Arabidopsis thaliana*," *Genetics*, vol. 161, no. 1, pp. 325–332, 2002.

[62] T. D. Brooks, B. S. Bushman, W. P. Williams, M. D. McMullen, and P. M. Buckley, "Genetic basis of resistance to fall armyworm (Lepidoptera: Noctuidae) and southwestern corn borer (Lepidoptera: Crambidae) leaf-feeding damage in maize," *Journal of Economic Entomology*, vol. 100, no. 4, pp. 1470–1475, 2007.

[63] W. P. Williams, F. M. Davis, P. M. Buckley, P. A. Hedin, G. T. Baker, and D. S. Luthe, "Factors associated with resistance to fall armyworm (Lepidoptera: Noctuidae) and Southwestern corn borer (Lepidoptera: Crambidae) in corn at different vegetative stages," *Journal of Economic Entomology*, vol. 91, no. 6, pp. 1471–1480, 1998.

[64] H. R. Boerma and D. R. Walker, "Discovery and utilization of QTLs for insect resistance in soybean," *Genetica*, vol. 123, no. 1-2, pp. 181–189, 2005.

[65] J. Wang, W. K. Song, W. B. Zhang, C. Y. Liu, G. H. Hu, and Q. S. Chen, "Meta-analysis of insect-resistance QTLs in soybean," *Yi Chuan*, vol. 31, no. 9, pp. 953–961, 2009.

[66] S. D. Tanksley, N. D. Young, A. H. Paterson, and M. W. Bonierbale, "RFLP mapping in plant breeding: new tools for an old science," *BioTechnology*, vol. 7, pp. 257–264, 1989.

[67] B. G. Rector, J. N. All, W. A. Parrott, and H. R. Boerma, "Quantitative trait loci for antibiosis resistance to corn earworm in soybean," *Crop Science*, vol. 40, no. 1, pp. 233–238, 2000.

[68] A. H. Paterson, "What has QTL mapping taught us about plant domestication?" *New Phytologist*, vol. 154, no. 3, pp. 591–608, 2002.

[69] M. Guo, S. Yang, M. Rupe et al., "Genome-wide allele-specific expression analysis using Massively Parallel Signature Sequencing (MPSS) Reveals cis- and trans-effects on gene expression in maize hybrid meristem tissue," *Plant Molecular Biology*, vol. 66, no. 5, pp. 551–563, 2008.

[70] A. L. Kahler, M. I. Morris, and R. W. Allard, "Gene triplication and fixed heterozygosity in diploid wild barley," *Journal of Heredity*, vol. 72, no. 6, pp. 374–376, 1981.

[71] A. B. Schooler, "Wild barley hybrids," *Journal of Heredity*, vol. 54, no. 3, pp. 130–132, 1963.

[72] C. A. Suneson, "Genetic diversity—a protection against plant diseases and insects," *Agronomy Journal*, vol. 52, pp. 319–321, 1960.

[73] R. W. Michelmore and B. C. Meyers, "Clusters of resistance genes in plants evolve by divergent selection and a birth-and-death process," *Genome Research*, vol. 8, no. 11, pp. 1113–1130, 1998.

[74] D. P. Singh and A. Singh, *Disease and Insect Resistance in Plants*, Science Publisher, Enfield, NH, USA, 2005.

[75] K. Zhao, M. Wright, J. Kimball et al., "Genomic diversity and introgression in *O. sativa* reveal the impact of domestication and breeding on the rice genome," *PloS ONE*, vol. 5, no. 5, Article ID e10780, 2010.

[76] A. Bonin, P. Taberlet, C. Miaud, and F. Pompanon, "Explorative genome scan to detect candidate loci for adaptation along a gradient of altitude in the common frog (*Rana temporaria*)," *Molecular Biology and Evolution*, vol. 23, no. 4, pp. 773–783, 2006.

[77] C. R. Yesudas, H. Sharma, and D. A. Lightfoot, "Identification of QTL in soybean underlying resistance to herbivory by Japanese beetles (*Popillia japonica*, Newman)," *Theoretical and Applied Genetics*, vol. 121, no. 2, pp. 353–362, 2010.

[78] B. Ordas, R. A. Malvar, R. Santiago, and A. Butron, "QTL mapping for Mediterranean corn borer resistance in European flint germplasm using recombinant inbred lines," *BMC Genomics*, vol. 11, no. 1, article 174, 2010.

[79] D. Fujita, K. Doi, A. Yoshimura, and H. Yasui, "A major QTL for resistance to green rice leafhopper (*Nephotettix cincticeps* Uhler) derived from African rice (*Oryza glaberrima* Steud.)," *Breeding Science*, vol. 60, no. 4, pp. 336–341, 2010.

[80] D. Cipollini and M. Heil, "Costs and benefits of induced resistance to herbivores and pathogens in plants," in *CAB Reviews: Perspectives in Agriculture, Veterinary Science, Nutrition and Natural Resources*, p. 10, 2010.

[81] D. Walters and M. Heil, "Costs and trade-offs associated with induced resistance," *Physiological and Molecular Plant Pathology*, vol. 71, no. 1–3, pp. 3–17, 2007.

[82] M. Heil, "Ecological costs of induced resistance," *Current Opinion in Plant Biology*, vol. 5, no. 4, pp. 345–350, 2002.

[83] M. Heil and I. T. Baldwin, "Fitness costs of induced resistance: emerging experimental support for a slippery concept," *Trends in Plant Science*, vol. 7, no. 2, pp. 61–67, 2002.

[84] N. D. Young and S. D. Tanksley, "RFLP analysis of the size of chromosomal segments retained around the Tm-2 locus of tomato during backcross breeding," *Theoretical and Applied Genetics*, vol. 77, no. 3, pp. 353–359, 1989.

[85] A. C. Zeven, D. R. Knott, and R. Johnson, "Investigation of linkage drag in near isogenic lines of wheat by testing for seedling reaction to races of stem rust, leaf rust and yellow rust," *Euphytica*, vol. 32, no. 2, pp. 319–327, 1983.

[86] W. Schuch, J. M. Kanczler, D. Robertson et al., "Fruit quality characteristics of transgenic tomato fruit with altered polygalacturonase activity," *HortScience*, vol. 26, pp. 1517–1520, 1991.

[87] S. H. Strauss, "Genomics, genetic engineering, and domestication of crops," *Science*, vol. 300, no. 5616, pp. 61–62, 2003.

[88] P. Christou, T. Capell, A. Kohli, J. A. Gatehouse, and A. M. R. Gatehouse, "Recent developments and future prospects in insect pest control in transgenic crops," *Trends in Plant Science*, vol. 11, no. 6, pp. 302–308, 2006.

[89] N. Ferry, M. G. Edwards, J. Gatehouse, T. Capell, P. Christou, and A. M. R. Gatehouse, "Transgenic plants for insect pest control: a forward looking scientific perspective," *Transgenic Research*, vol. 15, no. 1, pp. 13–19, 2006.

[90] S. M. High, M. B. Cohen, Q. Y. Shu, and I. Altosaar, "Achieving successful deployment of *Bt* rice," *Trends in Plant Science*, vol. 9, no. 6, pp. 286–292, 2004.

[91] M. Qaim and D. Zilberman, "Yield effects of genetically modified crops in developing countries," *Science*, vol. 299, no. 5608, pp. 900–902, 2003.

[92] T. Vaughn, T. Cavato, G. Brar et al., "A method of controlling corn rootworm feeding using a *Bacillus thuringiensis* protein expressed in transgenic maize," *Crop Science*, vol. 45, no. 3, pp. 931–938, 2005.

[93] C. James, "Global status of commercialized biotech/GM crops: 2009," ISAAA Brief 41 ISAAA, http://www.isaaa.org .

[94] F. J. Perlak, R. W. Deaton, T. A. Armstrong et al., "Insect resistant cotton plants," *Bio/Technology*, vol. 8, no. 10, pp. 939–943, 1990.

[95] C. N. Stewart, M. J. Adang, J. N. All et al., "Genetic transformation, recovery, and characterization of fertile soybean

transgenic for a synthetic *Bacillus thuringiensis* cryIAc gene," *Plant Physiology*, vol. 112, no. 1, pp. 121–129, 1996.

[96] S. H. Diehn, W. L. Chiu, E. Jay De Rocher, and P. J. Green, "Premature polyadenylation at multiple sites within a *Bacillus thuringiensis* toxin gene-coding region," *Plant Physiology*, vol. 117, no. 4, pp. 1433–1443, 1998.

[97] F. J. Perlak, R. L. Fuchs, D. A. Dean, S. L. McPherson, and D. A. Fischhoff, "Modification of the coding sequence enhances plant expression of insect control protein genes," *Proceedings of the National Academy of Sciences of the United States of America*, vol. 88, no. 8, pp. 3324–3328, 1991.

[98] B. Chaudhary, *Development of transgenic lines in cotton (Gossypium hirsutum L. cv. Coker 310FR) for insect resistance and marker gene removal [Ph.D. thesis]*, Department of Genetics, University of Delhi South Campus, New Delhi, India, 2006.

[99] B. Chaudhary, S. Kumar, K. V. S. K. Prasad, G. S. Oinam, P. K. Burma, and D. Pental, "Slow desiccation leads to high-frequency shoot recovery from transformed somatic embryos of cotton (*Gossypium hirsutum* L. cv. Coker 310 FR)," *Plant Cell Reports*, vol. 21, no. 10, pp. 955–960, 2003.

[100] Y. E. Dunaevsky, E. N. Elpidina, K. S. Vinokurov, and M. A. Belozersky, "Protease inhibitors in improvement of plant resistance to pathogens and insects," *Molecular Biology*, vol. 39, no. 4, pp. 608–613, 2005.

[101] L. L. Murdock and R. E. Shade, "Lectins and protease inhibitors as plant defenses against insects," *Journal of Agricultural and Food Chemistry*, vol. 50, no. 22, pp. 6605–6611, 2002.

[102] J. Huang, R. Hu, S. Rozelle, and C. Pray, "Plant science: insect-resistant GM rice in farmers' fields: assessing productivity and health effects in China," *Science*, vol. 308, no. 5722, pp. 688–690, 2005.

[103] S. B. Maqbool, S. Riazuddin, N. T. Loc, A. M. R. Gatehouse, J. A. Gatehouse, and P. Christou, "Expression of multiple insecticidal genes confers broad resistance against a range of different rice pests," *Molecular Breeding*, vol. 7, no. 1, pp. 85–93, 2001.

[104] R. E. Jackson, J. R. Bradley, and J. W. Van Duyn, "Performance of feral and *Cry1Ac*-selected *Helicoverpa zea* (Lepidoptera: Noctuidae) strains on transgenic cottons expressing one or two *Bacillus thuringiensis* ssp. kurstaki proteins under greenhouse conditions," *Journal of Entomological Science*, vol. 39, no. 1, pp. 46–55, 2004.

[105] E. Fitches, M. G. Edwards, C. Mee et al., "Fusion proteins containing insect-specific toxins as pest control agents: snowdrop lectin delivers fused insecticidal spider venom toxin to insect haemolymph following oral ingestion," *Journal of Insect Physiology*, vol. 50, no. 1, pp. 61–71, 2004.

[106] R. A. De Maagd, A. Bravo, C. Berry, N. Crickmore, and H. E. Schnepf, "Structure, diversity, and evolution of protein toxins from spore-forming entomopathogenic bacteria," *Annual Review of Genetics*, vol. 37, pp. 409–433, 2003.

[107] C. G. Yu, M. A. Mullins, G. W. Warren, M. G. Koziel, and J. J. Estruch, "The *Bacillus thuringiensis* vegetative insecticidal protein Vip3A lyses midgut epithelium cells of susceptible insects," *Applied and Environmental Microbiology*, vol. 63, no. 2, pp. 532–536, 1997.

[108] R. W. Kurtz, A. McCaffery, and D. O'Reilly, "Insect resistance management for Syngenta's VipCot transgenic cotton," *Journal of Invertebrate Pathology*, vol. 95, no. 3, pp. 227–230, 2007.

[109] S. H. Syed, *Genetic transformation of cotton with galanthus nivalis agglutinin (GNA) gene [Ph.D. thesis]*, University of the Punjab, Lahore, Pakistan, 2002.

[110] X. Fan, X. Shi, J. Zhao, R. Zhao, and Y. Fan, "Insecticidal activity of transgenic tobacco plants expressing both *Bt* and CpTI genes on cotton bollworm (*Helicoverpa armigera*)," *Chinese Journal of Biotechnology*, vol. 15, no. 1, pp. 1–5, 1999.

[111] G. D. Yang, Z. Zhu, Y. O. Li, and Z. J. Zhu, "Transformation of *Bt*-CpTi fusion protein gene to cabbage (*Brassica oleracea* var. capitata) mediated by *Agrobacterium tumefaciens* and particle bombardment," *Shi Yan Sheng Wu Xue Bao*, vol. 35, no. 2, pp. 117–122, 2002.

[112] R. E. Jackson, J. R. Bradley, and J. W. Van Duyn, "Field performance of transgenic cottons expressing one or two *Bacillus thuringiensis* endotoxins against bollworm, *Helicoverpa zea* (Boddie)," *Journal of Cotton Science*, vol. 7, no. 3, pp. 57–64, 2003.

[113] R. A. De Maagd, A. Bravo, and N. Crickmore, "How *Bacillus thuringiensis* has evolved specific toxins to colonize the insect world," *Trends in Genetics*, vol. 17, no. 4, pp. 193–199, 2001.

[114] E. Schnepf, N. Crickmore, J. Van Rie et al., "*Bacillus thuringiensis* and its pesticidal crystal proteins," *Microbiology and Molecular Biology Reviews*, vol. 62, no. 3, pp. 775–806, 1998.

[115] R. Karlova, M. Weemen-Hendriks, S. Naimov, J. Ceron, S. Dukiandjiev, and R. A. De Maagd, "*Bacillus thuringiensis* δ-endotoxin Cry1Ac domain III enhances activity against *Heliothis virescens* in some, but not all *Cry1-Cry1Ac* hybrids," *Journal of Invertebrate Pathology*, vol. 88, no. 2, pp. 169–172, 2005.

[116] P. Singh, M. Kumar, C. Chaturvedi, D. Yadav, and R. Tuli, "Development of a hybrid δ-endotoxin and its expression in tobacco and cotton for control of a polyphagous pest *Spodoptera litura*," *Transgenic Research*, vol. 13, no. 5, pp. 397–410, 2004.

[117] S. Naimov, S. Dukiandjiev, and R. de Maagd, "A hybrid *Bacillus thuringiensis* delta-endotoxin gives resistance against a coleopteran and a lepidopteran pest in transgenic potato," *Plant Biotechnology*, vol. 1, pp. 51–57, 2003.

[118] W. Manyangarirwa, M. Turnbull, G. S. McCutcheon, and J. P. Smith, "Gene pyramiding as a *Bt* resistance management strategy: how sustainable is this strategy?" *African Journal of Biotechnology*, vol. 5, no. 10, pp. 781–785, 2006.

[119] S. L. Bates, J. Z. Zhao, R. T. Roush, and A. M. Shelton, "Insect resistance management in GM crops: past, present and future," *Nature Biotechnology*, vol. 23, no. 1, pp. 57–62, 2005.

[120] R. T. Roush, "Two-toxin strategies for management of insecticidal transgenic crops: can pyramiding succeed where pesticide mixtures have not?" *Philosophical Transactions of the Royal Society B*, vol. 353, no. 1376, pp. 1777–1786, 1998.

[121] B. E. Tabashnik, "Evolution of resistance to *Bacillus thuringiensis*," *Annual Review of Entomology*, vol. 39, pp. 47–79, 1994.

[122] E. S. Sachs, J. H. Benedict, J. F. Taylor, D. M. Stelly, S. K. Davis, and D. W. Altman, "Pyramiding CryIA(b) insecticidal protein and terpenoids in cotton to resist tobacco budworm (Lepidoptera: Noctuidae)," *Environmental Entomology*, vol. 25, no. 6, pp. 1257–1266, 1996.

[123] J. J. Coombs, D. S. Douches, W. Li, E. J. Grafius, and W. L. Pett, "Combining engineered (*Bt-cry3A*) and natural resistance mechanisms in potato for control of Colorado potato beetle," *Journal of the American Society for Horticultural Science*, vol. 127, no. 1, pp. 62–68, 2002.

[124] D. Walker, H. Roger Boerma, J. All, and W. Parrott, "Combining Cry1Ac with QTL alleles from PI 229358 to improve soybean resistance to lepidopteran pests," *Molecular Breeding*, vol. 9, no. 1, pp. 43–51, 2002.

10

Application of Phosphoproteomics to Find Targets of Casein Kinase 1 in the Flagellum of *Chlamydomonas*

Jens Boesger, Volker Wagner, Wolfram Weisheit, and Maria Mittag

Institute of General Botany and Plant Physiology, Friedrich Schiller University Jena, Am Planetarium 1, 07743 Jena, Germany

Correspondence should be addressed to Maria Mittag, m.mittag@uni-jena.de

Academic Editor: Jaroslav Doležel

The green biflagellate alga *Chlamydomonas reinhardtii* serves as model for studying structural and functional features of flagella. The axoneme of *C. reinhardtii* anchors a network of kinases and phosphatases that control motility. One of them, Casein Kinase 1 (CK1), is known to phosphorylate the Inner Dynein Arm I1 Intermediate Chain 138 (IC138), thereby regulating motility. CK1 is also involved in regulating the circadian rhythm of phototaxis and is relevant for the formation of flagella. By a comparative phosphoproteome approach, we determined phosphoproteins in the flagellum that are targets of CK1. Thereby, we applied the specific CK1 inhibitor CKI-7 that causes significant changes in the flagellum phosphoproteome and reduces the swimming velocity of the cells. In the CKI-7-treated cells, 14 phosphoproteins were missing compared to the phosphoproteome of untreated cells, including IC138, and four additional phosphoproteins had a reduced number of phosphorylation sites. Notably, inhibition of CK1 causes also novel phosphorylation events, indicating that it is part of a kinase network. Among them, Glycogen Synthase Kinase 3 is of special interest, because it is involved in the phosphorylation of key clock components in flies and mammals and in parallel plays an important role in the regulation of assembly in the flagellum.

1. Introduction

Eukaryotic cilia or flagella are microtubule-based organelles that are highly conserved in protein composition and structural organization from protozoa to mammals. They are structurally characterized by nine microtubular doublets surrounding two central microtubular singlets [1]. Substructures like dynein arms and radial spokes are associated with the axoneme and important for motility in the flagellum. Matrix proteins that are not tightly associated with the flagellar membrane or the axoneme serve diverse functions in the flagellum and can be involved in intraflagellar transport [2].

Since many years, the green biflagellate alga *Chlamydomonas reinhardtii*, whose genome has been sequenced, is used as a model to study flagella structure, assembly, formation, and motility [3]. *C. reinhardtii* uses flagella for motility in aqueous environments, for attaching to surfaces and for cell-cell recognition during mating. A proteomic analysis of *Chlamydomonas* flagella revealed more than 600 proteins [4] that include, for example, motor and signal transduction components as well as proteins with homologues associated with human diseases (e.g., polycystic kidney disease, retinal degeneration, hydrocephalus, or changes in the left-right symmetry of organs) collectively known as ciliopathies [5]. But in many cases, Flagellar Associated Proteins (FAPs) still have unknown function.

Among the proteins in the flagellum, 21 protein kinases and 11 protein phosphatases were found pointing to regulation by reversible protein phosphorylation in this organelle. Phosphorylation events on specific amino acids residues can affect protein function, its intracellular localization, its activity, and its affinity to interaction partners (for review see [6]). But the identification of substrates for kinases in the phosphoregulatory pathway is still a challenge. In *C. reinhardtii*, several proteomes and phosphoproteomes of subcellular compartments (reviewed in [7, 8]) were analyzed including environmentally modulated photosynthetic membranes [9], the eyespot [10], and the flagellum [11]. The flagellum phosphoproteome was first studied under physiological conditions without postincubation of isolated flagellar proteins with ATP to increase the phosphorylation status. 126 *in vivo* phosphorylation sites were found belonging to 32 different structural and motor proteins, several kinases, and proteins with protein interaction domains [11]. Furthermore, a dynamic phosphorylation pattern and clustering of phosphorylation sites were found in some cases, indicating the specific control of proteins by reversible phosphorylation in the flagellum. In another study, flagellum phosphoproteins were examined during flagella shortening. In this case, postincubation with ATP was undertaken. Thereby, half of the identified phosphoproteins were only detected in shortening flagella [12].

The axoneme of *Chlamydomonas* flagella anchors multiple inner arm dyneins and a network of kinases and phosphatases that control motility by reversible protein phosphorylation [13]. One of the involved flagellum kinases is Casein Kinase 1 (CK1) [14–16]. In pharmacological experiments using a specific CK1 inhibitor (CKI-7), it was shown that CK1 regulates dynein activity and flagellum motility by phosphorylation of the Inner Dynein Arm I1 Intermediate Chain 138 (IC138) [14, 15]. Moreover, silencing of CK1 results in alterations of circadian phototaxis (shortening of the period), defects in flagella formation, and in hatching of the daughter cells [17]. Interestingly, alterations in the expression of several other key players of the clock machinery of *C. reinhardtii* named Rhythm of Chloroplast (ROC) and a homologue of Constans (CrCO) have in parallel severe effects on hatching, flagella formation, and/or movement, underlining that these processes are interconnected in *C. reinhardtii* [17–19].

Regarding the multiple functions of CK1 in flagella formation and motility along with its regulatory role in the circadian system in *C. reinhardtii*, we were interested in the identification of CK1 targets in flagella beside IC138. In a comparative phosphoproteomic approach using wild-type cells with and without CKI-7 treatment, we determined the targets of CK1 in the flagellum. In the CKI-7-treated cells, several phosphoproteins were missing or were identified with a reduced number of phosphorylation sites, compared to untreated wild-type cells. Also novel phosphopeptides or additional phosphorylation sites of known phosphopeptides were identified in the CKI-7-treated cells, suggesting that CK1 is part of a signaling network in the flagellum.

2. Materials and Methods

Standard molecular biology methods were done according to [20].

2.1. Cell Culture. C. reinhardtii strain 137c (*nit1 nit2*) was used with whom the flagellar proteome and phosphoproteome were analyzed [4, 11]. Cells were grown in TAP medium [21] under a 12 h light-12 h dark cycle (LD 12 : 12) with a light intensity of $71 \mu E\,m^{-2}\,sec^{-1}$ (1 E = 1 mol of photons) at 24°C. The beginning of the light period is defined as time zero (LD0) and the beginning of the dark period is LD12. In some cases, cells were released after growth in LD into constant conditions (LL) of dim light ($15 \mu E\,m^{-2}\,sec^{-1}$).

2.2. Crude Extract Preparation and Immunodetection. Protein extracts were prepared as described previously [11]. The concentration of proteins was measured according to [22]. Immunoblots were done with antibodies against phospho-Ser (Qiagen) and phosphoThr (Cell Signaling Technology) according to the manufacturer's instructions. Polyclonal antibodies against the C-terminal part of CK1 (amino acids 131–333 out of 333; ID JGI Vs3: 137286) were also used [23]. For this, the C-terminal part of CK1 was expressed and purified from *E. coli* according to the Qiagen protocol. Antibodies were raised by the "Pineda-Antikörper-Service," Berlin, Germany. Immunoblots were done as described [11] using the polyclonal anti-CK1 antibody in a dilution of 1 : 5,000.

2.3. Densitometry Analysis. Quantifications were done with the Image Master 2D Elite (version 4.01) software from GE Healthcare (formerly Amersham Pharmacia Biotech).

2.4. Measurement of Swimming Velocity of C. reinhardtii Cells. Measurement of swimming velocity was done by using a hemocytometer and a differential interference contrast microscope with a total magnification of 400 including a personal computer with a digital video recording system to measure displacement versus time. The swimming velocity was determined manually by measuring the linear displacement of cells on the scale of the micrometer. 10 samples were measured to obtain the average velocity of a given sample.

2.5. Cell Growth, CKI-7 Treatment, Isolation of Flagella, Protein Digestion, and Enrichment of Phosphopeptides by Immobilized Metal Affinity Chromatography (IMAC). Cells were grown in a LD cycle and harvested at the end of the night (LD24) at a cell density of $2\text{-}3 \times 10^6$ cells mL^{-1} by centrifugation (700 ×g, 5 min, 4°C).Cells were resuspended in one-half volume of minimal medium [21] and then the culture was kept under dim light conditions for 29 h representing subjective day (LL29), before cells were harvested (700 ×g, 15 min, 4°C). In some cases, the CK1 inhibitor, CKI-7, (N-(2-Aminoethyl)-5-chloroisoquinoline-8-sulfonamide; Toronto Research Chemicals Inc.) [24], was added to the culture to a final concentration of $50 \mu M$ following the shift to LL conditions. Isolation of the matrix membrane axoneme fraction (MMA) of flagella, tryptic digestion of MMA proteins, and enrichment of flagellum phosphopeptides by IMAC were done as previously described [11].

2.6. Peptide Identification by Nano-Liquid Chromatography-Electrospray Ionization-Mass Spectrometry (nLC-ESI-MS) and Data Analysis. nLC-ESI-MS and data analysis were carried out as described before [11]. Briefly, phosphopeptides were subjected to nLC-ESI-MS using an UltiMate 3000 nano-HPLC (Dionex Corporation) with a flow rate of 300 nL min^{-1} coupled online with a linear ion trap ESI-MS (Finnigan LTQ, Thermo Electron Corp.). The instrument was run by the data-dependent neutral loss method, cycling between one full MS and MS/MS scans of the four most abundant ions. After each cycle, these peptide masses were excluded from the analysis for 10 sec. The detection of a neutral loss fragment (98, 49, or 32.66 Da) in the MS^2 scans triggered an MS^3 scan of the neutral loss ion representing the dephosphorylated peptide.

Data analysis was done using the Proteome Discoverer software (Version 1.0) from Thermo Electron Corp. including the SEQUEST algorithm [25]. The software parameters were set to detect a modification of 79.96 Da in Ser, Thr, or Tyr in MS^2 and MS^3 spectra. For the database searches with MS^3 data, modifications of -18.00 Da on Ser and Thr residues representing the neutral loss were additionally used. Further, detection of a modification of 16 Da on Met representing its oxidized form was enabled and carboxyamidomethylation of Cys residues was enabled as a static modification. Peptide mass tolerance was set to 1.5 Da in MS mode. In MS^2 and MS^3 modes, fragment ion tolerance was set up to 1 Da. The parameters for all database searches were set to achieve a false discovery rate (FDR) of not more than 1% for each individual analysis. Data were searched against the flagellar proteome database [4] (http://labs.umassmed.edu/chlamyfp/index.php). Additionally, NCBI and the Joint Genome Institute *C. reinhardtii* databases (Version 2 and Version 3) were used for data evaluation.

3. Results

3.1. The Effects of the CK1 Inhibitor CKI-7 on the Phosphorylation Pattern of Flagellum Proteins and the Swimming Velocity of C. reinhardtii.
CK1 was found in the proteome of the flagellum [4] and was also shown immunologically to be enriched in flagella in wild-type strain SAG 73.72 [17]. For the comparative phosphoproteome analysis, flagella were isolated from strain 137c along with the dibucaine method [11]. We first examined the enrichment of CK1 in flagella of 137c using the applied conditions by immunodetection along with anti-CK1 antibodies (Figures 1(a) and 1(b)). Levels of CK1 were significantly enriched in the flagella fraction, especially compared to cell bodies lacking flagella. Thus, the procedure used for identification of the phosphoproteome maintains the enrichment of CK1 in flagella and is thus suited to screen for its targets.

In the next step, we examined to what degree the CK1-specific inhibitor, CKI-7 [24], which was already used for studying CK1 in *C. reinhardtii* [15], influences the phosphorylation pattern of flagellum proteins. Therefore, we grew cells with and without CKI-7 treatment, respectively, and compared the flagellum phosphoproteins from both aliquots by immunodetection with antiphosphoSer antibodies (Figure 1(c)). As expected, several phosphorylated protein bands were reduced to a significant extent or absent in the CKI-7-inhibited cells (Figure 1(c), labeled with "−"). At the same time, some phosphoprotein bands were stronger (Figure 1(c), labeled with "+"). These data show that inhibition of CK1 has a dual effect. On the one hand, the phosphorylation of CK1 targets drops strongly down or is fully stopped by its inhibition; on the other hand, inactive

CK1 seems to lead to the activation of other kinases resulting in the phosphorylation of other proteins.

As mentioned before, flagellum kinases affect motility. We also studied if the inhibition by CKI-7 results in changes in swimming velocity. To analyze the swimming behavior, we compared the swimming velocity of the *C. reinhardtii* strain 137c with cells that were cultivated with CKI-7 as described (see Section 2). Cells were spotted on a hemocytometer and the swimming velocity was measured using a differential interference contrast microscope including a personal computer with a video recording system (see Section 2). The assay revealed that the swimming speed of CK1-inhibited cells is significantly lower (75.6 μm/s; $\pm$4,1 SEM) compared to untreated cells (122.2 μm/s; $\pm$2.5 SEM) (Figure 1(d)). These data show that CK1-mediated phosphorylation events in flagella influence motility and swimming speed of *C. reinhardtii* cells.

3.2. The Flagellum Phosphoproteome of CKI-7-Treated Cells.
The targets of CK1 in the flagellum are of high interest with regard to flagella formation as well as for clock control events. They are largely unknown. An exception is IC138 that is suggested as a direct target of CK1 based on experimental data (summarized in [25]).

In a next step, the direct and indirect targets of CK1 were analyzed by a functional proteome approach. For that purpose, we compared the already existing phosphoproteome [11] with one investigated exactly under the same conditions with the single exception that CK1 is inhibited. Since strong silencing of CK1 by RNAi results in defects in flagella formation, flagellum material cannot be obtained in a significant amount from such strains [17]. Therefore, inhibition of CK1 with CKI-7 was used. Cells were grown under a light-dark cycle and the inhibitor was added for a 29 h period right at the moment when the cells were released to constant dim light. LL29 was also used as harvesting time point in the previous analysis [11].

We avoided to add high amounts of ATP to isolated flagella and to postincubate them at elevated temperatures to induce kinase activities *in vitro*, as done in another study [12]. We found that this treatment leads to severe phosphorylation events that include most likely phosphorylation steps that would not take place *in vivo* under physiological conditions See Supplemental Figures 1(a), 1(b) in Supplementary Material available online at doi: 1155/2012/581460.

The further analysis of the phosphoproteome in CKI-7-treated cells was carried out with the same procedure and criteria as applied before for the flagellum phosphoproteome [11]. Three independent isolations of flagella of CKI-7-inhibited cells were carried out and subjected to phosphopeptide purification along with liquid chromatography mass spectrometry (for details, see [11]). Previously identified phosphopeptides or phosphorylation sites within a phosphopeptide (listed in Table S1 in [11]) that had not been detected in any of the three analyses were considered to be either direct or indirect targets of CK1. The phosphoproteins to which these phosphopeptides belong are listed in Table 1. Novel phosphopeptides belonging to novel phosphoproteins that

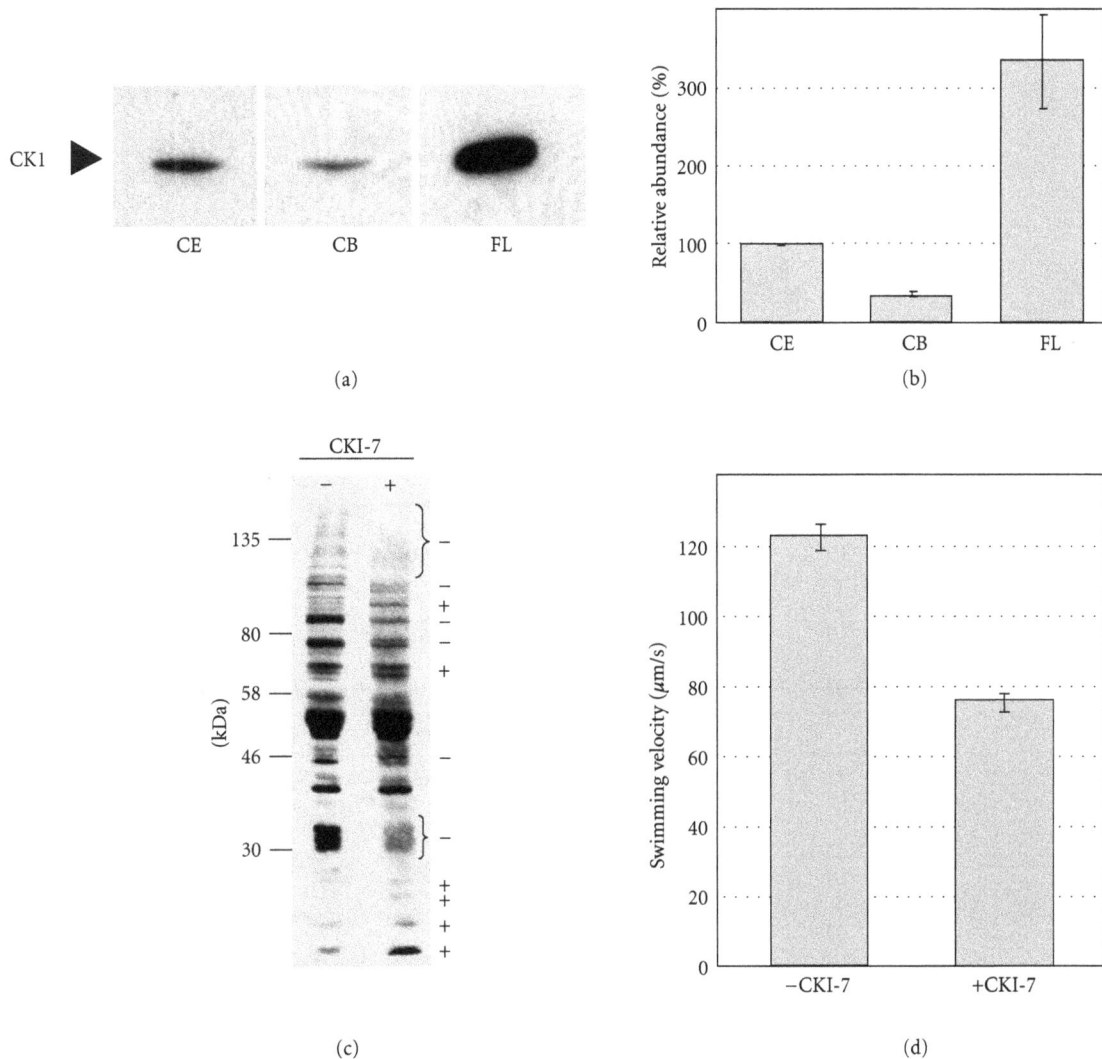

(a)

(b)

(c)

(d)

FIGURE 1: Enrichment of CK1 in flagella and the influence of CK1 inhibition on the phosphorylation status of flagellum proteins and swimming velocity of *C. reinhardtii* cells. (a) Cells were grown in TAP in a 12 h light-12 h dark cycle and then released to dim light (LL) according to Section 2. Cells were harvested at LL29 and flagella were isolated and a whole cell crude extract (CE), a flagellar extract (FL), and an extract from cell bodies lacking flagella (CB) were prepared. 25 μg proteins per fraction were separated by SDS-PAGE and analyzed by immunoblotting with anti-CK1 antibodies according to Section 2. (b) For quantifications, the amount of CK1 detected in the whole cell crude extract was set to 100%. Quantifications were done with three biological replicates using the ImageMaster 2D Elite Vs.4.01 software (GE Healthcare). (c) Changes in the phosphorylation pattern of flagellum proteins in cells treated with and without CKI-7. Cells were grown as described above (a) in the presence or absence of CKI-7 and harvested at LL29 before isolation of flagella. Proteins from the MMA fraction of the flagellum (25 μg each lane) were separated by 9% SDS-PAGE along with a molecular mass standard and immunoblotted with specific antibodies against phosphoSer according to Section 2. Changes in the phosphorylation status of proteins after CKI-7 treatment are indicated by "+" and "−" signs, respectively. (d) Swimming velocity of 137c cells in the absence (−CKI-7) or presence of CK1 inhibitor (+CKI-7). Cells were grown at 23°C in a LD cycle. Measurements of swimming velocity were done with a hemocytometer and a differential interference contrast microscope with a total magnification of 400 including a personal computer with a video recording system to measure displacement versus time ($n = 10$). Error bars represent the SEM.

had not been identified in the former analysis and additional phosphopeptides or phosphorylation sites of already identified phosphoproteins are listed in Table 2. Details about all newly identified peptides and phosphorylation sites can be found in Supplemental Table S1. In three cases, (TEKTIN, FAP18, and FAP262), all previous identified phosphorylation sites were detected again, but in some phosphopeptides with

different combinatory phosphorylation patterns (data not shown).

In the CKI-7-treated cells, phosphopeptides from 14 phosphoproteins were missing (Table 1). Four additional phosphoproteins were identified again but with a reduced number of phosphorylation sites. These are labeled by indices along with the missing sites in Table 1. Among

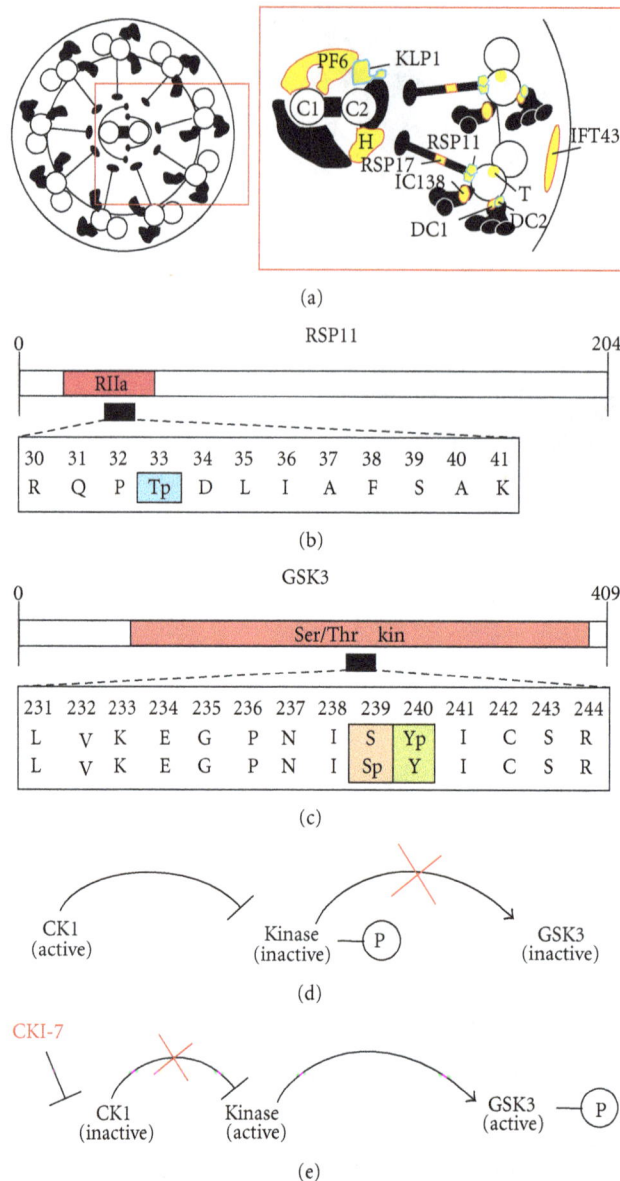

FIGURE 2: Analysis of CK1 targets in the flagellum. (a) Diagram of flagellum phosphoproteins in wild-type and CK1-inhibited cells. A cross-section of a flagellum from *C. reinhardtii* (left panel) and a more detailed view (red rectangle) are shown according to [11]. Structural phosphoproteins in CKI-7-inhibited cells, and such with a reduced number of phosphorylation sites are indicated in yellow color with a red frame. Novel phosphopeptides of structural proteins or additional phosphorylation sites of known phosphopeptides from structural phosphoproteins that were identified in the CKI-7-inhibited proteome are indicated by yellow color with a blue frame. Structural proteins with previously identified phosphopeptides, whose phosphorylation sites were detected again, are indicated in yellow without frame. Abbreviations are: C1 central pair projection (C1P), C2 central pair projection (C2P), PF6 protein (PF6), Hydin (H), Radial Spoke Protein17 (RSP17); Outer Dynein Arm Docking Complex (DC); Inner Dynein Arm Intermedite Chain138 (IC138), Tectin (T) as well as an Intraflagellar Transport Protein43 (IFT43). (b) and (c) Positions of identified phosphopeptides in the predicted domains of RSP11 (b) and GSK3 (c). Identified phosphopeptides are indicted by black boxes. The amino acid positions are mentioned. "p" indicates *in vivo* phosphorylation sites. RIIa, regulatory subunit of cAMP-Dependent Protein Kinase A; Ser/Thr Kin, Ser/Thr protein kinase catalytic domain. (d) and (e) Hypothetical model of GSK3 de-/activation via reversible phosphorylation triggered by CK1. (d) Regulatory signaling involves an additional kinase. The noninhibited active form of CK1 inactivates another still unknown kinase by phosphorylation. This kinase is needed in its active nonphosphorylated form for activating GSK3. (e) If CK1 is inhibited by CKI-7, the unknown kinase is not phosphorylated and thus active. This active kinase phosphorylates in turn GSK3, which is then activated.

TABLE 1: Phosphoproteins identified in 137c [11] whose phosphopeptides or phosphorylation sites are missing in CKI-7-treated cells.

Flagellar central pair-associated protein; PF6

Hydin-like protein; HYD3

Inner dynein arm I1 intermediate chain; IC138

Intraflagellar transport protein IFT43

Outer dynein arm docking complex subunit 1[a,b]; ODA-DC1, ODA3

Radial spoke protein 17; RSP17

FAP59[c]; RecF/RecN/SMC N-terminal domain

FAP116[a,d,e]; microtubule-binding protein MIP-T3 domain

FAP190[a,f]; sterile alpha motif

FAP228; callose synthase-like protein; 1,3-beta-glucan synthase component

FAP230; ankyrin repeats; ion transport protein domain

FAP254; putative ankyrin-like protein

FAP288; EF hand

FAP1[a,g]

FAP93

FAP147

FAP184

FAP263

The function of depicted proteins is given as determined by NCBI BLASTp, along with their conserved domains.
[a]Not all previously identified peptides (listed in Table S1 in [11]) are present in the CKI-7-treated cells.
[b]Variants of peptide TISGADTPEEVLAYWEGLK with the phosphorylation sites Thr-345, Ser-347, and Thr-351 as well as variants of peptide ILGYTGS-DVEEEEPESEEETEEEANKDDGVVDR with the phosphorylation sites Tyr-697 and Ser-709 are missing.
[c]Predicted functional domains are present only in the Vs3 model.
[d]Vs2 model differs significantly from Vs3 model.
[e]The phosphorylation site Ser-255 in peptide SASPGGEDPLNKSGSAAPK is missing.
[f]Variants of peptide STSSIGGGYSEPVGSDGEGSDAASAKPR with phosphorylation sites on Ser-370, Ser-375 and Ser-379 are missing.
[g]The phosphorylation site Ser-55 in peptide SRGSFQEGQAMVR is missing.

TABLE 2: Additional phosphopeptides/phosphorylation sites in CKI-7-treated cells of either novel phosphoproteins or phosphoproteins that were already identified in 137c [11].

Phosphoproteins only present in CKI-7-treated 137c cells
Glycogen synthase kinase 3; GSK3
Kinesin-like protein; Kinesin motor domain, KIF9-like subgroup
Phosphoglucomutase
Radial spoke protein 11; RSP11; RIIa domain
S-Adenosylmethionine synthetase
FAP139[a]; TIGR02680 domain
FAP21[a]
FAP56
FAP75
FAP98
FAP129
FAP165[a]
FAP236
FAP241
FAP243 (Vs3 FAP183)[a]
Phosphoproteins found in CKI-7-treated cells with additional phosphopeptide(s) in comparison to 137c
Outer dynein arm docking complex protein 2; ODA-DC2
FAP33[a,b]; ankyrin repeats
FAP154[a]
FAP217
Phosphoproteins found in CKI-7-treated cells with the same peptide [11] but with additional phosphorylation site(s)
FAP39[a]; plasma membrane calcium transporting ATPase
MAK7[a]; mitogen activated protein kinase 7

The function of depicted proteins is given as determined by NCBI BLASTp, along with their conserved domains.
[a]Vs2 model differs significantly from Vs3 model.
[b]Predicted functional domains are present only in the Vs3 model.

these 18 phosphoproteins, six known structural proteins are present including IC138 that was suggested to be a direct target of CK1 [26]. All missing structural phosphoproteins as well as those with a reduced number of phosphorylation sites are indicated in yellow color with a red frame in Figure 2(a). Moreover, seven FAPs with conserved domains are affected in the CKI-7-treated phosphoproteome as well as five FAPs without any conserved domains.

Also novel phosphopeptides or additional phosphorylation sites of known phosphopeptides were identified in the proteome of CKI-7-treated cells, suggesting that CK1 is part of a signaling network in the flagellum. They belong to either 15 new phosphoproteins or six already known phosphoproteins (Table 2, Supplemental Table 1). Among them, some structural components are present, indicated by yellow color with a blue frame in Figure 2(a). Thereby, Radial Spoke Protein 11 (RSP11) is of special interest. It has an RIIa domain, which is a regulatory subunit of cAMP Dependent

Protein Kinase A (PKA) and bears a phosphorylation site (Figure 2(b)). Two other kinases were also found in this category. One of them is Glycogen Synthase Kinase 3 (GSK3). The level of active GSK3 is postulated to be regulated via phosphorylation of a conserved Tyr correlating with flagellar length [27]. Exactly this Tyr that is situated in the Ser/Thr kinase domain of GSK3 is phosphorylated as well as a Ser in its surroundings (Supplemental Table 1; Figure 2(c)). Notable GSK3 is also clock relevant, for example, in *Drosophila* [28]. Moreover, a Mitogen Activated Kinase, MAK7, was found with additional phosphorylation sites.

4. Discussion

The identification of targets of CK1 in the flagellum will help understanding flagella formation as well as clock control events related to flagella [17–19]. The fact that several phosphorylated flagellar protein bands disappear in CKI-7-treated cells suggests that CK1 has multiple targets in the flagellum. Among the 32 phosphoproteins of the flagellum, 14 were missing in the flagellum phosphoproteome when the CKI-7 inhibitor was used or represented with a reduced number of phosphorylation sites (four cases, Table 1). Missing phosphorylation sites cannot be automatically considered as direct targets of CK1. It could be that the phosphorylation of an amino acid residue by CK1 represents a trigger that then allows a consequent phosphorylation of another amino acid residue in the surroundings by another kinase. An example for consequent phosphorylation steps of different kinases is mentioned below and involves PKA, GSK3, and CK1. Also, CK1 may activate or deactivate another kinase by reversible phosphorylation. In the current study, the previously identified kinases along with their phosphorylation sites were found again [11]. Only in case of FAP262 that bears a Ser/Thr kinase domain, a different combinatory phosphorylation pattern was observed, which might be relevant. But it could also be that some of the missing phosphoproteins in the FAP category whose functions are not known may have kinase activity. Networks that consist of interconnected kinases along with protein phosphatases are not unusual in signaling. In line with this, we found also 21 new phosphoproteins along with novel phosphopeptides or phosphorylation sites, including three kinase-related proteins. The presence of new phosphorylation sites in flagella of CKI-7-inhibited cells was already predictable from the appearance of novel flagellar phosphoprotein bands detected with anti-phosphoSer antibodies (Figure 1(c)). In this category, we identified two phosphoproteins involved in carbohydrate and amino acid metabolism, respectively (Table 2). One of them, phosphoglucomutase, catalyzes the bidirectional conversion of glucose-1-phosphate to glucose-6-phosphate. Glucose-1-phosphate can be transferred into glycolysis by this way. The flagellum contains all enzymes of the late glycolytic pathway; they are able to generate ATP for direct use in the flagellum [4]. In mammals, the activity of phosphoglucomutase is regulated by phosphorylation [29]. The other metabolically relevant enzyme in this category is S-adenosylmethionine synthetase, a key enzyme of methionine metabolism. In rat liver, the activity of the S-adenosylmethionine synthetase is regulated by Protein Kinase C [30].

One of the direct targets of CK1 was suggested to be IC138, the Inner Dynein Arm I1 Intermediate Chain 138. It was shown that phosphorylation of IC138 correlates with the inhibition of dynein activity and that PKA beside CK1 as well as the Protein Phosphatases PP2A and PP1 are involved there (summarized in [26]). IC138 was identified in CK1 active cells with one phosphopeptide that is situated at its N-terminus including variable phosphorylation sites [11]. None of these phosphorylation sites were detected after CKI-7 treatment, underlining that IC138 is a direct and/or indirect target of CK1. A pharmacological analysis using CKI-7 revealed the impact of CK1 on IC138 phosphorylation [14]. This mechanism authorizes CK1 to regulate dynein activity and control flagellum motility. Also an analysis of mutants lacking the IC138 subcomplex revealed strains that swim forward with reduced swimming velocities [31, 32]. Interestingly, the swimming speed of the CKI-7-treated cells

was reduced to a similar degree in comparison to the mutant strains that are lacking IC138, suggesting that the generation of flagellum motility is regulated by a CK1-mediated phosphorylation of IC138 as suggested before [14, 15].

Another structural phosphoprotein previously identified with two phosphopeptides and variable phosphorylation sites is ODA-DC1. The outer dynein arm docking complex (ODA-DC), which is composed of three proteins, designated DC1, DC2, and DC3, is associated with microtubules and targets the outer dynein arms to its binding site on the flagellum axoneme [33]. In both previously identified phosphopeptides certain phosphorylation sites are missing in CKI-7-inhibited cells (Table 1; indices a, b) pointing out that they are CK1 targets. ODA-DC2 had been also identified in the previous study [11] with one phosphopeptide and variable phosphorylation sites, which were all found again in the current study. But now a novel phosphopeptide with phosphorylation on Ser-278 was present in CKI-7 cells, underlining that CK1 seems to be indirectly involved in regulating further kinases.

Radial spokes represent a major structural feature of 9 + 2 axonemes and they are essential for flagellum beating. Each radial spoke consists of a thin stalk, which is attached to the A-tubule of the axonemal doublet microtubules and a head projecting toward the central apparatus [34]. The radial spoke of C. reinhardtii is composed of at least 23 proteins, and not all of them have been characterized at the molecular level [35]. RSP17, which is located in the spoke stalk, was identified in the flagellum phosphoproteome analysis with two different phosphopeptides [11]. The absence of both phosphopeptides in CKI-7-treated cells suggests that RSP17 is at the same time a direct and/or indirect target of CK1. Functional domains in radial spoke proteins reveal their role in mediating signaling pathways. For instance, RSP11 consists of a regulatory subunit (RIIa) of PKA [35]. However, RSP11 lacks the cAMP-binding domains of the RII regulatory subunit. We could identify RSP11 in the CKI-7-treated cells as a new phosphoprotein with one in vivo phosphorylation site at Thr-35, which is located directly in the RIIa domain (Figure 2(b)). The interaction between RII and A-kinase anchoring protein motifs (AKAP) can be regulated by phosphorylation of RII [36, 37]. A pharmacological analysis using an inhibitor and the RII regulatory subunits had detected an axonemal PKA activity [38]. But PKA could not be found in the flagellar proteome in contrast to CK1, PP1, and PP2A [4]. Thus, it was hypothesized that C. reinhardtii could express a PKA with an unconventional structure [39]. The identified phosphorylation site within the RII subunit of RSP11 may be relevant in this context.

An additional flagellum kinase is GSK3 whose enzymatic activity is inhibited by lithium causing flagellar elongation [27]. It is known that GSK3 has a Tyr-phosphorylated, active form and is enriched in flagella. GSK3 is associated with the axoneme in a phosphorylation-dependent manner. The level of active GSK3 correlates with flagellar length [27]. We could identify the Tyr-240-phosphorylated GSK3 as well as a Ser-239-phosphorylated alternative in the CKI-7-treated cells (Figure 2(c)), suggesting that inhibition of CK1 causes activation of GSK3. Both *in vivo* phosphorylation sites are located in the catalytic kinase domain, which could play important roles in the regulation of the activity of GSK3 within signaling pathways. Notably, interplay between CK1 and GSK3 is known for Hedgehog signaling pathways [40]. Thereby, the Cubitus Interruptus (Ci-155) transcriptional activator is involved. Ci-155 proteolysis depends on phosphorylation at three sites of PKA. Then, these phosphoSer prime further phosphorylation at GSK3 and CK1 sites. This principle is a good example for consecutive phosphorylation steps of different kinases as mentioned before.

Several studies have shown that reversible phosphorylation of Tyr causes increases and decreases in GSK3 kinase activity, respectively [41, 42]. For the interplay of CK1 and GSK3 in the *C. reinhardtii* flagella, one can imagine a regulatory mechanism, involving, for example, an additional kinase. In a hypothetical model (Figure 2(d)), the noninhibited, active CK1 inactivates another kinase by phosphorylation, which is responsible for the activation of GSK3 by Tyr-phosphorylation. If CKI-7 inhibits CK1 (Figure 2(e)), the additional kinase can stay active, because it is not phosphorylated by CK1 and consequently GSK3 gets converted to the phosphorylated active form.

GSK3 plays also an important role in the regulation of circadian systems. Shaggy (SGG), for example, the *Drosophila* homologue of GSK3, is a central player in determining period length in flies by phosphorylation of clock components [43]. In mammals, GSK3 is proposed to phosphorylate Clock (CLK), which is a core transcription factor that is essential for circadian behavior. Phosphorylation of CLK controls its activity and degradation [44]. Especially kinases and phosphatases, which are relevant in regulating circadian clocks in other organisms, are well conserved in *Chlamydomonas* [45]. Interestingly, many output rhythms that can be measured like phototaxis, chemotaxis, and stickiness to glass and mating during the cell cycle involve flagella. It is remarkable that kinases like CK1 or GSK3 as well as phosphatases like PP1 and PP2A are physically located in the axoneme [4, 26]. This underlines the important regulatory function of these components in the flagellum regarding circadian rhythms.

Acknowledgments

The authors thank the Joint Genome Institute (JGI) in the USA and the Kazusa Institute in Japan for the free delivery of EST and genome sequences. This study was supported by the Deutsche Forschungsgemeinschaft (Grants Mi 373 to MM) and the BMBF (Project GoFORSYS, Grant no. 0315260A, work package to MM).

References

[1] G. J. Pazour and G. B. Witman, "The vertebrate primary cilium is a sensory organelle," *Current Opinion in Cell Biology*, vol. 15, no. 1, pp. 105–110, 2003.

[2] J. L. Rosenbaum and G. B. Witman, "Intraflagellar transport," *Nature Reviews Molecular Cell Biology*, vol. 3, no. 11, pp. 813–825, 2002.

[3] S. S. Merchant, S. E. Prochnik, O. Vallon et al., "The Chlamydomonas genome reveals the evolution of key animal and plant functions," *Science*, vol. 318, no. 5848, pp. 245–251, 2007.

[4] G. J. Pazour, N. Agrin, J. Leszyk, and G. B. Witman, "Proteomic analysis of a eukaryotic cilium," *The Journal of Cell Biology*, vol. 170, no. 1, pp. 103–113, 2005.

[5] W. F. Marshall, "The cell biological basis of ciliary disease," *The Journal of Cell Biology*, vol. 180, no. 1, pp. 17–21, 2008.

[6] J. Reinders and A. Sickmann, "Modificomics: posttranslational modifications beyond protein phosphorylation and glycosylation," *Biomolecular Engineering*, vol. 24, no. 2, pp. 169–177, 2007.

[7] N. Rolland, A. Atteia, P. Decottignies et al., "Chlamydomonas proteomics," *Current Opinion in Microbiology*, vol. 12, no. 3, pp. 285–291, 2009.

[8] V. Wagner, J. Boesger, and M. Mittag, "Sub-proteome analysis in the green flagellate alga Chlamydomonas reinhardtii," *Journal of Basic Microbiology*, vol. 49, no. 1, pp. 32–41, 2009.

[9] A. V. Vener, "Environmentally modulated phosphorylation and dynamics of proteins in photosynthetic membranes," *Biochimica et Biophysica Acta*, vol. 1767, no. 6, pp. 449–457, 2007.

[10] V. Wagner, K. Ullmann, A. Mollwo, M. Kaminski, M. Mittag, and G. Kreimer, "The phosphoproteome of a Chlamydomonas reinhardtii eyespot fraction includes key proteins of the light signaling pathway," *Plant Physiology*, vol. 146, no. 2, pp. 772–788, 2008.

[11] J. Boesger, V. Wagner, W. Weisheit, and M. Mittag, "Analysis of flagellar phosphoproteins from Chlamydomonas reinhardtii," *Eukaryotic Cell*, vol. 8, no. 7, pp. 922–932, 2009.

[12] J. Pan, B. Naumann-Busch, L. Wang et al., "Protein phosphorylation is a key event of flagellar disassembly revealed by analysis of flagellar phosphoproteins during flagellar shortening in Chlamydomonas," *Journal of Proteome Research*, vol. 10, no. 8, pp. 3830–3839, 2011.

[13] M. E. Porter and W. S. Sale, "The 9 + 2 axoneme anchors multiple inner arm dyneins and a network of kinases and phosphatases that control motility," *The Journal of Cell Biology*, vol. 151, no. 5, pp. F37–F42, 2000.

[14] P. Yang and W. S. Sale, "Casein kinase I is anchored on axonemal doublet microtubules and regulates flagellar dynein phosphorylation and activity," *The Journal of Biological Chemistry*, vol. 275, no. 25, pp. 18905–18912, 2000.

[15] A. Gokhale, M. Wirschell, and W. S. Sale, "Regulation of dynein-driven microtubule sliding by the axonemal protein kinase CK1 in Chlamydomonas flagella," *The Journal of Cell Biology*, vol. 186, no. 6, pp. 817–824, 2009.

[16] M. Wirschell, R. Yamamoto, L. Alford, A. Gokhale, A. Gaillard, and W. S. Sale, "Regulation of ciliary motility: conserved protein kinases and phosphatases are targeted and anchored in the ciliary axoneme," *Archives of Biochemistry and Biophysics*, vol. 510, no. 2, pp. 93–100, 2011.

[17] M. Schmidt, G. Geßner, M. Luff et al., "Proteomic analysis of the eyespot of Chlamydomonas reinhardtii provides novel insights into its components and tactic movements," *The Plant Cell*, vol. 18, no. 8, pp. 1908–1930, 2006.

[18] T. Matsuo, K. Okamoto, K. Onai, Y. Niwa, K. Shimogawara, and M. Ishiura, "A systematic forward genetic analysis identified components of the Chlamydomonas circadian system," *Genes and Development*, vol. 22, no. 7, pp. 918–930, 2008.

[19] G. Serrano, R. Herrera-Palau, J. M. Romero, A. Serrano, G. Coupland, and F. Valverde, "Chlamydomonas CONSTANS

and the evolution of plant photoperiodic signaling," *Current Biology*, vol. 19, no. 5, pp. 359–368, 2009.

[20] J. Sambrook and D. W. Russel, *Molecular Cloning: A Laboratory Manual*, Cold Spring Harbor Laboratory Press, Cold Spring Harbor, NY, USA, 2001.

[21] E. H. Harris, *The Chlamydomonas Sourcebook*, Academic Press, San Diego, Calif, USA, 1989.

[22] V. Neuhoff, K. Philipp, H. G. Zimmer, and S. Mesecke, "A simple, versatile, sensitive and volume-independent method for quantitative protein determination which is independent of other external influences," *Hoppe-Seyler's Zeitschrift für Physiologische Chemie*, vol. 360, no. 11, pp. 1657–1670, 1979.

[23] T. Schulze, S. Schreiber, D. Iliev et al., "The heme-binding protein SOUL3 of *Chlamydomonas reinhardtii* influences size and position of the eyespot," *Molecular Plant*. In press.

[24] F. Preuss, J. Y. Fan, M. Kalive et al., "*Drosophila doubletime* mutations which either shorten or lengthen the period of circadian rhythms decrease the protein kinase activity of casein kinase I," *Molecular and Cellular Biology*, vol. 24, no. 2, pp. 886–898, 2004.

[25] A. J. Link, J. Eng, D. M. Schieltz et al., "Direct analysis of protein complexes using mass spectrometry," *Nature Biotechnology*, vol. 17, no. 7, pp. 676–682, 1999.

[26] M. Wirschell, T. Hendrickson, and W. S. Sale, "Keeping an eye on I1:I1 dynein as a model for flagellar dynein assembly and regulation," *Cell Motility and the Cytoskeleton*, vol. 64, no. 8, pp. 569–579, 2007.

[27] N. F. Wilson and P. A. Lefebvre, "Regulation of flagellar assembly by glycogen synthase kinase 3 in *Chlamydomonas reinhardtii*," *Eukaryotic Cell*, vol. 3, no. 5, pp. 1307–1319, 2004.

[28] E. Harms, M. W. Young, and L. Saez, "CK1 and GSK3 in the *Drosophila* and mammalian circadian clock," *Novartis Foundation Symposium*, vol. 253, pp. 267–277, 2003.

[29] A. Gururaj, C. J. Barnes, R. K. Vadlamudi, and R. Kumar, "Regulation of phosphoglucomutase 1 phosphorylation and activity by a signaling kinase," *Oncogene*, vol. 23, no. 49, pp. 8118–8127, 2004.

[30] M. A. Pajares, C. Durán, F. Corrales, and J. M. Mato, "Protein kinase C phosphorylation of rat liver S-adenosylmethionine synthetase: dissociation and production of an active monomer," *Biochemical Journal*, vol. 303, no. 3, pp. 949–955, 1994.

[31] T. W. Hendrickson, C. A. Perrone, P. Griffin et al., "IC138 is a WD-repeat dynein intermediate chain required for light chain assembly and regulation of flagellar bending," *Molecular Biology of the Cell*, vol. 15, no. 12, pp. 5431–5442, 2004.

[32] K. E. VanderWaal, R. Yamamoto, K. Wakabayashi et al., "bop5 mutations reveal new roles for the IC138 phosphoprotein in the regulation of flagellar motility and asymmetric waveforms," *Molecular Biology of the Cell*, vol. 22, no. 16, pp. 2862–2874, 2011.

[33] S. Takada, C. G. Wilkerson, K. I. Wakabayashi, R. Kamiya, and G. B. Witman, "The outer dynein arm-docking complex: composition and characterization of a subunit (Oda1) necessary for outer arm assembly," *Molecular Biology of the Cell*, vol. 13, no. 3, pp. 1015–1029, 2002.

[34] A. M. Curry and J. L. Rosenbaum, "Flagellar radial spoke: a model molecular genetic system for studying organelle assembly," *Cell Motility and the Cytoskeleton*, vol. 24, no. 4, pp. 224–232, 1993.

[35] P. Yang, D. R. Diener, C. Yang et al., "Radial spoke proteins of *Chlamydomonas* flagella," *Journal of Cell Science*, vol. 119, part 6, pp. 1165–1174, 2006.

[36] G. Keryer, Z. Luo, J. C. Cavadore, J. Erlichman, and M. Bornens, "Phosphorylation of the regulatory subunit of type IIβ cAMP-dependent protein kinase by cyclin B/p34(cdc2) kinase impairs its binding to microtubule-associated protein 2," *Proceedings of the National Academy of Sciences of the United States of America*, vol. 90, no. 12, pp. 5418–5422, 1993.

[37] S. Manni, J. H. Mauban, C. W. Ward, and M. Bond, "Phosphorylation of the cAMP-dependent protein kinase (PKA) regulatory subunit modulates PKA-AKAP interaction, substrate phosphorylation, and calcium signaling in cardiac cells," *The Journal of Biological Chemistry*, vol. 283, no. 35, pp. 24145–24154, 2008.

[38] A. R. Gaillard, L. A. Fox, J. M. Rhea, B. Craige, and W. S. Sale, "Disruption of the A-kinase anchoring domain in flagellar radial spoke protein 3 results in unregulated axonemal cAMP-dependent protein kinase activity and abnormal flagellar motility," *Molecular Biology of the Cell*, vol. 17, no. 6, pp. 2626–2635, 2006.

[39] E. H. Harris, *The Chlamydomonas Sourcebook*, vol. 3, Academic Press, San Diego, Calif, USA, 2009.

[40] M. A. Price and D. Kalderon, "Proteolysis of the Hedgehog signaling effector Cubitus interruptus requires phosphorylation by Glycogen Synthase Kinase 3 and Casein Kinase 1," *Cell*, vol. 108, no. 6, pp. 823–835, 2002.

[41] L. Kim, J. Liu, and A. R. Kimmel, "The novel tyrosine kinase ZAK1 activates GSK3 to direct cell fate specification," *Cell*, vol. 99, no. 4, pp. 399–408, 1999.

[42] H. Murai, M. Okazaki, and A. Kikuchi, "Tyrosine dephosphorylation of glycogen synthase kinase-3 is involved in its extracellular signal-dependent inactivation," *FEBS Letters*, vol. 392, no. 2, pp. 153–160, 1996.

[43] S. Panda, J. B. Hogenesch, and S. A. Kay, "Circadian rhythms from flies to human," *Nature*, vol. 417, no. 6886, pp. 329–335, 2002.

[44] M. L. Spengler, K. K. Kuropatwinski, M. Schumer, and M. P. Antoch, "A serine cluster mediates BMAL1-dependent CLOCK phosphorylation and degradation," *Cell Cycle*, vol. 8, no. 24, pp. 4138–4146, 2009.

[45] M. Mittag, S. Kiaulehn, and C. H. Johnson, "The circadian clock in *Chlamydomonas reinhardtii*. What is it for? What is it similar to?" *Plant Physiology*, vol. 137, no. 2, pp. 399–409, 2005.

11

A Cotton-Fiber-Associated Cyclin-Dependent Kinase A Gene: Characterization and Chromosomal Location

Weifan Gao,[1] Sukumar Saha,[2] Din-Pow Ma,[1] Yufang Guo,[3] Johnie N. Jenkins,[2] and David M. Stelly[4]

[1] *Department of Biochemistry, Molecular Biology, Entomology, and Plant Pathology, Mississippi State University, Mississippi State, MS 39762, USA*
[2] *USDA/ARS Crop Science Research Laboratory, P.O. Box 5367, Mississippi State, MS 39762, USA*
[3] *Department of Plant and Soil Sciences, Mississippi State University, Mississippi State, MS 39762, USA*
[4] *Department of Soil and Crop Sciences, Texas A&M University, College Station, TX 77845, USA*

Correspondence should be addressed to Din-Pow Ma, dm1@ra.msstate.edu

Academic Editor: Akhilesh Kumar Tyagi

A cotton fiber cDNA and its genomic sequences encoding an A-type cyclin-dependent kinase (*GhCDKA*) were cloned and characterized. The encoded GhCDKA protein contains the conserved cyclin-binding, ATP binding, and catalytic domains. Northern blot and RT-PCR analysis revealed that the *GhCDKA* transcript was high in 5–10 DPA fibers, moderate in 15 and 20 DPA fibers and roots, and low in flowers and leaves. GhCDKA protein levels in fibers increased from 5–15 DPA, peaked at 15 DPA, and decreased from 15 t0 20 DPA. The differential expression of *GhCDKA* suggested that the gene might play an important role in fiber development. The *GhCDKA* sequence data was used to develop single nucleotide polymorphism (SNP) markers specific for the *CDKA* gene in cotton. A primer specific to one of the SNPs was used to locate the *CDKA* gene to chromosome 16 by deletion analysis using a series of hypoaneuploid interspecific hybrids.

1. Introduction

Cotton fibers are unicellular seed trichomes differentiated from the outer integument of a developing seed. The regulation of cell division is thus an important aspect of fiber initiation and development. About 25% of commercial cotton ovule epidermal cells stops division and develops to produce fibers [1]. It has been reported that the cell cycle in fiber cells is arrested in the G1 phase during the early stages of fiber development [2]. A central role in the regulation of the cell division is played by cyclin-dependent kinases (CDKs) and their regulatory cyclin subunits [3–5]. Eleven types of cyclins (A, B, C, D, H, CycJ18, L, T, U, SDS (solo dancers), and P) have been identified in plants [6, 7]. Plant CDKs, identified in 23 species of algae, gymnosperms, and angiosperms, contain three functional domains: an ATP-binding domain, a cyclin-binding domain, and a catalytic domain. They are classified into five types (A, B, C, D, and E) based on their sequence differences in the cyclin-binding domain [8]. The A-type CDK (CDKA) proteins are characterized by the presence

of the PSTAIRE motif, which is essential for cyclin binding [9]. Plant CDKAs, but not CDKBs, have been shown to complement yeast CDK mutants [10–13], suggesting that plant CDKAs are functional homologues of the yeast CDK. Plant CDKAs not only control cell cycle progression from the G1 to S phase and from the G2 to M phase [5, 14] but also participate in cell proliferation and maintenance of cell division competence in differentiated tissues during development [15]. Since the *CDKA* gene is expressed in both dividing and differential tissues [15, 16], it has been suggested that the gene is involved in both cell division and differentiation [17, 18].

To dissect the possible functional role of CDKA in fiber cell differentiation and development, we have cloned and characterized a fiber *CDKA* cDNA and its corresponding genomic sequences. The expression levels of the *CDKA* transcript and the CDKA protein were also determined in elongating cotton fibers from 5 to 20 DPA ovules and other tissues. The *CDKA* sequence data was then used to develop single nucleotide polymorphism (SNP) markers specific for the

CDKA gene(s) in cotton. Lastly, a primer specific to one of the SNPs was used with single primer extension technology to locate the *CDKA* gene to chromosome 16 by deletion analysis using a series of hypoaneuploid interspecific hybrids.

2. Materials and Methods

2.1. Cloning of Fiber GhCDKA cDNA. Two degenerate primers (CDK1: 5′-ATHGGDGARGGHACHTAYGG-3′ and CDK2: 5′-CKATCWATCARYARRTTYTG-3′) (H: A + C + T, D: A + G + T, R: A + G, Y: C + T, K: G + T, W: A + T) designed from the conserved ATP-binding and catalytic domains of plant *CDKA* genes were used for PCR to amplify cDNA with homology to the *CDKA* gene using total cDNA from a cotton (*Gossypium hirsutum* L. cv. DES119) fiber cDNA library as template. The cDNA library was constructed using 10 DPA (days post-anthesis) fiber RNA with a Marathon cDNA amplification kit (BD Biosciences, San Jose, CA, USA). A 383 bp DNA fragment was amplified, purified using a QIAEX II gel extraction kit (Qiagen), cloned into pGEM-T Easy Vector (Promega), and sequenced with an ABI PRISM 310 Genetic Analyzer. The DNA sequencing data was analyzed using the BLAST program (NCBI) and LASERGENE software (DNASTAR). Analysis of the sequencing data showed that the 383 bp DNA fragment encoded an A-type CDK. Two gene specific primers CDKC-1 (5′GGCGTTGTTTATAAGGCTCGTGATCGTG-3′) and CDKC-2 (5′CATTCCTTTATCAAATTCTCCGTG-GTG-3′) were designed from the 383 bp DNA fragment and used to amplify a full-length *GhCDKA* cDNA by the Rapid Amplification of cDNA Ends (RACE) method with the Marathon cDNA Amplification kit. In the 3′ RACE reaction, CDKC-1 and the adaptor primer AP1 (5′-CCA-TCCTAATACGACTCACTATAGGGC-3′, 10 μM) were used in the first PCR, and CDKC-2 and the adaptor primer AP2 (5′-ACTCACTATAGGGCTCGAGCGGC-3′) were used in the second (nested) PCR. The 5′ RACEs were also performed as 3′ RACEs, except that primers CDK5-1 (5′-GACACTTTCTCAGGAAGATAGTTG-3′) and CDKC-3 (5′-CCCTATGAGAGTGACAATAAGCAATG-3′) were used in the first and second RACE amplifications, respectively. A full-length *GhCDKA* cDNA was assembled using the 5′ and 3′ RACE products and subsequently confirmed by PCR using *Pfu* DNA polymerase (Stratagene).

2.2. Isolation of the Genomic Sequence of the GhCDKA Gene. Two primers CDKC-1 and CDK5-1 were used in LA (long and accurate) PCR to amplify DES119 genomic DNA with the Takara LA PCR kit ver.2.1. The PCR was conducted with an initial denaturation at 94°C for 4 min, followed by 30 cycles at 94°C for 30 sec and 68°C for 4 min and a final extension at 68°C for 5 min. A 7547 bp DNA fragment containing the *GhCDKA* gene was amplified. The PCR product was gel purified and cloned, and both DNA strands are sequenced as described above.

The 5′ and 3′ flanking regions of the *GhCDKA* gene were amplified using a PCR-based genomic DNA walking method and inverse PCR. Genomic walking was conducted by amplifying the adaptor-ligated genomic libraries using gene-specific primers GSR-1 (5′-TGAGTTGTGCAGTGAAGT-GCATTG-3′) and GSR-2 (5′-CTCTAATTGCAGTGCTAG-GTACAC-3′). The self-ligated genomic DNA (previously restricted with *Hin*d III) was used as template in the inverse PCR amplification with primers GSF (5′-TCTGGAAGC-GGAAAGAAGCA-3′) and GSR-1 and LA *Taq* DNA polymerase (see Figure 1 in Supplementary Material available online at doi:10.1155/2012/613812.)

2.3. Expression Analyses of the GhCDKA Gene. Total RNA (10 μg) isolated from various cotton tissues were electrophoresed in a formaldehyde/agarose gel, transferred onto a nylon membrane, and fixed by UV-crosslinking. A 618 bp DNA fragment corresponding to the C-terminal and 3′-UTR region of the *GhCDKA* cDNA was amplified by PCR using two primers CDKC-2 and CDK5-1, labeled with [α-^{32}P] dCTP with the random priming labeling method, and used as a probe for Northern hybridization. After hybridization, the membrane was stringently washed and exposed to X-ray film for autoradiography. The relative *GhCDKA* transcript levels were determined by the ratio of radioactive intensity of hybridized band of the 1.2 kb *GhCDKA* mRNA to the EtBr stained 28S rRNA using the program of Scion Image for Windows (Scion Corporation). The *GhCDKA* transcript level was also determined by RT-PCR. First strand cDNA, labeled by [α-^{32}P] dCTP, was synthesized with SuperScript II reverse transcriptase (Invitrogen) using oligo-dT primer and total RNA (2 μg) isolated from flowers, leaves, roots, and 5, 10, and 15 DPA fibers. An equal amount of the synthesized first strand cDNA (based on scintillation counting) from different samples was serially diluted to 1x, 5x, 10x, 20x with sterile distilled water and used as template for PCR amplification with primers CDKC-2 and CDK5-1. Five microliters of the PCR products was analyzed by electrophoresis in a 1% agarose gel.

For Western analysis, 70 μg of total protein extracted from cotton flowers, leaves, and fibers (5, 10, 15, and 20 DPA) with a modified method of Barent and Elthon [19] was vacuum dried and resuspended in SDS-PAGE sample buffer (12 mM Tris-HCl, pH 6.8, 5% (v/v) glycerol, 0.4% (w/v) SDS, 1% (v/v) β-mercaptoethanol, 0.02% (w/v) bromophenol blue). The samples were heat denatured, separated by 12% SDS-PAGE, and transferred onto a nitrocellulose membrane. Immunodetection of the GhCDKA protein was carried out with an ECL Western blotting system (GE Healthcare) using rabbit anti-PSTAIRE (Santa Cruz Biotechnology) as primary antibody and anti-rabbit IgG-horseradish peroxidase conjugate (GE Healthcare) as secondary antibody.

2.4. SNP Analyses and Chromosomal Location of the GhCDKA Gene. Genomic DNAs were extracted from young leaves of CMD-01 (TM-1, *G. hirsutum*), CMD-02 (3–79, *G. barbadense*), CMD11 (*G. tomentosum*), CMD-3 (*G. arboreum*), and CMD-5 (*G. raimondii*) using a DNeasy Plant mini kit (Qiagen). These *Gossypium* genotypes have been widely used for the screening and preliminary characterization of cotton microsatellite markers [20]. The genomic DNA samples were amplified by *pfu* DNA polymerase with two primers (CDKP3, 5′-GGCTGGTTATGTTGTGGTAGTACTG-3′ (nt-913 to -889)); and CDKP4, 5′-GTGCAGCTCCACCAG-ACGAGAAG-3′ (nt-1 to -23)) designed from 5′-flanking region upstream of the start codon ATG of the *GhCDKA* gene. The amplified PCR products were gel purified, cloned, and then sequenced. The sequence of DES 119 (*G. hirsutum*) was then aligned with those of TM-1 (*G. hirsutum*), 3–79 (*G. barbadense*), and CMD11 (*G. tomentosum*) using the Clustal method (DNASTAR software) for SNP identification.

The chromosomal location of the *CDKA* gene was determined by following the overall strategy of Liu et al. [21] using hypoaneuploid chromosome substitution stocks (BC_0F_1) and a euploid $BC_5F_1S_1$ chromosome substitution line of TM-1 disomic for the chromosome 16 of *G. barbadense* [22]. The monotelodisomics included telosomes 1Lo, 2Lo, 2sh, 3Lo, 3sh, 4sh, 5Lo, 6Lo, 7Lo, 7sh, 9Lo, 11Lo, 14Lo, 15Lo, 15sh, 16sh, 16Lo, 18Lo, 18sh, 20Lo, 22sh, 25Lo, and 26sh, where Lo = long arm and sh = short arm. Monosomes included chromosomes 1, 2, 3, 4, 6, 7, 9, 10, 12, 17, 18, 20, 23, and 25. Each interspecific hybrid is expectedly heterozygous for all polymorphisms between the two parents, except those rendered hemizygous by the monosome- or telosome-defined deficiency. At hemizygous loci, the *G. hirsutum* allele is expectedly absent and only the *G. barbadense* allele is present. The telosomes expectedly lack all or most of the opposing arm, for example, an F_1 plant monotelodisomic for 6Lo will be hemizygous for *G. barbadense* polymorphisms in the short arm distal to the telosome breakpoint. We used cytologically identified BC5-derived inbred euploid backcross substitution line for chromosome 16 of *G. barbadense* in *G. hirsutum* in lieu of an available monosomic BC_0F_1 plant. The disomic chromosome substitution line is euploid but has one pair of chromosome 16 from *G. barbadense* line 3–79, whereas the other 25 chromosome pairs are largely or completely derived from TM-1.

A SNP primer (5′-GCCCAACTATAGAAATGAAA-3′) designed based on a single nucleotide differences in the sequences between the lines among the three *Gossypium* species (*G. hirsutum*, *G. barbadense*, and *G. tomentosum*) was used to screen SNP markers of the genetic stocks with the ABI Prism SNaPshot multiplex kit following the method of Buriev et al. [23]. Briefly, the *pfu*-amplified PCR products were incubated with SAP and *Exo* I (5 units of SAP and 2 units of *Exo* I for 15 μL PCR product) at 37°C for 1 hr followed by 75°C for 15 min. The PCR mixture contained 5 μL of SnaPshot Multiplex Ready Reaction Mix, 3 μL of purified PCR product, 1 μL of SNP primer (10 μM), and 1 μL of distilled water. The thermal cycle reaction was carried out with 25 cycles of 96°C, 10 sec, 50°C, 5 sec, and 60°C, 30 sec. After treated with SAP, 0.5 μL of SnaPshot product was mixed with 0.5 μL of size standard and 9 μL of Hi-Di formamide denatured at 95°C for 5 min and then run onto a 3100 Genetic Analyzer (Applied Biosystems).

3. Results

3.1. Cloning and Characterization of GhCDKA Gene. A 383 bp DNA fragment was amplified by PCR from a 10 DPA cotton fiber cDNA library using two degenerate primers designed from the conserved ATP-binding and catalytic domains of plant A-type *CDK* genes. BLAST searching in GenBank Databases indicated that the 383 bp cDNA encoded a protein with extensive homology to plant A-type CDKs. A full-length fiber *CDKA* cDNA (1211 bp), named *GhCDKA*, was subsequently cloned by 5′ and 3′ RACEs using gene-specific primers designed from the 383 bp fragment. The *GhCDKA* gene and its 5′ flanking region (9675 bp) (Supplementary Figure 1) were cloned by genomic walking and inverse PCR. *GhCDKA* encodes a protein of 294 aa with a predicted molecular mass of 34 kDa. The protein contained three conserved functional domains of CDK proteins: an ATP-binding domain, a cyclin-binding domain, and a catalytic domain. The GhCDKA protein also had the conserved PSTAIRE motif found in A-type CDKs in the cyclin binding domain. Comparisons of the cDNA and genomic sequences revealed that the *GhCDKA* gene contained 9 exons and 8 introns with 7 introns located within the coding region and one intron at the 5′ UTR region (Supplementary Figure 1). The *GhCDKA* gene had the same number and sizes of exons and the same number of introns as the *Arabidopsis CDKA; 1* gene (*AtCDKA; 1*, Genbank GI: 18408695), but the sizes of introns were much larger than those of *Arabidopsis* (Figure 1). The alignment of aa sequences of CDKA proteins from cotton (GhCDKA) and ten other plant species, including *Populus tremula* x *Populus tremuloides* (PtCDKA), *Helianthus annuus* (HaCDKA), *Picea abies* (PaCDKA), *Solanum lycopersicon* (LeCDKA; 1), *Pinus contorta* (PncCDKA), *Chenopodium rubrum* (CrCDKA), *Helianthus tuberosus* (HtCDKA), *Antirrhinum majus* (AmCDKA), *Nicotiana tobacum* (NtCDKA), and *Arabidopsis thaliana* (AtCDKA; 1) revealed that GhCDKA was 91.5–94.2% identical to PtCDKA, PaCDKA, HaCDKA, LeCDKA; 1, CrCDKA, PncCDKA, HtCDKA, AmCDKA, and NtCDK and 86.7% identical to AtCDKA; 1 (data not shown). Phylogenetic analysis of aa sequences of the 11 plant CDKA proteins indicated that GhCDKA was distant to AtCDKA; 1 but closer to the other nine CDKAs (Figure 2).

3.2. Expression of the GhCDKA Gene. The mRNA abundance of the *GhCDKA* gene was analyzed by Northern blot with total RNA isolated from flowers, leaves, roots, and fibers at different developmental stages (5, 10, 15, and 20 DPA). The 618 bp DNA fragment corresponding to the C-terminal and 3′-UTR region of *GhCDKA* cDNA (Supplementary Figure 1) was amplified by PCR with two primers CDKC-2 and CDK5-1 and used as a probe for Northern hybridization. Northern blotting had been performed three times, and the results were similar as shown in Figure 3(a), a 1.2 kb *GhCDKA* mRNA band was detected in all tissues. The *GhCDKA* transcript levels were high in 5 and 10 DPA fibers, moderate in 15 and 20 DPA fibers and roots, and low in flowers and leaves. The *GhCDKA* transcript level was also determined by RT-PCR. As shown in Figure 3(b), the amounts of 618 bp PCR products amplified with the primers CDKC-2 and CDK5-1 were proportional to the first strand cDNA input. The RT-PCR results indicated that transcript levels of the *GhCDKA*

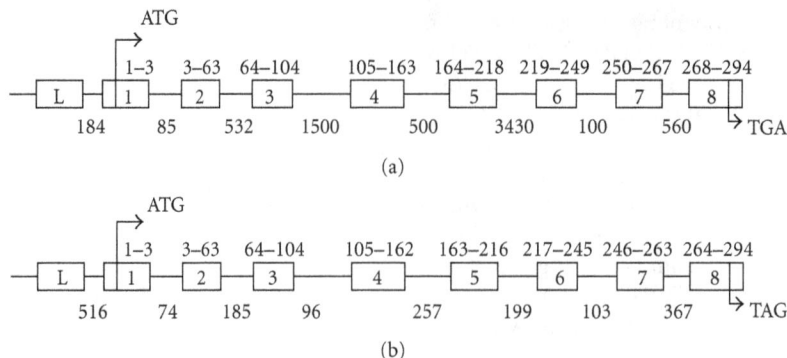

FIGURE 1: Diagrammatic comparison of cotton *GhCDKA* gene (a) with *Arabidopsis thaliana AtCDKA; 1* Gene ((b), Genbank GI: 18408695). ATG represents the start codon. TGA and TAG are stop codons. The exons containing the coding regions are boxed (numbers 1–8). The exons located in the 5'-UTR region are represented by the L boxes. The positions of codons at the 5' and 3' ends within exons are indicated. Intron sizes are indicated under the intron lines.

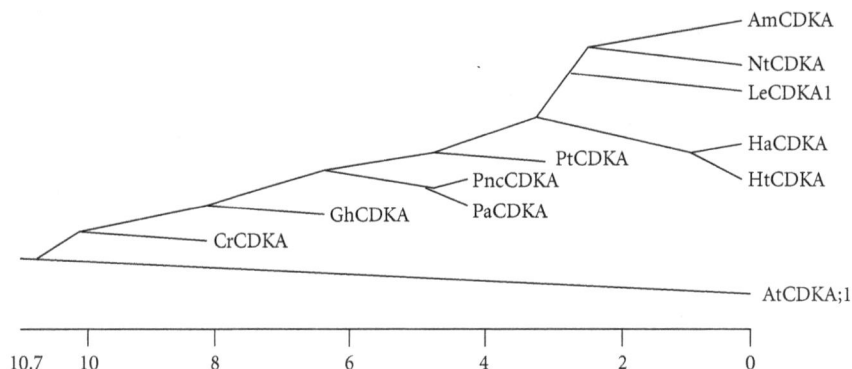

FIGURE 2: Phylogenetic analysis of eleven plant CDKA proteins. The phylogenetic tree was constructed based on amino acidsequences of 11 plant CDKAs using the Clustal method. The eleven CDKA proteins are GhCDKA (*Gossypium hirsutum*); PtCDKA (*Populus tremula x Populus tremuloides*); HaCDKA (*Helianthus annuus*); PaCDKA (*Picea abies*); LeCDKA1 (*Solanum lycopersicum*); PncCDKA (*Pinus contorta*); CrCDKA (*Chenopodium rubrum*); HtCDKA (*Helianthus tuberosus*); AmCDKA (*Antirrhinum majus*); NtCDKA (*Nicotiana tobacum*); AtCDKA; 1 (*Arabidopsis thaliana*).

gene were high in 5 and 10 DPA fibers, moderate in 15 DPA fibers and roots, and low in flowers and leaves. The RT-PCR result was consistent with Northern analyses.

Total protein isolated from 5, 10, 15, and 20 DPA cotton fibers, flowers, and leaves was separated by SDS-PAGE, electroblotted onto a nitrocellulose membrane, and probed with anti-PSTAIRE antibody. Western analysis showed that the antibody recognized a 34 kDa protein in all cotton tissues (Figure 4). The GhCDKA protein was present in a moderate level in leaves but low in flowers. The GhCDKA protein in fibers increased from 5 DPA, peaked at 15 DPA, and decreased from 15 to 20 DPA. The Western and Northern results suggest that the *GhCDKA* gene is differentially expressed and developmentally regulated.

3.3. *Identification of SNP in GhCDKA*. Analyses of PCR-amplified products from TM-1 (*G. hirsutum*), 3–79 (*G. barbadense*), CMD11 (*G. tomentosum*), CMD-5 (*G. raimondii*), and CMD-12 (*G. mustellinum*) by agarose gel electrophoresis revealed that the products were 0.9 kb in size and not discernibly polymorphic (data not shown). Genomic DNA

of CMD-3 (*G. arboreum*) did not yield an amplified product with *CDKA*-specific primers, although this DNA was amplified with other control primers (unpublished information).

The 0.9 kb 5' flanking sequence of the *CDKA* gene amplified from genomic DNA of CMD-01 (TM-1, *G. hirsutum*), CMD-02 (3–79, *G. barbadense*), and CMD-11 (*G. tomentosum*), respectively, was aligned with *G. hirsutum* var. DES 119 (Figure 5) for SNP identification. The incidence of SNP was about 1% in the -1 to -913 nt region of the *CDKA* gene. Specifically, we observed two indels, four transversions and three transitions type of mutation in the 5' flanking sequences of the *CDKA* gene (Figure 5). Two SNP occurred between the two *G. hirsutum* genotypes and six SNP occurred between *G. barbadense* and *G. hirsutum*. Results suggested that a putative *CDKA* locus with at least four different haplotype variants was present in the tetraploid cotton species.

3.4. *SNP Marker*. To develop a primer for a potentially scorable SNP marker, we targeted a deletion (G) site at nucleotide position 769 (Figure 5), as it distinguished the

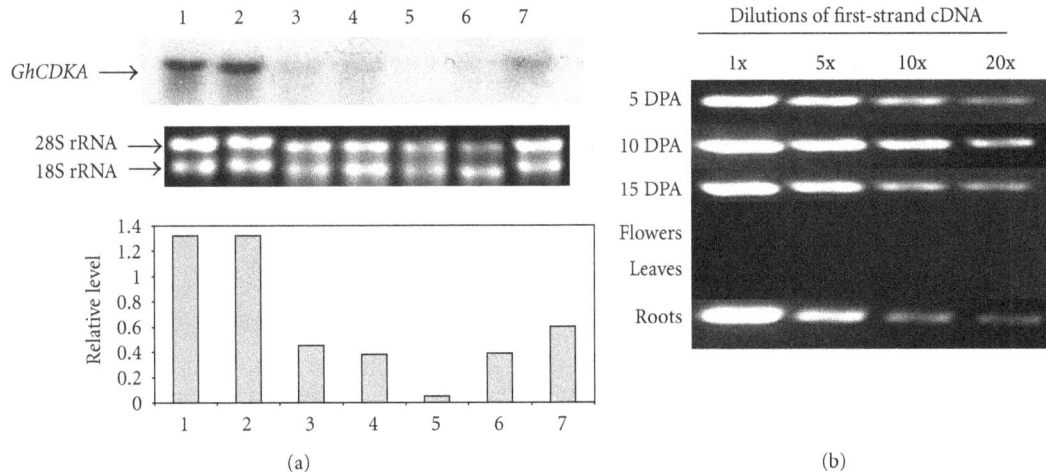

FIGURE 3: (a) Northern analysis of *GhCDKA* expression in different cotton tissues. Ten μg of total RNA from fibers (5, 10, 15, and 20 DPA, lanes 1–4), flowers (lane 5), leaves (lane 6), and roots (lane 7) was electrophoresed on an agarose gel, transferred onto a nylon membrane, and hybridized with a ³²P-labeled *GhCDKA* cDNA. Two EtBr-stained rRNA bands indicate that an equal amount of total RNA was loaded for each sample. The relative *GhCDKA* transcript levels were determined by the ratio of hybridized intensity of the 1.2 kb *GhCDKA* mRNA to the EtBr stained 28S rRNA band using the program of Scion Image for Windows (Scion Corporation). (b) RT-PCR analysis of *GhCDKA* mRNA. Total RNA from leaves, flowers, roots, and 5, 10, and 15 DPA fibers was used as template in generating first strand cDNA. Each cDNA was made 1x, 5x, 10x, and 20x dilutions and used as template for PCR amplification with two *GhCDKA* gene specific primers: CDKC-2 and CDK5-1.

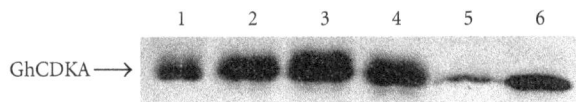

FIGURE 4: Western blot analysis of GhCDKA. Seventy μg of total protein was subjected to SDS-PAGE, blotted onto a nitrocellulose membrane, and probed with anti-PSTAIRE antibody. Proteins samples are from 5 DPA (lane 1), 10 DPA (lane 2), 15 DPA (lane 3), and 20 DPA (lane 4) fibers, flowers (lane 5), and leaves (lane 6).

3–79 *CDKA* sequence from those of the other tetraploids. The sequence of this specific SNP primer was 5'-GCCCAA-CTATAGAAATGAAA-3'. Two SNPs corresponding to the TM-1 (*G. hirsutum*) and 3–79 (*G. barbadense*) alleles were identified by the single primer extension technology and designated here as *CDKA^cg* (black) and *CDKA^at* (green) (Figure 6). F₁ hybrids between TM-1 and 3–79 exhibited peaks of both alleles, showing codominance. Our results also detected the presence of *CDKA^cg* allele in *G. tomentosum* and the presence of both *CDKA^cg* and *CDKA^at* alleles in the diploid species of *G. raimondii* (D5). We did not find the presence of any other bases except G or T as SNP markers specific to this SNP primer, suggesting that this locus was biallelic. We did not find the presence of any *CDKA*-specific SNP marker using the genomic DNA of *G. arboreum* (A2) species, suggesting the absence of any such locus in *G. arboreum* specific to the SNP primer or a major change in the primer annealing site of this marker in *G. arboreum*. This result was concordant with the absence of amplified products specific to *CDKA* gene in *G. arboreum* (A2) species, confirming the absence of the *CDKA* gene in *G. arboreum* (A2).

3.5. Chromosomal Location. Electropherograms revealed two peaks and thus heterozygosity for *CDKA^at* and *CDKA^cg* alleles in all of the hypoaneuploid chromosome substitution F₁ plants, except one (Figure 6). The single exception was the monotelodisomic Te16sh, which lacks all or most of the long arm of chromosome 16 and possessed the 3–79 allele, *CDKA^at*, but lacked the TM-1 allele, *CDKA^cg*. Similar results were observed for the disomic backcrossed chromosome 16 substitution line CSB 16 showing the presence of only the 3–79 alleles.

4. Discussion

As a first step toward understanding the mechanisms of fiber cell division and differentiation, a fiber cDNA, *GhCDKA*, and its corresponding gene have been cloned and characterized. The deduced aa sequence of GhCDKA shows high identity (more than 86%) to the CDKAs from 10 diverse plant species. The alignment of the 11 plant CDKAs indicates that they all contain 294 aa residues (except for 302 aa in AmCDKA) and their three functional domains (ATP-binding, cyclin-binding, and catalytic) have identical aa sequences (data not shown). These results indicate that A-type CDKs are highly conserved in higher plants. Comparisons of the cotton *CDKA* gene with the *Arabidopsis cdc2 A* (*CDKA; 1*) gene revealed that both genes contain 7 introns within their ORFs (Figure 1). Although the two *CDKA* genes encode proteins with identical molecular mass, the intron sizes of the two genes are quite different. It will be interesting to examine whether there are any differences in transcriptional regulation or RNA splicing between the two genes. A genome-wide analysis of cell cycle genes indicated that a single *CDKA* gene (*AtCDKA: 1*) exists in *Arabidopsis*

```
DES119 (G. hirsutum)  GGCTGGTTATGTTGTGGTAGTACTGTCAATATTAACTGTGACTGTTTTAGCTATGCATATGTCTTATTGTGATAGCTTAT 80
TM1 (G. hirsutum)     ------------------------------------------------------------------------------- 80
379 (G. barbadense)   ------------------------------------------------------------------------------- 80
CMD11 (G. tomentosum) ------------------------------------------------------------------------------- 80

DES119 (G. hirsutum)  TTTATACACAAATTGATGTAACCATTTAAAATTACTCACGAGAAATTTCATTTTATTTTGCTTCCTCATTCTTGTTGGTA 160
TM1 (G. hirsutum)     ------------------------------------------------------------------------------- 160
379 (G. barbadense)   ---------------------------------------------------------------------A------- 160
CMD11 (G. tomentosum) ------------------------------------------------------------------------------- 160

DES119 (G. hirsutum)  ATGATGATAACATTTTATTTTAACAATTGAAATTTAAGTTGGTGACGTTTATGATTGCGTTCTATAAATTTAAAACTATT 240
TM1 (G. hirsutum)     ----------------------------------------------C-------------------------------- 240
379 (G. barbadense)   -----------------------------------C--------C---------------------------------- 240
CMD11 (G. tomentosum) --------------------------------------------C--------------------------------- 240

DES119 (G. hirsutum)  ATTAATAAAGAAGATGGTGCCCCAATTTTCAGCATTTTTATTTCGCCTAAAATTATAGACTTTGTTGTAAAGATCATAGT 320
TM1 (G. hirsutum)     ------------------------------------------------------------------------------- 320
379 (G. barbadense)   ------------------------------------------------------------------------------- 320
CMD11 (G. tomentosum) ---------------------T-------------------------------------------------------- 320

DES119 (G. hirsutum)  TTCGAGGAACGATGTATTAATGGATGACATAAATACTAATATTTATTTTTTATTTTTTGGTACAATTTAATTTTGAATTT 400
TM1 (G. hirsutum)     ------A------------------------------------------------------------------------- 400
379 (G. barbadense)   ------A------------------------------------------------------------------------- 400
CMD11 (G. tomentosum) ------A------------------------------------------------------------------------- 400

DES119 (G. hirsutum)  TATCCAGTTAATATTATTGTCTAAGATAAAAAAATACCATAAAAATATTATTTGG CAGAAAAGAAAATATCGGAAGGGAG 479
TM1 (G. hirsutum)     ------------------------------------------------------- --------------------- 479
379 (G. barbadense)   -----------------------------------------------------------G------------------ 480
CMD11 (G. tomentosum) -----------------------------------------------------------G------------------ 480

DES119 (G. hirsutum)  AGCGAGAGATAAGTCAACTCAAAATGGGTCACAACGAGTCTCTCGGTTTTACTTTTAGGGTTGATGCTTGAGAGCATCAT 559
TM1 (G. hirsutum)     ------------------------------------------------------------------------------- 559
379 (G. barbadense)   ------------------------------------------------------------------------------- 560
CMD11 (G. tomentosum) ----------------------------------------A------------------------------------- 560

DES119 (G. hirsutum)  TTCTTACTCTAAGTACTTAAATAGAAATAAGAGATACATACATCTCGTCTCTACCTCCTCTCCTCTCCTCAACTTACTCA 639
TM1 (G. hirsutum)     ------------------------------------------------------------------------------- 639
379 (G. barbadense)   ------------------------------------------------------------------------------- 640
CMD11 (G. tomentosum) ------------------------------------------------------------------------------- 640

DES119 (G. hirsutum)  CTCTCTCGCTTGAAATCAAATCCGACTGTCACTTGAGGTACTTCTCTTCTCACATTGCATTTTACTTCTATTGATTTTAT 719
TM1 (G. hirsutum)     ------------------------------------------------------------------------------- 719
379 (G. barbadense)   ------------------------------------------------------------------------------- 720
CMD11 (G. tomentosum) ------------------------------------------------------------------------------- 720

DES119 (G. hirsutum)  TTTCTCTTGTTTGAATTGGATGCATTCTTATTCATTTGCTTGCTTCCCTGTTTCATTTCTATAGTTGGGCTTTTCAAGGA 799
TM1 (G. hirsutum)     ------------------------------------------------------------------------------- 799
379 (G. barbadense)   ------------------------------------------------------------------------------- 799
CMD11 (G. tomentosum) ------------------------------------------------------------------------------- 800

DES119 (G. hirsutum)  TTACAATCAGTTCATATTTCCTAGATTCATGTTTGAGATGAGTTAATTGTTTTTTTCCCTCTTTGTTTTAGCTTATTTGT 879
TM1 (G. hirsutum)     ------------------------------------------------------------------------------- 879
379 (G. barbadense)   -----------C------------------------------------------------------------------- 879
CMD11 (G. tomentosum) ------------------------------------------------------------------------------- 880

DES119 (G. hirsutum)  ACATTTTGATTCTTCTCGTCTGGTGGAGCTGCAC                                          913
TM1 (G. hirsutum)     ----------------------------------                                          913
379 (G. barbadense)   ----------------------------------                                          913
CMD11 (G. tomentosum) ----------------------------------                                          914
```

FIGURE 5: Alignment of 5′ flanking sequences of *CDKA* gene of DES119 (*G. hirsutum*), TM-1 (*G. hirsutum*), 3–79 (*G. barbadense*), and CMD11 (*G. tomentosum*) showing the presence of several SNPs. The arrow indicated the position and the direction of the SNP primer specific to the *CDKA* gene.

thaliana [24]. In contrast, multiple copies of two genes (*LeCDKA1* and *LeCDKA2*) encoding A-type CDKs have been found in tomato [25]. *Nicotiana tabacum* contains a single copy of the *CDKA* gene (*NtCDKA*) and at least one gene similar to *NtCDKA* in the genome [26]. In this study, Southern analysis revealed that one or two copies of the *GhCDKA* gene are present in cotton (*Gossypium hirstum*) (data not shown). *Gossypium hirstum* is a tetraploid plant which contains A and D genomes. Further work is needed to determine whether the *GhCDKA* gene is located in the A or D or both genomes.

The *Arabidopsis* and rice *CDKA* genes have been shown to be expressed not only in dividing tissues of root apex but also in differentiated tissues, such as, sclerenchyma, pericycle, and parenchyma of the vascular cylinder [15, 16]. These results suggest that A-type CDKs are involved not only in cell division but also in cell differentiation which is important to the integration of cell division and differentiation in meristems to produce new organs during plant development. In contrast, no *cdc2* (*CDKA*) transcripts have been detected in differentiated adult tissues of chicken and *Drosophila* [27, 28]. These findings suggest that plant CDKAs may have different functions from those of animals. The *Arabidopsis CDKA; 1* gene (*AtCDKA; 1*) has been shown to participate in trichome morphogenesis and development [29]. Fiber cells grown *in planta* do not divide after initiation; however, some

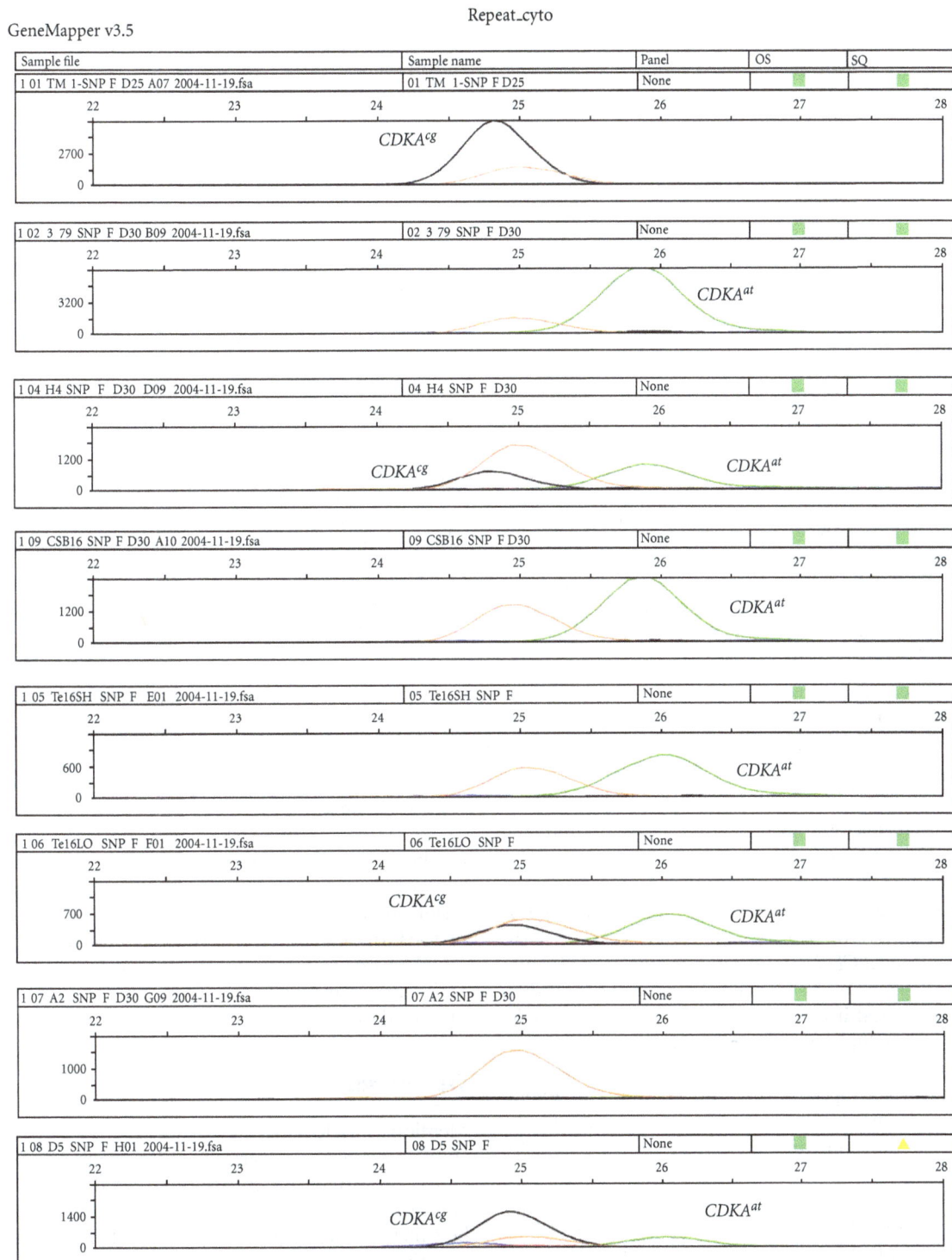

FIGURE 6: The electropherograms of two allelic SNPs, designated here as *CDKA^cg* (black) and *CDKA^at* (green), that corresponded to the polymorphism between *G. hirsutum* inbred TM-1 and *G. barbadense*-doubled haploid 3–79. Genomic dosage profiles are shown for [A] TM-1; [B] 3–79; [C–E] three hypoaneuploid-interspecific *G. hirsutum* x *G. barbadense* F₁ hybrids, [C] lacking *G. hirsutum* chromosome 4 (H4), [D] a monotelodisomic-16sh F1 (Te16SH) lacking most of the long arm of *G. hirsutum* chromosome 16, [E] a monotelodisomic 16 Lo F₁ (Te16LO) lacking most of the short arm of *G. hirsutum* chromosome-16; [F] a backcross disomic substitution plant (CSB-16) in which chromosome 16 of 3–79 has replaced the TM-1 chromosome 16; [G] a *G. arboreum* (A2 species); and [H] a *G. raimondii* plant (D5 species). Electropherograms revealing both peaks that indicate heterozygosity for both parental SNPs, as in the H4 interspecific F1 hybrid, indicate the locus is not in this chromosome. In contrast, absence of the *G. hirsutum* SNP *CDKA^cg* from CSB 16 and Te16SH hybrids and its presence in Te16LO concordantly indicate that the SNP marker is located in the long arm of the chromosome 16.

fiber cells can divide under *in vitro* conditions [1]. These observations suggest that fiber cells retain the competence to divide after initiation. In this study, the *GhCDKA* gene has been shown to be strongly expressed in elongated fibers (Figure 3). Western analysis revealed that the fiber GhCDKA protein level increased from 5 DPA, peaked at 15 DPA, and remained at a high level at 20 DPA (Figure 4), which coincided with primary and secondary cell wall syntheses during fiber development. The expression analysis results suggest that GhCDKA may play a role in fiber development. The low *GhCDKA* transcript level versus the high amount of GhCDKA protein in 20 DPA fibers suggests the possible existence of posttranscriptional regulation of the *GhCDKA* gene. In *Arabidopsis*, the transcript and protein levels of AtCDKB; 1 (but not AtCDKA; 1) have been shown to correlate with cell division rate [30].

Duplications through polyploidization and/or segmental duplication and retrotransposon activity have contributed extensively to the extant genomes of the Malvaceae, including those of *Gossypium* [31–33]. The normal plant cell cycle process is distinguished by a S phase (a round of DNA replication) followed by M phase which are separated by two gap phases (G1 and G2). Previous studies demonstrated that some plant cells followed a different cell cycle mode including endoreduplication where cells undergo iterative DNA replications without any subsequent cytokinesis [34]. Endoreduplication is usually considered to provide a mechanism for increasing cell size [35] and involved modulating the levels of CDKA activity [36, 37]. Cotton fibers are unique cells and they are differentiated from epidermal cells of the ovule. Regulation of cell cycle genes during the very early stages of fiber development triggered some specific epidermal cells in the ovule to stop cell division and then elongate into fiber cells. Previous experiments using 5-aminouracil (5-AU), an inhibitor of DNA replication, demonstrated that cotton fiber cells were arrested at the G1 stage [2]. Our results on Northern blot and RT-PCR analysis revealed that the *GhCDKA* transcript was high in 5–10 DPA fibers and moderate in 15 and 20 DPA fibers. Further studies are needed to reveal if GhCDKA is a regulator of cell cycle and DNA endoreduplication in fiber cells. Duplicated loci pose significant challenges in virtually all aspects of genomics research, including specific gene mapping in tetraploid cotton [23]. Locus-specific markers are thus particularly important for addressing these challenges, and the means to develop them are crucial to the advancement of structural genomics. One possible solution for marker development is to exploit sequence conservation of a specific gene and identify the gene in a locus-specific manner. The *CDK* gene is of special interest because of its possible importance to cotton fiber development, which entails major modifications of cell division and growth. Although cotton is clearly of polyploid origin, agarose gel analyses of amplified PCR product(s) from diverse taxa of cotton genomic DNAs using primers from conserved *CDKA* sequence regions showed no size polymorphisms. Such a result could be due to uniformity across duplicated loci or the existence of just one locus. The predicament had led us to seek SNP markers that could be used to define cotton *CDK* gene(s) and alleles in a locus-specific manner. This approach may be generally applicable for SNP development in cotton and is of particular value for genes that are highly conserved.

Deficiency tests with interspecific hypoaneuploid F1s provide a quick and fairly robust means of localizing various types of loci to specific chromosomes and arms of cotton. When we examined the hypoaneuploid F_1 hybrids used here, all but one exhibited a heterozygous banding pattern of $CDKA^{at}$ and $CDKA^{cg}$ alleles, suggesting that the *CDKA* locus was in any of respective chromosomes or chromosome arms. However, although $CDKA^{at}$ was present in the monotelodisomic Te16Lo-interspecific hybrid, it was differentially absent from the quasi-isogenic Te16sh hybrid. These results concordantly localized the *CDKA* gene to the long arm of chromosome 16. In lieu of a monosomic-interspecific F1 hybrid, we examined DNA from a euploid disomic backcross (BC_5S_n) substitution line, CS-B16 [38]. The disomic chromosome substitution line is euploid but has one pair of chromosome 16 from *G. barbadense* line 3–79, whereas the other 25 chromosome pairs are largely or completely derived from TM-1. Accordingly, CS-B16 is expectedly devoided of TM-1 chromosome-16 alleles, homozygous for all loci in the alien (3–79) chromosome-16 and also homozygous for TM-1 alleles at nearly all (~99%) other loci of the genome. We observed that only the 3–79 $CDKA^{at}$ allele is present in CS-B16, strongly supporting the finding from the monotelodisomic interspecific F1 plants that the *CDKA* gene is located on chromosome 16. Our results on the chromosomal location of *CDKA* SNP marker on chromosome 16 were concordant with the cytogenetic evidence on the origin of chromosome 16 from an ancestral D genome diploid species [39].

The identification of SNP markers enables facile tracking of the *CDKA* gene in cotton, and this gene has been successfully mapped in the long arm of chromosome 16. Our results indicate that single-primer extension technology can be used to identify SNP markers in cotton genes, including the 5′-upstream region of the genes and thus facilitate the mapping and investigation of candidate genes for their effects on fiber development.

Disclaimer

Mention of trademark or proprietary product does not constitute a guarantee or warranty of the product by the United States Department of Agriculture and does not imply its approval to the exclusion of other products that may also be suitable. The nucleotide sequence of *GhCDKA* has been submitted to GenBank and assigned an accession number EU006765.

Acknowledgments

The authors appreciate the help of Mr. Douglas Dollar, Dr. Chuan Fu, and Ms. Lillie Hendrix, for assisting in the experiments. The authors also thank Drs. Rowena Kelly of USDA/ARS, Mississippi State, Ibrokhim Abdurakhmonov,

Institute of Genetics and Plant Experimental Biology, Academy of Sciences of Uzbekistan, and Ramesh Kantety and Govind Sharma of Alabama A&M University for their critical evaluation of the paper and helpful comments. The authors also acknowledge the financial help from Plant Genome program of USDA/NRI. This paper was approved for publication as no.12145 of Mississippi Agricultural and Forestry Experiment Station (MAFES).

References

[1] J. Van't Hof and S. Saha, "Cotton fibers can undergo cell division," *American Journal of Botany*, vol. 84, no. 9, pp. 1231–1235, 1997.

[2] S. Saha and J. Van't Hof, "Cotton fiber cells are arrested at G1 stage," *Journal of New Seeds*, vol. 7, no. 1, pp. 1–8, 2005.

[3] W. Dewitte and J. A. H. Murray, "The plant cell cycle," *Annual Review of Plant Biology*, vol. 54, pp. 235–264, 2003.

[4] D. Inzé and L. De Veylder, "Cell cycle regulation in plant development," *Annual Review of Genetics*, vol. 40, pp. 77–105, 2006.

[5] V. Mironov, L. De Veylder, M. Van Montagu, and D. Inzé, "Cyclin-dependent kinases and cell division in plants—the nexus," *Plant Cell*, vol. 11, no. 4, pp. 509–522, 1999.

[6] G. Wang, H. Kong, Y. Sun et al., "Genome-wide analysis of the cyclin family in arabidopsis and comparative phylogenetic analysis of plant cyclin-like proteins," *Plant Physiology*, vol. 135, no. 2, pp. 1084–1099, 2004.

[7] J. A. Torres Acosta, J. de Almeida Engler, J. Raes et al., "Molecular characterization of *Arabidopsis* PHO80-like proteins, a novel class of CDKA;1-interacting cyclins," *Cellular and Molecular Life Sciences*, vol. 61, no. 12, pp. 1485–1497, 2004.

[8] J. Joubès, C. Chevalier, D. Dudits et al., "CDK-related protein kinases in plants," *Plant Molecular Biology*, vol. 43, no. 5-6, pp. 607–620, 2000.

[9] B. Ducommun, P. Brambilla, M. A. Felix, B. R. Franza, E. Karsenti Jr., and G. Draetta, "Cdc2 phosphorylation is required for its interaction with cyclin," *EMBO Journal*, vol. 10, no. 11, pp. 3311–3319, 1991.

[10] J. Colasanti, M. Tyers, and V. Sundaresan, "Isolation and characterization of cDNA clones encoding a functional p34cdc2 homologue from *Zea mays*," *Proceedings of the National Academy of Sciences of the United States of America*, vol. 88, no. 8, pp. 3377–3381, 1991.

[11] P. C. Ferreira, A. S. Hemerly, R. Villarroel, M. Van Montagu, and D. Inzé, "The *Arabidopsis* functional homolog of the p34cdc2 protein kinase," *Plant Cell*, vol. 3, no. 5, pp. 531–540, 1991.

[12] T. Hirayama, Y. Imajuku, T. Anai, M. Matsui, and A. Oka, "Identification of two cell-cycle-controlling cdc2 gene homologs in *Arabidopsis thaliana*," *Gene*, vol. 105, no. 2, pp. 159–165, 1991.

[13] H. Hirt, A. Páy, L. Bögre, I. Meskiene, and E. Heberle-Bors, "Cdc2MsB, a cognate *cdc2* gene from alfalfa, complements the G1/S but not the G2/M transition of budding yeast cdc28 mutants," *Plant Journal*, vol. 4, no. 1, pp. 61–69, 1993.

[14] Z. Magyar, T. Mészáros, P. Miskolczi et al., "Cell cycle phase specificity of putative cyclin-dependent kinase variants in synchronized alfalfa cells," *Plant Cell*, vol. 9, no. 2, pp. 223–235, 1997.

[15] M. Umeda, C. Umeda-Hara, M. Yamaguchi, J. Hashimoto, and H. Uchimiya, "Differential expression of genes for cyclin-dependent protein kinases in rice plants," *Plant Physiology*, vol. 119, no. 1, pp. 31–40, 1999.

[16] A. S. Hemerly, P. Ferreira, J. De Almeida Engler, M. Van Montagu, G. Engler, and D. Inze, "Cdc2a expression in *Arabidopsis* is linked with competence for cell division," *Plant Cell*, vol. 5, no. 12, pp. 1711–1723, 1993.

[17] M. Yamaguchi, H. Kato, S. Yoshida, S. Yamamura, H. Uchimiya, and M. Umeda, "Control of *in vitro* organogenesis by cyclin-dependent kinase activities in plants," *Proceedings of the National Academy of Sciences of the United States of America*, vol. 100, no. 13, pp. 8019–8023, 2003.

[18] S. Adachi, N. Nobusawa, and M. Umeda, "Quantitative and cell type-specific transcriptional regulation of A-type cyclin-dependent kinase in *Arabidopsis thaliana*," *Developmental Biology*, vol. 329, no. 2, pp. 306–314, 2009.

[19] R. L. Barent and T. E. Elthon, "Two-dimensional gels: an easy method for large quantities of proteins," *Plant Molecular Biology Reporter*, vol. 10, no. 4, pp. 338–344, 1992.

[20] J. Z. Yu, R. Cantrell, R. Kohel et al., "Establishment of the standardized cotton microsatellite database (CMD) panel," in *Proceedings of the Beltwide Cotton Improvement Conference*, pp. 5–8, National Cotton Council, San Antonio, Tex, USA, January 2004.

[21] S. Liu, S. Saha, D. Stelly, B. Burr, and R. G. Cantrell, "Chromosomal assignment of microsatellite loci in cotton," *Journal of Heredity*, vol. 91, no. 4, pp. 326–331, 2000.

[22] S. Saha, J. Wu, J. N. Jenkins et al., "Effect of chromosome substitutions from *Gossypium barbadense* L. 3-79 into *G. hirsutum* L. TM-1 on agronomic and fiber traits," *Journal of Cotton Science*, vol. 8, no. 3, pp. 162–169, 2004.

[23] Z. T. Buriev, S. Saha, I. Y. Abdurakhmonov et al., "Clustering, haplotype diversity and locations of *MIC-3*: a unique root-specific defense-related gene family in Upland cotton (*Gossypium hirsutum* L.)," *Theoretical and Applied Genetics*, vol. 120, no. 3, pp. 587–606, 2010.

[24] K. Vandepoele, J. Raes, L. De Veylder, P. Rouzé, S. Rombauts, and D. Inzé, "Genome-wide analysis of core cell cycle genes in Arabidopsis," *Plant Cell*, vol. 14, no. 4, pp. 903–916, 2002.

[25] J. Joubès, T. H. Phan, D. Just et al., "Molecular and biochemical characterization of the involvement of cyclin-dependent kinase a during the early development of tomato fruit," *Plant Physiology*, vol. 121, no. 3, pp. 857–869, 1999.

[26] Y. Y. Setiady, M. Sekine, N. Hariguchi, H. Kouchi, and A. Shinmyo, "Molecular cloning and characterization of a cDNA clone that encodes a Cdc2 homolog from *Nicotiana tabacum*," *Plant and Cell Physiology*, vol. 37, no. 3, pp. 369–376, 1996.

[27] W. Krek and E. A. Nigg, "Structure and developmental expression of the chicken CDC2 kinase," *EMBO Journal*, vol. 8, no. 10, pp. 3071–3078, 1989.

[28] C. F. Lehner and P. H. O'Farrell, "Drosophila cdc2 homologs: a functional homolog is coexpressed with a cognate variant," *EMBO Journal*, vol. 9, no. 11, pp. 3573–3581, 1990.

[29] Y. Imajuku, Y. Ohashi, T. Aoyama, K. Goto, and A. Oka, "An upstream region of the *Arabidopsis thaliana* CDKA;1 (CDC2aAt) gene directs transcription during trichome development," *Plant Molecular Biology*, vol. 46, no. 2, pp. 205–213, 2001.

[30] C. Richard, C. Granier, D. Inzé, and L. De Veylder, "Analysis of cell division parameters and cell cycle gene expression during the cultivation of *Arabidopsis thaliana* cell suspensions," *Journal of Experimental Botany*, vol. 52, no. 361, pp. 1625–1633, 2001.

[31] G. Moore, "Cereal chromosome structure, evolution, and pairing," *Annual Review Plant Physiology and Plant Molecular Biology*, vol. 51, pp. 195–222, 2000.

[32] M. Thangavelu, A. B. James, A. Bankier, G. J. Bryan, P. H. Dear, and R. Waugh, "Happy mapping in plant genome: reconstruction and analysis of a high-resolution physical map of a 1.9 Mbp region of *Arabidopsis thaliana* chromosome 4," *Plant Biotechnology Journal*, vol. 1, no. 1, pp. 23–31, 2003.

[33] B. E. Pfeil, C. L. Brubaker, L. A. Craven, and M. D. Crisp, "Paralogy and orthology in the malvaceae rpb2 gene family: Investigation of gene duplication in Hibiscus," *Molecular Biology and Evolution*, vol. 21, no. 7, pp. 1428–1437, 2004.

[34] J. T. Leiva-Neto, G. Grafi, P. A. Sabelli et al., "A dominant negative mutant of cyclin-dependent kinase A reduces endoreduplication but not cell size or gene expression in maize endosperm," *Plant Cell*, vol. 16, no. 7, pp. 1854–1869, 2004.

[35] K. Sugimoto-Shirasu and K. Roberts, "Big it up: endoreduplication and cell-size control in plants," *Current Opinion in Plant Biology*, vol. 6, no. 6, pp. 544–553, 2003.

[36] B. A. Larkins, B. P. Dilkes, R. A. Dante, C. M. Coelho, Y. M. Woo, and Y. Liu, "Investigating the hows and whys of DNA endoreduplication," *Journal of Experimental Botany*, vol. 52, no. 355, pp. 183–192, 2001.

[37] N. Gonzalez, F. Gévaudant, M. Hernould, C. Chevalier, and A. Mouras, "The cell cycle-associated protein kinase WEE1 regulates cell size in relation to endoreduplication in developing tomato fruit," *Plant Journal*, vol. 51, no. 4, pp. 642–655, 2007.

[38] D. M. Stelly, S. Saha, D. A. Raska, J. N. Jenkins, J. C. McCarty Jr., and O. A. Gutiérrez, "Registration of 17 upland (*Gossypium hirsutum*) cotton germplasm lines disomic for different *G. barbadense* chromosome or arm substitutions," *Crop Science*, vol. 45, pp. 2663–2665, 2005.

[39] J. E. Endrizzi, E. L. Turcotte, and R. J. Kohel, "Quantitative genetics, cytology and cytogenetics," in *Cotton*, R. J. Kohel and C. F. Lewis, Eds., pp. 81–129, American Society of Agronomy, Madison, Wis, USA, 1984.

12

Chromosomal Location of *HCA1* and *HCA2*, Hybrid Chlorosis Genes in Rice

Katsuyuki Ichitani,[1] Yuma Takemoto,[1] Kotaro Iiyama,[1] Satoru Taura,[2] and Muneharu Sato[1]

[1] *Faculty of Agriculture, Kagoshima University, 1-21-24 Korimoto, Kagoshima, Kagoshima 890-0065, Japan*
[2] *Institute of Gene Research, Kagoshima University, 1-21-24 Korimoto, Kagoshima, Kagoshima 890-0065, Japan*

Correspondence should be addressed to Katsuyuki Ichitani, ichitani@agri.kagoshima-u.ac.jp

Academic Editor: Yunbi Xu

Many postzygotic reproductive barrier forms have been reported in plants: hybrid weakness, hybrid necrosis, and hybrid chlorosis. In this study, linkage analysis of the genes causing hybrid chlorosis in F_2 generation in rice, *HCA1* and *HCA2*, was performed. *HCA1* and *HCA2* are located respectively on the distal regions of the short arms of chromosomes 12 and 11. These regions are known to be highly conserved as a duplicated chromosomal segment. The molecular mechanism causing F_2 chlorosis deduced from the location of the two genes was discussed. The possibility of the introgression of the chromosomal segments encompassing *HCA1* and/or *HCA2* was also discussed from the viewpoint of Indica-Japonica differentiation.

1. Introduction

Many post-zygotic reproductive barrier forms have been reported in plants [1]: hybrid weakness, hybrid necrosis, and hybrid chlorosis. The latter has been observed often in the F_1 generation from crosses among wheat (*Triticum aestivum* L.) and its relatives [2–6]. This phenomenon resulted from the complementary action of a pair of dominant genes. Research for distribution of these genes contributed greatly to the study of the origin of wheat.

Hybrid chlorosis in F_2 generation has been reported only in rice (*Oryza sativa* L.) [7] and interspecific crosses among *Melilotus* species [8]. Sato et al. [7] incidentally found a case of hybrid chlorosis in the F_2 population from a cross between two Japanese native cultivars: J-147 and J-321. Its first symptom was discoloration of the second or third leaf (Figure 1). The yellowish part expanded gradually. Then the whole plant died within 20 days [9], yielding no seed. The phenomenon was caused by a set of mutually independent duplicated recessive genes, named *hca-1* and *hca-2* by Sato and Morishima [9]. According to the new gene nomenclature system for rice [10], we changed our description of the gene symbols, as shown in Table 1.

Rice is classified into two types: Indica-type and Japonica-type. Sato and Morishima [9] examined the distribution of *HCA1* and *HCA2*. The experimentally obtained results can be summarized as follows. (1) The *hca2-1* gene is widely distributed in native Japonica-type cultivars, whereas many Indica-type cultivars carry its dominant allele, *Hca2-2*. (2) J-147 carries *hca1-1*. This gene is probably rare because the occurrence of F_2 chlorosis has not been reported in crosses between Taichung 65, which carries *hca2-1*, and many cultivars except for J-147. The mode of inheritance differs between wheat hybrid chlorosis and that of rice, but the distribution of causal genes is related to varietal differentiation in both cases.

We are interested in genes conferring the post-zygotic reproductive barrier in rice, and we have mapped these genes in the rice genome with the aid of DNA markers [11–14]. We produced hybrids from crosses between J-147 and several cultivars to verify the results of Sato and Morishima [9]. From them, we incidentally found chlorotic plants in the F_2 population from the cross between J-147 and a Philippine Indica-type cultivar IR24. We have never seen chlorotic plants in F_2 population from the cross between IR24 and rice cultivars except J-147. Moreover, no reports in

TABLE 1: Gene symbols frequently used in this study according to the new gene nomenclature system for rice [10].

| | Gene symbol | | | |
	Sato and Morishima (1988) [9]	This study	Gene full name	Cultivars harboring chlorosis-causing gene
Locus/gene	*hca-1*	*HCA1*	*HYBRID CHLOROSIS A1*	
Recessive allele	*hca-1*	*hca1-1*	*hybrid chlorosis a1-1*	J-147
Dominant allele	*hca-1*[+]	*Hca1-2*	*Hybrid chlorosis a1-2*	
Locus/gene	*hca-2*	*HCA2*	*HYBRID CHLOROSIS A2*	
Recessive allele	*hca-2*	*hca2-1*	*hybrid chlorosis a2-1*	Akihikari, Asominori IR24, Milyang 23 Many Japonica-type cultivars [9]
Dominant allele	*hca-2*[+]	*Hca2-2*	*Hybrid chlorosis a2-2*	

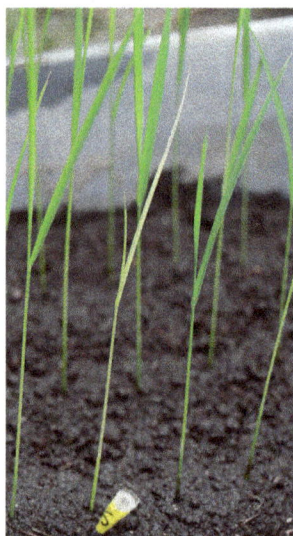

FIGURE 1: Hybrid chlorosis caused by *hca1-1* and *hca2-1*. Seedlings in an F_4 line from the cross between J-147 and IR24 are shown 10 days after sowing date. A chlorotic plant is located at the center. The neighboring green plants are normal.

the literature describe chlorotic plants in an F_2 population from the cross with IR24. These facts indicate that IR24 carries *hca2-1*. In general, much DNA polymorphism exists between Japonica-type and Indica-type cultivars. This report describes linkage analysis of *HCA1* and *HCA2* using progeny from the F_2 population between J-147 and IR24 assisted by DNA markers. Then the molecular mechanism causing F_2 chlorosis, as deduced from the location of the two genes, is discussed.

2. Materials and Methods

2.1. Plant Materials. Five rice cultivars were used for this study: J-147, Akihikari, Asominori, Milyang 23, and IR24. Dr. Sato of the Research Institute for Humanity and Nature provided J-147. Dr. Atsushi Yoshimura of Kyushu University provided Asominori and IR24 for this study. Dr. Yoshimichi Fukuta of Japan International Research Center for Agricultural Sciences provided Akihikari and Milyang

23. Akihikari and Milyang 23 are the parents of a set of recombinant inbred (RI) lines developed by Fukuta et al. [15]. Asominori and IR24 are the parents of another set of RI lines developed by Tsunematsu et al. [16]. No report in the relevant literature describes the appearance of chlorotic plants in progeny from a cross between Asominori and IR24, or from a cross between Akihikari and Milyang 23, although hybrid breakdown phenomena were reported for the cross between Asominori and IR24 [17, 18], and in the cross between Akihikari and Milyang 23 [19]. J-147, Akihikari, and Asominori are generally categorized as Japonica-type, whereas Milyang 23 and IR24 are generally categorized as Indica-type. J-147 was crossed with four cultivars: Akihikari, Asominori, Milyang 23, and IR24. F_2 populations from the above cross combinations were grown in nurseries. The plant spacing was 3×3 cm.

Chlorotic plants were segregated in the F_2 population from the cross between J-147 and IR24 (see Section 3). The normal plants were transplanted to a paddy field in the experimental farm of Kagoshima University to harvest self-pollinated seeds. Approximately 80 plants in each of 16 F_3 lines were grown in the nursery for the segregation of chlorotic plants. Among them, normal plants in the F_3 lines in which chlorotic plants segregated were transplanted in the same way as the F_2 generation. Approximately 80 plants in each F_4 line were also grown in the nursery for segregation of the chlorotic plants.

2.2. Linkage Analysis of HCA1 and HCA2. F_4 lines in which only *HCA1* or *HCA2* gene was expected to segregate were subjected to linkage analysis using DNA markers. Preliminary analysis using a small number of plants detected the approximate locations of *HCA1* and *HCA2*. Then, we selected the F_4 lines segregating chlorotic plants in which one locus is fixed for a recessive chlorosis-causing allele, whereas the other locus is heterozygous and the heterozygous chromosomal region encompasses the locus that is extended most. These lines were used for construction of linkage map of *HCA1* or *HCA2*. Linkage analysis was conducted using a computer program (MapDisto ver. 1.7; Lorieux [20]). Map distances were estimated using the Kosambi function [21].

After the linkage analysis, the F_2 populations from the cross between J-147 and IR24 were grown again. DNA was

J147 (Japonica-type)   ×   IR24 (Indica-type)
hca1-1/hca1-1 Hca2-2/Hca2-2   *Hca1-2/Hca1-2 hca2-1/hca2-1*

F_1
Hca1-2/hca1-1 Hca2-2/hca2-1

Self-pollinated seeds from normal plants F_2
were harvested to produce the F_3 generation

Self-pollinated seeds from normal plants in F_3 The distal regions of the short arm of
F_3 lines in which chlorotic plants were   chromosomes 11 and 12 are highly
segregated were harvested to produce F_4 conserved as a duplicated chromosomal
generation   segment [24, 25, 26]

F_4

TF$_4$ 27–10   TF$_4$ 29-30
Chlorotic plants were all homozygous of   Chlorotic plants were all homozygous of
J147 allele at the RM27421 locus   IR24 allele at the E30794 locus
HCA1 was mapped on the distal end of the   *HCA2* was mapped on the distal end of
short arm of chromosome 12   the short arm of chromosome 11

TF$_4$ 33–21   TF$_4$ 23–19
Construction of linkage map around *HCA1* locus   Construction of linkage map around *HCA2* locus
(Figure 3): cosegregation of *HCA1* and RM27404   (Figure 3): cosegregation of *HCA2* and RM25969
was detected   was detected

Back to F_2

Genotypes at the RM25969 and RM27404 loci were analyzed for all 503 F_2 plants (Table 5)

hca1-1 and *hca2-1* were sufficient to cause chlorosis in F_2 population from the cross between J147and IR24

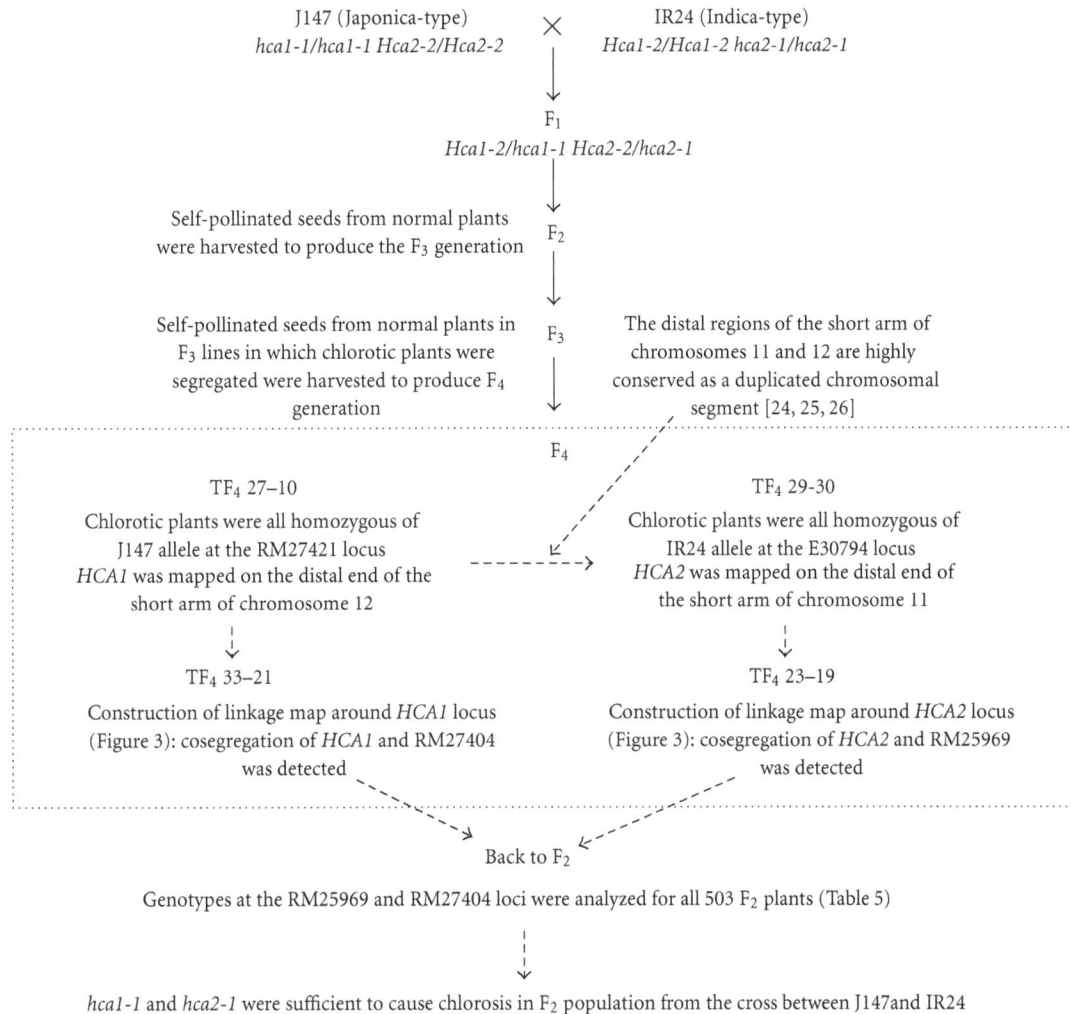

FIGURE 2: Flow chart showing the breeding of the plant materials for mapping *HCA1* and *HCA2* and for verifying that *hca1-1* and *hca2-1* were sufficient to cause hybrid chlorosis in the F_2 population from the cross between J147 and IR24. Arrows with a solid line indicate the flow of generation of plant materials. Arrows with a dotted line indicate the flow of information on linkage between chlorosis genes and DNA markers.

extracted from each plant and subject to genotyping of the most closely linked DNA markers with *HCA1* and *HCA2* loci to verify that *hca1-1* and *hca2-1* were sufficient to cause hybrid chlorosis in the F_2 population from the cross between J-147 and IR24.

Figure 2 shows a flow chart depicting the breeding of the plant materials for mapping *HCA1* and *HCA2* and for verification that *hca1-1* and *hca2-1* were sufficient to cause hybrid chlorosis in the F_2 population from the cross between J-147 and IR24.

2.3. DNA Marker Analysis. The DNA of plant materials except for the F_2 population was extracted using the process explained by Dellaporta et al. [22] with some modifications. The DNA of the F_2 population was extracted according to the experimental protocols of the Rice Genome Project (RGP) (http://rgp.dna.affrc.go.jp/E/rgp/protocols/index.html, written in Japanese) with some modifications [14]. The PCR conditions for indel and SSR markers used for this study were 95°C for 10 min, 40 cycles of 94°C for 30 s, 55°C for 30 s, and 72°C for 30 s with subsequent final extension of 72°C for 1 min. The PCR mixture (5 μL) contained 1 μl of template DNA, 200 mM of each dNTP, 0.2 μM of primers, 0.25 units of Taq polymerase (AmpliTaq Gold; Applied BioSystems), and 1 × buffer containing MgCl$_2$. The PCR products were analyzed using electrophoresis in 10% (29 : 1) polyacrylamide gel with subsequent ethidium bromide staining. Then they were viewed under ultraviolet light irradiation. Most PCR-based DNA markers used for this study have already been published. Some primer pairs did not perform well. Therefore, we redesigned them. Details of PCR-based DNA marker design were reported in our previous papers [13, 14].

3. Results

The F_2 populations of the four cultivars with J-147 all produced both normal plants and chlorotic plants (Table 2).

TABLE 2: Segregation for normal and chlorotic plants in the four F_2 populations.

| Cross combination | No. of plants | | | χ^2 (15:1) | P |
	Normal	Chlorotic	Total		
J-147 × Milyang 23	56	1	57	1.966	0.161
J-147 × Akihikari	59	1	60	2.151	0.142
J-147 × IR24	62	5	67	0.168	0.682
J-147 × Asominori	60	2	62	0.968	0.325

The ratios of normal plants and chlorotic plants were fitted to 15:1, the expected ratio deduced from the segregation of two independent recessive genes. These results indicated that J-147 carries hca1-1 gene and that Akihikari, Asominori, Milyang 23, and IR24 carry the hca2-1 gene.

Indica-type cultivars and Japonica-type cultivars have frequently shown DNA polymorphism between them. This is true for Asominori and IR24, and for Akihikari and Milyang 23. Using the DNA polymorphism between the two pair of cultivars, DNA marker-based linkage maps were constructed [15, 16]. Results of our recent study indicated much polymorphism between Indica-type cultivars and J-147, and little polymorphism between Japonica-type cultivars and J-147 [14]. Therefore, the progeny from the cross between Indica-type cultivars and J-147 were more suitable for mapping genes. We selected the progeny from the cross between IR24 and J-147 as a mapping population for HCA1 and HCA2 because the linkage map constructed from the cross between Asominori and IR24 covered almost the whole genome, whereas that constructed from Akihikari and Milyang 23 had some large gaps, suggesting that some chromosomal regions exist with no DNA polymorphism.

Seeds of normal plants in the F_2 population were harvested to produce the F_3 generation. The segregation of chlorotic plants was examined for 16 F_3 lines from each F_2 plant. Only normal plants appeared in 10 lines. Chlorotic plants segregated in six lines. In the six lines, the ratios of normal plants: chlorotic plants were all fitted to 15:1. In the F_3 generation, the expected ratio of lines fixed for normal plants, those showing 3 normal: 1 chlorotic segregation, and others showing a 15 normal: 1 chlorotic segregation, was 7:4:4. However, no lines showing 3 normal: 1 chlorotic segregation appeared. Then, six F_3 lines in which segregated chlorotic plants were transplanted to a paddy field, and the seeds of normal plants were harvested to produce F_4 generation.

The segregation of chlorotic plants was examined for 211 F_4 lines from each F_3 plant. The line named TF$_4$ 27-10 was a progeny from an F_3 line TF$_3$ 27. In TF$_4$ 27-10, 70 normal plants and 10 chlorotic plants segregated, showing the maximum chlorotic plant ratio among the 10 TF$_3$ 27-derived F_4 lines tested in our first experiment in the F_4 generation and significantly deviated from 15:1 ratio. Another line named TF$_4$ 23-5 was the progeny of an F_3 line TF$_3$ 23. In TF$_4$ 23-5, 80 normal plants and 7 chlorotic plants segregated, fitted to a 15:1 ratio. We performed a preliminary linkage analysis using a bulked DNA composed of 20 normal plants from TF$_4$ 27-10 and another bulked DNA composed of 20 normal

plants from TF$_4$ 23-5, and 39 DNA marker scattered on all the 12 chromosomes. Because of successive self-pollination, the ratio of heterozygous chromosomal region reduced to approximately 0.25 in the F_4 generation. HCA1 and/or HCA2 are expected to be located on heterozygous chromosomal region of the lines in which chlorotic plants segregated. Among the DNA markers, four showed heterozygosity in both lines. Then DNA from six chlorotic plants in both F_4 lines was analyzed individually. Possible linkage was observed between chlorosis and KGS1739 [14], a DNA marker located on the short arm of chromosome 12: no homozygotes of IR24 allele were detected. This result suggests that HCA1 was located on the chromosomal region. Then we analyzed 7 chlorotic plant and 54 normal plants in TF$_4$ 27-10 line individually. Results showed that the cosegregation of HCA1 and the more distal marker RM27421 [23] was detected: all 7 chlorotic plants in this line were homozygous for J-147 allele at the RM27421 locus, whereas all 54 normal plants were heterozygous or homozygous for the IR24 allele at the locus.

Based on RFLP-based linkage analysis using the F_2 population from the cross between Nipponbare and Kasalath, Nagamura et al. [24] reported that the distal regions of the short arms of rice chromosomes 11 and 12 are highly conserved as a duplicated chromosomal segment. Antonio et al. [25] confirmed the high degree of conservation of duplicated segments in these regions using four other mapping populations. Moreover, Wu et al. [26] generated physical maps covering most of the duplicated regions. RM27421 was within the duplicated region on chromosome 12 (Figure 3). These experimentally obtained results and the fact that the hybrid chlorosis was controlled by a couple of recessive duplicate genes led us to the idea that HCA2 might be located on the distal region of the short arm of chromosome 11. Our preliminary analysis using an F_4 line named TF$_4$ 29-30 showed the cosegregation of HCA2 and E30794, an STS marker designed by RGP (http://rgp.dna.affrc.go.jp/E/publicdata/caps/index.html), located on the distal region of the short arm of chromosome 11: all 20 chlorotic plants in this line were homozygous for IR24 allele at the E30794 locus, whereas all 68 normal plants were heterozygous or homozygous for J-147 allele at the locus.

To construct linkage maps around the HCA1 and HCA2 loci, we selected F_4 lines in which one locus is fixed for a recessive chlorosis-causing allele and the other locus is on the heterozygous region extending farthest. Using these lines (HCA1: TF$_4$ 33-21, HCA2: TF$_4$ 23-19, see Figure 2),

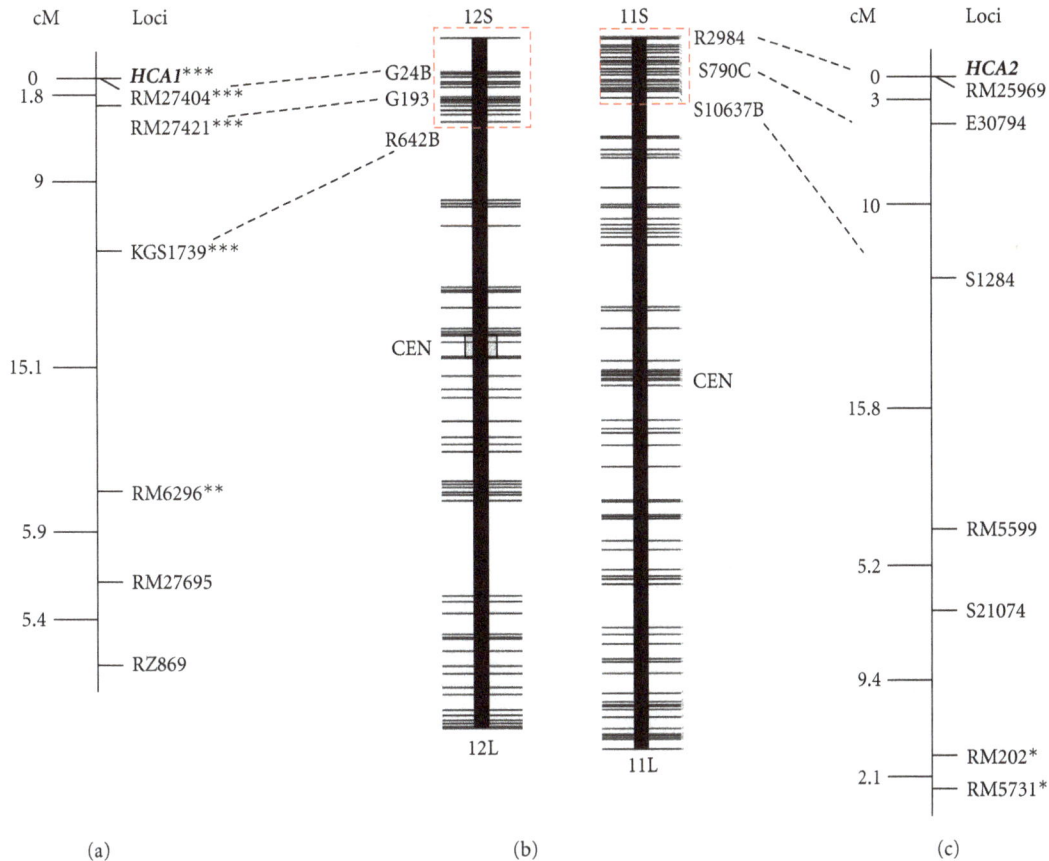

FIGURE 3: Linkage map showing the respective locations of *HCA1* and *HCA2* on rice chromosomes 12 and 11. (a) Linkage map of *HCA1* on rice chromosome 12 constructed from an F$_4$ line derived from the cross between J-147 and IR24 ($n = 115$). (b) RFLP framework maps of rice chromosomes 12 and 11 referred from Harushima et al. [27]. Duplicated genomic regions identified by Wu et al. [26] are enclosed by dotted red squares. (c) Linkage map of *HCA2* on rice chromosome 11constructed from another F$_4$ line derived from the cross between J-147 and IR24 ($n = 119$). DNA markers which are located near each other on Nipponbare pseudomolecules are connected with dotted lines. *, **, and *** respectively denote that the gene segregation was significantly deviated from the expected ratio (1 : 2 : 1 or 3 : 1) at 0.05, 0.01, and 0.001 levels.

we conducted linkage analysis of each gene using the DNA markers shown in Table 3. The DNA marker sources were the International Rice Genome Sequencing Project [23], RGP, McCouch et al. [28], Chen et al. [29], Monna et al. [30], and Ichitani et al. [14]. The linkage analysis of *HCA1* gene and six DNA markers using 13 chlorotic plants and 102 normal plants in TF$_4$ 33-21 showed that *HCA1* was located at the distal end of the short arm of chromosome 12 and that it cosegregated with RM27404 (Figure 3): all 13 chlorotic plants in this line were homozygous for J-147 allele at the RM27404 locus, whereas all 102 normal plants were heterozygous or homozygous for IR24 allele at the locus. The ratio of 102 : 13 did not fit to the expected ratio 3 : 1. Cosegregation of *HCA1* and RM27404 and distorted segregation of the linked DNA markers (Table 4) showed that the distorted segregation of *HCA1* resulted from gametophytic reproductive barrier gene(s), often found in the cross between Indica-type and Japonica-type (see Section 4).

The linkage analysis of *HCA2* gene and seven DNA markers using 23 chlorotic plants and 96 normal plants in

TF$_4$ 23-19 showed that *HCA2* was located at the distal end of the short arm of chromosome 11 and that it cosegregated with RM25969 (Figure 3): all 23 chlorotic plants in this line were homozygous for IR24 allele at the RM25969 locus, whereas 96 normal plants were heterozygous or homozygous for J-147 allele at the locus. The ratio of 23 chlorotic plants : 96 normal plants fitted the expected ratio 1 : 3. The segregation of the tightly linked DNA markers of *HCA2* also fitted the expected ratio 1 : 2 : 1 (Table 4). RM202 and RM5731, showing distorted segregation, were inherited independently of *HCA2*. Harushima et al. [27] constructed an often-cited restriction fragment length polymorphism (RFLP) marker-based high-density linkage map for rice, in which some RFLP markers have been sequenced. Based on the Nipponbare genome sequence, the relation between our linkage map and that by Harushima et al. [27] is shown in Figure 3.

Using F$_4$ generation, we located *HCA1* and *HCA2* genes on the rice genome. However, the ratio of normal plants to chlorotic plants was often distorted from the expected one; that is, the chlorotic plants were often significantly fewer than

Table 3: Primer sequences designed and used for mapping *HCA1* and *HCA2* loci.

Marker name	Kind of DNA marker		Primer sequences (5′–3′)	Chromosome	Position From	To	Source
RM25969	SSR	F	TAAATTTGGTTGTCTACGCATGG	11	364229	364418	[23]
		R	CTGCTCCAGATTAGGAGCCAG				redesigned in this study
E30794	Indel	F	TCTGCCTATGTATTTGTGCGTTAAT	11	679545	679744	RGP
		R	AAGTAACACAACGAAGGAGCAAC				redesigned in this study
S1284	Indel	F	ACATTCAACTGATCACAGCC	11	2186184	2186392	RGP
		R	AGCTCTGCACTAGGATGATG				
RM5599	SSR	F	AATTTTGTGCTGTTGTTGAA	11	3810354	3810494	[28]
		R	CTCACAATATCACCATCCAC				
S21074	Indel	F	TGCTATAGGTGGTGGTATGC	11	5588871	5589049	RGP
		R	TTTCAAGCTGACGAACGATG				
RM202	SSR	F	CCAGCAAGCATGTCAATGTA	11	9050541	9050718	[29]
		R	CAGATTGGAGATGAAGTCCTCC				
RM5731	SSR	F	CCTCCACCCTAAGCTTCTCC	11	10005222	10005423	[28]
		R	CGATGCACCTAGCGCATC				
RM27404	SSR	F	GCAGCGATTGGAGGTGGAA	12	204766	204863	[23]
		R	GACCGTGCCATCTTGTCCAG				redesigned in this study
RM27421	SSR	F	TCAACTCCATTCTACTCTCTACCC	12	468463	468538	[23]
		R	GCTGCTGGTACTTCGTAGAGG				
KGS1739	Indel	F	AGAGACGCAGGAGCTGCTTA	12	1999528	1999818	[14, 30]
		R	CATGACCCTTCTATGGCAATTAT				
RM6296	SSR	F	CCCACGTTTCTCTTGTCCTC	12	3200580	3200734	[28]
		R	TCTTGCCTCGCTAGGGTTAG				
RM27695	SSR	F	CTATAAAGAGTCCGAGGGGTATTGT	12	4828865	4829000	[23]
		R	GGGAGAGGATGTGAATGAAGG				redesigned in this study
RZ869	Indel	F	TTTTGGTATTTGCTGCATGG	12	7739507	7739753	RGP
		R	TATCAATCACCCCCAACCTC				

expected. Therefore, we again produced the F_2 generation from the cross between J-147 and IR24 to confirm that *hca1-1* and *hca2-1* are sufficient to cause chlorosis in F_2 generation with the aid of tightly linked DNA markers.

A total of 503 F_2 plants were classified into 481 normal plants and 22 chlorotic plants. The ratio 481 : 22 looked slightly skewed towards normal plants but fitted the expected ratio 15 : 1 ($\chi^2 = 3.033$, $P = 0.08$). Genotypes at the

RM25969 and RM27404 loci were analyzed for all F_2 plants (Table 5). No homozygotes of J-147 allele at the RM27404 locus and IR24 allele at the RM25969 locus were present in normal plants. In contrast, all chlorotic plants were homozygotes of J-147 allele at RM27404 locus and IR24 allele at the RM25969 locus. This result indicated that *hca1-1* and *hca2-1* were sufficient to cause chlorosis in F_2 population from the cross between J-147 and IR24. Segregation of

TABLE 4: Segregation of DNA markers linked with *HCA1* and *HCA2* in the mapping population of these genes from the cross between J-147 and IR24.

Marker	Genotype[a]			n	χ^2 (1:2:1)	P
	J	H	I			
HCA1 (chr. 12)						
RM27404	13	59	43	115	15.730	<0.001
RM27421	12	59	44	115	17.887	<0.001
KGS1739	11	61	43	115	18.235	<0.001
RM6296	15	62	38	115	9.904	0.007
RM27695	19	59	37	115	5.713	0.057
RZ869	21	61	33	115	2.930	0.231
HCA2 (chr. 11)						
RM25969	31	65	23	119	2.092	0.351
E30794	28	68	23	119	2.849	0.241
S1284	27	68	24	119	2.580	0.275
RM5599	33	61	25	119	1.151	0.562
S21074	36	63	20	119	4.714	0.095
RM202	42	58	19	119	8.966	0.011
RM5731	43	55	21	119	8.815	0.012

[a] I, H, and J respectively denote homozygote for IR24 allele, and heterozygote and homozygote for J-147 allele.

TABLE 5: Segregation of DNA markers linked with *HCA1* and *HCA2* in the F_2 population ($n = 503$) derived from the cross between J-147 and IR24. Numbers of chlorotic plants are shown in parentheses.

		Genotypes for RM27404[a] on chr. 12			Total	χ^2 (1:2:1)	P
		I	H	J			
Genotypes	I	38 (0)	60 (0)	22 (22)	120		
for	H	94 (0)	139 (0)	37 (0)	270	2.917	0.233
RM25969	J	37 (0)	66 (0)	10 (0)	113		
on chr. 11							
	Total	169	265	69			
	χ^2 (1:2:1)		41.211				
	P		<0.0001				

[a] I, H, and J respectively denote homozygote for IR24 allele and heterozygote, and homozygote for J-147 allele.

RM25969 fitted the expected ratio 1:2:1, although that of RM27404 did not fit the expected ratio: homozygotes of IR24 allele were more numerous than expected, and homozygotes of J-147 allele were much fewer than expected. This result is consistent with that of F_4 generation. The distorted segregation of chlorotic plants in F_3 generation was thought to result from abnormal segregation at the chromosomal region around the *HCA1* locus.

4. Discussion

In this study, *HCA1* and *HCA2* were located on the respective distal regions of the short arms of chromosomes 12 and 11. These regions are known to be highly conserved as a duplicated chromosomal segment [24–26]. According to the Rice Genome Annotation Project (http://rice.plantbiology.msu.edu/index.shtml) [31], 30 genes with known functions are located in common in the areas of interest of chromosomes 11 and 12. Therefore, *HCA1* and *HCA2* are thought to be mutually homoeologous genes, and the loss-of-function mutations of these genes are thought to be the cause of hybrid chlorosis. Based on the symptom, the causal genes of hybrid chlorosis should be involved in chlorophyll synthesis or chloroplast metabolism.

Recently, two papers reported the reciprocal disruption of duplicate genes causing reproductive barrier in interspecific and intraspecific rice crosses. Yamagata et al. [32] reported that the reciprocal loss of duplicated gene encoding mitochondrial ribosomal protein L27 causes hybrid pollen sterility in F_1 hybrids of *O. sativa* and *O. glumaepatula*. Mizuta et al. [33] reported disruption of duplicated genes *DOPPELGANGER1* (*DPL1*) and *DOPPELGANGER2* (*DPL2*) causing pollen sterility: independent disruption of *DPL1* and *DPL2* occurred, respectively, in Indica-type and Japonica-type. DPLs encode highly conserved, plant-specific small proteins. In *Arabidopsis thaliana*, recessive embryo lethality is caused by a disruption of duplicated histidinol-phosphate amino-transferase genes encoding for a protein that catalyzes an important pathway leading to histidine incorporated into proteins [34]. The hybrid chlorosis described in this study might offer a new example of a reproductive barrier caused by disruption of duplicate genes.

The segregation of *HCA2* and its tightly linked DNA marker RM27404 was significantly distorted from the expected ratio: the frequencies of Japonica-type alleles were smaller than expected. The segregation distortion of the distal end of the short arm of chromosome 12 was reported in the populations derived from the crosses between Indica-type cultivars and Japonica-type cultivars: IR24 and Asominori [16], and Milyang 23 and Akihikari [35]. The peak of distortion was at an RFLP marker XNpb193 (= G193), and the frequencies of Japonica-type alleles were smaller than expected in both crosses. Therefore, the segregation distortion observed in this study might result from the same genetic factor(s).

Haplotype analysis around reproductive barrier genes might shed new light on varietal differentiation. Kuboyama et al. [13] performed haplotype analysis around the HWC2 locus. Carriers of the weakness-inducing allele Hwc2-1, most of which are categorized as temperate Japonica-type, share the same haplotype in the 200 kb region between the two DNA markers, KGC4M5 and KGC4M52. The carriers of hwc2-2 have different haplotypes, most of which are distinct from those of Hwc2-1 carriers. These results suggested that Hwc2-1 diffused in temperate Japonica-types, dragging adjacent genes with it. The hca2-1 gene is mainly carried by Japonica-type cultivars. Therefore, the haplotype analysis of HCA2 might engender new findings related to Japonica-Indica differentiation. We are undertaking haplotype analysis of HCA2 using a core collection of world rice [36] and a minicollection of Japanese rice landrace [37].

Before the experiment, we expected that Akihikari and Asominori, both generally classified as Japonica-type, carry hca2-1 gene whereas Milyang 23 and IR24, both generally classified as Indica-type, carry the wild type Hca2-2. Based on that expectation, we produced an experimental design in which J-147 was crossed with a set of RI lines so that linkage between HCA2 and DNA markers was detectable by combining the genotype of each RI line of DNA markers on the whole genome, which had been analyzed by the breeders of the RI lines, and the segregation of chlorotic plants in the F_2 population between J-147 and each RI line. The experimental design using RI lines for mapping a reproductive-barrier-related gene was successful in a hybrid weakness gene HWC2 [11]. However, contrary to our expectation, all four cultivars carry hca2-1 gene. Milyang 23 is a descendant of IR24. Therefore, hca2-1 gene of Milyang 23 might derive from IR24. Zhao et al. [38] used a 1,536 SNP panel genotyped across 395 rice diverse accessions to study genomic-wide pattern of polymorphism, to characterize population structure, and to infer the introgression history, revealing that most accessions exhibit some degree of admixture, with many individuals within a population sharing the same introgressed segment because of artificial selection. Therefore, a high probability exists that the HCA2 locus and its surrounding chromosomal region of IR24 were introgressed from a Japonica-type cultivar. However, another possibility exists: loss-of-function mutation occurred at the HCA2 locus independently in Indica-type cultivars. The origin of Hca2-2 allele of J-147 is also interesting. Ichitani et al. [14] reported that J-147 shares the same banding patterns of 38 out of 39 PCR-based DNA markers scattered on the whole genome with the three cultivars generally classified as temperate Japonica-type. This result indicates that J-147 can be categorized as temperate Japonica-type. A DNA marker S1284 (Table 3, Figure 3), located at 2.2 Mbp from distal end of short arm of chromosome 11 (IRGSP pseudomolecules Build05), clearly discriminates Japonica-type and Indica-type, and J-147 carries the Japonica-type allele [14]. The Hca2-2 allele of J-147 might also be the result of the introgression of small chromosomal segment of the distal end of the short arm of chromosome 11.

To identify the causal genes, we are undertaking closer linkage analysis and linkage disequilibrium analysis of both HCA1 and HCA2 genes. Identification of causal genes will contribute to the study of rice varietal differentiation, gene duplication, and chlorophyll synthesis or chloroplast metabolism.

Abbreviations

RGP: Rice genome project
PCR: Polymerase chain reaction
SSR: Simple sequence repeat marker
RI: Recombinant inbred
Indel: Insertion/deletion
RFLP: Restriction fragment length polymorphism.

Acknowledgments

The authors gratefully acknowledge Dr. Lisa Monna of the Plant Genome Center for valuable technical advice on DNA markers. They are also grateful to Dr. Atsushi Yoshimura of Kyushu University, Dr. Yoshimichi Fukuta of Japan International Research Center for Agricultural Sciences, and Dr. Yo-Ichiro Sato of Research Institute for Humanity and Nature for their kind provision of rice cultivars. They thank Ms. Yuki Shirata for technical assistance. This work was funded by the Ministry of Education, Culture, Sports, Science and Technology, Japan.

References

[1] K. Bomblies and D. Weigel, "Hybrid necrosis: autoimmunity as a potential gene-flow barrier in plant species," Nature Reviews Genetics, vol. 8, no. 5, pp. 382–393, 2007.

[2] K. Tsunewaki, "Gene analysis on chlorosis of the hybrid, Triticum aestivum var. Chenese Spring × T. macha var. subletschchumicum, and its bearing on the genetic basis of necrosis and chlorosis," Japanese Journal of Genetics, vol. 41, no. 6, pp. 413–426, 1966.

[3] K. Tsunewaki and J. Hamada, "A new type of hybrid chlorosis found in tetraploid wheats," Japanese Journal of Genetics, vol. 43, no. 4, pp. 279–288, 1968.

[4] K. Tsunewaki, "Distribution of necrosis genes in wheat V. Triticum macha, T. spelta and T. vavilovii," Japanese Journal of Genetics, vol. 46, no. 2, pp. 93–101, 1971.

[5] K. Tsunewaki and Y. Nakai, "Consideration on the origin and speciation of four groups of wheat from the distribution of necrosis and chlorosis genes," in Proceedings of the 4th International Wheat Genetics Symposium, pp. 123–129, 1973.

[6] T. Kawahara, "Genetic analysis of Cs chlorosis in tetraploid wheats," Japanese Journal of Genetics, vol. 68, no. 2, pp. 147–153, 1993.

[7] Y. I. Sato, S. Matsuura, and K. Hayashi, "The genetic basis of hybrid chlorosis found in a cross between two Japanese native cultivars," *Rice Genetics Newsletter*, vol. 1, p. 106, 1984.

[8] Y. Sano and F. Kita, "Reproductive barriers distributed in *melilotus* species and their genetic bases," *Canadian Journal of Genetics and Cytology*, vol. 20, no. 2, pp. 275–289, 1978.

[9] Y. I. Sato and H. Morishima, "Distribution of the genes causing F_2 chlorosis in rice cultivars of the Indica and Japonica types," *Theoretical and Applied Genetics*, vol. 75, no. 5, pp. 723–727, 1988.

[10] S. R. McCouch and Committee on Gene Symbolization, Nomenclature and Linkage, Rice Genetics Cooperative (CGSNL), "Gene nomenclature system for rice," *Rice*, vol. 1, pp. 72–84, 2008.

[11] K. Ichitani, Y. Fukuta, S. Taura, and M. Sato, "Chromosomal location of *Hwc2*, one of the complementary hybrid weakness genes, in rice," *Plant Breeding*, vol. 120, no. 6, pp. 523–525, 2001.

[12] K. Ichitani, K. Namigoshi, M. Sato et al., "Fine mapping and allelic dosage effect of *Hwc1*, a complementary hybrid weakness gene in rice," *Theoretical and Applied Genetics*, vol. 114, no. 8, pp. 1407–1415, 2007.

[13] T. Kuboyama, T. Saito, T. Matsumoto et al., "Fine mapping of *HWC2*, a complementary hybrid weakness gene, and haplotype analysis around the locus in rice," *Rice*, vol. 2, pp. 93–103, 2009.

[14] K. Ichitani, T. Taura, T. Tezuka, Y. Okiyama, and T. Kuboyama, "Chromosomal location of *HWA1* and *HWA2*, complementary weakness genes in rice," *Rice.*, vol. 4, no. 2, pp. 29–38, 2011.

[15] Y. Fukuta, K. Tamura, H. Sasahara, and T. Fukuyama, "Genetic and breeding analysis using molecular markers. 18. Variations of gene frequency and the RFLP map of the hybrid population derived from the cross between the rice variety, Milyang 23 and Akihikari," *Breeding Research*, vol. 1, supplement 2, p. 176, 1999 (Japanese).

[16] H. Tsunematsu, A. Yoshimura, Y. Harushima et al., "RFLP framework map using recombinant inbred lines in rice," *Breeding Science*, vol. 46, no. 3, pp. 279–284, 1996.

[17] T. Kubo and A. Yoshimura, "Genetic basis of hybrid breakdown in a Japonica/Indica cross of rice, *Oryza sativa* L.," *Theoretical and Applied Genetics*, vol. 105, no. 6-7, pp. 906–911, 2002.

[18] T. Kubo and A. Yoshimura, "Epistasis underlying female sterility detected in hybrid breakdown in a Japonica-Indica cross of rice (*Oryza sativa* L.)," *Theoretical and Applied Genetics*, vol. 110, no. 2, pp. 346–355, 2005.

[19] M. Yokoo, S. Saito, T. Higashi, and S. Matsumoto, "Use of a Korean rice cultivar Milyang 23 for improving Japanese rice," *Breeding Science*, vol. 44, pp. 219–222, 1994 (Japanese).

[20] M. Lorieux, "MapDisto, A free user-friendly program for computing genetic maps," Computer demonstration (P958) given at the Plant and Animal Genome 15th conference, San Diego, Calif, USA, 2007, http://mapdisto.free.fr/.

[21] D. Kosambi, "The estimation of map distance from recombination values," *Annals of Eugenics*, vol. 12, no. 3, pp. 172–175, 1944.

[22] S. L. Dellaporta, J. Wood, and J. B. Hicks, "A plant DNA minipreparation: version II," *Plant Molecular Biology Reporter*, vol. 1, no. 4, pp. 19–21, 1983.

[23] T. Sasaki, "International Rice Genome Sequencing Project," "The map-based sequence of the rice genome," *Nature*, vol. 436, no. 7052, pp. 793–800, 2005.

[24] Y. Nagamura, T. Inoue, B. A. Antonio et al., "Conservation of duplicated segments between rice chromosomes 11 and 12," *Breeding Science*, vol. 45, no. 3, pp. 373–376, 1995.

[25] B. A. Antonio, T. Inoue, H. Kajiya et al., "Comparison of genetic distance and order of DNA markers in five populations of rice," *Genome*, vol. 39, no. 5, pp. 946–956, 1996.

[26] J. Wu, N. Kurata, H. Tanoue et al., "Physical mapping of duplicated genomic regions of two chromosome ends in rice," *Genetics*, vol. 150, no. 4, pp. 1595–1603, 1998.

[27] Y. Harushima, M. Yano, A. Shomura et al., "A high-density rice genetic linkage map with 2275 markers using a single F_2 population," *Genetics*, vol. 148, no. 1, pp. 479–494, 1998.

[28] S. R. McCouch, L. Teytelman, Y. Xu et al., "Development and mapping of 2240 new SSR markers for rice (*Oryza sativa* L.)," *DNA Research*, vol. 9, no. 6, pp. 199–207, 2002.

[29] X. Chen, S. Temnykh, Y. Xu, Y. G. Cho, and S. R. McCouch, "Development of a microsatellite framework map providing genome-wide coverage in rice (*Oryza sativa* L.)," *Theoretical and Applied Genetics*, vol. 95, no. 4, pp. 553–567, 1997.

[30] L. Monna, R. Ohta, H. Masuda, A. Koike, and Y. Minobe, "Genome-wide searching of single-nucleotide polymorphisms among eight distantly and closely related rice cultivars (*Oryza sativa* L.) and a wild accession (*Oryza rufipogon* Griff.)," *DNA Research*, vol. 13, no. 2, pp. 43–51, 2006.

[31] S. Ouyang, W. Zhu, J. Hamilton et al., "The TIGR Rice Genome Annotation Resource: improvements and new features," *Nucleic Acids Research*, vol. 35, no. 1, pp. D883–D887, 2007.

[32] Y. Yamagata, E. Yamamoto, K. Aya et al., "Mitochondrial gene in the nuclear genome induces reproductive barrier in rice," *Proceedings of the National Academy of Sciences of the United States of America*, vol. 107, no. 4, pp. 1494–1499, 2010.

[33] Y. Mizuta, Y. Harushima, and N. Kurata, "Rice pollen hybrid incompatibility caused by reciprocal gene loss of duplicated genes," *Proceedings of the National Academy of Sciences of the United States of America*, vol. 107, no. 47, pp. 20417–20422, 2010.

[34] D. Bikard, D. Patel, C. Le Metté et al., "Divergent evolution of duplicate genes leads to genetic incompatibilities within *A. thaliana*," *Science*, vol. 323, no. 5914, pp. 623–626, 2009.

[35] Y. Fukuta, H. Sasahara, K. Tamura, and T. Fukuyama, "RFLP linkage map included the information of segregation distortion in a wide-cross population between Indica and Japonica rice (*Oryza sativa* L.)," *Breeding Science*, vol. 50, no. 2, pp. 65–72, 2000.

[36] Y. Kojima, K. Ebana, S. Fukuoka, T. Nagamine, and M. Kawase, "Development of an RFLP-based rice diversity research set of germplasm," *Breeding Science*, vol. 55, no. 4, pp. 431–440, 2005.

[37] K. Ebana, Y. Kojima, S. Fukuoka, T. Nagamine, and M. Kawase, "Development of mini core collection of Japanese rice landrace," *Breeding Science*, vol. 58, no. 3, pp. 281–291, 2008.

[38] K. Zhao, M. Wright, J. Kimball et al., "Genomic diversity and introgression in *O. sativa* reveal the impact of domestication and breeding on the rice genome," *PLoS One*, vol. 5, no. 5, article e10780, 2010.

A Bayesian Framework for Functional Mapping through Joint Modeling of Longitudinal and Time-to-Event Data

Kiranmoy Das,[1,2,3] **Runze Li,**[2] **Zhongwen Huang,**[4,5] **Junyi Gai,**[4,5] **and Rongling Wu**[2,3]

[1] Department of Statistics, Temple University, Philadelphia, PA 19122, USA
[2] Department of Statistics, The Pennsylvania State University, University Park, PA 16802, USA
[3] Center for Statistical Genetics, The Pennsylvania State University, Hershey, PA 17033, USA
[4] Department of Agronomy, Henan Institute of Science and Technology, Xinxiang 453003, China
[5] National Center for Soybean Improvement, National Key Laboratory of Crop Genetics and Germplasm Enhancement, Soybean Research Institute, Nanjing Agricultural University, Nanjing 210095, China

Correspondence should be addressed to Kiranmoy Das, kiranmoy.das@temple.edu

Academic Editor: Pierre Sourdille

The most powerful and comprehensive approach of study in modern biology is to understand the whole process of development and all events of importance to development which occur in the process. As a consequence, joint modeling of developmental processes and events has become one of the most demanding tasks in statistical research. Here, we propose a joint modeling framework for functional mapping of specific quantitative trait loci (QTLs) which controls developmental processes and the timing of development and their causal correlation over time. The joint model contains two submodels, one for a developmental process, known as a longitudinal trait, and the other for a developmental event, known as the time to event, which are connected through a QTL mapping framework. A nonparametric approach is used to model the mean and covariance function of the longitudinal trait while the traditional Cox proportional hazard (PH) model is used to model the event time. The joint model is applied to map QTLs that control whole-plant vegetative biomass growth and time to first flower in soybeans. Results show that this model should be broadly useful for detecting genes controlling physiological and pathological processes and other events of interest in biomedicine.

1. Introduction

To study biology, a classic approach is dimension reduction in which a biological phenomenon or process is dissected into several discrete features over time and space. Most efforts in the past decades have been made to understand biological details of individual features and then use knowledge from each feature to draw an inference about biology as a whole. There has been increasing recognition of the limitation of this approach because it fails to detect a rule that governs the transition from one feature to next, thus leading to a significant loss of information behind the development of a biological trait. More recently, tremendous developments in statistics and computer science have enabled scientists to model and compute the dynamic behavior of a biological phenomenon and construct a comprehensive view

of how a cell, tissue, or organ grows and develops across the time-space scale.

A statistical dynamic model, called functional mapping, is one of the products of such developments [1, 2]. The merit of functional mapping lies in its biological relevance to study the tempo-spatial pattern of change for the trait and further predict the physiological or pathological status of trait phenotype. Functional mapping has proven to be powerful for elucidating the dynamic genetic architecture of complex phenotypic traits by identifying when specific genes (known as quantitative trait loci or QTLs) involved turn on and turn off and how long they are expressed in a time course. With the advent of new automatic techniques that collect dynamic data in a cost-effective way, functional mapping can be anticipated to play an increasingly important role in shedding light on the genetic control mechanisms of complex traits or diseases.

The statistical foundation of functional mapping is longitudinal data analysis or functional data analysis. There has been a considerable body of literature on statistical modeling of time-varying mean and covariance structure using various parametric, nonparametric, and semiparametric methods [3–7]. A joint mean-covariance model was proposed by Pourahmadi [8, 9], which shows some advantages over modeling the mean and covariance separately. Since the publication of the pioneering work by Laird and Ware [10], random effects model have been extensively used for longitudinal data analysis [11]. All these statistical approaches have been incorporated into functional mapping [12, 13], aiming to provide the most parsimonious estimates of QTL effects for a given data set. A Bayesian algorithm for functional mapping has been proposed recently by Liu and Wu [14].

The complexity of biology lies in the fact that no biological trait is isolated, rather every trait is affected by other traits through genes and environmental factors. For example, when a plant grows into a particular stage, reproductive behavior, such as flowering, starts to emerge as one of the important events in plant development. The time to first flower is highly associated with the amount of vegetative growth, depending on the environment where the plant is grown. Likewise, the time to recurrence of prostate cancer in humans is related with dynamic changes of prostate specific antigen level. How to jointly model longitudinal and time-to-event data within functional mapping has become an important issue for studying the common genetic basis of these processes and predicting events based on longitudinal traits.

Simultaneous modeling of longitudinal traits and time to events has been an active area in biostatistics during the past twenty years. A linear random effects model and EM estimation approach are proposed by Henderson et al. [15] for joint modeling. Guo and Carlin [16] made a comparative study between separate models and a joint model, showing that a joint model is more powerful when there is a strong correlation between the trait and the event. Wang and Taylor [17] developed a Bayesian method and MCMC algorithm for joint modeling of longitudinal and event time data and applied their algorithm on AIDS data. A review article by Tsiatis and Davidian [18] nicely summarizes the recent developments for such joint modeling.

By simply estimating the correlation between longitudinal traits and event time, Lin and Wu [19] developed a first model that connects these two aspects within functional mapping. However, they developed a likelihood-based framework where the covariance structure for the longitudinal trait was modelled by the known AR(1), and model parameters were estimated using maximum likelihood estimation. Taking advantage of event models, such as semiparametric Cox proportional hazard model, Weibull model, accelerated failure time (AFT) model, we here propose a sophisticated model for joint modeling of longitudinal trait and time to event to locate the QTLs which control the event via a dynamic trait. The detection of those QTLs that are common to these types of traits may help to

prevent or accelerate the outcome by genetic approaches. Our model is constructed with a Bayesian paradigm and model parameters are estimated by the MCMC algorithm. Local polynomials are used to model the mean trajectory and generalized-linear-model- (GLM-) based approach is used to model the covariance matrix. The model is validated using a real example in which whole-plant biomass as a longitudinal trait measured at a series of discrete time points and the time to first flower as a time-to-event are jointly modeled through functional mapping. The statistical properties of the model applied to estimate QTL temporal effects in this example and its practical usefulness are investigated by simulation studies.

2. Joint Modeling Framework

2.1. Model for the Longitudinal Trait. Genetic mapping should be based on a segregating population, such as the backcross, F_2, or recombinant inbred lines (RILS), initiated with two inbred lines each carrying an alternative allele. An RIL population is generated by self-crossing the hybrids of the two inbred lines continuously for 7-8 successive generations, which leads to two homozygous genotypes for alternative alleles at each locus. Methods for other designs can be derived similarly. Suppose a backcross has n progeny which is genotyped to construct a linkage map, aiming at locating putative QTLs that trigger significant effects on a longitudinal trait and its associated event. For each progeny, the trait is measured repeatedly at T different time points and a time-to-event is also recorded. At a specific time point t, the phenotypic value of the trait for progeny i affected by a putative QTL can be expressed by a linear model as follows:

$$y_i(t) = \sum_{j=1}^{2} x_{ij} u_j(t) + r_i(t) + e_i(t), \qquad (1)$$

where x_{ij} is an indicator variable for a possible QTL genotype of progeny i and defined as 1 if a particular QTL genotype j is indicated and 0 otherwise ($j = 1$ for QTL genotype QQ and 2 for genotype qq), $u_j(t)$ is the mean phenotypic value of QTL genotype j for progeny i at time t, $r_i(t)$ is the subject specific random effect, and $e_i(t)$ is the residual error assumed to follow a normal distribution with mean zero and covariance matrix Σ.

The central theme of functional mapping is to model the mean and covariance structures for the longitudinal trait efficiently. Here, we model the mean vector by polynomial function and the covariance matrix by an approach that guarantees the positive definiteness of the estimated covariance matrix. Without loss of generality, assume the response vector for progeny i, $\mathbf{y}_i = (y_i(1), \ldots, y_i(T))$, has mean 0 and covariance matrix Σ. The response at time t, $y_i(t)$, can be predicted by its predecessors as the follows:

$$y_i(t) = \sum_{t'=1}^{t-1} \phi_{t,t'} y_i(t') + \epsilon_i(t), \qquad (2)$$

where $\phi_{t,t'}$ is the corresponding regression coefficient, $\epsilon_i(t)$ is the prediction error for progeny i with mean = 0, and $\sigma^2(t)$ is its variance. Assuming that $\epsilon_i(t)$'s are uncorrelated (Pourahmadi [8]), we get $\text{cov}(\epsilon_i) = D$, a diagonal matrix with $\sigma^2(t)$ being the tth diagonal element, where $\epsilon_i = (\epsilon_i(1), \ldots, \epsilon_i(T))'$ is the vector of prediction errors. Hence, the matrix representation of the above autoregression becomes

$$\epsilon_i = \mathbf{M} \mathbf{y}_i, \tag{3}$$

where $\mathbf{M}$ is a lower triangular matrix with 1's in diagonal elements and $-\phi_{t,t'}$ in the (t, t')th position. The above equation simply gives

$$\text{cov}(\epsilon_i) = \mathbf{M} \, \text{cov}(\mathbf{y}_i) \mathbf{M}^T = \mathbf{M} \Sigma \mathbf{M}^T = \mathbf{D}, \tag{4}$$

which is related to the modified Cholesky decomposition of Σ [20].

Equation (4) will be considered as the basis for modeling the covariance structure, since this guarantees the estimated covariance matrix to be positive definite. Following Pourahmadi [8, 9], we model the mean vector, unconstrained variance parameters $\log \sigma^2(t)$, and dependence parameter $\phi_{t,t'}$, using a polynomial function of a particular order, expressed as

$$u_j(t) = \beta_{j0} + \beta_{j1}t + \beta_{j2}t^2 + \cdots + \beta_{jr}t^r, \tag{5}$$

$$r_i(t) = \theta_{i0} + \theta_{i1}t + \theta_{i2}t^2 + \cdots + \theta_{im}t^m, \tag{6}$$

$$\log \sigma_t^2 = \eta_0 + \eta_1 t + \eta_2 t^2 + \cdots + \eta_g t^g, \tag{7}$$

$$\phi_{t,t'} = \delta_0 + \delta_1(t - t') + \delta_2(t - t')^2 + \cdots + \delta_h(t - t')^h, \tag{8}$$
$$(t' = 1, 2, \ldots, t - 1).$$

The optimal (r, m, g, h) is determined from the information criteria (AIC/BIC). We note that different genotypes are assumed to have the same covariance structure but different means. Note that the above method of modeling the covariance structure for a longitudinal response is more robust than the traditional first-order autoregressive (AR(1)) or compound symmetry (CS) structure since real data might not show a parametric dependence structure. We refer to the proposed approach as GLM-based approach to estimate the covariance matrix.

Denote $\boldsymbol{\beta}_j = (\beta_{j0}, \beta_{j1}, \ldots, \beta_{jr})$ for QTL genotype j and $\boldsymbol{\theta}_i = (\theta_{i0}, \theta_{i1}, \ldots, \theta_{im})$ for subject i. Then, the conditional mean function for progeny i carrying QTL genotype j ($j = 1$ or 2) for the given subject specific random effect ($\boldsymbol{\theta}_i$) can be expressed as

$$\boldsymbol{\mu}_{ij} = \mathbf{X}_i^{(r)} \boldsymbol{\beta}_j^T + \mathbf{X}_i^{(m)} \boldsymbol{\theta}_i^T, \tag{9}$$

where

$$\mathbf{X}_i^{(r)} = \begin{bmatrix} 1 & t_{i1} & t_{i1}^2 & \cdots & t_{i1}^r \\ 1 & t_{i2} & t_{i2}^2 & \cdots & t_{i2}^r \\ \vdots & \vdots & \vdots & \ddots & \vdots \\ 1 & t_{i\tau} & t_{i\tau}^2 & \cdots & t_{i\tau}^r \\ \vdots & \vdots & \vdots & \ddots & \vdots \\ 1 & t_{iT} & t_{iT}^2 & \cdots & t_{iT}^r \end{bmatrix}. \tag{10}$$

Assume the vectors of subject specific random effects $\boldsymbol{\theta}_i$ follow m-variate normal distribution with mean 0 and covariance matrix $\sigma^2 I_m$ and they are independent of the residual errors. Note that under this assumption, $\mathbf{y}_i | \boldsymbol{\theta}_i$ will follow $\text{MVN}(\mathbf{X}_i^{(r)} \boldsymbol{\beta}_j^T + \mathbf{X}_i^{(m)} \boldsymbol{\theta}_i^T, \Sigma)$, and the marginal distribution of $\mathbf{y}_i$ will be $\text{MVN}(\mathbf{X}_i^{(r)} \boldsymbol{\beta}_j^T, \Sigma + \sigma^2 \mathbf{X}_i^{(m)} \mathbf{X}_i^{(m)T})$.

2.2. Model for the Event Time. We use s_i to denote the event time of progeny i. Since in the current situation, the event time is recorded for all progeny; no progeny is censored. Assuming a Cox proportional hazard model for this event, we get for progeny i,

$$\lambda_i(t) = \lambda_0(t) \exp \left(\gamma \mu_{ij}(t) \right), \tag{11}$$

where $\lambda_0(t)$ denotes the baseline hazard at time t, $\mu_{ij}(t)$ is the mean longitudinal trait at time t for given $\boldsymbol{\theta}_i$ when progeny i is of QTL genotype j and the regression coefficient γ represents the effect of the trait on the event time. The survival function for progeny i can be expressed in terms of the hazard function as $s_i(t) = \exp\left[-\int_0^t \lambda_i(u) du\right]$.

The longitudinal model described above is linked to the hazard model by γ. If $\gamma = 0$, then the event is independent of the trait, and hence we should better fit separate models for the trait and the event. However, when γ is different from zero, a joint model performs better than the separate models [17]. For simplicity, the baseline hazard is assumed to be a step function, $\lambda_0(t) = \lambda_{0k}$ over a partition of the observed time scale $[0, \max(s_i)]$ into K (possibly evenly spaced) intervals, $(t_0^\lambda = 0, t_1^\lambda, t_2^\lambda, \ldots, t_K^\lambda)$. The value of K is usually not too large, possibly smaller than 10.

2.3. Likelihood for the Joint Model. Since the QTL genotype of a progeny is unknown, we use a mixture model to describe the likelihood of the progeny in terms of its possible underlying QTL genotypes [21]. The joint likelihood of unknown parameters $\boldsymbol{\Theta}$ given the longitudinal trait $\mathbf{y} = (\mathbf{y}_i)_{i=1}^n$ and event time $\mathbf{s} = (\mathbf{s}_i)_{i=1}^n$ for all n progeny can be expressed as

$$L(\boldsymbol{\Theta} \mid \mathbf{y}, \mathbf{s}) = \prod_{i=1}^n \left(\sum_{j=1}^2 \omega_{j|i} [\pi(\mathbf{y}_i, s_i \mid Q_i = j, \boldsymbol{\theta}_i)] \right)$$
$$= \prod_{i=1}^n \left(\sum_{j=1}^2 \omega_{j|i} \left\{ [f(\mathbf{y}_i \mid Q_i = j, \boldsymbol{\theta}_i)] \right. \right.$$
$$\left. \left. \times \left[\lambda_i(s_i) \exp\left(-\int_0^{s_i} \lambda_i(u) du \right) \right] \right\} \right), \tag{12}$$

where $\pi(\cdot)$ denotes the joint density of the longitudinal trait and event time; $f(.)$ denotes a multivariate normal with QTL genotype-specific mean $\boldsymbol{\mu}_{ij}$ modeled as (9) and covariance matrix Σ; hazard function $\lambda_i(s_i)$ is modeled as (11); and $\omega_{j|i}$ denotes the conditional probability of QTL genotype j given that the marker information of progeny i and Q_i is the QTL genotype for the i-th subject.

The QTL genotype is inferred from marker genotypes of the linkage map. Let $M_i = (M_{i1}, \ldots, M_{im})$ be the m-marker genotypes for progeny i, D^* the position of the putative QTL measured by its distance from the very first marker of an ordered linkage group, and D_k the distances between marker 1 and k. Assume that the QTL is located between marker k and $k+1$. Then, the conditional probability of QTL genotype j given the genotype of these two markers that flank the QTL is expressed as

$$\omega_{j|i} = \mathrm{Prob}(Q_i = j \mid D^*, M_{ik}, M_{i(k+1)}, D_k, D_{k+1}). \quad (13)$$

Note that, given the QTL locations D^*, D_k, and D_{k+1}, one can compute d_1, the distance of the QTL from marker k and d_2, the distance of the QTL from marker $k+1$ [22]. Using the Haldane map function, one can compute recombination fractions between marker k and QTL (r_1), between QTL and marker $k+1$ (r_2), and between markers k and $k+1$ (r) as follows:

$$r_1 = \frac{1}{2}\left(1 - e^{-2d_1}\right), \qquad r_2 = \frac{1}{2}\left(1 - e^{-2d_2}\right),$$
$$r = \frac{1}{2}\left(1 - e^{-2d}\right), \quad (14)$$

where $d = D_{k+1} - D_k$ is the distance between marker k and $k+1$. Wu et al. [22] provide a procedure for deriving the conditional probabilities of QTL genotypes given marker interval genotypes for the backcross, F$_2$, and RIL populations, respectively.

Unknown parameters Θ in likelihood (12) contain QTL genotype-specific parameters $\boldsymbol{\beta}_j$, σ^2, the parameters that model the variance structure and dependence structure $\boldsymbol{\eta} = (\eta_0, \eta_1, \dots, \eta_g)$ and $\boldsymbol{\delta} = (\delta_0, \delta_1, \dots, \delta_h)$, as shown in models (7) and (8), respectively, the effect of the longitudinal trait on the event γ, QTL position D^* and the baseline hazards λ_{0k}.

2.4. Posterior Distribution and Sampling Procedure. We derive a Bayesian approach for estimating the unknown parameters. This will first need to specify the prior distributions for the parameters and, given the data and the priors, derive the posterior distribution over all the unknown parameters. For β_j, we place a multivariate normal prior with zero mean and covariance matrix Σ_β. An inverse gamma prior with parameters (α_1, α_2) is considered for σ^2. We consider a uniform prior on γ and uniform $(0, D_m)$ prior for the parameter D^*. Independent Gamma (a, b) priors are taken for λ_{0k}, for $k = 1, \dots, K$. Priors for $\boldsymbol{\eta}$ and $\boldsymbol{\delta}$ are taken as MVN$(0, \Sigma_\eta)$ and MVN$(0, \Sigma_\delta)$, respectively.

With the above priors and likelihood function, we have the joint posterior distribution for the parameters. In this case, it is quite straightforward to get the full conditional posterior distributions. Assume that the priors are independent for different parameters. Thus, we get the posterior density of β, σ^2, $\boldsymbol{\gamma}$, D^*, $\boldsymbol{\eta}$, $\boldsymbol{\delta}$, $\boldsymbol{\lambda_0}$ as

$$\pi(\beta, \sigma^2, \boldsymbol{\gamma}, D^*, \boldsymbol{\eta}, \boldsymbol{\delta}, \boldsymbol{\lambda_0} \mid \mathbf{y}, \mathbf{s}) \propto \pi(\mathbf{y}, \mathbf{s} \mid \beta, \sigma^2, \boldsymbol{\gamma}, D^*, \boldsymbol{\eta}, \boldsymbol{\delta}, \boldsymbol{\lambda_0})$$
$$\times \pi(\beta)\pi(\sigma^2)\pi(\boldsymbol{\gamma})\pi(D^*)\pi(\boldsymbol{\eta})\pi(\boldsymbol{\delta})\pi(\boldsymbol{\lambda_0}). \quad (15)$$

Assuming that priors for different genotypes are independent, we can express the above posterior distribution as

$$\pi(\beta, \sigma^2, \gamma, D, \boldsymbol{\eta}, \boldsymbol{\delta}, \boldsymbol{\lambda_0} \mid \mathbf{y}, \mathbf{s})$$
$$\propto \pi(\mathbf{y}, \mathbf{s} \mid \beta, \sigma^2, \gamma, D^*, \boldsymbol{\eta}, \boldsymbol{\delta}, \boldsymbol{\lambda_0})$$
$$\times \left[\prod_{j=1}^{2} \pi(\boldsymbol{\beta}_j)\right] \pi(\sigma^2)\pi(\gamma)\pi(D^*)\pi(\boldsymbol{\eta})\pi(\boldsymbol{\delta})\pi(\boldsymbol{\lambda_0}). \quad (16)$$

The full conditional distributions for the model parameters, as derived in the Appendix, are used to estimate the parameters using the MCMC algorithm. Note that the full conditional distribution for β_j is expressed as a product of a normal distribution term which comes from the longitudinal trait and two other terms from the hazard model. To update β_j, therefore, we use a Metropolis-Hastings (MH) algorithm with a normal proposal density since it is a part of its posterior distribution. We also note that in the full conditionals of η and δ, normal distribution coming from the longitudinal part of the data is the main determinant. Hence for η and δ, we consider normal proposals with the current value of the parameter as the mean and covariance matrix as Σ_η and Σ_δ, respectively. Selection of a good proposal density for γ is a bit tricky and we follow the recommendation given by Wang and Taylor [17]. By evaluating several choices for a good proposal, we consider a normal distribution with mean as the current state of the parameter and a suitable standard deviation in such a way that the proposed density gets well mixed with the target distribution (acceptance rate between 0.25 to 0.40). Because of conjugacy, we can directly simulate from the full conditional of λ_{0k}'s.

The parameter D^* which specifies the location of the QTL is updated following the idea of Satagopan et al. [23] by using the MH algorithm. A new value of D^*, which we denote by $D^{*\mathrm{new}}$, is generated from Uniform $(\max(0, D^* - \psi), \min(D^* + \psi, D_m))$, where ψ is the tuning parameter. Denote this proposed distribution by $q(D^*, D^{*\mathrm{new}})$. The proposed value will be considered as the new value of the chain with probability

$$\alpha(D^*, D^{*\mathrm{new}})$$
$$= \min\left[1, \frac{\pi(D^{*\mathrm{new}} \mid y, s, \beta, \gamma, \eta, \delta, \lambda_0)q(D^{*\mathrm{new}}, D^*)}{\pi(D^* \mid y, s, \beta, \gamma, \eta, \delta, \lambda_0)q(D^*, D^{*\mathrm{new}})}\right]. \quad (17)$$

We note that $\pi(D^{*\mathrm{new}} \mid \dots) \propto \pi(Q \mid D^{*\mathrm{new}}, M)\pi(D^{*\mathrm{new}}) = \prod_{i=1}^{n} \pi(Q_i \mid D^{*\mathrm{new}}, M_i)\pi(D^{*\mathrm{new}})$ and, similarly, $\pi(D^* \mid \dots) = \prod_{i=1}^{n} \pi(Q_i \mid D^*, M_i)\pi(D^*)$.

Because of the independence among n progeny, Q is updated by updating for each Q_i separately. For each progeny i, the full conditional density is in the form of a multinomial with the following cell probabilities:

$$\Pi(Q_i = j \mid y_i, s_i) = \frac{\pi(Q_i = j)\pi(y_i, s_i \mid Q_i = j)}{\sum_{j'=1}^{2} \pi(Q_i = j')\pi(y_i, s_i \mid Q_i = j')}. \quad (18)$$

We can sample the QTL genotype directly from this full conditional density at each cycle. Details of the estimation procedure can be found in Satagopan et al. [23].

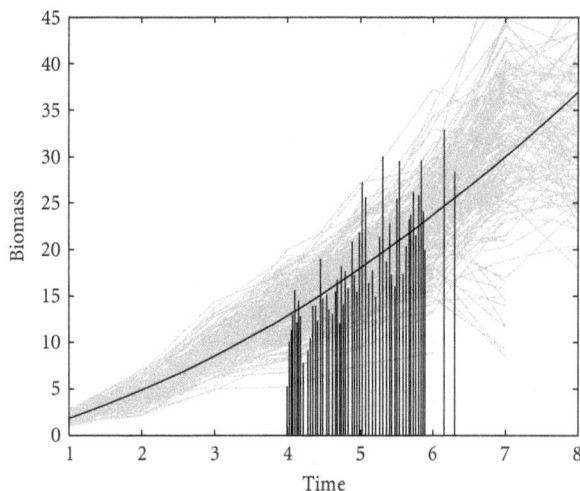

FIGURE 1: Whole-plant biomass growth trajectories for 184 soybean RILs. The time to first flower is indicated by a vertical line on each biomass growth curve. The black curve is the mean growth trajectory.

3. Application

3.1. Material and Analysis. The new model was applied to analyze a real data set for QTL mapping in soybeans. The mapping population contains 184 RILs derived from two cultivars, Nannong 1138-2 and Kefeng no. 1. A genetic linkage map of this population was first established by Zhang et al. [24] with 452 makers including RFLP, SSR, EST distributed among the 21 linkage group. This map was recently updated by adding some new SSR makers and dumping some unreliable markers. The new map contains 834 molecular makers covering a length of 2,308 cM in 24 linkage groups, with an average genetic distance of 2.85 cM between adjacent markers. Those markers with missing information were excluded from the analysis, leading to a total of 780 markers involved in our analysis.

The plants and their parents were grown in a sample lattice design with two replicates at Jiangpu Soybean Experiment Station, Nanjing Agricultural University, China. After 20 days of seedling emergence, plant biomass (in gms.) were measured once every 5–10 days until most plants stopped height growth. A total of 8 measurements were taken for the biomass and the time to get the first flower in that growing season was also recorded for each plant. Figure 1 shows the raw data, both the trait (biomass) and the event time for 184 plants.

Prior distributions for the model parameters were taken as follows. For genotype specific fixed effect β_j, a multivariate normal prior was used with zero mean and a diagonal covariance matrix with all diagonal elements 100. For σ^2, we took $IG(3, 1)$ prior which has mean = 0.5 with small variance (0.25). A uniform prior $U(-3.0, -0.1)$ was taken for γ following Wang and Taylor [17]. Observed time scale for the event time data was partitioned into 5 parts; that is, we took $K = 5$ and considered independent gamma $(0.04, 1.0)$ priors for $\lambda_{01}, \ldots, \lambda_{05}$, following Wang and Taylor

[17]. Uniform prior was taken for D^*. For η and δ, we considered multivariate normal priors with zero means and diagonal covariance matrices with diagonal elements 30 and 20, respectively. To investigate the effect of prior distributions on the estimation method, we did a sensitivity analysis. Considering different sets of priors, we fitted our model many times and it turned out the estimation is almost insensitive to the choice of priors. Hence the choice of our priors (even with huge variances for β, η, and δ) does not affect much the estimation of the model parameters.

We fitted our joint model as described in Section 2, by MCMC sampling. We ran chains 120,000 times. To remove the effect of the starting values, we excluded first 20,000 burn-in iterations. With the remaining 100,000 iterations, we estimated the posterior distributions and the parameters were estimated by the posterior mean and also calculated the sample standard deviations for the posterior densities. For the MH algorithm, the acceptance rates were 0.31, 0.26, 0.25, and 0.30 for σ^2, γ, η, and δ, respectively.

Since our model is complex, we perform several standard diagnostic tests to assess the convergence of the Markov chains. First, we use the method proposed by Brooks and Gelman [25]. Considering five different chains with different starting points and discarding the burn-in iterations, we computed multivariate potential scale reduction factors (MPSRF) to assess the convergence of the chains. Starting points for the model parameters were drawn from the respective priors. The computed values of this statistic get stabilized near 1 after 60,000 iterations for our model parameters, which indicates convergence of the chains.

Second, we perform Geweke test which compares the earlier part of the markov chain to the later part for assessing convergence. After deleting the burn-in iterations, from the remaining 100,000 iterations, we take out two subsequences; the first 50,000 and the last 50,000 iterations. Also consistent spectral density estimates at zero frequency are calculated to compute the z-scores. The calculated Pvalues for our model parameters are above 0.18, which indicates a small absolute z-scores assessing the convergence of our chains. A detailed discussion of this method can be found in Geweke [26].

Finally, we perform the Heidelberger and Welch test as proposed by Heidelberger and Welch [27]. This test has two parts: a stationary test and a half-width test. Our chains after deletion of the burn-in iterations, pass the stationary test. To assess whether the number of iterations is adequate to estimate the parameters accurately, we calculate relative half-width (RHW). We consider the default α value (0.05) and predetermined tolerance value is taken as 0.1. For all our model parameters, the calculated RHW is in between 0.045 and 0.081. This indicates that we have enough iterations to estimate the model parameters with 95% confidence under tolerance of 0.1.

3.2. Results. Figure 1 is the growth trajectories of whole-plant biomass over time for all RILs, in which the times to first flower for each RIL are also indicated. There are great variability found for these two traits among RILs. In the exploratory data analysis, we computed BIC values for different orders for the triplet (r, m, g, h), showing that

TABLE 1: BIC values for selecting the optimum (r, m, g, h) to model the mean-covariance structures for whole-plant biomass growth trajectories and the time to first flower in soybeans.

(r, m, g, h)	BIC
$(3, 3, 3, 2)$	7.18
$(3, 2, 3, 2)$	4.15
$(3, 3, 2, 2)$	4.26
$(3, 2, 2, 2)$	3.91
$(2, 2, 2, 2)$	**2.03**
$(2, 3, 2, 2)$	2.87
$(2, 2, 3, 2)$	3.01
$(2, 3, 3, 2)$	4.26
$(3, 3, 3, 3)$	4.66
$(3, 2, 3, 3)$	3.17
$(3, 3, 2, 3)$	3.93
$(3, 2, 2, 3)$	4.05
$(2, 2, 2, 3)$	3.34
$(2, 3, 2, 3)$	5.16
$(2, 3, 2, 3)$	4.88
$(2, 3, 3, 3)$	3.16

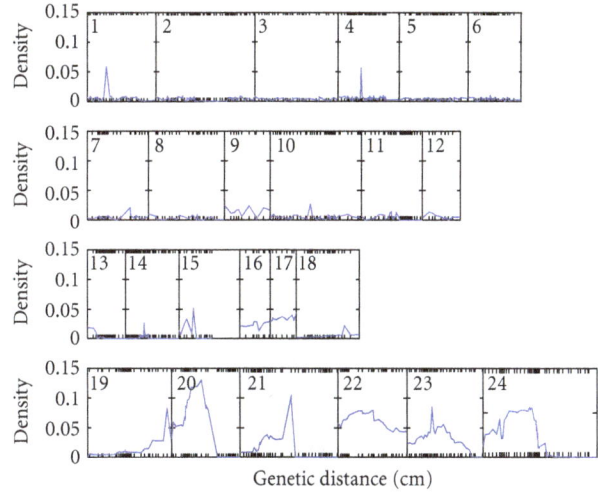

FIGURE 2: Marginal posterior plot for QTL locations over 24 linkage groups. Marker locations are indicated by ticks on the x-axis.

the BIC value is smallest for the order $(2, 2, 2, 2)$, that is, second order polynomials can best fit the the mean, variance, and dependence structures (Table 1). By scanning for the existence of QTLs over the genetic linkage map, we obtained the posterior distribution of the model parameters and estimate marginal posterior distributions of the QTL locations (D^*) for all 24 linkage groups (Figure 2).

We observed posterior peaks in linkage groups 1, 4, 15, 19, 20, 21, and 23. To draw inference about the existence of a putative QTL in each of these groups, we computed the Bayes Factor (BF), defined as

$$BF = \frac{P(\mathbf{Y}, \mathbf{S} \mid \kappa = 0)}{P(\mathbf{Y}, \mathbf{S} \mid \kappa = 1)}, \quad (19)$$

where κ denotes the number of QTLs in that particular group. Following Jeffrey's scale, the BFS with value smaller than 1 gives strong evidence against the null hypothesis and higher than 10 gives enough evidence for the null hypothesis. Note that in this case, we are testing the existence of no QTL (null) versus the existence of one QTL (alternative) for each linkage group. Here, no QTL in a group means $\beta_1 = \beta_2$, for that group. So, in order to compute BF we run our MCMC twice, first under the null and then under the alternative. The calculated BF for the above 7 linkage groups were 0.2781, 11.274, 13.493, 10.610, 0.5913, 11.475, and 0.7953, respectively, implying the existence of QTL in groups 1, 20, and 23. Table 2 provides genotype-specific mean parameters for the QTLs located on linkage groups 1, 20, and 23, along with their 95 percent credible intervals (C.I.).

Since the nature of our model is complex and our estimation is based on MCMC, we perform posterior predictive check for the aforementioned 7 linkage groups. We simulate observations to get the posterior predictive distribution. Let Θ be the set of all model parameters and $\mathbf{D}^{rep} = (\mathbf{y}^{rep}, \mathbf{s}^{rep})$ be the replicated data. Then given the data $\mathbf{D} = (\mathbf{y}, \mathbf{s})$, the posterior predictive distribution of $\mathbf{D}^{rep}$ is given by $p(\mathbf{D}^{rep} \mid \mathbf{D}) = \int p(\mathbf{D}^{rep} \mid \Theta) p(\Theta \mid \mathbf{D}) d\Theta$.

One can simulate from the posterior predictive distribution using the following two steps. First from the posterior distributions of the model parameters, simulate m values (vectors) of Θ. Next for each value of Θ, simulate a value (vector) $\mathbf{D}^{rep}$ from the likelihood. The m values (vectors) of $\mathbf{D}^{rep}$ drawn in this way will essentially come from posterior predictive distribution $p(\mathbf{D}^{rep} \mid \mathbf{D})$.

We simulate 100 draws ($m=100$) from the posterior predictive distribution and then apply proposed joint analysis and estimate the model parameters using MCMC as described earlier. We compute BF for each of those 7 linkage groups by considering the problem of testing the existence of no QTL (null) versus the existence of one QTL (alternative). Table 3 shows the average BF (with estimated SE) for each linkage group and the existence of QTL in groups 1, 20, and 23 is quite evident.

Heritability (broad-sense) for the traits is estimated from the data, as the proportion of phenotypic variance attributable to genetic variance. The estimated heritability in our case is 32.6%. Also we compute the percentage of variance explained by three identified QTLs. It turns out the QTLs identified in linkage groups 1, 20, and 23 explain 6.8%, 14.3% and 11.4% of the total variance, respectively.

We show the marginal posterior plots with 95% credible intervals for the parameter γ in Figure 3. Note that for all three groups, the estimates and the confidence intervals are in the negative part which indicates a negative relationship between the trait and the event time. Biologically this is sensible since the plants with higher body mass will take less time to have the first flower compared to the plants with lower body masses.

Figure 4 illustrates genotypic differences in whole-plant biomass trajectory and the time to first flower for the three

TABLE 2: Estimates of the parameters that describe genotype-specific biomass growth trajectories and QTL locations on linkage groups 1, 20, and 23, with 95% credible intervals.

Parameter	Group 1		Group 20		Group 23	
	estimate	C.I.	estimate	C.I.	estimate	C.I.
β_{10}	−0.4762	(−0.5081, −0.4442)	−1.1524	(−1.1641, −1.1405)	0.4190	(0.3897, 0.4484)
β_{11}	3.3214	(3.3023, 3.3404)	2.6829	(2.6463, 2.7194)	0.5971	(0.5566, 0.6377)
β_{12}	0.1548	(0.1363, 0.1731)	0.2295	(0.2020, 0.2570)	0.5438	(0.5176, 0.5701)
β_{20}	−5.4762	(−5.4788, −5.4734)	−3.1004	(−3.0231, −2.9768)	−2.3571	(−2.3701, −2.3442)
β_{21}	8.4464	(8.4381, 8.4546)	5.3250	(5.3159, 5.3340)	4.4660	(4.4294, 4.5028)
β_{22}	−0.4702	(−0.5011, −0.4392)	−0.0250	(−0.0618, 0.0118)	0.0410	(0.0390, 0.0432)
Marker Interval	Sat-356–B30T		GNE097b–A199H		LC4-4T–Sat-280	
D^*	30.810	(29.1934, 31.7150)	49.600	(48.7515, 50.0726)	39.472	(38.8143, 40.1863)

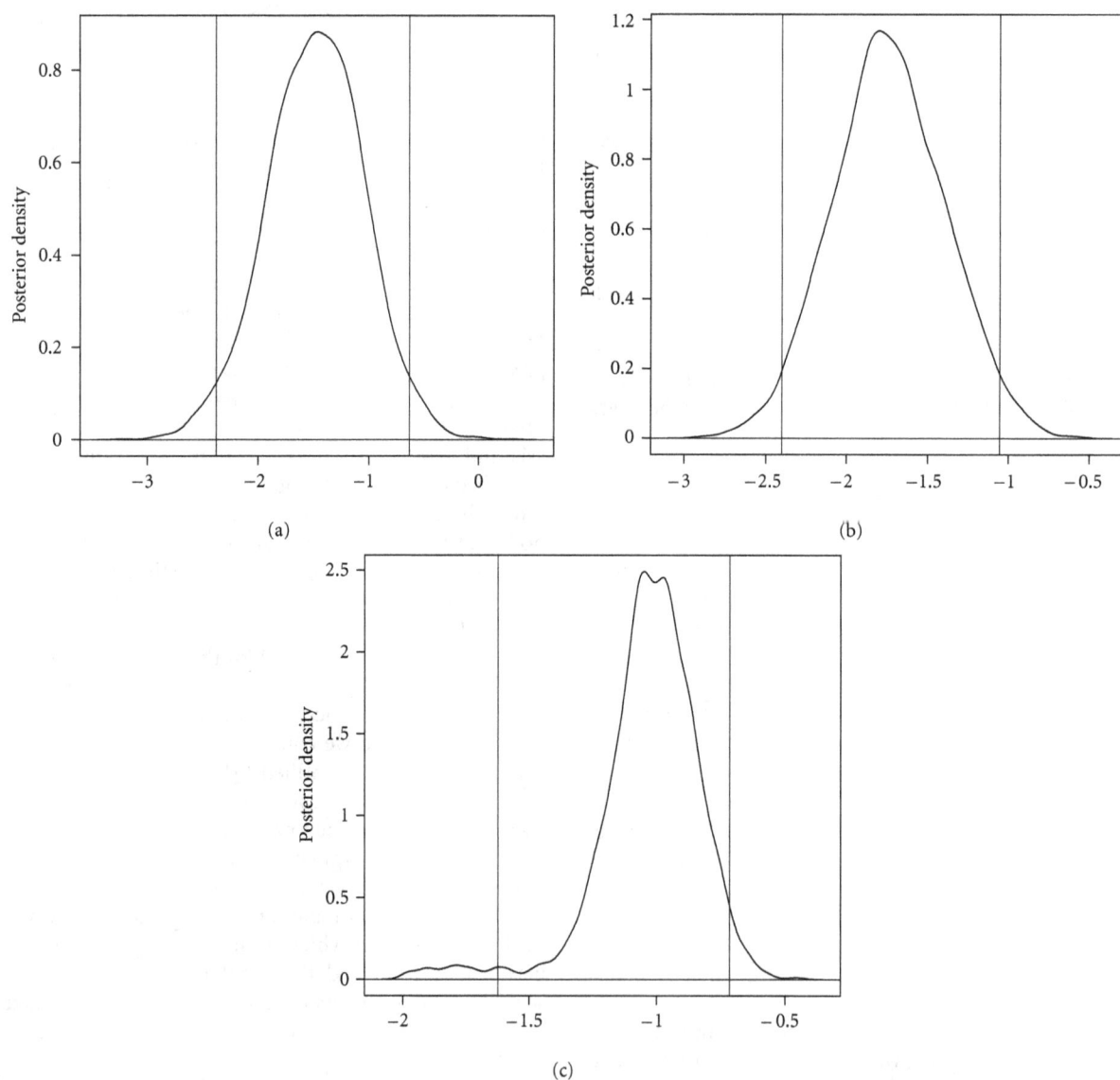

(a)

(b)

(c)

FIGURE 3: Marginal posterior plot for γ for linkage groups 1 (a), 20 (b), and 23 (c).

TABLE 3: Posterior predictive check for 7 linkage groups.

Linkage group	BF (from actual data)	BF with SE (from posterior predictive distribution)
1	0.2781	0.2659 (0.15)
4	11.274	12.086 (1.76)
15	13.493	12.962 (2.17)
19	10.610	11.138 (1.56)
20	0.5913	0.6281 (0.73)
21	11.475	12.183 (1.18)
23	0.7953	0.7682 (0.89)

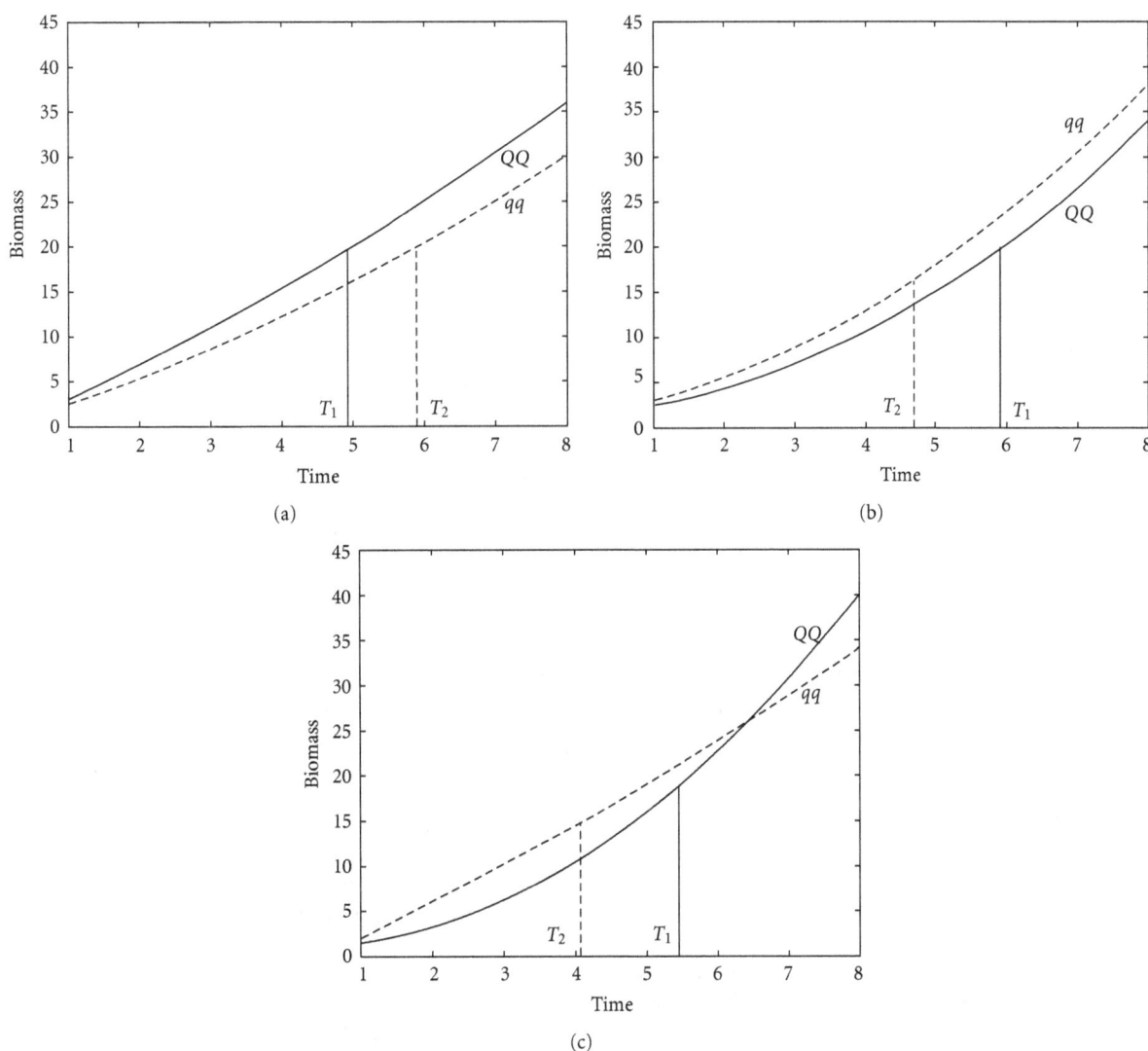

FIGURE 4: Whole-plant biomass growth trajectories and times to first flower (T_1 and T_2) for two different genotypes at each of the QTLs detected on linkage groups 1 (a), 20 (b), and 23 (c). Genotypes QQ inherit two alleles from parent Nannong 1138-2, whereas genotype qq inherits two alleles from parent Kefeng no. 1.

QTLs detected. At the QTL on linkage group 1, the allele (Q) inherited from parent Nannong 1138-2 leads to increased biomass growth and earlier flowering than the allele (q) from parent Kefeng no. 1. The inverse pattern is observed for the QTL on linkage group 20. The QTL on linkage group 23 alters its direction of genetic effect. Affected by the first two QTLs, poor vegetative biomass growth in plants stimulates early flowering (Figures 4(a) and 4(b)). Yet, the QTL on linkage group 23 makes the fast-growing genotype to flower earlier than the slow-growing genotype (Figure 4(c)).

TABLE 4: Simulation results for genotypic-mean parameters and QTL locations under different covariance structures, AR(1)-, CS- and GLM-based approach.

Parameter	Actual value	AR(1)		CS		GLM-based approach	
		Estimate	MCSE	Estimate	MCSE	Estimate	MCSE
β_{10}	−1.1524	−1.1872	0.0871	−1.0982	0.1094	−1.1667	0.0361
β_{11}	2.6829	2.5391	0.0502	2.6103	0.0495	2.7001	0.0103
β_{12}	0.2295	0.2288	0.0805	0.2301	0.0302	0.2291	0.0113
β_{20}	−3.1004	−3.1204	0.0307	−3.093	0.1025	−3.1255	0.1011
β_{21}	5.3250	5.3140	0.0291	5.2998	0.1130	5.3433	0.0405
β_{22}	−0.0250	−0.0241	0.1302	−0.0257	0.1035	−0.0244	0.0603
D^*	43.00	39.32	2.3561	40.61	1.5694	42.48	1.1572

4. Simulation Study

We performed simulation studies to study the statistical properties of the joint model. We assumed an RIL design of 200 progeny and simulated 11 evenly spaced markers on a linkage group of length 100 cM. A QTL is located at 43 cM from the very first marker of the linkage group. To reflect a practical problem, we used parameter estimates of the soybean QTL detected in linkage group 20 as true values to simulate the data, allowing the covariance structure. Time-dependent phenotypic values were assumed to follow a multivariate normal distribution and the event times were taken the same as the soybean data. To make a comparison, we analyzed the simulated data using our nonparametric GLM-based covariance structure and the traditional AR(1) and CS covariance structures.

The prior distributions for the model parameters were taken in the same way as discussed in Section 3.1. A uniform prior on (0,100) was considered for D^*. For each situation, we ran Markov chains 120,000 iterations and initial 20,000 burn-in iterations were discarded. Model parameters were estimated from the posterior distributions on the basis of remaining 100,000 iterations. The computed BF was 0.649 giving a strong evidence against the null hypothesis. Table 4 shows the means of the Bayesian estimates of model parameters with their respective Monte Carlo standard errors, (MCSE). It can be seen that the estimates are quite close to the actual values with a reasonably small standard errors which justifies the accuracy and precision of our estimation procedure. However, our GLM-based approach provides better estimation of parameters than AR(1) and CS-based approaches.

Figure 5 elucidates the marginal posterior plot for the QTL location under three different covariance structures. It is found that both AR(1)- and CS-based models provide the peaks at wrong locations, whereas GLM-based nonparametric covariance structure locates QTL more accurately in which case the length of the credible interval is narrower than those obtained from the former two structures. This provides numerical evidence that the proposed GLM-based model has better precision of QTL localization.

We perform further simulation studies to assess the reliability of BF in our data application. For each of the linkage groups 1, 20, and 23, we simulate data under the "null" model. As mentioned earlier, under the "null" model, $\beta_1 = \beta_2$. For group 1, we consider the null model $\beta_1 = \beta_2 = (-0.4762, 3, 3214, 0.1548)$, for group 20 it is $\beta_1 = \beta_2 = (-1.1524, 2.6829, 0.2295)$, and for group 23, our null model is given as $\beta_1 = \beta_2 = (0.4190, 0.5971, 0.5438)$. The computed BFS for these three groups are 34.55, 46.79, and 41.86, respectively, suggesting strong evidence for the null.

5. Discussion

Tools to reveal the secret of life should reflect the dynamic nature of life. More recently, a series of statistical models have been developed to map quantitative trait loci (QTLs) that control the dynamic process of a complex trait [1, 2, 22, 28]. These so-called functional mapping models integrate mathematical aspects of biological processes into a statistical framework derived to map complex trait QTLs and have proved to be useful for detecting and identifying genes and genetic interactions involved in quantitative genetic variation for plant height, plant rooting ability, and animal body mass. Functional mapping is also flexible to incorporate complex biological phenomena, such as genotype-environment interactions and allometric scaling providing powerful means for addressing biological questions of fundamental importance.

In this paper, we develop a new version of functional mapping that can map QTLs for developmental events affected by organismic growth trajectories in time. This version is benefited from recent statistical developments for joint modeling of longitudinal traits and event time [16–18, 29]. In the presence of strong correlation between a longitudinal trait and event, a joint model performs better than submodels separately for a single trait. In the joint modeling framework for these two types of traits, we applied a GLM-based approach to model the covariance structure and local polynomials for the mean curves. Bayesian estimation method using the MCMC algorithm was used since it is computationally much simpler than a likelihood-based approach. Simulation results show the effectiveness of GLM-based covariance model compared to traditional parametric compound symmetry or autoregressive structure.

Our joint model, embedded within functional mapping, promotes the study of testing how QTLs pleiotropically

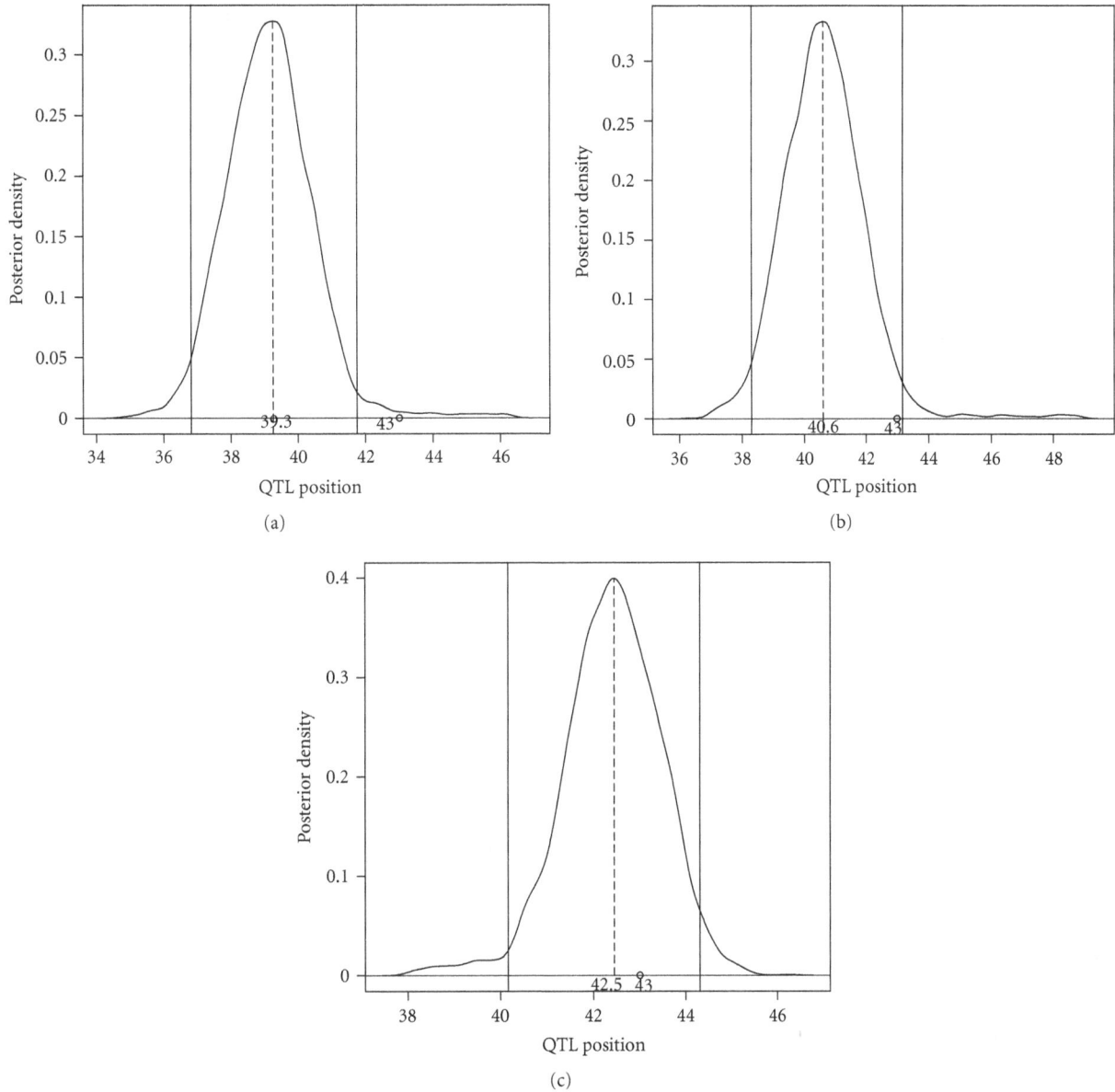

FIGURE 5: The Bayesian estimate of QTL location (indicated by dash vertical lines) from simulation studies under different covariance structures, AR(1) (a), CS (b), and nonparametric (c). The true QTL location is at 43 cM from the very first marker of the linkage group.

affect different biological processes and how one trait is predicted by other traits through genetic information. The application of the new model to soybean mapping data does not only validate its usefulness and utilization, but also gains new insight into the genetic and developmental regulation of trait correlations in plants. There is no doubt that the new model can be modified to study the genetic associations between HIV dynamics and the time to death as well as prostate specific antigen change and the time to recurrence of prostate. However, there is much room for modifying this model. First, to clearly describe our idea, we assume one QTL at a time for trait control. Epistatic interactions between multiple QTLs may play an important role in trait development as well as in correlations between

longitudinal traits and events. Second, from a dynamic systems perspective, we need to model dynamic correlations among multiple longitudinal traits and multiple events. Third, with the availability of efficient genotyping techniques, our model should accommodate a high-dimension model selection scheme to identify significant genetic variants from a flood of marker data.

Appendix

Denote $\beta_{-j} = (\beta_{j'}: j' = 1, 2; j' \neq j)$, $S(t) = \exp(-\int_0^t \lambda(u)du)$, $h(t) = (1, t, t^2, \ldots, t^r)$, and $g(t) = (1, t, t^2, \ldots, t^m)$. Let m_j be the number of progeny that carries QTL genotype j. We derive the full conditional distributions for the unknown parameters as

$$\pi\left(\boldsymbol{\beta}_j \mid \mathbf{y}, \mathbf{s}, \sigma^2, \gamma, D, \boldsymbol{\eta}, \boldsymbol{\delta}, \lambda_0, \boldsymbol{\beta}_{-j}\right)$$

$$\propto \pi\left(\boldsymbol{\beta}_j\right)\pi\left(\mathbf{y}, \mathbf{s} \mid \boldsymbol{\beta}, \sigma^2, \gamma, D, \boldsymbol{\eta}, \boldsymbol{\delta}, \lambda_0\right)$$

$$\propto \pi\left(\boldsymbol{\beta}_j\right)\pi\left(\mathbf{s} \mid \mathbf{y}, \boldsymbol{\beta}, \sigma^2, \gamma, D, \boldsymbol{\eta}, \boldsymbol{\delta}, \lambda_0\right)$$

$$\times \pi\left(\mathbf{y} \mid \boldsymbol{\beta}, \sigma^2, \gamma, D, \boldsymbol{\eta}, \boldsymbol{\delta}, \lambda_0\right)$$

$$\propto \exp\left[-\frac{1}{2}\boldsymbol{\beta}_j\boldsymbol{\Sigma}_\beta^{-1}\boldsymbol{\beta}_j^T - \frac{1}{2}\sum_{i=1}^{m_j}\left(\mathbf{y}_i - \mathbf{X}_i^{(r)}\boldsymbol{\beta}_j^T\right)^T\right.$$

$$\left.\times\boldsymbol{\Sigma}^{-1}\left(\mathbf{y}_i - \mathbf{X}_i^{(r)}\boldsymbol{\beta}_j^T\right)\right] \times \left[\exp\left(\gamma\sum_{i=1}^{m_j}h_i(s_i)\boldsymbol{\beta}_j^T\right)\right]$$

$$\times\left[\prod_{i=1}^{m_j}S_i(s_i)\right]$$

$$= \exp\left[-\frac{1}{2}\boldsymbol{\beta}_j\boldsymbol{\Sigma}_\beta^{-1}\boldsymbol{\beta}_j^T - \frac{1}{2}\sum_{i=1}^{m_j}\left(\mathbf{y}_i - \mathbf{X}_i^{(r)}\boldsymbol{\beta}_j^T\right)^T\right.$$

$$\left.\times\boldsymbol{\Sigma}^{-1}\left(\mathbf{y}_i - \mathbf{X}_i^{(r)}\boldsymbol{\beta}_j^T\right)\right] \times \left[\exp\left(\gamma\sum_{i=1}^{m_j}h_i(s_i)\boldsymbol{\beta}_j^T\right)\right]$$

$$\times\left[\prod_{i=1}^{m_j}S_i(s_i)\right]$$

$$\propto \exp\left[\boldsymbol{\beta}_j\boldsymbol{\Sigma}_\beta^{-1}\boldsymbol{\beta}_j^T + \sum_{i=1}^{m_j}\left(\boldsymbol{\beta}_j\mathbf{X}_i^{(r)T}\boldsymbol{\Sigma}^{-1}\mathbf{X}_i^{(r)}\boldsymbol{\beta}_j^T\right)\right.$$

$$\left.-2\sum_{i=1}^{m_j}\left(\mathbf{y}_i^T\boldsymbol{\Sigma}^{-1}\mathbf{X}_i^{(r)}\boldsymbol{\beta}_j^T\right)\right]\left[\exp\left(\gamma\sum_{i=1}^{m_j}h_i(s_i)\boldsymbol{\beta}_j^T\right)\right]$$

$$\times\left[\prod_{i=1}^{m_j}S_i(s_i)\right]$$

$$= \exp\left[\boldsymbol{\beta}_j\left(\boldsymbol{\Sigma}_\beta^{-1}+\sum_{i=1}^{m_j}\mathbf{X}_i^{(r)T}\boldsymbol{\Sigma}^{-1}\mathbf{X}_i^{(r)}\right)\boldsymbol{\beta}_j^T - 2\sum_{i=1}^{m_j}\left(\mathbf{y}_i^T\boldsymbol{\Sigma}^{-1}\mathbf{X}_i^{(r)}\boldsymbol{\beta}_j^T\right)\right]$$

$$\times\left[\exp\left(\gamma\sum_{i=1}^{m_j}h_i(s_i)\boldsymbol{\beta}_j^T\right)\right]\left[\prod_{i=1}^{m_j}S_i(s_i)\right]. \tag{A.1}$$

Hence, we have

$$\pi\left(\boldsymbol{\beta}_j \mid \cdot\right) : \mathrm{MVN}\left[\boldsymbol{\beta}_j, M_\beta, V_\beta\right]$$

$$\times\left[\exp\left(\gamma\sum_{i=1}^{m_j}h_i(s_i)\boldsymbol{\beta}_j^T\right)\right]\left[\prod_{i=1}^{m_j}S_i(s_i)\right], \tag{A.2}$$

where

$$M_\beta = \left[\left(\boldsymbol{\Sigma}_\beta^{-1}+\sum_{i=1}^{m_j}\mathbf{X}_i^{(r)T}\boldsymbol{\Sigma}^{-1}\mathbf{X}_i^{(r)}\right)^{-1}\left(\sum_{i=1}^{m_j}\mathbf{X}_i^{(r)T}\boldsymbol{\Sigma}^{-1}\mathbf{y}_i\right)\right],$$

$$V_\beta = \left(\boldsymbol{\Sigma}_\beta^{-1}+\sum_{i=1}^{m_j}\mathbf{X}_i^{(r)T}\boldsymbol{\Sigma}^{-1}\mathbf{X}_i^{(r)}\right)^{-1}. \tag{A.3}$$

Similarly, the full conditional distributions for the other parameters can be derived as follows:

$$\pi(\gamma \mid \cdot) \propto \left[\exp\left(\gamma\sum_{j=1}^{2}\sum_{i=1}^{m_j}\left(h_i(s_i)\boldsymbol{\beta}_j^T + g_i(s_i)\boldsymbol{\theta}_i^T\right)\right)\right]$$

$$\times\left[\prod_{i=1}^{n}S_i(s_i)\right].$$

$$\pi(\boldsymbol{\eta} \mid \cdot) \propto \left[\prod_{i=1}^{n}\det\left(\boldsymbol{\Sigma}\right)^{-(1/2)}\right]\mathrm{f}$$

$$\times\exp\left[-\frac{1}{2}\boldsymbol{\eta}'\boldsymbol{\Sigma}_\eta^{-1}\boldsymbol{\eta} - \frac{1}{2}\sum_{j=1}^{2}\sum_{i=1}^{m_j}\left(\mathbf{y}_i - \mathbf{X}_i^{(r)}\boldsymbol{\beta}_j^T\right.\right.$$

$$\left.\left.- \mathbf{X}_i^{(m)}\boldsymbol{\theta}_i^T\right)^T\boldsymbol{\Sigma}^{-1}\left(\mathbf{y}_i - \mathbf{X}_i^{(r)}\boldsymbol{\beta}_j^T - \mathbf{X}_i^{(m)}\boldsymbol{\theta}_i^T\right)\right].$$

$$\pi(\boldsymbol{\delta} \mid \cdot) \propto \left[\prod_{i=1}^{n}\det\left(\boldsymbol{\Sigma}\right)^{-(1/2)}\right]$$

$$\times\exp\left[-\frac{1}{2}\boldsymbol{\delta}'\boldsymbol{\Sigma}_\delta^{-1}\boldsymbol{\delta} - \frac{1}{2}\sum_{j=1}^{2}\sum_{i=1}^{m_j}\left(\mathbf{y}_i - \mathbf{X}_i^{(r)}\boldsymbol{\beta}_j^T\right.\right.$$

$$\left.\left.- \mathbf{X}_i^{(m)}\boldsymbol{\theta}_i^T\right)^T\boldsymbol{\Sigma}^{-1}\left(\mathbf{y}_i - \mathbf{X}_i^{(r)}\boldsymbol{\beta}_j^T - \mathbf{X}_i^{(m)}\boldsymbol{\theta}_i^T\right)\right].$$

$$\pi(\sigma^2 \mid \cdot) \propto \left[\prod_{i=1}^{n}\det\left(\boldsymbol{\Sigma} + \sigma^2\mathbf{X}_i^{(m)}\mathbf{X}_i^{(m)T}\right)^{-(1/2)}\right]$$

$$\times(\sigma^2)^{-\alpha_1-1}\exp\left(-\frac{\alpha_2}{\sigma^2}\right)$$

$$\times\exp\left[-\frac{1}{2}\sum_{j=1}^{2}\sum_{i=1}^{m_j}\left(\mathbf{y}_i - \mathbf{X}_i^{(r)}\boldsymbol{\beta}_j^T\right)^T\right.$$

$$\left.\times\left(\boldsymbol{\Sigma} + \sigma^2\mathbf{X}_i^{(m)}\mathbf{X}_i^{(m)T}\right)^{-1}\left(\mathbf{y}_i - \mathbf{X}_i^{(r)}\boldsymbol{\beta}_j^T\right)\right].$$

$$\pi(\lambda_{0k} \mid \cdot) \propto \mathrm{Gamma}\left[d_k + 1, \phi_{\lambda_{0k}}^{-1}\right]\pi(\lambda_{0k})$$

$$= \mathrm{Gamma}\left(a + d_k + 1, \left[\phi_{\lambda_{0k}} + \frac{1}{b}\right]^{-1}\right), \tag{A.4}$$

where d_k is the number of events in the interval $(t_{k-1}^\lambda, t_k^\lambda]$ and

$$\phi_{\lambda_{0k}} = \sum_{i:s_i\geq t_k^\lambda}\int_{t_{k-1}^\lambda}^{t_k^\lambda}\exp\left[\gamma\left(h(t)\boldsymbol{\beta}_j^T + g(t)\boldsymbol{\theta}_i^T\right)\right]dt$$

$$+ \sum_{i:t_{k-1}^\lambda<s_i\leq t_k^\lambda}\int_{t_{k-1}^\lambda}^{s_i}\exp\left[\gamma\left(h(t)\boldsymbol{\beta}_j^T + g(t)\boldsymbol{\theta}_i^T\right)\right]dt. \tag{A.5}$$

Acknowledgments

This work is partially supported by NSF/IOS-0923975, Changjiang Scholars Award, "Thousand-Person Plan" Award, the China National Key Basic Research Program (2006CB1017, 2009CB1184, 2010CB125906), the China National Hightech R&D Program (2006AA100104), the

Natural Science Foundation of China (30671266), and the China MOE 111 Project (B08025). R.Li was supported by an NIDA Grant P50-DA10075 and an NNSF of China Grant 11028103. The content is solely the responsibility of the authors and does not necessarily represent the official views of the NIDA, or the NIH.

References

[1] C. X. Ma, G. Casella, and R. Wu, "Functional mapping of quantitative trait loci underlying the character process: a theoretical framework," *Genetics*, vol. 161, no. 4, pp. 1751–1762, 2002.

[2] R. Wu and M. Lin, "Opinion: functional mapping—how to map and study the genetic architecture of dynamic complex traits," *Nature Reviews Genetics*, vol. 7, no. 3, pp. 229–237, 2006.

[3] J. Fan and R. Li, "New estimation and model selection procedures for semiparametric modeling in longitudinal data analysis," *Journal of the American Statistical Association*, vol. 99, no. 467, pp. 710–723, 2004.

[4] J. Fan, T. Huang, and R. Li, "Analysis of longitudinal data with semiparametric estimation of covariance function," *Journal of the American Statistical Association*, vol. 102, no. 478, pp. 632–641, 2007.

[5] J. O. Ramsay and B. W. Silverman, *Functional Data Analysis*, Springer, New York, NY, USA, 2nd edition, 2005.

[6] F. Yao, H. G. Müller, and J. L. Wang, "Functional data analysis for sparse longitudinal data," *Journal of the American Statistical Association*, vol. 100, no. 470, pp. 577–590, 2005.

[7] F. Yao, H. G. Muller, and J. L. Wang, "Functional linear regression analysis for longitudinal data," *Annals of Statistics*, vol. 33, no. 6, pp. 2873–2903, 2005.

[8] M. Pourahmadi, "Joint mean-covariance models with applications to longitudinal data: unconstrained parameterisation," *Biometrika*, vol. 86, no. 3, pp. 677–690, 1999.

[9] M. Pourahmadi, "Maximum likelihood estimation of generalised linear models for multivariate normal covariance matrix," *Biometrika*, vol. 87, no. 2, pp. 425–435, 2000.

[10] N. M. Laird and J. H. Ware, "Random-effects models for longitudinal data," *Biometrics*, vol. 38, no. 4, pp. 963–974, 1982.

[11] M. Davidian and D. M. Giltinan, "Nonlinear models for repeated measurement data: an overview and update," *Journal of Agricultural, Biological, and Environmental Statistics*, vol. 8, no. 4, pp. 387–419, 2003.

[12] J. S. Yap, J. Fan, and R. Wu, "Nonparametric modeling of longitudinal covariance structure in functional mapping of quantitative trait loci," *Biometrics*, vol. 65, no. 4, pp. 1068–1077, 2009.

[13] K. Das, J. Li, Z. Wang et al., "A dynamic model for genome-wide association studies," *Human Genetics*, vol. 129, no. 6, pp. 629–639, 2011.

[14] T. Liu and R. L. Wu, "A bayesian algorithm for functional mapping of dynamic complex traits," *Algorithms*, vol. 2, pp. 667–691, 2009.

[15] R. Henderson, P. Diggle, and A. Dobson, "Joint modelling of longitudinal measurements and event time data," *Biostatistics*, vol. 1, pp. 465–480, 2000.

[16] X. Guo and B. P. Carlin, "Separate and joint modeling of longitudinal and event time data using standard computer packages," *American Statistician*, vol. 58, no. 1, pp. 16–24, 2004.

[17] Y. Wang and J. M. G. Taylor, "Jointly modeling longitudinal and event time data with application to acquired immunodeficiency syndrome," *Journal of the American Statistical Association*, vol. 96, no. 455, pp. 895–905, 2001.

[18] A. A. Tsiatis and M. Davidian, "Joint modeling of longitudinal and time-to-event data: an overview," *Statistica Sinica*, vol. 14, no. 3, pp. 809–834, 2004.

[19] M. Lin and R. Wu, "A joint model for nonparametric functional mapping of longitudinal trajectory and time-to-event," *BMC Bioinformatics*, vol. 7, article 138, 2006.

[20] H. J. Newton, *IMESLAB: A Time Series Analysis Laboratory*, T. Wadsworth and Brooks/Coles, Pacific Grove, Calif, USA, 1988.

[21] E. S. Lander and S. Botstein, "Mapping mendelian factors underlying quantitative traits using RFLP linkage maps," *Genetics*, vol. 121, no. 1, p. 185, 1989.

[22] R. Wu, W. Hou, Y. Cui et al., "Modeling the genetic architecture of complex traits with molecular markers," *Recent Patents on Nanotechnology*, vol. 1, no. 1, pp. 41–49, 2007.

[23] J. M. Satagopan, B. S. Yandell, M. A. Newton, and T. C. Osborn, "A Bayesian approach to detect quantitative trait loci using Markov chain Monte Carlo," *Genetics*, vol. 144, no. 2, pp. 805–816, 1996.

[24] W. K. Zhang, Y. J. Wang, G. Z. Luo et al., "QTL mapping of ten agronomic traits on the soybean (Glycine max L. Merr.) genetic map and their association with EST markers," *Theoretical and Applied Genetics*, vol. 108, no. 6, pp. 1131–1139, 2004.

[25] S. P. Brooks and A. Gelman, "General methods for monitoring convergence of iterative simulations," *Journal of Computational and Graphical Statistics*, vol. 7, no. 4, pp. 434–455, 1998.

[26] J. Geweke, "Evaluating the accuracy of sampling-based approaches to calculating posterior moments," in *Bayesian Statistics*, vol. 4, Clarendon Press, Oxford, UK, 1992.

[27] P. Heidelberger and P. D. Welch, "Simulation run length control in the presence of an initial transient," *Operations Research*, vol. 31, no. 6, pp. 1109–1144, 1983.

[28] R. Wu, C. X. Ma, M. Lin, Z. Wang, and G. Casella, "Functional mapping of quantitative trait loci underlying growth trajectories using a transform-both-sides logistic model," *Biometrics*, vol. 60, no. 3, pp. 729–738, 2004.

[29] D. Rizopoulos, "Dynamic predictions and prospective accuracy in joint models for longitudinal and time-to-event data," *Biometrics*, vol. 67, no. 3, pp. 819–829, 2011.

14

SNP Markers and Their Impact on Plant Breeding

Jafar Mammadov,[1] Rajat Aggarwal,[1] Ramesh Buyyarapu,[1] and Siva Kumpatla[2]

[1] Department of Trait Genetics and Technologies, Dow AgroSciences LLC, 9330 Zionsville Road, Indianapolis, IN 46268, USA
[2] Department of Biotechnology Regulatory Sciences, Dow AgroSciences LLC, 9330 Zionsville Road, Indianapolis, IN 46268, USA

Correspondence should be addressed to Jafar Mammadov, jamammadov@dow.com

Academic Editor: Ian Bancroft

The use of molecular markers has revolutionized the pace and precision of plant genetic analysis which in turn facilitated the implementation of molecular breeding of crops. The last three decades have seen tremendous advances in the evolution of marker systems and the respective detection platforms. Markers based on single nucleotide polymorphisms (SNPs) have rapidly gained the center stage of molecular genetics during the recent years due to their abundance in the genomes and their amenability for high-throughput detection formats and platforms. Computational approaches dominate SNP discovery methods due to the ever-increasing sequence information in public databases; however, complex genomes pose special challenges in the identification of informative SNPs warranting alternative strategies in those crops. Many genotyping platforms and chemistries have become available making the use of SNPs even more attractive and efficient. This paper provides a review of historical and current efforts in the development, validation, and application of SNP markers in QTL/gene discovery and plant breeding by discussing key experimental strategies and cases exemplifying their impact.

1. Introduction

Allelic variations within a genome of the same species can be classified into three major groups that include differences in the number of tandem repeats at a particular locus [microsatellites, or simple sequence repeats (SSRs)] [1], segmental insertions/deletions (InDels) [2], and single nucleotide polymorphisms (SNPs) [3]. In order to detect and track these variations in the individuals of a progeny at DNA level, researchers have been developing and using genetic tools called molecular markers [4]. Although SSRs, InDels, and SNPs are the three major allelic variations discovered so far, a plethora of molecular markers were developed to detect the polymorphisms that resulted from these three types of variation [5]. Evolution of molecular markers has been primarily driven by the throughput and cost of detection method and the level of reproducibility [6]. Depending on detection method and throughput, all molecular markers can be divided into three major groups: (1) low-throughput, hybridization-based markers such as restriction fragment length polymorphisms (RFLPs) [4]; (2) medium-throughput, PCR-based markers that include random amplification of polymorphic DNA (RAPD) [7], amplified fragment length polymorphism (AFLP) [8], SSRs [9]; (3) high-throughput (HTP) sequence-based markers: SNPs [3]. In late eighties, RFLPs were the most popular molecular markers that were widely used in plant molecular genetics because they were reproducible and codominant [10]. However, the detection of RFLPs was an expensive, labor- and time-consuming process, which made these markers eventually obsolete. Moreover, RFLP markers were not amenable to automation. Invention of PCR technology and the application of this method for the rapid detection of polymorphisms overthrew low-throughput RFLP markers, and new generation of PCR-based markers emerged in the beginning of nineties. RAPD, AFLP, and SSR markers are the major PCR-based markers that research community has been using in various plant systems. RAPDs are able to simultaneously detect polymorphic loci in various regions of a genome [11]. However, they are anonymous and the level of their reproducibility is very low due to the non-specific binding of short, random primers. Although AFLPs are anonymous too, the level of their reproducibility and sensitivity is very high owing to the longer +1 and

+3 selective primers and the presence of discriminatory nucleotides at 3′ end of each primer. That is why AFLP markers are still popular in molecular genetics research in crops with little to zero reference genome sequence available [12]. However, AFLP markers did not find widespread application in molecular breeding owing to the lengthy and laborious detection method, which was not amenable to automation either. Therefore, it was not surprising that soon after the discovery of SSR markers in the genome of a plant, they were declared as "markers of choice" [13], because SSRs were able to eliminate all drawbacks of the above-mentioned DNA marker technologies. SSRs were no longer anonymous; they were highly reproducible, highly polymorphic, and amenable to automation. Despite the cost of detection remaining high, SSR markers had pervaded all areas of plant molecular genetics and breeding in late 90s and the beginning of 21st century. However, during the last five years, the hegemony of medium-throughput SSRs was eventually broken by SNP markers. First discovered in human genome, SNPs proved to be universal as well as the most abundant forms of genetic variation among individuals of the same species [14]. Although SNPs are less polymorphic than SSR markers because of their biallelic nature, they easily compensate this drawback by being abundant, ubiquitous, and amenable to high- and ultra-high-throughput automation. However, despite these obvious advantages, there were only a limited number of examples of application of SNP markers in plant breeding by 2009 [15]. In this paper, we tried to summarize the recent progress in the utility of SNP markers in plant breeding.

2. SNP Discovery in Complex Plant Genomes

While SNP discovery in crops with simple genomes is a relatively straightforward process, complex genomes pose serious obstacles for the researchers interested in developing SNPs. One of the major problems is the highly repetitive nature of the plant genomes [16]. Prior to the emergence of next-generation sequencing (NGS) technologies, researchers used to rely on different experimental strategies to avoid repetitive portions of the genome. These include discovery of SNPs experimentally by resequencing of unigene-derived amplicons using Sanger's method [17] and *in silico* SNP discovery through the mining of SNPs within EST databases followed by PCR-based validation [18]. Although these approaches allowed the detection of gene-based SNPs, their frequency is generally low in conserved genic regions, and they were unable to discover SNPs located in low-copy noncoding regions and intergenic spaces. Additionally, amplicon resequencing was an expensive and labor-intensive procedure [15]. As many crops are ancient tetraploids with mosaics of scattered duplicated regions [19], *in silico* and experimental mining of EST databases resulted in the discovery of a large number of nonallelic SNPs that represented paralogous sequences and were suboptimal for application in molecular breeding [20]. Recent emergence of NGS technologies such as 454 Life

Sciences (Roche Applied Science, Indianapolis, IN), HiSeq (Illumina, San Diego, CA), SOLiD and Ion Torrent (Life Technologies Corporation, Carlsbad, CA) has eliminated the problems associated with low throughput and high cost of SNP discovery [21]. Transcriptome resequencing using NGS technologies allows rapid and inexpensive SNP discovery within genes and avoids highly repetitive regions of a genome [22]. This methodology was successfully applied in several plant genomes, including maize [23], canola [24], eucalyptus [25], sugarcane [26], tree species [27], wheat [28], avocado [29], and black currant [30]. Originally developed for human disease diagnostic research, the NimbleGen sequence capture technology (Roche Applied Science, IN) [31] brought the detection of gene-based SNPs in plants into higher throughput and coverage level [32]. This technology consists of exon sequence capture and enrichment by microarray followed by NGS for targeted resequencing. Similar in-solution target capture technologies, such as Agilent SureSelect, are also commercially available for genome/exome mining studies. However, this technology would be efficient only for crops with available reference genome sequence or large transcriptome (EST) datasets, since the design of capture probes requires these reference resources.

Despite the attractiveness of SNP discovery via transcriptome or exome resequencing, this process is targeted, focusing solely on coding regions. It is obvious that the availability of SNPs within coding sequences is a very powerful tool for molecular geneticists to detect a causative mutation [33]. However, often QTL are located in noncoding regulatory sequences such as enhancers or locus control regions, which could be located several megabases away from genes within intergenic spaces [34]. Discovery of SNPs located within those regulatory elements via transcriptome or exon sequencing is limited. In order to discover SNPs in a genome-wide fashion and avoid repetitive and duplicated DNA, it is very important to employ genome complexity reduction techniques coupled with NGS technologies. Several genome complexity reduction techniques have been developed over the years, including High Cot selection [35], methylation filtering [36], and microarray-based genomic selection [37]. These techniques mainly reduce the number of repetitive sequences but lack the power to recognize and eliminate duplicated sequences, which cause the detection of false-positive SNPs. Unlike the above-mentioned techniques, recently developed genome complexity reduction technologies such as Complexity Reduction of Polymorphic Sequences (CRoPS) (Keygene N.V., Wageningen, The Netherlands) [38] and Restriction Site Associated DNA (RAD) (Floragenics, Eugene, OR, USA) [39] are computationally well equipped and capable of filtering out duplicated SNPs. These systems were successfully applied to discover SNPs in crops with [40] and without reference genome sequences [41].

Although several complexity reduction approaches are being developed to generate data from NGS platforms, it is often challenging to identify candidate SNPs in polyploid crops species such as potato, tobacco, cotton, canola, and wheat. In general, minor allele frequency could be used as a measure to identify candidate SNPs in diploid species [42]. However, in polyploid crops, you often find loci that are polymorphic within a single genotype due to the presence of either homoeologous loci from the individual subgenomes (homoeologous SNPs) or paralogous loci from duplicated regions of the genome. Such false positive SNPs are not useful for genetic mapping purposes and often lead to a lower validation rate during assays. Successful SNP validation in allopolyploids depends upon differentiation of the sequence variation classes [43]. Use of haplotype information beside the allelic frequency would help to identify homologous SNPs (true SNPs) from those of homoeologous loci (false positives). Bioinformatic programs such as HaploSNPer [44] would facilitate identification of candidate loci for assay design purposes in polyploid crops. Elimination of homoeologous loci for the assay design process would improve the validation rate. Such approaches could also be extended to other complex and highly repetitive diploid genomes such as barley. Complexity reduction approaches, combined with sophisticated computational tools, would expedite SNP discovery and validation efforts in polyploids.

Although CRoPS and RAD technologies are powerful tools to detect SNPs in genome-wide fashion, they can hardly be called HTP, because on an average only ~1,000 SNPs pass stringent quality control [40]. While these numbers are enough to generate genetic linkage maps of reasonable saturation and carry out preliminary QTL mapping, they are not adequate to implement genome-wide association studies (GWAS). Depending on the rate of linkage disequilibrium decay, GWAS might require several million genetic landmarks. From this point of view, genotyping-by-sequencing (GBS) technique offers many more opportunities. Discovery of a large number of SNPs using GBS was demonstrated in maize [45] and sorghum [46]. GBS not only increases the sequencing throughput by several orders of magnitude but also has multiplexing capabilities [47]. To eliminate a large portion of repetitive sequences, a type II restriction endonuclease, ApeKI, is applied to digest DNA prior to sequencing to generate reduced representation libraries (genome complexity reduction component), which are further subject to sequencing [47]. In polyploid crops, GBS might be challenging, but the associated complexity reduction methods could be used for SNP discovery. For discovery purposes, the availability of a reference genome is not an absolute requirement to implement GBS approach. However, in organisms that do not have a reference genome, GBS-derived SNPs must be validated using one of the techniques that are described in the following section, which might dramatically increase per marker price. Validation needs to be done primarily to discard paralogous SNPs. For organisms with a reference genome sequence, the validation step is replaced by *in silico* mapping of the sequenced fragments to the genome. Although GBS has the potential to discover several million SNPs, one of the major drawbacks of this technique is large numbers of missing data. To solve this problem, computational biologists developed data imputation models such as BEAGLE v3.0.2 [48] and IMPUTE v2 [49], to bring imputed data as close as possible to the real data [50, 51].

3. SNP Validation and Modern Genotyping Platforms and Chemistries

The availability of reference sequence and sophisticated software does not always guarantee that the discovered SNP can be converted into a valid marker. In order to insure that the discovered SNP is a Mendelian locus, it has to be validated. The validation of a marker is the process of designing an assay based on the discovered polymorphism and then genotyping a panel of diverse germplasm and segregating population. Compared to the collection of unrelated lines, a segregating population is more informative as a validation panel because it allows the inspection of the discriminatory ability and segregation patterns of a marker which helps the researcher to understand whether it is a Mendelian locus or a duplicated/repetitive sequence that escaped the software filter [40].

The most popular HTP assays/chemistries and genotyping platforms that are currently being used for SNP validation are Illumina's BeadArray technology-based Golden Gate (GG) [52] and Infinium assays [53], Life Technologies' TaqMan [54] assay coupled with OpenArray platform (TaqMan OpenArray Genotyping system, Product bulletin), and KBiosciences' Competitive Allele Specific PCR (KASPar) combined with the SNP Line platform (SNP Line XL; http://www.kbioscience.co.uk). These modern genotyping assays and platforms differ from each other in their chemistry, cost, and throughput of samples to genotype and number of SNPs to validate. The choice of chemistry and genotyping platform depends on many factors that include the length of SNP context sequence, overall number of SNPs to genotype, and finally the funds available to the researcher, because most of these chemistries still remain cost intensive. Comparative analyses of these four genotyping assays and platforms were described in Kumpatla et al. [55].

Though all genotyping chemistries and platforms are applicable to generate genotypic data in polyploid crops, analysis of SNP calls is somewhat challenging in polyploids due to multiallele combinations in the genotypes. SNPs in polyploid species can be broadly classified as simple SNPs, hemi-SNPs, and homoeo-SNPs. Here, we describe simple, hemi-, and homoeo-SNPs using an example of allele calls in tetraploid and diploid cotton species (Figure 1). Genomes of tetraploid cotton species, *Gossypium hirsutum* (AD$_1$) and *G. barbadense* (AD$_2$), consist of two subgenomes A and D, where A genome was derived from diploid progenitors, such as *G. herbaceum* (A$_1$) and *G. arboreum* (A$_2$), and D genome resulted from another diploid progenitor *G. raimondii* (D$_5$). Simple, or true SNPs are markers that detect allelic variation between homologous loci of the same subgenome of two tetraploid samples. For example, in Figure 1(a), a SNP marker clearly detects polymorphism within A subgenomes of *G. hirsutum* (AD$_1$) and *G. barbadense* (AD$_2$) and separates samples into homozygous A (blue) and B (red) clusters. This marker does not discriminate polymorphism in D subgenome, because the D genome allele is absent there (pink dot in *G. raimondii*). In contrast to simple SNPs, hemi-SNPs detect allelic variation in the homozygous state in one sample and the heterozygous state in the other sample.

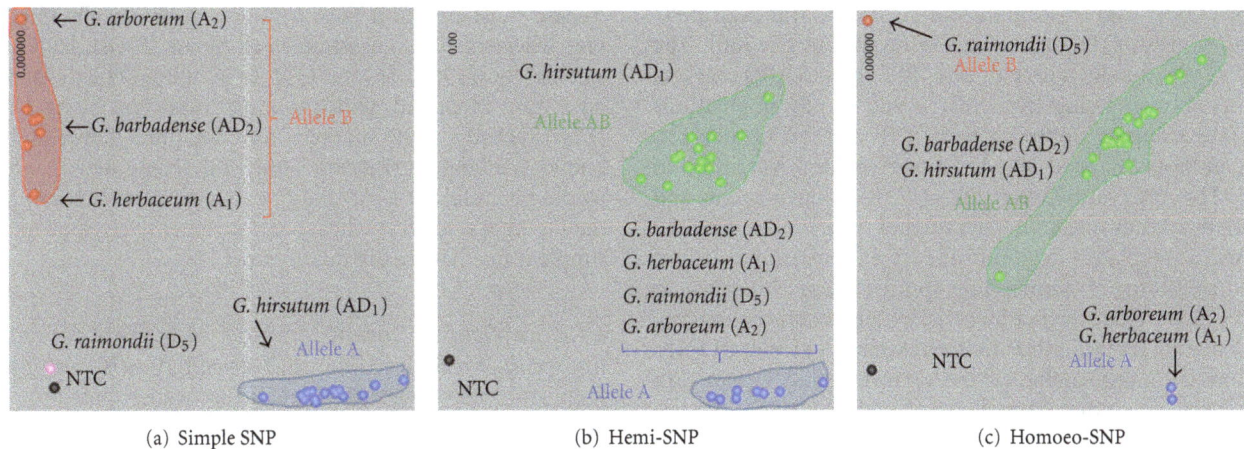

FIGURE 1: Segregation patterns of simple, hemi-, and homoeo-SNPs assayed using KASPar chemistry across tetraploid cotton species [*G. hirsutum* (AD₁) and *G. barbadense* (AD₂)] and their diploid progenitors [*G. arboreum*, *G. herbaceum* (A subgenome), and *G. raimondii* (D-subgenome)]. (a) Simple, or true SNP detects allelic variation between homologous loci of A subgenome of *G. hirsutum* (AD₁) and *G. barbadense* (AD₂). (b) Hemi-SNPs detect allelic variation in homozygous state in *G. barbadense* (AD₂) and heterozygous state in *G. hirsutum* (AD₁). (c) Homoeo-SNP detects homoelogous and/or paralogous loci both in A and D subgenomes, which are monomorphic between *G. barbadense* (AD₂) and *G. hirsutum* (AD₁). Allele calls depicted in blue, red, green, pink, and black represent alleles A, B, and AB, no amplification or missing locus, and no template control (NTC), respectively.

In Figure 1(b), SNP marker detects both alleles (A and B) in *G. hirsutum* (heterozygous green cluster) and one allele A in *G. barbadense* (a homozygous blue cluster) and could be *vice versa*. Homoeo-SNPs detect homoeologous and possibly paralogous loci both in A and D subgenomes and result in monomorphic loci in tetraploid species (right image). In Figure 1(c) A genome progenitors (*G. herbaceum and G. arboreum*) had allele A (blue) and D genome progenitor (*G. raimondii*) had allele B (red), but both tetraploid species (*G. hirsutum and G. barbadense*) were grouped into heterozygous AB (green) cluster. As homoeo-SNPs can detect paralogous loci, the diploid progenitors both have different alleles.

Simple SNPs as well as hemi-SNPs are useful markers for genetic mapping and diversity screening studies. Simple SNPs segregate like the markers in diploids in most of the mapping populations and would account for approximately 10–30% of total polymorphic SNPs in various polyploid crop species. Hemi-SNPs form a major category (30–60%) of polymorphic SNPs in a polyploid crop species and could be used for genetic mapping purposes in F₂, RIL, and DH populations. Homoeo-SNPs are of lesser value for mapping purposes as most of the genotypes result in heterologous loci due to polymorphism between the homoeologous genomes or duplicated loci within each of the polyploid genotypes [56].

4. Application of SNP Markers in Gene/QTL Discovery

4.1. Biparental Approach. Genetic mapping studies involve genetic linkage analysis, which is based on the concept of genetic recombination during meiosis [57]. This encompasses developing genetic linkage maps following genotyping of individuals in segregating populations with DNA markers covering the genome of that organism. Since their discovery in the 1980s, DNA-based markers have been widely used in developing saturated genetic linkage maps as well as for the mapping and discovery of genes/QTL. With the large-scale availability of the sequence information and development of HTP technologies for SNP genotyping, SNP markers have been increasingly used for QTL mapping studies. This is primarily, because SNPs are highly abundant in the genomes and, therefore, they can provide the highest map resolution compared to other marker systems [58, 59]. A review of the selected examples of QTL and gene discovery using SNP markers is presented below.

4.1.1. Examples in Rice. A recent study on QTL analysis in rice for yield and three-yield-component traits, number of tillers per plant, number of grains per panicle, and grain weight compared a SNP-based map to that of a previous RFLP/SSR-based QTL map generated using the same mapping population [42]. Using the ultra-high-density SNP map, the authors showed that this map had more power and resolution relative to the RFLP/SSR map. This was clearly evident by the analysis of the two main QTL for grain weight, *kgw3a* (*GS3*) and *kgw5* (*GW5/qSW5*). Using the SNP bin map, *GW5/qSW5* QTL for grain width was accurately narrowed down to a 123 kb region as compared to the 12.4 Mb region based on the RFLP/SSR genetic map. Likewise, *GS3* QTL for grain length was mapped to a 197 kb interval in comparison to 6 Mb region with the RFLP/SSR genetic map. Beside the power and the resolution, maps based on high-density SNP markers are also highly suitable for fine mapping and cloning of QTL and at times SNPs on these maps are also functionally associated with the natural variation in the trait. In another QTL mapping project, SNP and InDel markers were used to fine map *qSH1* gene, a major QTL of seed shattering trait in rice [60]. The QTL were initially detected using RFLP and RAPD markers on F2 plants. Using large BC$_4$F$_2$ and BC$_3$F$_2$ populations in fine mapping approach with SNP and InDel markers, the authors mapped the functional natural variation to a 612 bp interval between the QTL flanking markers and discovered only one SNP. They further showed that this SNP in the 5′ regulatory region of the *qSH1* gene caused loss of seed shattering. Fine mapping approach was also taken to positionally clone the rice bacterial blight resistance gene *xa5*, by isolating the recombination breakpoints to a pair of SNPs followed by sequencing of the corresponding 5 kb region [61]. Several studies have shown that the SNPs and InDels are highly abundant and present throughout the genome in various species including plants [62–64]. SNP genotyping is a valuable tool for gene mapping, map-based cloning, and marker assisted selection (MAS) in crops [65]. A study was conducted to assess the feasibility of SNPs and InDels as DNA markers in genetic analysis and marker-assisted breeding in rice by analyzing these sequence polymorphisms in the genomic region containing *Piz* and *Piz-t* rice blast resistance genes and developing PCR-based SNP markers [65]. The authors discovered that SNPs were abundant in the *Piz* and *Piz-t* (averaging one SNP every 248 bp), while InDels were much lower. This dense distribution of SNPs helped in developing SNP markers in the vicinity of these genes. Advancements in rice genomics have led to mapping and cloning of several genes and QTL controlling agronomically important traits, enabled routine use of SNP markers for MAS, gene pyramiding, and marker-assisted breeding (MAB) [66–68].

4.1.2. Examples in Maize.

SNP markers have facilitated the dissection of complex traits such as flowering time in maize. Using a set of 5000 RILs, which represent the nested association mapping (NAM) population and genotyping with 1,200 SNP markers, the authors discovered that the genetic architecture of flowering time is controlled by small additive QTL rather than a single large-effect QTL [69]. The same NAM population was used for mapping resistance to northern leaf blight disease [70]. Twenty-nine QTL were discovered and candidate genes were identified with genome-wide NAM approach using 1.6 million SNPs. Proprietary SNP markers developed by companies are being predominantly used in their private breeding programs. A study from Pioneer Hi-Bred International Inc. reported identifying a high-oil QTL (*qHO6*) affecting maize seed oil and oleic acid contents. This QTL encodes an acyl-CoA:diacylglycerol acyltransferase (*DGAT1-2*), which catalyzes the final step of oil synthesis [71].

4.1.3. Examples in Wheat.

Recent advances in wheat genomics have led to the implementation of high-density SNP genotyping in wheat [72–75]. Gene-based SNP markers were developed for *Lr34/Yr18/Pm38* locus that confers resistance to leaf rust, stripe rust, and powdery mildew diseases [76]. These markers serve as efficient tools for MAS and MAB of

disease resistant wheat lines. Another economically important wheat disease, Fusarium head blight (FHB), has been extensively studied. Several QTL controlling FHB resistance have been identified, with the most important being *Fhb1* [77]. Recently, SNP markers were mapped between the known flanking markers for *Fhb1* [78]. These new markers would be useful for MAS and fine mapping towards cloning the *Fhb1* gene. MAS in wheat has been extensively applied for simple traits that are difficult to score [79].

4.1.4. Examples in Soybean.

In order to improve the effectiveness of MAS and clone soybean aphid resistance gene, *Rag1*, fine mapping was done to accurately position the gene, which was previously mapped to a 12 cM interval [80]. The authors mapped the gene between two SNP markers that corresponded to a physical distance of 115 kb and identified several candidate genes. Similarly, another aphid resistance gene, *Rag2*, originally mapped to a 10 cM interval, was fine mapped to a 54 kb interval using SNP markers that were developed by resequencing of target intervals and sequence-tagged sites [81]. In another study that used a similar approach, the authors identified SNP markers tightly linked to a QTL conferring resistance to southern root-knot nematode by developing these SNP markers from the bacterial artificial chromosome (BAC) ends and SSR-containing genomic DNA clones [82]. In all of these examples the main idea behind the identification of closely linked SNP markers was to enhance the efficiency and cost effectiveness through MAS and increase the resolution within the target locus.

4.1.5. Examples in Other Crops.

In a study conducted in canola to map the *fad2* and *fad3* gene, single nucleotide mutations were identified by sequencing the genomic clones of these genes and subsequently SNP markers were developed [83]. Allele-specific PCR assays were developed to enable direct selection of desirable *fad2* and *fad3* alleles in marker-assisted trait introgression and breeding. In barley, SNP markers were identified that were linked to a covered smut resistance gene, *Ruh.7H*, by using high-resolution melting (HRM) technique [84]. In sugar beet, an anchored linkage map based on AFLP, SNP, and RAPD markers was developed to map QTL for *Beet necrotic yellow vein virus* resistance genes, *Rz4* [85] and *Rz5* [86]. A consensus genetic map based on EST-derived SNPs was developed for cowpea that would be an important resource for genomic and QTL mapping studies in this crop [87]. In one of the post-genomic era studies in 2002, the fine mapping and map-based cloning approaches were used to clone the *VTC2* gene in Arabidopsis [88]. The authors fine mapped the gene interval from ~980 kb region to a 20 kb interval with SNP and InDel markers. Additional nine candidate genes were identified in that interval and subsequently the underlying mutation was discovered. Although only a few examples that demonstrate the application of SNP markers in QTL mapping and genomic studies have been mentioned here, several other studies have been published in this area. Recent advances in HTP genotyping technologies and sequence information will further pave the way for rapid identification of causative variations and cloning of QTL of interest for use in MAB.

4.2. Genome-Wide Association Study Approach. GWAS is increasingly becoming a popular tool for dissecting complex traits in plants [89–92]. The idea behind GWAS is to genotype a large number of markers distributed across the genome so that the phenotype or the functional alleles will be in LD with one or few markers that could then be used in the breeding program. However, due to limited extent of LD, a greater number of markers are required for sufficient power to detect linkage between the marker and the underlying phenotypic variation. Several studies on association mapping in plants have been published and reviewed in the past [89, 90, 92, 93]. A few selected examples on the GWAS and candidate gene association (CGA) studies that utilized SNP markers are described below.

The successful use and first time demonstration of the power of GWAS was through the identification of a putative gene associated with a QTL in maize [94]. In that study, a single locus with major effect on oleic acid was mapped to a 4 cM genetic interval by using SNP haplotypes at 8,590 loci. The authors identified a fatty acid desaturase gene, *fad2*, at ~2 kb from one of the associated markers, and this was considered a likely causative gene. With the discovery of millions of SNPs in maize and the availability of tools such as NAM populations, GWAS was effectively applied to dissect the genetic architecture of leaf traits and it was also shown that variations at the *liguleless* genes contributed to more upright leaf phenotype [95]. Utility of the GWAS approach was demonstrated in barley through the mapping of a QTL for spot blotch disease resistance [96]. Using the diversity array technology (DArT) and SNP markers, the authors identified several QTL, some of which were not identified for this trait earlier. Another variant of the association mapping method is the CGA, where the association between one or few gene candidate loci and the trait of interest is tested. Using this approach 24 gene candidates were analyzed for association with the field resistance to late blight disease in potato and plant maturity. Nine SNPs were identified to be associated with maturity corrected resistance, explaining 50% of the genetic variance of this trait [97]. Two SNPs at the *allene oxide synthase 2* (*StAOS2*) gene locus were associated with the largest effect on the trait of interest. A GWAS approach was also successfully applied to understand the genetic architecture of complex diseases such as northern and southern corn leaf blights [70, 98]. Although the number of papers dedicated to the application of GWAS to reveal the genetic basis of agronomic traits is growing, the practical utility of minor QTL in molecular breeding is yet to be shown. As GWAS requires large number of molecular markers, the utility of GWAS in dissection of molecular basis of traits in polyploid crops such as canola, wheat, and cotton has been fairly limited due to the insufficient number of polymorphic markers and the absence of reference genome. However, recently developed associative transcriptomics method has

a potential to overcome the above-mentioned shortages [99]. Harper et al. [99] leveraged differentially expressed transcriptome sequences to develop molecular markers in tetraploid crop *Brassica napus* and associated them with glucosinolate content variation in seeds. Due to the precision of this method, scientists were able to correlate specific deletions in canola genome with two QTL controlling the trait. Annotation of deleted regions revealed the orthologs of the transcription factor HAG1, which controlled aliphatic glucosinolate biosynthesis in *A. thaliana*. This research work gives an optimism on successful application of GWAS in polyploid crops.

5. Implementation of SNP Markers in Plant Breeding

Due to the availability of HTP SNP detection and validation technologies, the development of SNP markers becomes a routine process, especially in crops with reference genome. How has that influenced the application of SNP markers in plant breeding? In a review article, Xu and Crouch [100] indicated fairly low number of articles dedicated to the marker assisted selection for the 1986–2005 period. The combination of three key phrases ("marker-assisted selection" AND "SNP" AND "plant breeding"), indeed, shows only 637 articles at Google Scholar for that period. However, similar search for the period, spanning 2006 through 2012, demonstrates almost sevenfold (~4,560) increase in the number of articles indicating the application of SNPs in MAS. A vast majority of those publications are from public sector and primarily describe mapping QTL using SNPs and state the potential usefulness of those markers in MAS without any experimental support for that. For most of those research studies, QTL mapping is the final destination and further application of those markers in actual MAS leading to the development of varieties seldom happens. Fairly low impact of academic research in the MAS-based variety development can be explained by the lack of funding to complete the entire marker development pipeline (MDP), which can be long term and cost intensive. MDP includes several steps such as (1) population development, (2) initial QTL mapping, (3) QTL validation (testing in several locations and years and implementing fine mapping), and (4) marker validation (development of inexpensive but HTP and automation amenable assays) [101]. Every step of the development of markers linked to QTL is associated with numerous constraints, which may take several years and substantial funding to resolve. However, since 2006, there have been a few success stories about the development of varieties using SNPs in publications derived from academic research, including the development of submergence-tolerant rice cultivars [102], rice cultivars with improved eating, cooking, and sensory quality [103], leaf rust resistant wheat variety "Patwin" [104], and maize cultivar with low phytic acid [105]. Although the private sector does not normally release details of its breeding methodologies to the public, several papers published by Monsanto [106, 107], Pioneer Hi-bred [71], Syngenta [108], and Dow AgroSciences [109] indicate that commercial organizations are the main drivers in the application of SNP markers in MAS [110].

Current MAS strategies fit the breeding programs for the traits that are highly heritable and governed by a single gene or one major QTL that explains a large portion of the phenotypic variability. In reality, most of the agronomic traits such as yield, drought and heat tolerance, nitrogen and water use efficiency, and fiber quality in cotton have complex inheritance that is controlled by multiple QTL with minor effect. Use of one of those minor QTL in MAS will be inefficient because of its negligible effect on phenotype.

The MAS scheme using paternity testing has recently been proposed to address challenges associated with selection gains that can be achieved in outbred forage crops [111]. Paternity testing, a nonlinkage-based MAS scheme, improves selection gains by increasing parental control in the selection gain equation. The authors demonstrated paternity testing MAS in three red clover breeding populations by using permutation-based truncation selection for a biomass-persistence index trait and achieved paternity-based selection gains that were greater than double the selection gains based on maternity alone. The paternity was determined by using a small set (11) of SSR markers. SNP markers can also be used for paternity testing, but one would require a relatively larger number of SNP loci [112].

Meuwissen et al. [113] described a new methodology in plant breeding called genomic selection (GS) that was intended to solve problems related to MAS of complex traits. This methodology also applies molecular markers but in a different fashion in both diploid and polyploid crop species. Unlike MAS, in GS markers are not used for tracking a trait. In GS high-density marker coverage is needed to potentially have all QTL in LD with at least one marker. Then the comprehensive information on all possible loci, haplotypes, and marker effects across the entire genome is used to calculate genomic estimated breeding value (GEBV) of a particular line in the breeding population.

GS of superior lines can be carried out within any breeding population. In order to enable successful GS, the experimental population must be identified. The population should not be necessarily derived from bi-parental cross but must be representative of selection candidates in the breeding program to which GS will be applied [114]. The experimental population must be genotyped with a large number of markers. Taking into account the low cost of sequencing, the best choice is the GBS implementation, which will yield maximum number of polymorphisms. The sequence of the two events, that is, phenotypic and genotypic data collection, is arbitrary and can be done in parallel. When both phenotypic and genotypic data are ready, one can start "training" molecular markers [115]. In order to train the GS model, the effect of each marker is calculated computationally. The effect of a marker is represented by a number with a positive or negative sign that indicates the positive or negative effect, respectively, of a particular locus to the phenotype. When the effects of all markers are known, they are considered "trained" and ready to assess any breeding population different from the experimental one for the same trait. Availability of trained GS model does not require the collection of phenotypic data from new breeding populations. The same set of "trained" markers will be used to genotype a new breeding population. Based on genotypic data, the known effects of each marker will be summed and GEBV of each line will be calculated. The higher the GEBV value of an individual line, the more the chances that this line will be selected and advanced in the breeding cycle. Thus, GS using high-density marker coverage has a potential to capture QTL with major and minor effects and eliminate the need to collect phenotypic data in all breeding cycles. Also, the application of GS was demonstrated to reduce the number of breeding cycles and increase the annual gain [114]. One of the problems of GS is the level of GEBV accuracy. Simulation studies based on simulated and empirical data demonstrated that GEBV accuracy could be within 0.62–0.85. Heffner et al. [114] used previously reported GEBV accuracy of 0.53 and reported three- and twofold annual gain in maize and winter barley, respectively. The obvious advantages of GS over traditional MAS have been successfully proven in animal breeding [116]. Rapid evolution of sequencing technologies and HTP SNP genotyping systems are enabling generation and validation of millions of markers, giving a "cautious optimism" for successful application of GS in breeding for complex traits [117–120].

6. Conclusion

SNP markers have become extremely popular in plant molecular genetics due to their genome-wide abundance and amenability for high- to ultra-high-throughput detection platforms. Unlike earlier marker systems, SNPs made it possible to create saturated, if not, supersaturated genetic maps, thereby enabling genome-wide tracking, fine mapping of target regions, rapid association of markers with a trait, and accelerated cloning of gene/QTL of interest. On the flip side, there are some challenges that need to be addressed or overcome while using SNPs. For example, the biallelic nature of SNPs needs to be compensated by discovering and using a larger number of SNPs to arrive at the same or higher power as that of earlier-generation molecular markers. This could be cost prohibitive depending on the crop and the sequence resources available for that genome. Working with polyploid crops is another challenge where useful SNPs are only a small percentage of the total available polymorphisms. Creative strategies need to be employed to generate a reasonable number of SNPs in those species. The use of SNP markers in MAB programs has been growing at a faster pace and so is the development of technologies and platforms for the discovery and HTP screening of SNPs in many crops. SNP chips are currently available for several crops; however, one disadvantage is that these readily available chips are made based on SNPs discovered from certain genotypes and, therefore, may not be ideal for projects utilizing unrelated genotypes. This necessitates creation of multiple chips or the usage of technologies that permit design flexibility but are economical. Although GBS creates great opportunities to discover a large number of SNPs at lower per sample cost within the genotypes of interest, the lack of adequate computational capabilities such as reliable data imputation algorithms and powerful computers allowing quick processing and the storage of a large amount of sequencing data becomes a major bottleneck. Despite certain disadvantages or challenges, it is clear that SNP markers, in combination with genomics and other next-generation technologies, have been accelerating the pace and gains of plant breeding.

Acknowledgments

The authors would like to thank Drs. Shunxue Tang and Peizhong Zheng of the Trait Genetics and Technologies Department of Dow AgroSciences (DAS) and Raghav Ram of the IP Portfolio Development Department of DAS for careful review of the paper and the DAS Seeds and Traits R & D leaders Drs. David Meyer and Steve Thompson for general support and help.

References

[1] J. L. Weber and P. E. May, "Abundant class of human DNA polymorphisms which can be typed using the polymerase chain reaction," *American Journal of Human Genetics*, vol. 44, no. 3, pp. 388–396, 1989.

[2] R. Ophir and D. Graur, "Patterns and rates of indel evolution in processed pseudogenes from humans and murids," *Gene*, vol. 205, no. 1-2, pp. 191–202, 1997.

[3] D. G. Wang, J. B. Fan, C. J. Siao et al., "Large-scale identification, mapping, and genotyping of single-nucleotide polymorphisms in the human genome," *Science*, vol. 280, no. 5366, pp. 1077–1082, 1998.

[4] D. Botstein, R. L. White, M. Skolnick, and R. W. Davis, "Construction of a genetic linkage map in man using restriction fragment length polymorphisms," *American Journal of Human Genetics*, vol. 32, no. 3, pp. 314–331, 1980.

[5] P. K. Gupta, R. K. Varshney, P. C. Sharma, and B. Ramesh, "Molecular markers and their applications in wheat breeding," *Plant Breeding*, vol. 118, no. 5, pp. 369–390, 1999.

[6] R. Bernardo, "Molecular markers and selection for complex traits in plants: learning from the last 20 years," *Crop Science*, vol. 48, no. 5, pp. 1649–1664, 2008.

[7] J. Welsh and M. McClelland, "Fingerprinting genomes using PCR with arbitrary primers," *Nucleic Acids Research*, vol. 18, no. 24, pp. 7213–7218, 1990.

[8] P. Vos, R. Hogers, M. Bleeker et al., "AFLP: a new technique for DNA fingerprinting," *Nucleic Acids Research*, vol. 23, no. 21, pp. 4407–4414, 1995.

[9] H. J. Jacob, K. Lindpaintner, S. E. Lincoln et al., "Genetic mapping of a gene causing hypertension in the stroke-prone spontaneously hypertensive rat," *Cell*, vol. 67, no. 1, pp. 213–224, 1991.

[10] E. S. Lander and S. Botstein, "Mapping mendelian factors underlying quantitative traits using RFLP linkage maps," *Genetics*, vol. 121, no. 1, p. 185, 1989.

[11] J. G. K. Williams, A. R. Kubelik, K. J. Livak, J. A. Rafalski, and S. V. Tingey, "DNA polymorphisms amplified by arbitrary primers are useful as genetic markers," *Nucleic Acids Research*, vol. 18, no. 22, pp. 6531–6535, 1990.

[12] Z. Zhang, X. Guo, B. Liu, L. Tang, and F. Chen, "Genetic diversity and genetic relationship of Jatropha curcas between China and Southeast Asian revealed by amplified fragment length polymorphisms," *African Journal of Biotechnology*, vol. 10, no. 15, pp. 2825–2832, 2011.

[13] W. Powell, G. C. Machray, and J. Proven, "Polymorphism revealed by simple sequence repeats," *Trends in Plant Science*, vol. 1, no. 7, pp. 215–222, 1996.

[14] S. Ghosh, P. Malhotra, P. V. Lalitha, S. Guha-Mukherjee, and V. S. Chauhan, "Novel genetic mapping tools in plants: SNPs and LD-based approaches," *Plant Science*, vol. 162, no. 3, pp. 329–333, 2002.

[15] M. W. Ganal, T. Altmann, and M. S. Röder, "SNP identification in crop plants," *Current Opinion in Plant Biology*, vol. 12, no. 2, pp. 211–217, 2009.

[16] B. C. Meyers, S. V. Tingey, and M. Morgante, "Abundance, distribution, and transcriptional activity of repetitive elements in the maize genome," *Genome Research*, vol. 11, no. 10, pp. 1660–1676, 2001.

[17] S. I. Wright, I. V. Bi, S. C. Schroeder et al., "Evolution: the effects of artificial selection on the maize genome," *Science*, vol. 308, no. 5726, pp. 1310–1314, 2005.

[18] J. Batley, G. Barker, H. O'Sullivan, K. J. Edwards, and D. Edwards, "Mining for single nucleotide polymorphisms and insertions/deletions in maize expressed sequence tag data," *Plant Physiology*, vol. 132, no. 1, pp. 84–91, 2003.

[19] A. Pratap, S. Gupta, J. Kumar, and R. Solanki, "Soybean," *Technological Innovations in Major World Oil Crops*, vol. 1, pp. 293–321, 2012.

[20] I. Y. Choi, D. L. Hyten, L. K. Matukumalli et al., "A soybean transcript map: gene distribution, haplotype and single-nucleotide polymorphism analysis," *Genetics*, vol. 176, no. 1, pp. 685–696, 2007.

[21] E. R. Mardis, "The impact of next-generation sequencing technology on genetics," *Trends in Genetics*, vol. 24, no. 3, pp. 133–141, 2008.

[22] O. Morozova and M. A. Marra, "Applications of next-generation sequencing technologies in functional genomics," *Genomics*, vol. 92, no. 5, pp. 255–264, 2008.

[23] W. B. Barbazuk, S. J. Emrich, H. D. Chen, L. Li, and P. S. Schnable, "SNP discovery via 454 transcriptome sequencing," *The Plant Journal*, vol. 51, no. 5, pp. 910–918, 2007.

[24] M. Trick, Y. Long, J. Meng, and I. Bancroft, "Single nucleotide polymorphism (SNP) discovery in the polyploid *Brassica napus* using Solexa transcriptome sequencing," *Plant Biotechnology Journal*, vol. 7, no. 4, pp. 334–346, 2009.

[25] E. Novaes, D. R. Drost, W. G. Farmerie et al., "High-throughput gene and SNP discovery in *Eucalyptus grandis*, an uncharacterized genome," *BMC Genomics*, vol. 9, article 312, 2008.

[26] P. C. Bundock, F. G. Eliott, G. Ablett et al., "Targeted single nucleotide polymorphism (SNP) discovery in a highly polyploid plant species using 454 sequencing," *Plant Biotechnology Journal*, vol. 7, no. 4, pp. 347–354, 2009.

[27] T. L. Parchman, K. S. Geist, J. A. Grahnen, C. W. Benkman, and C. A. Buerkle, "Transcriptome sequencing in an ecologically important tree species: assembly, annotation, and marker discovery," *BMC Genomics*, vol. 11, no. 1, article 180, 2010.

[28] K. Lai, C. Duran, P. J. Berkman et al., "Single nucleotide polymorphism discovery from wheat next-generation sequence data," *Plant Biotechnology Journal*, vol. 10, no. 6, pp. 743–749, 2012.

[29] D. Kuhn, "Design of an Illumina Infinium 6k SNPchip for genotyping two large avocado mapping populations," in *Proceedings of the 20th Conference on Plant and Animal Genome*, San Diego, CA, January 2012.

[30] J. R. Russell, M. Bayer, C. Booth et al., "Identification, utilisation and mapping of novel transcriptome-based markers from blackcurrant (*Ribes nigrum*)," *BMC Plant Biology*, vol. 11, article 147, 2011.

[31] E. Hodges, Z. Xuan, V. Balija et al., "Genome-wide in situ exon capture for selective resequencing," *Nature Genetics*, vol. 39, no. 12, pp. 1522–1527, 2007.

[32] N. M. Springer, K. Ying, Y. Fu et al., "Maize inbreds exhibit high levels of copy number variation (CNV) and presence/absence variation (PAV) in genome content," *PLoS Genetics*, vol. 5, no. 11, Article ID e1000734, 2009.

[33] R. K. Varshney, "Gene-based marker systems in plants: high throughput approaches for marker discovery and genotyping," in *Molecular Techniques in Crop Improvement*, S. M. Jain and D. S. Brar, Eds., pp. 119–142, 2009.

[34] A. Dean, "On a chromosome far, far away: LCRs and gene expression," *Trends in Genetics*, vol. 22, no. 1, pp. 38–45, 2006.

[35] Y. Yuan, P. J. SanMiguel, and J. L. Bennetzen, "High-Cot sequence analysis of the maize genome," *The Plant Journal*, vol. 34, no. 2, pp. 249–255, 2003.

[36] J. Emberton, J. Ma, Y. Yuan, P. SanMiguel, and J. L. Bennetzen, "Gene enrichment in maize with hypomethylated partial restriction (HMPR) libraries," *Genome Research*, vol. 15, no. 10, pp. 1441–1446, 2005.

[37] D. T. Okou, K. M. Steinberg, C. Middle, D. J. Cutler, T. J. Albert, and M. E. Zwick, "Microarray-based genomic selection for high-throughput resequencing," *Nature Methods*, vol. 4, no. 11, pp. 907–909, 2007.

[38] N. J. van Orsouw, R. C. J. Hogers, A. Janssen et al., "Complexity reduction of polymorphic sequences (CRoPS): a novel approach for large-scale polymorphism discovery in complex genomes," *PLoS ONE*, vol. 2, no. 11, Article ID e1172, 2007.

[39] N. A. Baird, P. D. Etter, T. S. Atwood et al., "Rapid SNP discovery and genetic mapping using sequenced RAD markers," *PLoS ONE*, vol. 3, no. 10, Article ID e3376, 2008.

[40] J. A. Mammadov, W. Chen, R. Ren et al., "Development of highly polymorphic SNP markers from the complexity reduced portion of maize (*Zea mays* L.) genome for use in marker-assisted breeding," *Theoretical and Applied Genetics*, vol. 121, no. 3, pp. 577–588, 2010.

[41] Y. Chutimanitsakun, R. W. Nipper, A. Cuesta-Marcos et al., "Construction and application for QTL analysis of a Restriction Site Associated DNA (RAD) linkage map in barley," *BMC Genomics*, vol. 12, article 4, 2011.

[42] H. Yu, W. Xie, J. Wang et al., "Gains in QTL detection using an ultra-high density SNP map based on population sequencing relative to traditional RFLP/SSR markers," *PLoS ONE*, vol. 6, no. 3, Article ID e17595, 2011.

[43] A. Bus, J. Hecht, B. Huettel, R. Reinhardt, and B. Stich, "High-throughput polymorphism detection and genotyping in *Brassica napus* using next-generation RAD sequencing," *BMC Genomics*, vol. 13, no. 1, p. 281, 2012.

[44] J. Tang, J. A. M. Leunissen, R. E. Voorrips, C. G. van der Linden, and B. Vosman, "HaploSNPer: a web-based allele and SNP detection tool," *BMC Genetics*, vol. 9, article 23, 2008.

[45] A. Narechania, M. A. Gore, E. S. Buckler et al., "Large-scale discovery of gene-enriched SNPs," *The Plant Genome*, vol. 2, no. 2, pp. 121–133, 2009.

[46] J. C. Nelson, S. Wang, Y. Wu et al., "Single-nucleotide polymorphism discovery by high-throughput sequencing in sorghum," *BMC Genomics*, vol. 12, article 352, 2011.

[47] R. J. Elshire, J. C. Glaubitz, Q. Sun et al., "A robust, simple genotyping-by-sequencing (GBS) approach for high diversity species," *PLoS ONE*, vol. 6, no. 5, Article ID e19379, 2011.

[48] S. R. Browning and B. L. Browning, "Rapid and accurate haplotype phasing and missing-data inference for whole-genome association studies by use of localized haplotype clustering," *American Journal of Human Genetics*, vol. 81, no. 5, pp. 1084–1097, 2007.

[49] B. N. Howie, P. Donnelly, and J. Marchini, "A flexible and accurate genotype imputation method for the next generation of genome-wide association studies," *PLoS Genetics*, vol. 5, no. 6, Article ID e1000529, 2009.

[50] X. Huang, X. Wei, T. Sang et al., "Genome-wide association studies of 14 agronomic traits in rice landraces," *Nature Genetics*, vol. 42, no. 11, pp. 961–967, 2010.

[51] J. Marchini and B. Howie, "Genotype imputation for genome-wide association studies," *Nature Reviews Genetics*, vol. 11, no. 7, pp. 499–511, 2010.

[52] J. B. Fan, A. Oliphant, R. Shen et al., "Highly parallel SNP genotyping," *Cold Spring Harbor Symposia on Quantitative Biology*, vol. 68, pp. 69–78, 2003.

[53] F. J. Steemers and K. L. Gunderson, "Whole genome genotyping technologies on the BeadArray platform," *Biotechnology Journal*, vol. 2, no. 1, pp. 41–49, 2007.

[54] K. J. Livak, S. J. A. Flood, J. Marmaro, W. Giusti, and K. Deetz, "Oligonucleotides with fluorescent dyes at opposite ends provide a quenched probe system useful for detecting PCR product and nucleic acid hybridization," *Genome Research*, vol. 4, no. 6, pp. 357–362, 1995.

[55] S. P. Kumpatla, R. Buyyarapu, I. Y. Abdurakhmonov, and J. A. Mammadov, "Genomics-assisted plant breeding in the 21st century: technological advances and progress," in *Plant Breeding*, I. Y. Abdurakhmonov, Ed., pp. 131–184.

[56] R. Buyyarapu, R. Ren, S. Kumpatla et al., "In silico discovery and validation of SNP markers for molecular breeding in cotton," in *Proceedings of the 19th Conference on Plant & Animal Genome*, San Diego, Calif, USA, January 2011.

[57] S. D. Tanksley, "Mapping polygenes," *Annual Review of Genetics*, vol. 27, pp. 205–233, 1993.

[58] D. Bhattramakki, M. Dolan, M. Hanafey et al., "Insertion-deletion polymorphisms in 3′ regions of maize genes occur frequently and can be used as highly informative genetic markers," *Plant Molecular Biology*, vol. 48, no. 5-6, pp. 539–547, 2002.

[59] E. S. Jones, H. Sullivan, D. Bhattramakki, and J. S. C. Smith, "A comparison of simple sequence repeat and single nucleotide polymorphism marker technologies for the genotypic analysis of maize (*Zea mays* L.)," *Theoretical and Applied Genetics*, vol. 115, no. 3, pp. 361–371, 2007.

[60] S. Konishi, T. Izawa, S. Y. Lin et al., "An SNP caused loss of seed shattering during rice domestication," *Science*, vol. 312, no. 5778, pp. 1392–1396, 2006.

[61] A. S. Iyer and S. R. McCouch, "The rice bacterial blight resistance gene xa5 encodes a novel form of disease resistance," *Molecular Plant-Microbe Interactions*, vol. 17, no. 12, pp. 1348–1354, 2004.

[62] E. Drenkard, B. G. Richter, S. Rozen et al., "A simple procedure for the analysis of single nucleotide polymorphism facilitates map-based cloning in Arabidopsis," *Plant Physiology*, vol. 124, no. 4, pp. 1483–1492, 2000.

[63] K. Garg, P. Green, and D. A. Nickerson, "Identification of candidate coding region single nucleotide polymorphisms in

165 human genes using assembled expressed sequence tags," *Genome Research*, vol. 9, no. 11, pp. 1087–1092, 1999.

[64] S. Nasu, J. Suzuki, R. Ohta et al., "Search for and analysis of single nucleotide polymorphisms (SNPS) in rice (*Oryza sativa, Oryza rufipogon*) and establishment of SNP markers," *DNA Research*, vol. 9, no. 5, pp. 163–171, 2002.

[65] K. Hayashi, N. Hashimoto, M. Daigen, and I. Ashikawa, "Development of PCR-based SNP markers for rice blast resistance genes at the *Piz* locus," *Theoretical and Applied Genetics*, vol. 108, no. 7, pp. 1212–1220, 2004.

[66] M. Ashikari and M. Matsuoka, "Identification, isolation and pyramiding of quantitative trait loci for rice breeding," *Trends in Plant Science*, vol. 11, no. 7, pp. 344–350, 2006.

[67] K. K. Jena and D. J. Mackill, "Molecular markers and their use in marker-assisted selection in rice," *Crop Science*, vol. 48, no. 4, pp. 1266–1276, 2008.

[68] R. K. Varshney, D. A. Hoisington, and A. K. Tyagi, "Advances in cereal genomics and applications in crop breeding," *Trends in Biotechnology*, vol. 24, no. 11, pp. 490–499, 2006.

[69] E. S. Buckler, J. B. Holland, P. J. Bradbury et al., "The genetic architecture of maize flowering time," *Science*, vol. 325, no. 5941, pp. 714–718, 2009.

[70] J. A. Poland, P. J. Bradbury, E. S. Buckler, and R. J. Nelson, "Genome-wide nested association mapping of quantitative resistance to northern leaf blight in maize," *Proceedings of the National Academy of Sciences of the United States of America*, vol. 108, no. 17, pp. 6893–6898, 2011.

[71] P. Zheng, W. B. Allen, K. Roesler et al., "A phenylalanine in DGAT is a key determinant of oil content and composition in maize," *Nature Genetics*, vol. 40, no. 3, pp. 367–372, 2008.

[72] E. Akhunov, C. Nicolet, and J. Dvorak, "Single nucleotide polymorphism genotyping in polyploid wheat with the Illumina GoldenGate assay," *Theoretical and Applied Genetics*, vol. 119, no. 3, pp. 507–517, 2009.

[73] A. M. Allen, G. L. Barker, S. T. Berry et al., "Transcript-specific, single-nucleotide polymorphism discovery and linkage analysis in hexaploid bread wheat (*Triticum aestivum* L.)," *Plant Biotechnology Journal*, vol. 9, no. 9, pp. 1086–1099, 2011.

[74] A. Bérard, M. C. Le Paslier, M. Dardevet et al., "High-throughput single nucleotide polymorphism genotyping in wheat (*Triticum* spp.)," *Plant Biotechnology Journal*, vol. 7, no. 4, pp. 364–374, 2009.

[75] M. O. Winfield, P. A. Wilkinson, A. M. Allen et al., "Targeted re-sequencing of the allohexaploid wheat exome," *Plant Biotechnology Journal*, vol. 10, no. 6, pp. 733–742, 2012.

[76] E. S. Lagudah, S. G. Krattinger, S. Herrera-Foessel et al., "Gene-specific markers for the wheat gene *Lr34/Yr18/Pm38* which confers resistance to multiple fungal pathogens," *Theoretical and Applied Genetics*, vol. 119, no. 5, pp. 889–898, 2009.

[77] H. Buerstmayr, T. Ban, and J. A. Anderson, "QTL mapping and marker-assisted selection for Fusarium head blight resistance in wheat: a review," *Plant Breeding*, vol. 128, no. 1, pp. 1–26, 2009.

[78] A. N. Bernardo, H. Ma, D. Zhang, and G. Bai, "Single nucleotide polymorphism in wheat chromosome region harboring *Fhb1* for Fusarium head blight resistance," *Molecular Breeding*, vol. 29, no. 2, pp. 477–488, 2012.

[79] P. K. Gupta, P. Langridge, and R. R. Mir, "Marker-assisted wheat breeding: present status and future possibilities," *Molecular Breeding*, vol. 26, no. 2, pp. 145–161, 2010.

[80] K. S. Kim, S. Bellendir, K. A. Hudson et al., "Fine mapping the soybean aphid resistance gene *Rag1* in soybean," *Theoretical and Applied Genetics*, vol. 120, no. 5, pp. 1063–1071, 2010.

[81] K. S. Kim, C. B. Hill, G. L. Hartman, D. L. Hyten, M. E. Hudson, and B. W. Diers, "Fine mapping of the soybean aphid-resistance gene *Rag2* in soybean PI 200538," *Theoretical and Applied Genetics*, vol. 121, no. 3, pp. 599–610, 2010.

[82] B. K. Ha, R. S. Hussey, and H. R. Boerma, "Development of SNP assays for marker-assisted selection of two southern root-knot nematode resistance QTL in soybean," *Crop Science*, vol. 47, no. 2, pp. S73–S82, 2007.

[83] X. Hu, M. Sullivan-Gilbert, M. Gupta, and S. A. Thompson, "Mapping of the loci controlling oleic and linolenic acid contents and development of *fad2* and *fad3* allele-specific markers in canola (*Brassica napus* L.)," *Theoretical and Applied Genetics*, vol. 113, no. 3, pp. 497–507, 2006.

[84] A. Lehmensiek, M. W. Sutherland, and R. B. McNamara, "The use of high resolution melting (HRM) to map single nucleotide polymorphism markers linked to a covered smut resistance gene in barley," *Theoretical and Applied Genetics*, vol. 117, no. 5, pp. 721–728, 2008.

[85] M. K. Grimmer, S. Trybush, S. Hanley, S. A. Francis, A. Karp, and M. J. C. Asher, "An anchored linkage map for sugar beet based on AFLP, SNP and RAPD markers and QTL mapping of a new source of resistance to Beet necrotic yellow vein virus," *Theoretical and Applied Genetics*, vol. 114, no. 7, pp. 1151–1160, 2007.

[86] M. K. Grimmer, T. Kraft, S. A. Francis, and M. J. C. Asher, "QTL mapping of BNYVV resistance from the WB258 source in sugar beet," *Plant Breeding*, vol. 127, no. 6, pp. 650–652, 2008.

[87] W. Muchero, N. N. Diop, P. R. Bhat et al., "A consensus genetic map of cowpea [*Vigna unguiculata* (L) Walp.] and synteny based on EST-derived SNPs," *Proceedings of the National Academy of Sciences of the United States of America*, vol. 106, no. 43, pp. 18159–18164, 2009.

[88] G. Jander, S. R. Norris, S. D. Rounsley, D. F. Bush, I. M. Levin, and R. L. Last, "Arabidopsis map-based cloning in the post-genome era," *Plant Physiology*, vol. 129, no. 2, pp. 440–450, 2002.

[89] I. Y. Abdurakhmonov and A. Abdukarimov, "Application of association mapping to understanding the genetic diversity of plant germplasm resources," *International Journal of Plant Genomics*, vol. 2008, Article ID 574927, 2008.

[90] D. Hall, C. Tegström, and P. K. Ingvarsson, "Using association mapping to dissect the genetic basis of complex traits in plants," *Briefings in Functional Genomics and Proteomics*, vol. 9, no. 2, pp. 157–165, 2010.

[91] S. Myles, J. Peiffer, P. J. Brown et al., "Association mapping: critical considerations shift from genotyping to experimental design," *Plant Cell*, vol. 21, no. 8, pp. 2194–2202, 2009.

[92] M. Gore, E. S. Buckler, J. Yu, and C. Zhu, "Status and prospects of association mapping in plants," *The Plant Genome*, vol. 1, no. 1, pp. 5–20, 2008.

[93] J. A. Rafalski, "Association genetics in crop improvement," *Current Opinion in Plant Biology*, vol. 13, no. 2, pp. 174–180, 2010.

[94] A. Beló, P. Zheng, S. Luck et al., "Whole genome scan detects an allelic variant of *fad2* associated with increased oleic acid levels in maize," *Molecular Genetics and Genomics*, vol. 279, no. 1, pp. 1–10, 2008.

[95] F. Tian, P. J. Bradbury, P. J. Brown et al., "Genome-wide association study of leaf architecture in the maize nested

association mapping population," *Nature Genetics*, vol. 43, no. 2, pp. 159–162, 2011.

[96] J. K. Roy, K. P. Smith, G. J. Muehlbauer, S. Chao, T. J. Close, and B. J. Steffenson, "Association mapping of spot blotch resistance in wild barley," *Molecular Breeding*, vol. 26, no. 2, pp. 243–256, 2010.

[97] K. Pajerowska-Mukhtar, B. Stich, U. Achenbach et al., "Single nucleotide polymorphisms in the Allene Oxide Synthase 2 gene are associated with field resistance to late blight in populations of tetraploid potato cultivars," *Genetics*, vol. 181, no. 3, pp. 1115–1127, 2009.

[98] K. L. Kump, P. J. Bradbury, R. J. Wisser et al., "Genome-wide association study of quantitative resistance to southern leaf blight in the maize nested association mapping population," *Nature Genetics*, vol. 43, no. 2, pp. 163–168, 2011.

[99] A. L. Harper, M. Trick, J. Higgins et al., "Associative transcriptomics of traits in the polyploid crop species *Brassica napus*," *Nature Biotechnology*, vol. 30, no. 8, pp. 798–802, 2012.

[100] Y. Xu and J. H. Crouch, "Marker-assisted selection in plant breeding: from publications to practice," *Crop Science*, vol. 48, no. 2, pp. 391–407, 2008.

[101] B. C. Y. Collard and D. J. Mackill, "Marker-assisted selection: an approach for precision plant breeding in the twenty-first century," *Philosophical Transactions of the Royal Society B*, vol. 363, no. 1491, pp. 557–572, 2008.

[102] E. M. Septiningsih, A. M. Pamplona, D. L. Sanchez et al., "Development of submergence-tolerant rice cultivars: the Sub1 locus and beyond," *Annals of Botany*, vol. 103, no. 2, pp. 151–160, 2009.

[103] L. Jin, Y. Lu, Y. Shao et al., "Molecular marker assisted selection for improvement of the eating, cooking and sensory quality of rice (*Oryza sativa* L.)," *Journal of Cereal Science*, vol. 51, no. 1, pp. 159–164, 2010.

[104] M. Asif, T. Shaheen, N. Tabbasam, Y. Zafar, and A. H. Paterson, "Marker-assisted breeding in higher plants," *Alternative Farming Systems, Biotechnology, Drought Stress and Ecological Fertilisation*, vol. 6, pp. 39–76, 2011.

[105] R. Naidoo, G. M. F. Watson, J. Derera, P. Tongoona, and M. Laing, "Marker-assisted selection for low phytic acid (*lpa1-1*) with single nucleotide polymorphism marker and amplified fragment length polymorphisms for background selection in a maize backcross breeding programme," *Molecular Breeding*, vol. 30, pp. 1207–1217, 2012.

[106] S. R. Eathington, T. M. Crosbie, M. D. Edwards, R. S. Reiter, and J. K. Bull, "Molecular markers in a commercial breeding program," *Crop Science*, vol. 47, supplement 3, pp. S154–S163, 2007.

[107] M. L. Rosso, S. A. Burleson, L. M. Maupin, and K. M. Rainey, "Development of breeder-friendly markers for selection of *MIPS1* mutations in soybean," *Molecular Breeding*, vol. 28, no. 1, pp. 127–132, 2011.

[108] J. M. Ribaut and M. Ragot, "Marker-assisted selection to improve drought adaptation in maize: the backcross approach, perspectives, limitations, and alternatives," *Journal of Experimental Botany*, vol. 58, no. 2, pp. 351–360, 2007.

[109] R. Ren, B. A. Nagel, S. P. Kumpatla et al., "Maize Cytoplasmic Male Sterility (Cms) C-Type Restorer Rf4 Gene, Molecular Markers And Their Use," Google Patents, 2011.

[110] M. Ragot, M. Lee, E. Guimarães et al., "Marker-assisted selection in maize: current status, potential, limitations and perspertives from the private and public sectors," *Marker-Assisted Selection, Current Status and Future Perspectives in Crops, Livestock, Forestry and Fish*, pp. 117–150, 2007.

[111] H. Riday, "Paternity testing: a non-linkage based marker-assisted selection scheme for outbred forage species," *Crop Science*, vol. 51, no. 2, pp. 631–641, 2011.

[112] D. W. Gjertson, C. H. Brenner, M. P. Baur et al., "ISFG: recommendations on biostatistics in paternity testing," *Forensic Science International*, vol. 1, no. 3-4, pp. 223–231, 2007.

[113] T. H. E. Meuwissen, B. J. Hayes, and M. E. Goddard, "Prediction of total genetic value using genome-wide dense marker maps," *Genetics*, vol. 157, no. 4, pp. 1819–1829, 2001.

[114] E. L. Heffner, M. E. Sorrells, and J. L. Jannink, "Genomic selection for crop improvement," *Crop Science*, vol. 49, no. 1, pp. 1–12, 2009.

[115] Z. Shengqiang, J. C. M. Dekkers, R. L. Fernando, and J. L. Jannink, "Factors affecting accuracy from genomic selection in populations derived from multiple inbred lines: a barley case study," *Genetics*, vol. 182, no. 1, pp. 355–364, 2009.

[116] B. Hayes and M. Goddard, "Genome-wide association and genomic selection in animal breeding," *Genome*, vol. 53, no. 11, pp. 876–883, 2010.

[117] J. L. Jannink, A. J. Lorenz, and H. Iwata, "Genomic selection in plant breeding: from theory to practice," *Briefings in Functional Genomics and Proteomics*, vol. 9, no. 2, pp. 166–177, 2010.

[118] A. M. Mastrangelo, E. Mazzucotelli, D. Guerra, P. Vita, and L. Cattivelli, "Improvement of drought resistance in crops: from conventional breeding to genomic selection," *Crop Stress and Its Management*, pp. 225–259, 2012.

[119] M. D. V. Resende, M. F. R. Resende Jr., C. P. Sansaloni et al., "Genomic selection for growth and wood quality in *Eucalyptus*: capturing the missing heritability and accelerating breeding for complex traits in forest trees," *New Phytologist*, vol. 194, no. 1, pp. 116–128, 2012.

[120] Y. Zhao, M. Gowda, W. Liu et al., "Accuracy of genomic selection in European maize elite breeding populations," *Theoretical and Applied Genetics*, vol. 124, no. 4, pp. 769–776, 2012.

15

SNP Discovery through Next-Generation Sequencing and Its Applications

Santosh Kumar,[1] Travis W. Banks,[2] and Sylvie Cloutier[1, 3]

[1] *Department of Plant Science, University of Manitoba, Winnipeg, MB, Canada R3T 2N2*
[2] *Department of Applied Genomics, Vineland Research and Innovation Centre, Vineland Station, ON, Canada L0R 2E0*
[3] *Cereal Research Centre, Agriculture and Agri-Food Canada, Winnipeg, MB, Canada R3T 2M9*

Correspondence should be addressed to Sylvie Cloutier, sylvie.j.cloutier@agr.gc.ca

Academic Editor: Roberto Tuberosa

The decreasing cost along with rapid progress in next-generation sequencing and related bioinformatics computing resources has facilitated large-scale discovery of SNPs in various model and nonmodel plant species. Large numbers and genome-wide availability of SNPs make them the marker of choice in partially or completely sequenced genomes. Although excellent reviews have been published on next-generation sequencing, its associated bioinformatics challenges, and the applications of SNPs in genetic studies, a comprehensive review connecting these three intertwined research areas is needed. This paper touches upon various aspects of SNP discovery, highlighting key points in availability and selection of appropriate sequencing platforms, bioinformatics pipelines, SNP filtering criteria, and applications of SNPs in genetic analyses. The use of next-generation sequencing methodologies in many non-model crops leading to discovery and implementation of SNPs in various genetic studies is discussed. Development and improvement of bioinformatics software that are open source and freely available have accelerated the SNP discovery while reducing the associated cost. Key considerations for SNP filtering and associated pipelines are discussed in specific topics. A list of commonly used software and their sources is compiled for easy access and reference.

1. Introduction

Molecular markers are widely used in plant genetic research and breeding. Single Nucleotide Polymorphisms (SNPs) are currently the marker of choice due to their large numbers in virtually all populations of individuals. The applications of SNP markers have clearly been demonstrated in human genomics where complete sequencing of the human genome led to the discovery of several million SNPs [1] and technologies to analyze large sets of SNPs (up to 1 million) have been developed. SNPs have been applied in areas as diverse as human forensics [2] and diagnostics [3], aquaculture [4], marker assisted-breeding of dairy cattle [5], crop improvement [6], conservation [7], and resource management in fisheries [8]. Functional genomic studies have capitalized upon SNPs located within regulatory genes, transcripts, and Expressed Sequence Tags (ESTs) [9, 10]. Until recently large scale SNP discovery in plants was limited to maize,

Arabidopsis, and rice [11–15]. Genetic applications such as linkage mapping, population structure, association studies, map-based cloning, marker-assisted plant breeding, and functional genomics continue to be enabled by access to large collections of SNPs. *Arabidopsis thaliana* was the first plant genome sequenced [16] followed soon after by rice [17, 18]. In the year 2011 alone, the number of plant genomes sequenced doubled as compared to the number sequenced in the previous decade, resulting in currently, 31 and counting, publicly released sequenced plant genomes (http://www.phytozome.net/). With the ever increasing throughput of next-generation sequencing (NGS), *de novo* and reference-based SNP discovery and application are now feasible for numerous plant species.

Sequencing refers to the identification of the nucleotides in a polymer of nucleic acids, whether DNA or RNA. Since its inception in 1977, sequencing has brought about the field of genomics and increased our understanding of the organization and composition of plant genomes. Tremendous improvements in sequencing have led to the generation of large amounts of DNA information in a very short period of time [19]. The analyses of large volumes of data generated through various NGS platforms require powerful computers and complex algorithms and have led to a recent expansion of the bioinformatics field of research. This book chapter focuses on the *a priori* discovery of SNPs through NGS, bioinformatics tools and resources, and the various downstream applications of SNPs.

2. History and Evolution of Sequencing Technologies

2.1. Invention of Sequencing. In 1977, two sequencing methods were developed and published. The Sanger method is a sequencing-by-synthesis (SBS) method that relies on a combination of deoxy- and dideoxy-labeled chain terminator nucleotides [20]. The first complete genome sequencing, that of bacteriophage *phi X174*, was achieved that same year using this pioneering method [21]. The chemical modification followed by cleavage at specific sites method also published in 1977 [22] quickly became the less favored of the two methods because of its technical complexities, use of hazardous chemicals, and inherent difficulty in scale-up. In contrast, the Sanger method, for which Frederick Sanger was awarded his second Nobel Prize in chemistry in 1980, was quickly adopted by the biotechnology industry which implemented it using a broad array of chemistries and detection methods [19].

2.2. Sequencing Technologies. In the last decade, new sequencing technologies have outperformed Sanger-based sequencing in throughput and overall cost, if not quite in sequence length and error rate [23]. This section will focus on the three main NGS platforms as well as the two main third-generation sequencing (TGS) platforms, their throughput and relative cost. We made every effort to ensure the accuracy of the data at the time of submission. However, the cost and throughput of these sequencing platforms change rapidly and, as such, our analysis only represents a snapshot in time. The flux of innovation in this field imposes a need for constant assessment of the technologies' potentials and realignment of research goals.

2.2.1. Roche (454) Sequencing. Pyrosequencing was the first of the new highly parallel sequencing technologies to reach the market [24]. It is commonly referred to as 454 sequencing after the name of the company that first commercialized it. It is an SBS method where single fragments of DNA are hybridized to a capture bead array and the beads are emulsified with regents necessary to PCR amplifying the individually bound template. Each bead in the emulsion acts as an independent PCR where millions of copies of the original template are produced and bound to the capture beads which then serve as the templates for the subsequent sequencing reaction. The individual beads are deposited into a picotiter plate along with DNA polymerase, primers,

and the enzymes necessary to create fluorescence through the consumption of inorganic phosphate produced during sequencing. The instrument washes the picotiter plate with each of the DNA bases in turn. As template-specific incorporation of a base by DNA polymerase occurs, a pyrophosphate (PPi) is produced. This pyrophosphate is detected by an enzymatic luminometric inorganic pyrophosphate detection assay (ELIDA) through the generation of a light signal following the conversion of PPi into ATP [25]. Thus, the wells in which the current nucleotides are being incorporated by the sequencing reaction occurring on the bead emit a light signal proportional to the number of nucleotides incorporated, whereas wells in which the nucleotides are not being incorporated do not. The instrument repeats the sequential nucleotide wash cycle hundreds of times to lengthen the sequences. The 454 GS FLX Titanium XL$^+$ platform currently generates up to 700 MB of raw 750 bp reads in a 23 hour run. The technology has difficulty quantifying homopolymers resulting in insertions/deletions and has an overall error rate of approximately 1%. Reagent costs are approximately \$6,200 per run [26].

2.2.2. Illumina Sequencing. Illumina technology, acquired by Illumina from Solexa, followed the release of 454 sequencing. With this sequencing approach, fragments of DNA are hybridized to a solid substrate called a flow cell. In a process called bridge amplification, the bound DNA template fragments are amplified in an isothermal reaction where copies of the template are created in close proximity to the original. This results in clusters of DNA fragments on the flow cell creating a "lawn" of bound single strand DNA molecules. The molecules are sequenced by flooding the flow cell with a new class of cleavable fluorescent nucleotides and the reagents necessary for DNA polymerization [27]. A complementary strand of each template is synthesized one base at a time using fluorescently labeled nucleotides. The fluorescent molecule is excited by a laser and emits light, the colour of which is different for each of the four bases. The fluorescent label is then cleaved off and a new round of polymerization occurs. Unlike 454 sequencing, all four bases are present for the polymerization step and only a single molecule is incorporated per cycle. The flagship HiSeq2500 sequencing instrument from Illumina can generate up to 600 GB per run with a read length of 100 nt and 0.1% error rate. The Illumina technique can generate sequence from opposite ends of a DNA fragment, so called paired-end (PE) reads. Reagent costs are approximately \$23,500 per run [26].

2.2.3. Applied Biosystems (SOLiD) Sequencing. The SOLiD system was jointly developed by the Harvard Medical School and the Howard Hughes Medical Institute [28]. The library preparation in SOLiD is very similar to Roche/454 in which clonal bead populations are prepared in microreactors containing DNA template, beads, primers, and PCR components. Beads that contain PCR products amplified by emulsion PCR are enriched by a proprietary process. The DNA templates on the beads are modified at their 3′ end to allow attachment to glass slides. A primer is annealed to an adapter on the DNA template and a mixture of fluorescently tagged oligonucleotides is pumped into the flow cell. When the oligonucleotide matches the template sequence, it is ligated onto the primer and the unincorporated nucleotides are washed away. A charged couple device (CCD) camera captures the different colours attached to the primer. Each fluorescence wavelength corresponds to a particular dinucleotide combination. After image capture, the fluorescent tag is removed and new set of oligonucleotides are injected into the flow cell to begin the next round of DNA ligation [19]. This sequencing-by-ligation method in SOLiD-5500x1 platform generates up to 1,410 million PE reads of 75 + 35 nt each with an error rate of 0.01% and reagent cost of approximately $10,500 per run [26].

Although widely accepted and used, the NGS platforms suffer from amplification biases introduced by PCR and dephasing due to varying extension of templates. The TGS technologies use single molecule sequencing which eliminates the need for prior amplification of DNA thus overcoming the limitations imposed by NGS. The advantages offered by TGS technology are (i) lower cost, (ii) high throughput, (iii) faster turnaround, and (iv) longer reads [19, 29]. The TGS can broadly be classified into three different categories: (i) SBS where individual nucleotides are observed as they incorporate (Pacific Biosciences single molecule real time (SMART), Heliscope true single molecule sequencing (tSMS), and Life Technologies/Starlight and Ion Torrent), (ii) nanopore sequencing where single nucleotides are detected as they pass through a nanopore (Oxford/Nanopore), and (iii) direct imaging of individual molecules (IBM).

2.2.4. Helicos Biosciences Corporation (Heliscope) Sequencing.
Heliscope sequencing involves DNA library preparation and DNA shearing followed by addition of a poly-A tail to the sheared DNA fragments. These poly-A tailed DNA fragments are attached to flow cells through poly-T anchors. The sequencing proceeds by DNA extension with one out of 4 fluorescent tagged nucleotides incorporated followed by detection by the Heliscope sequencer. The fluorescent tag on the incorporated nucleotide is then chemically cleaved to allow subsequent elongation of DNA [30]. Heliscope sequencers can generate up to 28 GB of sequence data per run (50 channels) with maximum read length of 55 bp at ~99% accuracy [31]. The cost per run per channel is approximately $360.

2.2.5. Pacific Biosciences SMART Sequencing.
The Pacific Biosciences sequencer uses glass anchored DNA polymerases which are housed at the bottom of a zero-mode waveguide (ZMW). DNA fragments are added into the ZMW chamber with the anchored DNA polymerase and nucleotides, each labeled with a different colour fluorophore, and are diffused from above the ZMW. As the nucleotides circulate through the ZMW, only the incorporated nucleotides remain at the bottom of the ZMW while unincorporated nucleotides diffuse back above the ZMW. A laser placed below the ZMW excites only the fluorophores of the incorporated nucleotides as the ZMW entraps the light and does not allow

it to reach the unincorporated nucleotides above [32]. The Pacific Biosciences sequencers can generate up to 140 MB of sequences per run (per smart cell) with reads of 2.5 Kbp at ~85% accuracy. The cost per run per smart cell is approximately $600.

Among the TGS technologies, Pacific Biosciences SMART and Heliscope tSMS have been used in characterizing bacterial genomes and in human-disease-related studies [31]; however, TGS has yet to be capitalized upon in plant genomes. The Heliscope generates short reads (55 bp) which may cause ambiguous read mapping due to the presence of paralogous sequences and repetitive elements in plant genomes. The Pacific Biosciences reads have high error rates which limit their direct use in SNP discovery. However, their long reads offer a definite advantage to fill gaps in genomic sequences and, at least in bacterial genomes, NGS reads have proven capable of "correcting" the base call errors of this TGS technology [33–36]. Hybrid assemblies incorporating short (Illumina, SOLiD), medium (454/Roche), and long reads (Pac-Bio) have the potential to yield better quality reference genomes and, as such, would provide an improved tool for SNP discovery.

The choice of a sequencing strategy must take into account the research goals, ability to store and analyze data, the ongoing changes in performance parameters, and the cost of NGS/TGS platforms. Some key considerations include cost per raw base, cost per consensus base, raw and consensus accuracy of bases, read length, cost per read, and availability of PE or single end reads. The pre- and postprocessing protocols such as library construction [37] and pipeline development and implementation for data analysis [38] are also important.

2.3. RNA and ChIP Sequencing.
Genome-wide analyses of RNA sequences and their qualitative and quantitative measurements provide insights into the complex nature of regulatory networks. RNA sequencing has been performed on a number of plant species including *Arabidopsis* [39], soybean [40], rice [41], and maize [42] for transcript profiling and detection of splice variants. RNA sequencing has been used in *de novo* assemblies followed by SNP discovery performed in nonmodel plants such as *Eucalyptus grandis* [43], *Brassica napus* [44], and *Medicago sativa* [45].

RNA deep-sequencing technologies such as digital gene expression [46] and Illumina RNASeq [47] are both qualitative and quantitative in nature and permit the identification of rare transcripts and splice variants [48]. RNA sequencing may be performed following its conversion into cDNA that can then be sequenced as such. This method is, however, prone to error due to (i) the inefficient nature of reverse transcriptases (RTs) [49], (ii) DNA-dependent DNA polymerase activity of RT causing spurious second strand DNA [50], and (iii) artifactual cDNA synthesis due to template switching [51]. Direct RNA sequencing (DRS) developed by Helicos Biosciences Corporation is a high throughput and cost-effective method which eliminates the need for cDNA synthesis and ligation/amplification leading to improved accuracy [52].

Chromatin immunoprecipitation (ChIP) is a specialized sequencing method that was specifically designed to identify DNA sequences involved in *in vivo* protein DNA interaction [53]. ChIP-sequencing (ChIP-Seq) is used to map the binding sites of transcription factors and other DNA binding sites for proteins such as histones. As such, ChIP-Seq does not aid SNP discovery, but the availability of SNP data along with ChIP-Seq allows the study of allele-specific states of chromatin organization. Deep sequence coverage leading to dense SNP maps permits the identification of transcription factor binding sites and histone-mediated epigenetic modifications [54]. ChIP-Seq can be performed on serial analysis of gene expression (SAGE) tags or PE using Sanger, 454, and Illumina platforms [55, 56].

The DNA, RNA, and ChIP-Seq data is analysed using a reference sequence if available or, in the absence of such reference, it requires *de novo* assembly, all of which is performed using specialized software, algorithms, pipelines, and hardware.

3. Computing Resources for Sequence Assembly

The next-generation platforms generate a considerable amount of data and the impact of this with respect to data storage and processing time can be overlooked when designing an experiment. Bioinformatics research is constantly developing new software and algorithms, data storage approaches, and even new computer architectures to better meet the computation requirements for projects incorporating NGS. This chapter describes the state-of-the-art with respect to software for NGS alignment and analysis at the time of writing.

3.1. Software for Sequence Analysis. Both commercial and noncommercial sequence analysis software are available for Windows, Macintosh, and Linux operating systems. NGS companies offer proprietary software such as consensus assessment of sequence and variation (Cassava) for Illumina data and Newbler for 454 data. Such software tend to be optimized for their respective platform but have limited cross applicability to the others. Web-based portals such as Galaxy [57] are tailored to a multitude of analyses, but the requirement to transfer multigigabyte sequence files across the internet can limit its usability to smaller datasets. Commercially available software such as CLC-Bio (http://www.clcbio.com/) and SeqMan NGen (http://www.dnastar.com/t-sub-products-genomics-seqman-ngen.aspx) provide a friendly user interface, are compatible with different operating systems, require minimal computing knowledge, and are capable of performing multiple downstream analyses. However, they tend to be relatively expensive, have narrow customizability, and require locally available high computing power. A recent review by Wang et al. [58] recommends Linux-based programs because they are often free, not specific to any sequencing platform, and less computing power hungry and, as a consequence, tend to perform faster. Flexibility in the parameter's choice for read assembly is another major advantage. However,

most biologists are unfamiliar with Linux operating systems, its structure and command lines, thereby imposing a steep learning curve for adoption. Linux-based software such as Bowtie [59], BWA [60], and SOAP2/3 [61] have been used widely for the analysis of NGS data. Other software may not have gained broad acceptance but may have unique features worth noting. For reviews on NGS software, see Li and Homer [62], Wang et al. [58], and Treangen and Salzberg [63]. Characteristics of the most common NGS software and their attributes are listed in Table 1, and their download information can be found in Table 4.

3.2. Consideration for Software Selection. In selecting software for NGS data analysis one must consider, among other things, the sequencing platform, the availability of a reference genome, the computing and storage resources necessary, and the bioinformatics expertise available. Algorithms used for sequence analysis have matured significantly but may still require computing power beyond what is currently available in most genomics facilities and/or long processing time. For example, in aligning $2 \times 13,326,195$ paired-end reads (76 bp) from The Cancer Genome Atlas project (SRR018643) [64], SHRiMP [65] took 1,065 hrs with a peak memory footprint of 12 gigabytes to achieve the mapping of 81% of the reads to the human genome reference whereas Bowtie used 2.9 gigabytes of memory, a run time of 2.2 hrs but only achieved a 67% mapping rate [58]. Both time and memory become critical when dealing with a very large NGS dataset. Fast and memory efficient sequence mapping seems to be preferred over slower, memory demanding software even at the cost of a reduced mapping rate. It should be noted that a higher percentage of mapped reads is not a strict measure of quality because it may be indicative of a higher level of misaligned reads or reads aligned against repetitive elements, features that are not desirable [63].

In the absence of a reference genome, *de novo* assembly of a plant genome is achieved using sequence information obtained through a combination of Sanger and/or NGS of bacterial artificial chromosome (BAC) clones, or by whole genome shotgun (WGS) with NGS [66]. *De novo* assemblies are time consuming and require much greater computing power than read mapping onto a reference genome. The assembly accuracy depends in part on the read length and depth as well as the nature of the sequenced genome. The genomes of *Arabidopsis thaliana* [16], rice [67], and maize [68] were generated using a BAC-by-BAC approach while poplar [69], grape [70], and sorghum [71] genomic sequences were obtained through WGS. All genomes sequenced to date are fragmented to varying degrees because of the inability of sequencing technologies and bioinformatics algorithms to assemble through highly conserved repetitive elements. A list of current plant genome sequencing projects, their sequencing strategies, and status from standard draft to finished can be found in the review by Feuillet et al. [72].

Software programs such as Mira [73], SOAPdenovo [74], ABySS [75], and Velvet [76] have been used for *de novo* assembly. MIRA is well documented and can be readily

TABLE 1: List of most cited/used software for sequence assembly of NGS data. Source locations for these software are compiled in Table 4.

Name (current version)	Assembly type (algorithm)	Supported parameters				Output format	Platform
		Color space	Read length	Gapped alignment	Paired-end		
CLC-Bio[1]	Reference[2]	Yes	Arbitrary	Yes	Yes	CLC-Bio	Linux/Windows/Mac OS X
SeqMan NGen[1]	Reference[2]	Yes	Arbitrary	Yes	Yes	ACE, BAM	Windows/Mac OS X
NextGENe[1]	Reference[2]	Yes	Arbitrary	Yes	Yes	NextGENe	Windows/Mac OS X
Bowtie (2)	Reference (FM-index)	Yes	Arbitrary	Yes	Yes	SAM	Linux/Windows/Mac OS X
BWA	Reference (FM-index)	Yes	Arbitrary	Yes	Yes	SAM	Linux
SOAP (3)	Reference (FM-index)	Yes	Arbitrary	No	Yes	SOAP2/3	Linux
MAQ (0.6.6)	Reference (Hashing reads)	Yes	≤127	Yes	Yes	MAQ	Linux/Solaris/Mac OS X
Novoalign (2.07.07)	Reference (Hashing reference)	Yes	Arbitrary	Yes	Yes	SAM	Linux/Mac OS X
Mosaik (1.1.0018)	Reference (Hashing reference)	Yes	Arbitrary	Yes	Yes	SAM	Linux/Windows/Mac OS X/Solaris
SHRiMP (2.2.2)	Reference (Hashing reference)	Yes	Arbitrary	Yes	Yes	SAM	Linux/Mac OS X
Mira (3.4)	Reference[2]	Yes	Arbitrary	Yes	Yes	FASTA, ACE	Linux

[1]Commercial software. [2]Option for *de novo* assembly and modules included for variant calling.

customized, but it requires substantial computing memory and is not suited for large complex genomes. Of the freely available software, SOAPdenovo is one of the fastest read assembly programs and it uses a comparatively moderate amount of computing memory. The assembly generated by SOAPdenovo can be used for SNP discovery using SOAPsnp as implemented for the apple genome [77]. ABySS can be deployed on a computer cluster. It requires the least amount of memory and can be used for large genomes. Velvet requires the largest amount of memory. It can use mate-pair information to resolve and correct assembly errors.

4. SNP Discovery

The most common application of NGS is SNP discovery, whose downstream usefulness in linkage map construction, genetic diversity analyses, association mapping, and marker-assisted selection has been demonstrated in several species [78]. NGS-derived SNPs have been reported in humans [79], *Drosophila* [80], wheat [81, 82], eggplant [83], rice [84–86], *Arabidopsis* [87, 88], barley [14, 89], sorghum [90], cotton [91], common beans [78], soybean [92], potato [93], flax [94], *Aegilops tauschii* [95], alfalfa [96], oat [97], and maize [98] to name a few.

SNP discovery using NGS is readily accomplished in small plant genomes for which good reference genomes are available such as rice and *Arabidopsis* [86, 99]. Although SNP discovery in complex genomes without a reference genome such as wheat [81, 82], barley [14, 89], oat [97], and beans [78] can be achieved through NGS, several challenges remain in other nonmodel but economically important crops. The presence of repeat elements, paralogs, and incomplete or inaccurate reference genome sequences can create ambiguities in SNP calling [63]. NGS read mapping can also suffer from sequencing error (erroneous base calling) and misaligned reads. The following section focuses on programs tailored for SNP discovery and emphasizes some of the precautions and considerations to minimize erroneous SNP calling.

4.1. Software and Pipelines for SNP Discovery. In theory, a SNP is identified when a nucleotide from an accession read differs from the reference genome at the same nucleotide position. In the absence of a reference genome, this is achieved by comparing reads from different genotypes using *de novo* assembly strategies [95]. Read assembly files generated by mapping programs are used to perform SNP calling. In practice, various empirical and statistical criteria are used to call SNPs, such as a minimum and maximum number of reads considering the read depth, the quality score and the consensus base ratio for examples [95]. Thresholds for these criteria are adjusted based on the read length and the genome coverage achieved by the NGS data. In assemblies generated allowing single nucleotide variants and insertions/deletions (indels), a list of SNP and indel coordinates is generated and the read mapping results can be visualized using graphical user interface programs such as Tablet [100] (Figure 1), SNP-VISTA [101], or Savant [102] (refer to Table 4 for download

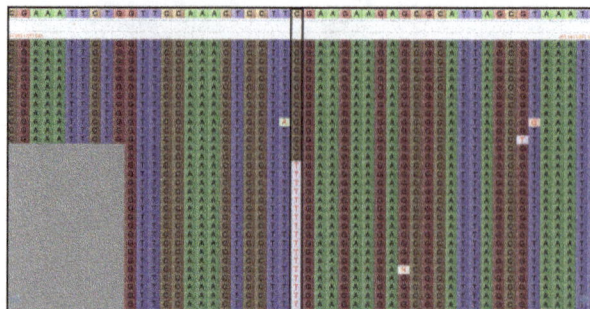

FIGURE 1: Graphical user interface of Tablet, an assembly visualization program, displays the reference genome on top and the mapped reads with color-coded SNPs on the bottom.

information). Tablet has a user-friendly interface and is widely used because it supports a wide array of commonly used file formats such as SAM, BAM, SOAP, ACE, FASTQ, and FASTA generated by different read assemblers such as Bowtie, BWA, SOAP, MAQ, and SeqMan NGen. It displays contig overview, coverage information, read names and it allows searching for specific coordinates on scaffolds.

Broadly used SNP calling software include Samtools [103], SNVer [104], and SOAPsnp [74]. Samtools is popular because of its various modules for file conversion (SAM to BAM and vice-versa), mapping statistics, variant calling, and assembly visualization. Recently, SOAPsnp has gained popularity because of its tight integration with SOAP aligner and other SOAP modules which are constantly upgraded and provide a one stop shop for the sequencing analysis continuum. Variant calling algorithms such as Samtools and SNVer can be used as stand-alone programs or incorporated into pipelines for SNP calling. Reviews of SNP calling software have been published [63, 105]. Some of the main features of the current commonly used software are listed in Table 2 (refer to Table 4 for download information).

4.2. SNP Discovery from Multiple Individuals and Complex Genomes. SNP discovery is more robust when multiple and divergent genotypes are used simultaneously, creating the necessary basis to capture the genetic variability of a species. Large parts of plant genomes consist of repetitive elements [106] which can cause spurious SNP calling by erroneous read mapping to paralogous repeat element sequences. In polyploid genomes such as cotton (allotetraploid), homoeologous sequences can cause similar misalignment [91]. Improved read assembly and filtering of SNPs become even more important factors for accurate SNP calling in these cases because they can mitigate the effects of errors caused by paralogs and homoeologs.

Read assembly algorithms such as Bowtie and SOAP as well as variant calling/genotyping softwares such as GATK [107] are rapidly evolving to accommodate an ever increasing number of reads, increased read length, nucleotide quality values, and mate-pair information of PE reads. Assembly programs such as Novoalign (http://www.novo-craft.com/main/index.php) and STAMPY [108], although

TABLE 2: Commonly used NGS variant calling software. Download information for these software is compiled in Table 4. A more comprehensive list of variant calling programs is available at http://seqanswers.com/wiki/Software/list.

Software	Multisample support	Reference	Features	Platform
Samtools	Yes	Aligned reads	Include computation of genotype likelihoods and variant calling	Linux
SOAPsnp	No	Variant database	Part of SOAP3 for variant calling	Linux
GATK	Yes	Aligned reads	Include variant caller, SNP filter, and SNP quality calibrator	Linux
SNVer	Yes	Aligned reads	Fast variant caller, assigning SNP significance based on read depth	Windows, Linux, Mac OS X
SHORE	Yes	Aligned reads	Variant calling based on reference sequence even from other species	Linux, Mac OS X
MaCH	Yes	Genotype likelihoods	Variant calling with or without LD information	Windows, Linux, Mac OSX
IMPUTE2	Yes	Candidate SNPs and genotype likelihoods	Variant calling and linkage map-based SNP imputation	Windows, Linux, Mac OS X

memory and time intensive, are highly sensitive for simultaneous mapping of short reads from multiple individuals [105].

SNP calls can be significantly improved using filtering criteria that are specific to the genome characteristics and the dataset. For instance, projects aimed at resequencing can compare different datasets from the same genotype and thus eliminate data with large discrepancies. This strategy identifies the most common sources of error and is applied in the 1000 genome project [109]. Reduced representation libraries (RRLs), that is, sequencing an enriched subset of a genome by eliminating a proportion of its repetitive fractions [79], reduce the probability of misalignments to repeats and thus potential downstream erroneous SNP calling. Filtering criteria that can improve SNP accuracy include (i) a minimum read depth (often ≥3 per genotype), (ii) >90% nucleotides within a genotype having identical call at a given position (~<10% sequencing error), (iii) a read depth ≤ mean of the sequence depth over the entire mapping assembly, (iv) the elimination of ribosomal DNA and other repetitive elements in the 50 nt flanking any SNP call, and (v) masking of homopolymer SNPs with a given base string length (often ≥2). Additionally, in polyploid species, separate assembly of homoeologs using stringent mapping parameters is often essential for genome-wide SNP identification to avoid spurious SNP calls caused by erroneous homoeologous read mapping [91].

4.3. SNP Validation. Prior to any SNP applications, the discovered SNPs must be validated to identify the true SNPs and get an idea of the percentage of potentially false SNPs resulting from an SNP discovery exercise. The need for validation arises because a proportion of the discovered SNPs could have been wrongly called for various reasons including those outlined above. SNP validation can be accomplished using a variety of material such as a biparental segregating population or a diverse panel of genotypes. Usually a small subset of the SNPs is used for validation through assays such as the Illumina Goldengate [110], KBiosciences Competitive AlleleSpecific-PCR SNP genotyping system (KASPar) (http://www.lgcgenomics.com/) or the High Resolution

Melting (HRM) curve analysis. Validation can serve as an iterative and informative process to modify and optimize the SNP filtering criteria to improve SNP calling. For example, a subset of 144 SNPs from a total of 2,113,120 SNPs were validated using the Goldengate assay on 160 accessions in apple [77]. Another example is illustrated in Figure 2 where a KASPar assay was performed on 92 genotypes from a segregating population illustrating the validation of a single "T/C" SNP in two distinct clusters. Other validation strategies used in nonmodel organisms are tabulated in Garvin et al. [111]. With the continuously competitive pricing of NGS, genotyping-by-sequencing (GBS) is becoming a viable SNP validation method. Either biparental segregating populations or a collection of diverse genotypes can be sequenced at a reasonable cost using indexing, that is, combining multiple independently tagged genotypes in a single NGS run to obtain genome-wide or reduced representation genome sequences at a lower coverage but potentially validating a much larger number of SNPs than the methods described above. Sequencing of segregating populations or diverse genotypes may also lead to the discovery of additional SNPs.

The two major factors affecting the SNP validation rate are sequencing and read mapping errors as discussed above. NGS platforms have different levels of sequencing accuracies, and this may be the most important factor determining the variation in the validation, from 88.2% for SOLiD followed by Illumina at 85.4% and Roche 454 at 71% [95]. The SNP validation rates can be improved using RRL for SNP discovery and choosing SNPs within the nonrepetitive sequences including predicted single copy genes and single copy repeat junctions shown to have high validation rates [95].

5. SNP Genotyping

SNP genotyping is the downstream application of SNP discovery to identify genetic variations. SNP applications include phylogenic analysis, marker-assisted selection, genetic mapping of quantitative trait loci (QTL), bulked segregant analysis, genome selection, and genome-wide association studies (GWAS). The number of SNPs and

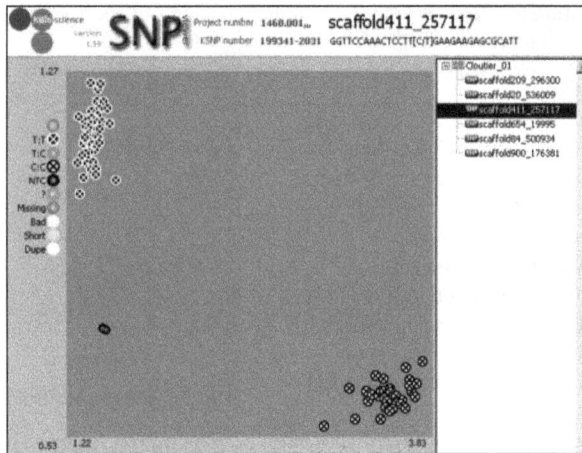

FIGURE 2: Validation of a T/C SNP by a KASPar assay (KBiosciences, Herts, England). Genotypes with a "T" are represented by black dots with a white cross clustered in the upper left and those with a "C" by white dots with a black cross in the bottom right cluster. The two black dots near the bottom left are negative controls. No heterozygous individuals were present in this population.

individuals to screen are of primary importance in choosing an SNP genotyping assay, though cost of the assay and/or equipment and the level of accuracy are also important considerations.

Illumina Goldengate is a commonly used genotyping assay because of its flexibility in interrogating 96 to 3,072 SNP loci simultaneously (http://www.illumina.com/). HRM analysis is suitable for a few to an intermediate number of SNPs and can be performed within a typical laboratory setting. KASPar and SNPline genotyping systems (http://www.lgcgenomics.com/) can be used for genotyping a few to thousands of SNPs in a laboratory setting. The SNPline system is available in SNPlite or SNPline XL versions to allow flexibility in sample number and SNP assays. The iPLEX Gold technology developed by Sequenom (http://www.sequenom.com/) is based on the MassARRAY system which uses primer extension chemistry and matrix-assisted laser desorption/ionisation-time of flight (MALDI-TOF) mass spectrometry for genotyping.

The iPLEX Gold system has gained acceptance due to its high precision and cost-effective implementation. High throughput chip-based genotyping assays such as the Affymetrix GeneChip arrays (http://www.affymetrix.com/estore/) and the Illumina BeadChips (http://www.illumina.com/) are capable of validating up to a million SNPs per reaction across an entire genome. Detailed analyses of SNP genotyping assays and their features are reviewed in Tsuchihashi and Dracopoli [112], Sobrino and Carracedo [113], Giancola et al. [114], Kim and Misra [115], Gupta et al. [116], and Ragoussis [117]. A list of the most commonly used genotyping assays describing the assay type, technology, throughput, multiplexing ability, and relative scalability can be found in Table 3.

Array-based technologies such as Infinium and Goldengate substantially improved SNP genotyping efficiency, but they are species-specific, expensive to design and require specific equipment and chemistry. PCR and primer extension technologies like KASPar and Taqman (http://www.lifetechnologies.com/global/en/home.html) are limited by their low SNP throughput but can be useful to assay a large number of genotypes with few SNPs. NGS technologies have become viable for genotyping studies and may offer advantages over other genotyping methods in cost and efficiency.

5.1. Genotyping-by-Sequencing (GBS). There have been a number of approaches developed that use complexity reduction strategies to lower the cost and simplify the discovery of SNP markers using NGS, RNA-Seq, complexity reduction of polymorphic sequences (CRoPS), restriction-site-associated DNA sequencing (RAD-Seq), and GBS [118]. Of these methodologies GBS holds the greatest promise to serve the widest base of plant researchers because of its ability to allow simultaneous marker discovery and genotyping with low cost and a simple molecular biology workflow. Briefly, GBS involves digesting the genome of each individual in a population to be studied with a restriction enzyme [119]. One unique and one common adapter are ligated to the fragments and a PCR is carried out which is biased towards amplifying smaller DNA fragments. The resulting PCR products are then pooled and sequenced using an Illumina platform. The amplicons are not fragmented so only the ends of the PCR products are sequenced. The unique adapter acts as an ID tag so sequencing reads can be associated with an individual. The technique can be applied to species with or without a reference genome. The choice of enzyme has an effect on the number of markers identified and the amount of sequence coverage required. The more frequent the restriction recognition site, the higher the number of fragments and therefore more potential markers. Use of more frequent cutters may necessitate greater amounts of sequencing depending on the application. Poland et al. [120] recently demonstrated the use of two restriction enzymes to perform GBS in bread wheat, a hexaploid genome.

GBS has the potential to be a truly revolutionary technology in the arena of plant genomics. It brings high density genotyping to the vast majority of plant species that, until now, have had almost no investment in genomics resources. With little capital investment requirement and an affordable per sample cost, all plant researchers now have powerful genomic and genetic methodologies available to them. Uses of GBS include applications in marker discovery, phylogenetics, bulked segregant analysis, QTL mapping in biparental lines, GWAS, and genome selection. GBS can also be applied to fine mapping in candidate gene discovery and be used to generate high-density SNP genetic maps to assist in *de novo* genome assembly. We predict tremendous advances in functional genomics and plant breeding from the implementation of GBS because it is truly a democratizing application for NGS in nonmodel plant systems.

TABLE 3: Commonly used genotyping platforms.

Name	Assay type	Technology	Throughput (samples)	Multiplexing	Relative scale (no. of SNP/no. of individuals)
Genechip	Hybridization	Oligo nucleotide array	96/5 days	Up to 18×10^6	Small/large
Infinium II	Hybridization	Bead array	Up to 128/5 days	Up to 13×10^6	Large/small-large
Goldengate	Primer extension-ligation	Bead array	172/3 days	Up to 3,072	Medium/large
iPlex	Primer extension	Mass spectrometry (MALDI-TOF)	3840/2.5 days	Up to 40	Medium/large
Taqman	PCR	Taqman probe	Up to 1536/day	Up to 256	Medium/medium
SNPlex	PCR	Capillary electrophoresis	Up to 1536/3 days	Up to 48	Medium/large
KASPar	PCR	FRET quenching oligos	Up to 96/day	—	Medium/large
Invader	Primer annealing/endonuclease digestion	FRET quenching oligos	Up to 384/day	Up to 200,000	Medium/large
HRM	PCR	Melting curve analysis	Up to 1536/day	—	Medium/large

6. Applications of SNPS

NGS and SNP genotyping technologies have made SNPs the most widely used marker for genetic studies in plant species such as *Arabidopsis* [121] and rice [122]. SNPs can help to decipher breeding pedigree, to identify genomic divergence of species to elucidate speciation and evolution, and to associate genomic variations to phenotypic traits [85]. The ease of SNP development, reasonable genotyping costs, and the sheer number of SNPs present within a collection of individuals allow an assortment of applications that can have a tremendous impact on basic and applied research in plant species.

6.1. SNPs in Genetic Mapping. A genetic map refers to the arrangement of traits, genes, and markers relative to each other as measured by their recombination frequency. Genetic maps are essential tools in molecular breeding for plant genetic improvement as they enable gene localization, map-based cloning, and the identification of QTL [123]. SNPs have greatly facilitated the production of much higher density maps than previous marker systems. SNPs discovered using RNA-Seq and expressed sequence tags (ESTs) have the added advantage of being gene specific [124]. Their high abundance and rapidly improving genotyping technologies make SNPs an ideal marker type for generating new genetic maps as well as saturating existing maps created with other markers. Most SNPs are biallelic thereby having a lower polymorphism information content (PIC) value as compared to most other marker types which are often multiallelic [125]. The limited information associated with their biallelic nature is greatly compensated by their high frequency, and a map of 700–900 SNPs has been found to be equivalent to a map of 300–400 simple sequence repeat (SSR) markers [125]. SNP-based linkage maps have been constructed in many economically important species such as rice [126], cotton [91] and *Brassica* [127]. The identification of candidate genes for flowering time in *Brassica* [127] and

maize [128] are practical examples of gene discovery through SNP-based genetic maps.

6.2. Genome-Wide Association Mapping. Association mapping (AM) panels provide a better resolution, consider numerous alleles, and may provide faster marker-trait association than biparental populations [129]. AM, often referred to as linkage disequilibrium (LD) mapping, relies on the nonrandom association between markers and traits [130]. LD can vary greatly across a genome. In low LD regions, high marker saturation is required to detect marker-trait association, hence the need for densely saturated maps. In general, GWASs require 10,000–100,000 markers applied to a collection of genotypes representing a broad genetic basis [130].

In the past few years, NGS technologies have led to the discovery of thousands, even millions of SNPs, and novel application platforms have made it possible to produce genome-wide haplotypes of large numbers of genotypes, making SNPs the ideal marker for GWASs. So far, 951 GWASs have been reported in humans (http://www.bing.com/search?q=www.genome.gov%2Fgwastudies%2F&src=ie9tr). In plants, such a study was first reported in *Arabidopsis* for flowering time and pathogen-resistance genes [131]. A GWAS performed in rice using ~3.6 million SNPs identified genomic regions associated with 14 agronomic traits [132]. The genetic structure of northern leaf blight, southern leaf blight, and leaf architecture was studied using ~1.6 million SNPs in maize [133–135]. SNP-based GWAS was also performed on species such as barley for which a reference genome sequence is not available [136]. Soto-Cerda and Cloutier [137] have reviewed the concepts, benefits, and limitations of AM in plants.

6.3. Evolutionary Studies. SSRs and mitochondrial DNA have been used in evolutionary studies since the early 1990s [138]. However, the biological inferences from results of these two

TABLE 4: Download information of software used for NGS data.

Software	Source
Bowtie	http://bowtie-bio.sourceforge.net/bowtie2/index.shtml
BWA	http://bio-bwa.sourceforge.net/
SOAP	http://soap.genomics.org.cn/soap3.html#down2
MAQ	http://sourceforge.net/projects/maq/
Novoalign	http://www.novocraft.com/main/index.php
CLC-Bio Genomics	http://www.clcbio.com/index.php?id=1240
SeqMan NGen	http://www.dnastar.com/t-products-seqman-ngen.aspx
NextGENe	http://softgenetics.com/NextGENe.html
Mosaik	http://bioinformatics.bc.edu/marthlab/Mosaik
SHRiMP	http://compbio.cs.toronto.edu/shrimp/
Mira	http://sourceforge.net/projects/mira-assembler/files/MIRA/stable/
Cassava	http://www.illumina.com/software/genome_analyzer_software.ilmn
Newbler	http://www.454.com/products/analysis-software/index.asp
Novoalign	http://www.novocraft.com/main/downloadpage.php
Tablet	http://bioinf.scri.ac.uk/tablet/
SNP-VISTA	http://genome.lbl.gov/vista/snpvista/
Samtools	http://sourceforge.net/projects/samtools/
Savant	http://genomesavant.com/savant/download.php
SOAPsnp	http://soap.genomics.org.cn/soapsnp.html
GATK	http://www.broadinstitute.org/gsa/wiki/index.php/ The_Genome_Analysis_Toolkit
SNver	http://snver.sourceforge.net/
MaCH	http://www.sph.umich.edu/csg/abecasis/MACH/
IMPUTE2	http://mathgen.stats.ox.ac.uk/impute/impute_v2.html# download_impute2
MEGA	http://www.megasoftware.net/
PHYLIP	http://evolution.genetics.washington.edu/phylip.html

marker types may be misinterpreted due to homoplasy, a phenomenon in which similarity in traits or markers occurs due to reasons other than ancestry, such as convergent evolution, evolutionary reversal, gene duplication, and horizontal gene transfer [139]. The advantage of SNPs over microsatellites and mitochondrial DNA resides in the fact that SNPs represent single base nucleotide substitutions and, as such, they are less affected by homoplasy because their origin can be explained by mutation models [140]. SNPs have been employed to quantify genetic variation, for individual identification, to determine parentage relatedness and population structure [138]. Seed shattering (or loss thereof) has been associated with an SNP through a GWAS aimed at unraveling the evolution of rice that led to its domestication [141]. SNPs have also been used to study the evolution of genes such as WAG-2 in wheat [142]. Algorithms such as neighbor-joining and maximum likelihood implemented in the PHYLIP [143] and MEGA [144] software are commonly used to generate phylogenetic trees.

The main advantage of SNPs is unquestionably their large numbers. As with all marker systems the researcher must be aware of ascertainment biases that exist in the panel of SNPs being used. These biases exist because SNPs are often developed from examining a small group of individuals and selecting the markers that maximize the amount of polymorphism that can be detected in the population used. This results in a collection of markers that sample only a fraction of the diversity that exists in the species but that are nevertheless used to infer relatedness and determine genetic distance for whole populations. Ideally, a set of SNP markers randomly distributed throughout the genome would be developed for each population studied. GBS moves us closer to this goal by incorporating simultaneous discovery of SNPs and genotyping of individuals. With this approach genome sample bias remains but can be mitigated by careful restriction enzyme selection.

7. Future Perspectives

SNP discovery incontestably made a quantum leap forward with the advent of NGS technologies and large numbers of SNPs are now available from several genomes including large and complex ones (see Section 4). Unlike model systems such as humans and *Arabidopsis*, SNPs from crop plants remain limited for the time being, but broad access to reasonable cost NGS promises to rapidly increase the production of reference genome sequences as well as SNP discovery. Many issues remain to be addressed, such as the ascertainment bias of popular biparental populations and the low validation rate of some array-based genotyping platforms [145]. The area of epigenetic regulation of various genome components can be better understood as accurate and deeper sequencing is achieved. RNA and ChIP-sequencing projects, similar to RNA-Seq in the nonmodel plant sweet cherry to identify SNPs and haplotypes [146], can be undertaken to study functional genomics. A great deal of knowledge that is still elusive about the noncoding and repetitive elements can be determined with the next wave of modern and efficient sequencing technologies.

The first (Sanger) and the second (next) generation sequencing technologies have enabled researchers to characterize DNA sequence variation, sequence entire genomes, quantify transcript abundance, and understand mechanisms such as alternative splicing and epigenetic regulation [29].

Numerous plant genomes are now sequenced at various levels of completion and many more are underway [72]. The NGS technologies have made SNP discovery affordable even in complex genomes and the technologies themselves have improved tremendously in the past decade. Improvements in TGS promise synergies with NGS technologies to further assist our understanding of plant genetics and genomics. NGS has revolutionized genomics-related research, and it is our belief that the NGS-enabled discoveries will continue in the next decade.

Acknowledgments

The authors are grateful to Andrzej Walichnowski for help with paper editing, Joanne Schiavoni for formatting, and Michael Shillinglaw for figure preparation. This chapter was written within the scope of the Genome Canada TUFGEN project, and support from all funding partners is gratefully acknowledged.

References

[1] K. A. Frazer, D. G. Ballinger, D. R. Cox et al., "A second generation human haplotype map of over 3.1 million SNPs," *Nature*, vol. 449, no. 7164, pp. 851–861, 2007.

[2] C. H. Brenner and B. S. Weir, "Issues and strategies in the DNA identification of World Trade Center victims," *Theoretical Population Biology*, vol. 63, no. 3, pp. 173–178, 2003.

[3] M. I. McCarthy, G. R. Abecasis, L. R. Cardon et al., "Genome-wide association studies for complex traits: consensus, uncertainty and challenges," *Nature Reviews Genetics*, vol. 9, no. 5, pp. 356–369, 2008.

[4] Z. J. Liu and J. F. Cordes, "DNA marker technologies and their applications in aquaculture genetics," *Aquaculture*, vol. 238, no. 1–4, pp. 1–37, 2004.

[5] L. R. Schaeffer, "Strategy for applying genome-wide selection in dairy cattle," *Journal of Animal Breeding and Genetics*, vol. 123, no. 4, pp. 218–223, 2006.

[6] H. Yu, W. Xie, J. Wang et al., "Gains in QTL detection using an ultra-high density SNP map based on population sequencing relative to traditional RFLP/SSR markers," *PLoS ONE*, vol. 6, no. 3, Article ID e17595, 2011.

[7] J. M. Seddon, H. G. Parker, E. A. Ostrander, and H. Ellegren, "SNPs in ecological and conservation studies: a test in the Scandinavian wolf population," *Molecular Ecology*, vol. 14, no. 2, pp. 503–511, 2005.

[8] C. T. Smith, C. M. Elfstrom, L. W. Seeb, and J. E. Seeb, "Use of sequence data from rainbow trout and Atlantic salmon for SNP detection in Pacific salmon," *Molecular Ecology*, vol. 14, no. 13, pp. 4193–4203, 2005.

[9] B. N. Chorley, X. Wang, M. R. Campbell, G. S. Pittman, M. A. Noureddine, and D. A. Bell, "Discovery and verification of functional single nucleotide polymorphisms in regulatory genomic regions: current and developing technologies," *Mutation Research*, vol. 659, no. 1-2, pp. 147–157, 2008.

[10] K. Faber, K. H. Glatting, P. J. Mueller, A. Risch, and A. Hotz-Wagenblatt, "Genome-wide prediction of splice-modifying SNPs in human genes using a new analysis pipeline called AASsites," *BMC Bioinformatics*, vol. 12, supplement 4, article S2, 2011.

[11] S. Atwell, Y. S. Huang, B. J. Vilhjálmsson et al., "Genome-wide association study of 107 phenotypes in *Arabidopsis thaliana* inbred lines," *Nature*, vol. 465, no. 7298, pp. 627–631, 2010.

[12] W. B. Barbazuk, S. J. Emrich, H. D. Chen, L. Li, and P. S. Schnable, "SNP discovery via 454 transcriptome sequencing," *Plant Journal*, vol. 51, no. 5, pp. 910–918, 2007.

[13] A. Ching, K. S. Caldwell, M. Jung et al., "SNP frequency, haplotype structure and linkage disequilibrium in elite maize inbred lines," *BMC Genetics*, vol. 3, article 19, 2002.

[14] T. J. Close, P. R. Bhat, S. Lonardi et al., "Development and implementation of high-throughput SNP genotyping in barley," *BMC Genomics*, vol. 10, article 582, 2009.

[15] X. Xu, X. Liu, S. Ge et al., "Resequencing 50 accessions of cultivated and wild rice yields markers for identifying agronomically important genes," *Nature Biotechnology*, vol. 30, no. 1, pp. 105–111, 2012.

[16] S. Kaul, H. L. Koo, J. Jenkins et al., "Analysis of the genome sequence of the flowering plant *Arabidopsis thaliana*," *Nature*, vol. 408, no. 6814, pp. 796–815, 2000.

[17] S. A. Goff, D. Ricke, T. H. Lan et al., "A draft sequence of the rice genome (*Oryza sativa* L. ssp. *japonica*)," *Science*, vol. 296, no. 5565, pp. 92–100, 2002.

[18] J. Yu, S. Hu, J. Wang et al., "A draft sequence of the rice genome (*Oryza sativa* L. ssp. *indica*)," *Science*, vol. 296, no. 5565, pp. 79–92, 2002.

[19] J. A. Shendure, G. J. Porreca, and G. M. Church, "Overview of DNA sequencing strategies," *Current Protocols in Molecular Biology*, chapter 7, no. 81, pp. 7.1.1–7.1.11, 2008.

[20] F. Sanger, S. Nicklen, and A. R. Coulson, "DNA sequencing with chain-terminating inhibitors," *Proceedings of the National Academy of Sciences of the United States of America*, vol. 74, no. 12, pp. 5463–5467, 1977.

[21] F. Sanger, G. M. Air, B. G. Barrell et al., "Nucleotide sequence of bacteriophage phiX174 DNA," *Nature*, vol. 265, no. 5596, pp. 687–695, 1977.

[22] A. M. Maxam and W. Gilbert, "A new method for sequencing DNA," *Proceedings of the National Academy of Sciences of the United States of America*, vol. 74, no. 2, pp. 560–564, 1977.

[23] M. Kircher and J. Kelso, "High-throughput DNA sequencing—concepts and limitations," *BioEssays*, vol. 32, no. 6, pp. 524–536, 2010.

[24] M. Ronaghi, M. Uhlén, and P. Nyrén, "A sequencing method based on real-time pyrophosphate," *Science*, vol. 281, no. 5375, pp. 363–365, 1998.

[25] M. Ronaghi, S. Karamohamed, B. Pettersson, M. Uhlén, and P. Nyrén, "Real-time DNA sequencing using detection of

pyrophosphate release," *Analytical Biochemistry*, vol. 242, no. 1, pp. 84–89, 1996.

[26] T. C. Glenn, "Field guide to next-generation DNA sequencers," *Molecular Ecology Resources*, vol. 11, no. 5, pp. 759–769, 2011.

[27] G. Turcatti, A. Romieu, M. Fedurco, and A. P. Tairi, "A new class of cleavable fluorescent nucleotides: synthesis and optimization as reversible terminators for DNA sequencing by synthesis," *Nucleic Acids Research*, vol. 36, no. 4, article e25, 2008.

[28] J. Shendure, G. J. Porreca, N. B. Reppas et al., "Molecular biology: accurate multiplex polony sequencing of an evolved bacterial genome," *Science*, vol. 309, no. 5741, pp. 1728–1732, 2005.

[29] E. E. Schadt, S. Turner, and A. Kasarskis, "A window into third-generation sequencing," *Human Molecular Genetics*, vol. 19, no. 2, pp. R227–R240, 2010.

[30] T. D. Harris, P. R. Buzby, H. Babcock et al., "Single-molecule DNA sequencing of a viral genome," *Science*, vol. 320, no. 5872, pp. 106–109, 2008.

[31] C. S. Pareek, R. Smoczynski, and A. Tretyn, "Sequencing technologies and genome sequencing," *Journal of Applied Genetics*, vol. 52, no. 4, pp. 413–435, 2011.

[32] J. Eid, A. Fehr, J. Gray et al., "Real-time DNA sequencing from single polymerase molecules," *Science*, vol. 323, no. 5910, pp. 133–138, 2009.

[33] S. Koren, M. C. Schatz, B. P. Walenz et al., "Hybrid error correction and de novo assembly of single-molecule sequencing reads," *Nature Biotechnology*, vol. 30, no. 7, pp. 693–700, 2012.

[34] F. Ribeiro, D. Przybylski, S. Yin et al., "Finished bacterial genomes from shotgun sequence data," *Genome Research*. In press.

[35] A. Bashir, A. A. Klammer, W. P. Robins et al., "A hybrid approach for the automated finishing of bacterial genomes," *Nature Biotechnology*, vol. 30, no. 7, pp. 701–707, 2012.

[36] X. Zhang, K. W. Davenport, W. Gu et al., "Improving genome assemblies by sequencing PCR products with PacBio," *BioTechniques*, vol. 53, no. 1, pp. 61–62, 2012.

[37] P. Kothiyal, S. Cox, J. Ebert, B. J. Aronow, J. H. Greinwald, and H. L. Rehm, "An overview of custom array sequencing," *Current Protocols in Human Genetics*, no. 61, chapter 7, pp. 7.17.1–17.17.11, 2009.

[38] J. D. McPherson, "Next-generation gap," *Nature Methods*, vol. 6, no. 11, supplement, pp. S2–S5, 2009.

[39] A. P. M. Weber, K. L. Weber, K. Carr, C. Wilkerson, and J. B. Ohlrogge, "Sampling the arabidopsis transcriptome with massively parallel pyrosequencing," *Plant Physiology*, vol. 144, no. 1, pp. 32–42, 2007.

[40] M. Libault, A. Farmer, T. Joshi et al., "An integrated transcriptome atlas of the crop model *Glycine max*, and its use in comparative analyses in plants," *Plant Journal*, vol. 63, no. 1, pp. 86–99, 2010.

[41] T. Lu, G. Lu, D. Fan et al., "Function annotation of the rice transcriptome at single-nucleotide resolution by RNA-seq," *Genome Research*, vol. 20, no. 9, pp. 1238–1249, 2010.

[42] W. B. Barbazuk, S. Emrich, and P. S. Schnable, "SNP mining from maize 454 EST sequences," *Cold Spring Harbor Protocols*. In press.

[43] E. Novaes, D. R. Drost, W. G. Farmerie et al., "High-throughput gene and SNP discovery in *Eucalyptus grandis*, an uncharacterized genome," *BMC Genomics*, vol. 9, article 312, 2008.

[44] M. Trick, Y. Long, J. Meng, and I. Bancroft, "Single nucleotide polymorphism (SNP) discovery in the polyploid *Brassica napus* using Solexa transcriptome sequencing," *Plant Biotechnology Journal*, vol. 7, no. 4, pp. 334–346, 2009.

[45] S. S. Yang, Z. J. Tu, F. Cheung et al., "Using RNA-Seq for gene identification, polymorphism detection and transcript profiling in two alfalfa genotypes with divergent cell wall composition in stems," *BMC Genomics*, vol. 12, no. 1, article 199, 2011.

[46] F. Ozsolak, D. T. Ting, B. S. Wittner et al., "Amplification-free digital gene expression profiling from minute cell quantities," *Nature Methods*, vol. 7, no. 8, pp. 619–621, 2010.

[47] Z. Wang, M. Gerstein, and M. Snyder, "RNA-Seq: a revolutionary tool for transcriptomics," *Nature Reviews Genetics*, vol. 10, no. 1, pp. 57–63, 2009.

[48] H. Xu, Y. Gao, and J. Wang, "Transcriptomic analysis of rice (*Oryza sativa*) developing embryos using the RNA-Seq technique," *PLoS ONE*, vol. 7, no. 2, Article ID e30646, 2012.

[49] J. D. Roberts, B. D. Preston, L. A. Johnston, A. Soni, L. A. Loeb, and T. A. Kunkel, "Fidelity of two retroviral reverse transcriptases during DNA-dependent DNA synthesis *in vitro*," *Molecular and Cellular Biology*, vol. 9, no. 2, pp. 469–476, 1989.

[50] U. Gubler, "Second-strand cDNA synthesis: mRNA fragments as primers," *Methods in Enzymology*, vol. 152, pp. 330–335, 1987.

[51] J. Cocquet, A. Chong, G. Zhang, and R. A. Veitia, "Reverse transcriptase template switching and false alternative transcripts," *Genomics*, vol. 88, no. 1, pp. 127–131, 2006.

[52] F. Ozsolak, A. R. Platt, D. R. Jones et al., "Direct RNA sequencing," *Nature*, vol. 461, no. 7265, pp. 814–818, 2009.

[53] M. J. Solomon, P. L. Larsen, and A. Varshavsky, "Mapping protein-DNA interactions *in vivo* with formaldehyde: evidence that histone H4 is retained on a highly transcribed gene," *Cell*, vol. 53, no. 6, pp. 937–947, 1988.

[54] T. S. Mikkelsen, M. Ku, D. B. Jaffe et al., "Genome-wide maps of chromatin state in pluripotent and lineage-committed cells," *Nature*, vol. 448, no. 7153, pp. 553–560, 2007.

[55] P. Ng, J. J. Tan, H. S. Ooi et al., "Multiplex sequencing of paired-end ditags (MS-PET): a strategy for the ultra-high-throughput analysis of transcriptomes and genomes," *Nucleic Acids Research*, vol. 34, no. 12, p. e84, 2006.

[56] G. Robertson, M. Hirst, M. Bainbridge et al., "Genome-wide profiles of STAT1 DNA association using chromatin immunoprecipitation and massively parallel sequencing," *Nature Methods*, vol. 4, no. 8, pp. 651–657, 2007.

[57] B. Giardine, C. Riemer, R. C. Hardison et al., "Galaxy: a platform for interactive large-scale genome analysis," *Genome Research*, vol. 15, no. 10, pp. 1451–1455, 2005.

[58] W. Wang, Z. Wei, T.-W. Lam, and J. Wang, "Next generation sequencing has lower sequence coverage and poorer SNP-detection capability in the regulatory regions," *Scientific Reports*, vol. 1, article 55, 2011.

[59] B. Langmead, C. Trapnell, M. Pop, and S. L. Salzberg, "Ultrafast and memory-efficient alignment of short DNA sequences to the human genome," *Genome Biology*, vol. 10, no. 3, article R25, 2009.

[60] H. Li and R. Durbin, "Fast and accurate short read alignment with Burrows-Wheeler transform," *Bioinformatics*, vol. 25, no. 14, pp. 1754–1760, 2009.

[61] R. Li, C. Yu, Y. Li et al., "SOAP2: an improved ultrafast tool for short read alignment," *Bioinformatics*, vol. 25, no. 15, pp. 1966–1967, 2009.

[62] H. Li and N. Homer, "A survey of sequence alignment algorithms for next-generation sequencing," *Briefings in Bioinformatics*, vol. 11, no. 5, Article ID bbq015, pp. 473–483, 2010.

[63] T. J. Treangen and S. L. Salzberg, "Repetitive DNA and next-generation sequencing: computational challenges and solutions," *Nature Reviews Genetics*, vol. 13, no. 1, pp. 36–46, 2012.

[64] R. McLendon, A. Friedman, D. Bigner et al., "Comprehensive genomic characterization defines human glioblastoma genes and core pathways," *Nature*, vol. 455, no. 7216, pp. 1061–1068, 2008.

[65] S. M. Rumble, P. Lacroute, A. V. Dalca, M. Fiume, A. Sidow, and M. Brudno, "SHRiMP: accurate mapping of short color-space reads," *PLoS Computational Biology*, vol. 5, no. 5, Article ID e1000386, 2009.

[66] S. Rounsley, P. R. Marri, Y. Yu et al., "De novo next generation sequencing of plant genomes," *Rice*, vol. 2, no. 1, pp. 35–43, 2009.

[67] T. Sasaki, "The map-based sequence of the rice genome," *Nature*, vol. 436, no. 7052, pp. 793–800, 2005.

[68] E. Pennisi, "Plant sciences: corn genomics pops wide open," *Science*, vol. 319, no. 5868, p. 1333, 2008.

[69] G. A. Tuskan, S. DiFazio, S. Jansson et al., "The genome of black cottonwood, *Populus trichocarpa* (Torr. & Gray)," *Science*, vol. 313, no. 5793, pp. 1596–1604, 2006.

[70] O. Jaillon, J. M. Aury, B. Noel et al., "The grapevine genome sequence suggests ancestral hexaploidization in major angiosperm phyla," *Nature*, vol. 449, no. 7161, pp. 463–467, 2007.

[71] A. H. Paterson, J. E. Bowers, R. Bruggmann et al., "The *Sorghum bicolor* genome and the diversification of grasses," *Nature*, vol. 457, no. 7229, pp. 551–556, 2009.

[72] C. Feuillet, J. E. Leach, J. Rogers, P. S. Schnable, and K. Eversole, "Crop genome sequencing: lessons and rationales," *Trends in Plant Science*, vol. 16, no. 2, pp. 77–88, 2011.

[73] B. Chevreux, T. Pfisterer, B. Drescher et al., "Using the miraEST assembler for reliable and automated mRNA transcript assembly and SNP detection in sequenced ESTs," *Genome Research*, vol. 14, no. 6, pp. 1147–1159, 2004.

[74] R. Li, Y. Li, X. Fang et al., "SNP detection for massively parallel whole-genome resequencing," *Genome Research*, vol. 19, no. 6, pp. 1124–1132, 2009.

[75] J. T. Simpson, K. Wong, S. D. Jackman, J. E. Schein, S. J. M. Jones, and I. Birol, "ABySS: a parallel assembler for short read sequence data," *Genome Research*, vol. 19, no. 6, pp. 1117–1123, 2009.

[76] D. R. Zerbino and E. Birney, "Velvet: algorithms for de novo short read assembly using de Bruijn graphs," *Genome Research*, vol. 18, no. 5, pp. 821–829, 2008.

[77] D. Chagné, R. N. Crowhurst, M. Troggio et al., "Genome-wide SNP detection, validation, and development of an 8K SNP array for apple," *PLoS ONE*, vol. 7, no. 2, Article ID e31745, 2012.

[78] A. J. Cortés, M. C. Chavarro, and M. W. Blair, "SNP marker diversity in common bean (*Phaseolus vulgaris* L.)," *Theoretical and Applied Genetics*, vol. 123, no. 5, pp. 827–845, 2011.

[79] D. Altshuler, V. J. Pollara, C. R. Cowles et al., "An SNP map of the human genome generated by reduced representation shotgun sequencing," *Nature*, vol. 407, no. 6803, pp. 513–516, 2000.

[80] J. Berger, T. Suzuki, K. A. Senti, J. Stubbs, G. Schaffner, and B. J. Dickson, "Genetic mapping with SNP markers in *Drosophila*," *Nature Genetics*, vol. 29, no. 4, pp. 475–481, 2001.

[81] A. M. Allen, G. L. Barker, S. T. Berry et al., "Transcript-specific, single-nucleotide polymorphism discovery and linkage analysis in hexaploid bread wheat (*Triticum aestivum* L.)," *Plant Biotechnology Journal*, vol. 9, no. 9, pp. 1086–1099, 2011.

[82] D. Trebbi, M. Maccaferri, P. de Heer et al., "High-throughput SNP discovery and genotyping in durum wheat (*Triticum durum* Desf.)," *Theoretical and Applied Genetics*, vol. 123, no. 4, pp. 555–569, 2011.

[83] L. Barchi, S. Lanteri, E. Portis et al., "Identification of SNP and SSR markers in eggplant using RAD tag sequencing," *BMC Genomics*, vol. 12, article 304, 2011.

[84] F. A. Feltus, J. Wan, S. R. Schulze, J. C. Estill, N. Jiang, and A. H. Paterson, "An SNP resource for rice genetics and breeding based on subspecies *Indica* and *Japonica* genome alignments," *Genome Research*, vol. 14, no. 9, pp. 1812–1819, 2004.

[85] K. L. McNally, K. L. Childs, R. Bohnert et al., "Genomewide SNP variation reveals relationships among landraces and modern varieties of rice," *Proceedings of the National Academy of Sciences of the United States of America*, vol. 106, no. 30, pp. 12273–12278, 2009.

[86] T. Yamamoto, H. Nagasaki, J. I. Yonemaru et al., "Fine definition of the pedigree haplotypes of closely related rice cultivars by means of genome-wide discovery of single-nucleotide polymorphisms," *BMC Genomics*, vol. 11, no. 1, article 267, 2010.

[87] G. Jander, S. R. Norris, S. D. Rounsley, D. F. Bush, I. M. Levin, and R. L. Last, "Arabidopsis map-based cloning in the postgenome era," *Plant Physiology*, vol. 129, no. 2, pp. 440–450, 2002.

[88] X. Zhang and J. O. Borevitz, "Global analysis of allele-specific expression in *Arabidopsis thaliana*," *Genetics*, vol. 182, no. 4, pp. 943–954, 2009.

[89] R. Waugh, J. L. Jannink, G. J. Muehlbauer, and L. Ramsay, "The emergence of whole genome association scans in barley," *Current Opinion in Plant Biology*, vol. 12, no. 2, pp. 218–222, 2009.

[90] J. C. Nelson, S. Wang, Y. Wu et al., "Single-nucleotide polymorphism discovery by high-throughput sequencing in sorghum," *BMC Genomics*, vol. 12, article 352, 2011.

[91] R. L. Byers, D. B. Harker, S. M. Yourstone, P. J. Maughan, and J. A. Udall, "Development and mapping of SNP assays in allotetraploid cotton," *Theoretical and Applied Genetics*, vol. 124, no. 7, pp. 1201–1214, 2012.

[92] D. L. Hyten, S. B. Cannon, Q. Song et al., "High-throughput SNP discovery through deep resequencing of a reduced representation library to anchor and orient scaffolds in the soybean whole genome sequence," *BMC Genomics*, vol. 11, no. 1, article 38, 2010.

[93] J. P. Hamilton, C. N. Hansey, B. R. Whitty et al., "Single nucleotide polymorphism discovery in elite north American potato germplasm," *BMC Genomics*, vol. 12, article 302, 2011.

[94] Y.-B. Fu and G. W. Peterson, "Developing genomic resources in two *Linum* species via 454 pyrosequencing and genomic reduction," *Molecular Ecology Resources*, vol. 12, no. 3, pp. 492–500, 2012.

[95] F. M. You, N. Huo, K. R. Deal et al., "Annotation-based genome-wide SNP discovery in the large and complex *Aegilops tauschii* genome using next-generation sequencing without a reference genome sequence," *BMC Genomics*, vol. 12, article 59, 2011.

[96] Y. Han, Y. Kang, I. Torres-Jerez et al., "Genome-wide SNP discovery in tetraploid alfalfa using 454 sequencing and high resolution melting analysis," *BMC Genomics*, vol. 12, p. 350, 2011.

[97] R. E. Oliver, G. R. Lazo, J. D. Lutz et al., "Model SNP development for complex genomes based on hexaploid oat using high-throughput 454 sequencing technology," *BMC Genomics*, vol. 12, no. 1, article 77, 2011.

[98] E. Jones, W. C. Chu, M. Ayele et al., "Development of single nucleotide polymorphism (SNP) markers for use in commercial maize (*Zea mays* L.) germplasm," *Molecular Breeding*, vol. 24, no. 2, pp. 165–176, 2009.

[99] S. Ossowski, K. Schneeberger, R. M. Clark, C. Lanz, N. Warthmann, and D. Weigel, "Sequencing of natural strains of *Arabidopsis thaliana* with short reads," *Genome Research*, vol. 18, no. 12, pp. 2024–2033, 2008.

[100] I. Milne, M. Bayer, L. Cardle et al., "Tablet-next generation sequence assembly visualization," *Bioinformatics*, vol. 26, no. 3, pp. 401–402, 2009.

[101] N. Shah, M. V. Teplitsky, S. Minovitsky et al., "SNP-VISTA: an interactive SNP visualization tool," *BMC Bioinformatics*, vol. 6, no. 1, article 292, 2005.

[102] M. Fiume, V. Williams, A. Brook, and M. Brudno, "Savant: genome browser for high-throughput sequencing data," *Bioinformatics*, vol. 26, no. 16, Article ID btq332, pp. 1938–1944, 2010.

[103] H. Li, B. Handsaker, A. Wysoker et al., "The sequence alignment/map format and SAMtools," *Bioinformatics*, vol. 25, no. 16, pp. 2078–2079, 2009.

[104] Z. Wei, W. Wang, P. Hu, G. J. Lyon, and H. Hakonarson, "SNVer: a statistical tool for variant calling in analysis of pooled or individual next-generation sequencing data," *Nucleic acids research*, vol. 39, no. 19, article e132, 2011.

[105] R. Nielsen, J. S. Paul, A. Albrechtsen, and Y. S. Song, "Genotype and SNP calling from next-generation sequencing data," *Nature Reviews Genetics*, vol. 12, no. 6, pp. 443–451, 2011.

[106] R. Ragupathy, R. Rathinavelu, and S. Cloutier, "Physical mapping and BAC-end sequence analysis provide initial insights into the flax (*Linum usitatissimum* L.) genome," *BMC Genomics*, vol. 12, article 217, 2011.

[107] A. McKenna, M. Hanna, E. Banks et al., "The genome analysis toolkit: a MapReduce framework for analyzing next-generation DNA sequencing data," *Genome Research*, vol. 20, no. 9, pp. 1297–1303, 2010.

[108] G. Lunter and M. Goodson, "Stampy: a statistical algorithm for sensitive and fast mapping of Illumina sequence reads," *Genome Research*, vol. 21, no. 6, pp. 936–939, 2011.

[109] R. M. Durbin, "A map of human genome variation from population-scale sequencing," *Nature*, vol. 467, no. 7319, pp. 1061–1073, 2010.

[110] J. B. Fan, M. S. Chee, and K. L. Gunderson, "Highly parallel genomic assays," *Nature Reviews Genetics*, vol. 7, no. 8, pp. 632–644, 2006.

[111] M. R. Garvin, K. Saitoh, and A. J. Gharrett, "Application of single nucleotide polymorphisms to non-model species: a technical review," *Molecular Ecology Resources*, vol. 10, no. 6, pp. 915–934, 2010.

[112] Z. Tsuchihashi and N. C. Dracopoli, "Progress in high throughput SNP genotyping methods," *Pharmacogenomics Journal*, vol. 2, no. 2, pp. 103–110, 2002.

[113] B. Sobrino and A. Carracedo, "SNP typing in forensic genetics: a review," *Methods in Molecular Biology*, vol. 297, pp. 107–126, 2005.

[114] S. Giancola, H. I. McKhann, A. Bérard et al., "Utilization of the three high-throughput SNP genotyping methods, the GOOD assay, Amplifluor and TaqMan, in diploid and polyploid plants," *Theoretical and Applied Genetics*, vol. 112, no. 6, pp. 1115–1124, 2006.

[115] S. Kim and A. Misra, "SNP genotyping: technologies and biomedical applications," *Annual Review of Biomedical Engineering*, vol. 9, pp. 289–320, 2007.

[116] P. K. Gupta, S. Rustgi, and R. R. Mir, "Array-based high-throughput DNA markers for crop improvement," *Heredity*, vol. 101, no. 1, pp. 5–18, 2008.

[117] J. Ragoussis, "Genotyping technologies for genetic research," *Annual Review of Genomics and Human Genetics*, vol. 10, pp. 117–133, 2009.

[118] J. W. Davey, P. A. Hohenlohe, P. D. Etter, J. Q. Boone, J. M. Catchen, and M. L. Blaxter, "Genome-wide genetic marker discovery and genotyping using next-generation sequencing," *Nature Reviews Genetics*, vol. 12, no. 7, pp. 499–510, 2011.

[119] R. J. Elshire, J. C. Glaubitz, Q. Sun et al., "A robust, simple genotyping-by-sequencing (GBS) approach for high diversity species," *PLoS ONE*, vol. 6, no. 5, Article ID e19379, 2011.

[120] J. A. Poland, P. J. Brown, M. E. Sorrells, and J.-L. Jannink, "Development of high-density genetic maps for barley and wheat using a novel two-enzyme genotyping-by-sequencing approach," *PLoS ONE*, vol. 7, no. 2, Article ID e32253, 2012.

[121] M. W. Horton, A. M. Hancock, Y. S. Huang et al., "Genome-wide patterns of genetic variation in worldwide *Arabidopsis thaliana* accessions from the RegMap panel," *Nature Genetics*, vol. 44, no. 2, pp. 212–216, 2012.

[122] G. K. Subbaiyan, D. L. E. Waters, S. K. Katiyar, A. R. Sadananda, S. Vaddadi, and R. J. Henry, "Genome-wide DNA polymorphisms in elite indica rice inbreds discovered by whole-genome sequencing," *Plant Biotechnology Journal*, vol. 10, no. 6, pp. 623–634, 2012.

[123] J. C. Nelson, "Methods and software for genetic mapping," in *The Handbook of Plant Genome Mapping*, pp. 53–74, Wiley-VCH, Weinheim, Germany, 2005.

[124] A. Rafalski, "Applications of single nucleotide polymorphisms in crop genetics," *Current Opinion in Plant Biology*, vol. 5, no. 2, pp. 94–100, 2002.

[125] L. Kruglyak, "The use of a genetic map of biallelic markers in linkage studies," *Nature Genetics*, vol. 17, no. 1, pp. 21–24, 1997.

[126] W. Xie, Q. Feng, H. Yu et al., "Parent-independent genotyping for constructing an ultrahigh-density linkage map based on population sequencing," *Proceedings of the National Academy of Sciences of the United States of America*, vol. 107, no. 23, pp. 10578–10583, 2010.

[127] F. Li, H. Kitashiba, K. Inaba, and T. Nishio, "A *Brassica rapa* linkage map of EST-based SNP markers for identification of candidate genes controlling flowering time and leaf morphological traits," *DNA Research*, vol. 16, no. 6, pp. 311–323, 2009.

[128] E. S. Buckler, J. B. Holland, P. J. Bradbury et al., "The genetic architecture of maize flowering time," *Science*, vol. 325, no. 5941, pp. 714–718, 2009.

[129] S. A. Flint-Garcia, J. M. Thornsberry, and S. B. Edward, "Structure of linkage disequilibrium in plants," *Annual Review of Plant Biology*, vol. 54, pp. 357–374, 2003.

[130] P. K. Gupta, S. Rustgi, and P. L. Kulwal, "Linkage disequilibrium and association studies in higher plants: present status and future prospects," *Plant Molecular Biology*, vol. 57, no. 4, pp. 461–485, 2005.

[131] M. J. Aranzana, S. Kim, K. Zhao et al., "Genome-wide association mapping in Arabidopsis identifies previously known flowering time and pathogen resistance genes," *PLoS Genetics*, vol. 1, no. 5, p. e60, 2005.

[132] X. Huang, X. Wei, T. Sang et al., "Genome-wide asociation studies of 14 agronomic traits in rice landraces," *Nature Genetics*, vol. 42, no. 11, pp. 961–967, 2010.

[133] K. L. Kump, P. J. Bradbury, R. J. Wisser et al., "Genome-wide association study of quantitative resistance to southern leaf blight in the maize nested association mapping population," *Nature Genetics*, vol. 43, no. 2, pp. 163–168, 2011.

[134] J. A. Poland, P. J. Bradbury, E. S. Buckler, and R. J. Nelson, "Genome-wide nested association mapping of quantitative resistance to northern leaf blight in maize," *Proceedings of the National Academy of Sciences of the United States of America*, vol. 108, no. 17, pp. 6893–6898, 2011.

[135] F. Tian, P. J. Bradbury, P. J. Brown et al., "Genome-wide association study of leaf architecture in the maize nested association mapping population," *Nature Genetics*, vol. 43, no. 2, pp. 159–162, 2011.

[136] R. K. Pasam, R. Sharma, M. Malosetti et al., "Genome-wide association studies for agronomical traits in a world wide spring barley collection," *BMC Plant Biology*, vol. 12, article 16, 2012.

[137] B. J. Soto-Cerda and S. Cloutier, "Association mapping in plant genomes," in *Genetic Diversity in Plants*, M. Çalişkan, Ed., pp. 29–54, InTech, 2012.

[138] P. A. Morin, G. Luikart, and R. K. Wayne, "SNPs in ecology, evolution and conservation," *Trends in Ecology and Evolution*, vol. 19, no. 4, pp. 208–216, 2004.

[139] P. W. Hedrick, "Perspective: highly variable loci and their interpretation in evolution and conservation," *Evolution*, vol. 53, no. 2, pp. 313–318, 1999.

[140] A. Vignal, D. Milan, M. SanCristobal, and A. Eggen, "A review on SNP and other types of molecular markers and their use in animal genetics," *Genetics Selection Evolution*, vol. 34, no. 3, pp. 275–305, 2002.

[141] S. Konishi, T. Izawa, S. Y. Lin et al., "An SNP caused loss of seed shattering during rice domestication," *Science*, vol. 312, no. 5778, pp. 1392–1396, 2006.

[142] O. Wei, Z. Peng, Y. Zhou, Z. Yang, K. Wu, and Z. Ouyang, "Nucleotide diversity and molecular evolution of the WAG-2 gene in common wheat (*Triticum aestivum* L.) and its relatives," *Genetics and Molecular Biology*, vol. 34, no. 4, pp. 606–615, 2011.

[143] J. D. Retief, "Phylogenetic analysis using PHYLIP," *Methods in Molecular Biology*, vol. 132, pp. 243–258, 2000.

[144] K. Tamura, J. Dudley, M. Nei, and S. Kumar, "MEGA4: Molecular Evolutionary Genetics Analysis (MEGA) software version 4.0," *Molecular Biology and Evolution*, vol. 24, no. 8, pp. 1596–1599, 2007.

[145] M. W. Ganal, T. Altmann, and M. S. Röder, "SNP identification in crop plants," *Current Opinion in Plant Biology*, vol. 12, no. 2, pp. 211–217, 2009.

[146] T. Koepke, S. Schaeffer, V. Krishnan et al., "Rapid gene-based SNP and haplotype marker development in non-model eukaryotes using 3'UTR sequencing," *BMC Genomics*, vol. 13, no. 1, article 18, 2012.

16

Evolutionary History of LTR Retrotransposon Chromodomains in Plants

Anton Novikov,[1] Georgiy Smyshlyaev,[2] and Olga Novikova[3, 4]

[1] Laboratory of Molecular Genetic Systems, Institute of Cytology and Genetics, Novosibirsk, 630090, Russia
[2] Department of Natural Sciences, Novosibirsk State University, Novosibirsk, 630090, Russia
[3] Department of Plant Pathology, University of Kentucky, Lexington, KY 40546, USA
[4] Department of Biological Sciences, University at Albany, Life Sciences Building 2061, 1400 Washington Avenue, Albany, NY 12222, USA

Correspondence should be addressed to Olga Novikova, novikova.olga.uky@gmail.com

Academic Editor: Jim Leebens-Mack

Chromodomain-containing LTR retrotransposons are one of the most successful groups of mobile elements in plant genomes. Previously, we demonstrated that two types of chromodomains (CHDs) are carried by plant LTR retrotransposons. Chromodomains from group I (CHD_I) were detected only in Tcn1-like LTR retrotransposons from nonseed plants such as mosses (including the model moss species *Physcomitrella*) and lycophytes (the *Selaginella* species). LTR retrotransposon chromodomains from group II (CHD_II) have been described from a wide range of higher plants. In the present study, we performed computer-based mining of plant LTR retrotransposon CHDs from diverse plants with an emphasis on spike-moss *Selaginella*. Our extended comparative and phylogenetic analysis demonstrated that two types of CHDs are present only in the *Selaginella* genome, which puts this species in a unique position among plants. It appears that a transition from CHD_I to CHD_II and further diversification occurred in the evolutionary history of plant LTR retrotransposons at approximately 400 MYA and most probably was associated with the evolution of chromatin organization.

1. Introduction

A chromodomain (CHD) is a protein domain involved in chromatin remodeling and the regulation of gene expression in eukaryotes (e.g., [1–3]). CHDs perform a wide range of functions, including chromatin targeting and interactions between different proteins, RNA and DNA [3]. There are two major groups of CHDs that are found in eukaryotic chromodomain-containing proteins. The so-called "classical" CHDs carry the characteristic chromo-box motif (Y/f)-(L/F/Y)-(L/I/V)-K-(W/y)-(k/r)-g (single-letter code, capital letters standing for the most prominent aminoacid) [4]. The "classical" CHDs are highly conserved among eukaryotes and are represented in a large number of proteins in many genomes. They are believed to have a similar three-dimensional structure, which consists of an N-terminal three-stranded β-barrel capped by a C-terminal helix [5].

Three conserved residues, Y24, W45, and Y48, are essential for aromatic pocket formation [6, 7].

The second group of CHDs, "shadow" chromodomains, is more variable and includes chromo-related domains, which are well conserved in their central region, but they deviate significantly in other regions. The majority of the "shadow" CHDs contain the conserved residue W45 and lack Y24 and Y48 [3, 4]. In comparison with the "classical" CHDs, the shadow chromodomains contain one helix at the N-terminus and another inserted before the C-terminal helix [8]. The best-known protein with both types of chromodomains is heterochromatin protein-1 (HP1). HP1 presence is a hallmark of constitutive heterochromatin in *Drosophila*, a condensed and highly repressive type of chromatin that organizes the repetitive pericentromeric DNA. This protein contains an N-terminal "classical" CHD and a C-terminal shadow chromo-related domain. The CHD of HP1 binds to histone

H3 dimethyl-K9 (H3K9me2) and histone H3 trimethyl-K9 (H3K9me3) to help establish transcriptionally silent heterochromatin [9–11].

The chromodomain has been found not only in eukaryotic functional proteins but also in diverse LTR retrotransposons, which are called chromodomain-containing Gypsy LTR retrotransposons or chromoviruses [12, 13]. Chromoviruses are the most widespread lineage of Gypsy LTR retrotransposons and are present in the genomes of fungi, plants, and vertebrates [13, 14]. Two distinct groups of retrotransposon CHDs have been described. Group I CHDs from retrotransposons (CHDs_I) are similar to "classical" CHDs of chromodomain-containing proteins [15]. This group of CHDs was found in diverse eukaryotic LTR retrotransposons, including fungal and vertebrate Gypsy elements, as well as in LTR retrotransposons from moss *Physcomitrella patens* and spike-moss *Selaginella moellendorffii*, which belong to the Tcn1 clade [13, 16–18]. The information on the role of CHDs_I in the retrotransposition of LTR retrotransposons is limited. The transposition activity of the MAGGY retrotransposon of the rice blast fungus *Magnaporthe oryzae* dramatically decreased with the loss or alteration of the chromodomain [19]. On the other hand, the chromointegrase of the Tf1 LTR retrotransposon from *Schizosaccharomyces pombe* that lacks the chromodomain demonstrated a significantly higher activity and a substantially reduced substrate specificity [20]. As was demonstrated recently, the MAGGY chromodomain interacts with H3K9me2 and H3K9me3 in a similar way compared to the HP1 "classical" chromodomain. It was proposed that chromodomains can target the integration of chromoviruses into heterochromatic regions [15].

Representatives of group II CHDs from retrotransposons (CHDs_II) lack the first conserved aromatic residue (Y24) and usually the third (Y48). Group II has only been identified in plant Gypsy LTR retrotransposons. Little is known about the activity and role of CHDs_II from plant retrotransposons. The mostly heterochromatic distribution of plant chromoviruses along with data describing the localization of chromodomain-YFP fused protein in heterochromatin can be used as indirect evidence for recognizing heterochromatin and directing the integration role of CHDs_II [15]. Nevertheless, the actual mechanisms with which these chromodomains act are still unknown.

Previously, we demonstrated that our knowledge of plant chromodomain-containing LTR retrotransposons is mostly limited to knowledge about seed plants (mostly angiosperms) [16–18]. An investigation of retrotransposons from nonseed plants could shed light on the evolutionary history of retrotransposons and their impact on the evolutionary history of plant genomes. For example, it is still not clear whether CHDs_I and CHDs_II were acquired independently by distinct lineages of LTR retrotransposons or whether they evolved from a common ancestor. The present survey of chromodomains from diverse plants with an emphasis on the spike-moss *Selaginella* demonstrated that a transition from CHDs_I to CHDs_II occurred in the evolutionary history of plants approximately 500–400 MYA. Moreover, several types of clade-specific CHDs_II were found in plants; sequence dissimilarities among these clade-specific CHDs

hypothesized to indicate functional differences. We examined the evolutionary constraints that shaped the diversity of CHDs_II in plants and demonstrated that positive selection contributed to the diversification of clade-specific LTR retrotransposon CHDs. We propose that the presence of CHDs_I or CHDs_II is related to the distribution of heterochromatin/euchromatin marks and molecular differences in these marks between distinct lineages of eukaryotes, such as fungi/metazoa and plants. Both the transition from CHDs_I to CHDs_II and the diversification of clade-specific CHDs reflect evolutionary changes that occurred in plant chromatin organization.

2. Results

2.1. Novel LTR Retrotransposons from Selaginella moellendorffii. Previously, we described SM-Tcn1 CHD-containing Gypsy LTR retrotransposons, which presumably appeared as a result of a horizontal transfer from fungi during the early evolution of plants [18]. Several other families of LTR retroelements were identified by performing BLASTN and TBLASTN searches of the *S. moellendorffii* Whole Genome Shotgun (WGS) database (http://genome.jgi-psf.org/Selmo1) using the SM-Tcn1 retrotransposon as well as previously described retrotransposons from other species as queries (see Section 4). The newly identified retrotransposons were classified as representatives of the same or different families based on the levels of their similarities. More than 80% identity at the nucleotide level is believed to be sufficient for the classification of retrotransposons in the same family [21]. An additional criterion to designate retrotransposon families is a minimum of 50% nucleotide identity in LTRs [22]. The exemplar element was retrieved or reconstructed based on copies that were available for each family. These retrotransposons were used for further classification based on comparative and phylogenetic analysis, which included known LTR retrotransposons from other plants. The result of this analysis is shown in Figure 1. In total, five diverse families of CHD-containing LTR retrotransposons were found in addition to the previously described SM-Tcn1 [18]. The five families were named SM1-Galahad and SM2-Galahad, SM-Fogey, SM-Diluvium, and SM-Cranky (Table 1).

SM1-Galahad and SM2-Galahad are closely related to each other and form a common clade with Galadriel-like retrotransposons from monocots and dicots, suggesting that this clade originated before nonvascular and vascular plants separated from a common ancestor, roughly 400 MYA [23]. Among the sequences that are available in the NCBI protein database based on BLASTP analysis, the retroelement that is most closely related to SM1-Galahad and SM2-Galahad is the Galadriel LTR retrotransposon from *Lycopersicon esculentum*, which was identified in the Cf-9 disease resistance gene cluster [24].

We did not identify a putative intact copy of SM1-Galahad in *S. moellendorffii* WGS; thus, we used a number of copies to obtain a consensus sequence. SM1-Galahad is highly repeated and represented by hundreds of copies per genome, which share 95% nucleotide similarity on average. Many copies contain large deletions and/or insertions.

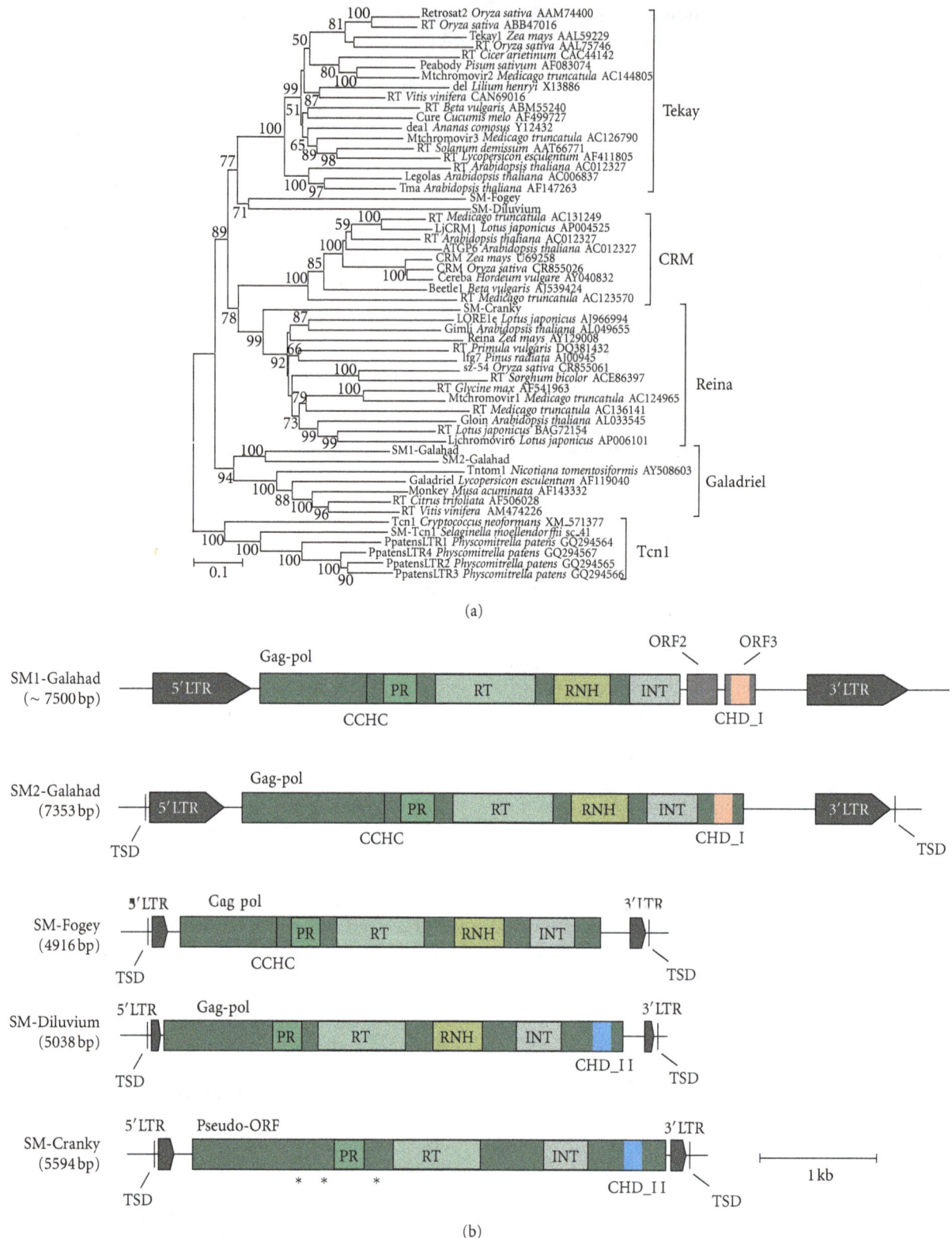

FIGURE 1: A neighbor-joining (NJ) phylogenetic tree based on multiple alignment of reverse transcriptase aminoacid sequences of LTR retrotransposons, including newly identified chromodomain-containing LTR retrotransposons from *Selaginella moellendorffii* (a) and the structural organization of novel retroelements from *Selaginella* (b). Statistical support was evaluated by bootstrapping (1000 replications); nodes with bootstrap values over 50% are shown. The plant-specific clades: Reina, CRM, Galadriel, and Tekay, as well as the Tcn1 clade, are indicated. The name of the host species and the accession number is indicated for the LTR elements that were taken from GenBank. Abbreviations: ORF: open reading frame, PR: aspartyl protease, RT: reverse transcriptase, RNH: ribonuclease H, INT: integrase, CHD_I and CHD_II: chromodomain group I and group II, CCHC: Zn-finger motif, TSD: target site duplications, and 5' and 3'LTRs: 5' and 3' long terminal repeats. The positions of stop-codons (SM-Cranky) are marked by asterisks.

TABLE 1: The characteristics of chromodomain-containing Gypsy LTR retrotransposons from *Selaginella moellendorffii*.

Element name	Element size [bp]	Intact copies	LTRs size [bp]	LTRs identity [%]	TSD	ITR
SM1-Galahad*	~7500	ND	1076/1146	90.2	ND	TG···CA
SM2-Galahad	7353	+	919	99.7	CCTAT···CCTAT	TGT···ACA
SM-Fogey	4916	+	233	98.7	TGCCC···TGCCC	TG···CA
SM-Diluvium	5038	+	112	97.3	GCGTA···GCGTA	TG···CA
SM-Cranky	5594	—	217	98.6	GTTTCT···GTTTCT	TG···CA

*SM1-Galahad was reconstructed based on a number of sequences; TSD: terminal site duplications; ITR: dinucleotide inverted repeat; ND: not detected.

The reconstructed SM1-Galahad is 7.5 kbp in length and carries LTRs that are more than 1 kbp in length and which possess a short inverted terminal repeat (TG···CA), typical for LTR retrotransposons and retroviruses [25]. They also contain polyguanine tracks that vary in length (from 11 to 25 bp) between different copies. A 13 bp primer binding site-like sequence (PBS) that is complementary to the 3′ end region of tRNA^{Met} is present downstream of the 5′LTR, and a polypurine tract (PPT) was detected immediately upstream of the 3′LTR (Figure 1(b)). The putative single open reading frame (ORF) of SM1-Galahad is 3885 bp in length. The hypothetical protein product of the ORF of SM1-Galahad (1295 aa) exhibits significant similarity to both Gag and Pol gene products of known retroelements, especially with those of Gypsy retrotransposons. A search for characteristic motifs within the polyprotein sequence identified several functional domains, such as the cysteine or Zn-finger motif with Cys-X$_2$-Cys-X$_4$-His-X$_4$-Cys (CCHC) composition, proteinase (PR), reverse transcriptase (RT), ribonuclease H (RNH), and core integrase (core Int), in the order indicated, confirming that SM1-Galahad belongs to the Ty3/Gypsy LTR retrotransposons. No chromodomain (CHD) was found at the C-terminal end of the Gag-Pol protein from the reconstructed SM1-Galahad. However, analysis of the DNA sequence downstream of the putative ORF revealed two additional short ORFs (316 bp and 309 bp in length). The 309-bp ORF3 encodes a protein (113 aa) containing a CHD group I (CHD_I) domain.

The full-length putatively intact SM2-Galahad was found in scaffold_24 (between position 1030224 and 1037576). This LTR retrotransposon is represented by hundreds of copies per genome. SM2-Galahad is a 7.4 kbp LTR retrotransposon with a single ORF (Figure 1(b)). LTRs are 919 bp in length and share 99.7% similarity. A 5 bp target site duplication (TSD) was also detected (CCTAT···CCTAT). An 11 bp PBS complementary to the tRNA^{Met} is located just downstream of the 5′LTR, and 13 bp PPT is located upstream of 3′LTR. The putative ORF (4731 bp) encodes a fused Gag-Pol protein (1577 aa), which carries several functional domains, including the CCHC Zn-finger motif, PR, RT, RNH, core Int, and CHD_I.

The complete retrotransposons SM-Fogey and SM-Diluvium are 4.9 kbp and 5 kbp long, respectively (Figure 1(b)). The full-length intact copies of SM-Fogey and SM-Diluvium can be found in scaffold_20 (position from 2127930 to 2132846) and scaffold_73 (position from 936919 to 941957). Target site duplications of five base pairs mark the integration of the intact SM-Fogey and SM-Diluvium in the genomic sequence. Both LTR retrotransposons contain four catalytic regions, PR, RT, RNH, and Int of the retroviral genes *gag*

and *pol*, in a continuous single open reading frame. SM-Diluvium also has a CHD group II (CHD_II), whereas SM-Fogey does not carry a chromodomain. SM-Fogey is characterized by 233 bp terminal repeats; in contrast, the LTRs of SM-Diluvium are only 112 bp in length. The LTRs of SM-Fogey and SM-Diluvium contain consensus short inverted terminal repeats (TG···CA), which are important for the integration of retroviral sequences [26]. The primer binding site (PBS), which is complementary to the 3′ end of tRNA^{Met} and has a one-nucleotide spacer next to the adjacent 5′ LTR, was identified in both retrotransposons. The PPT with the stretch of 14 purines for SM-Fogey and SM-Diluvium is located immediately before the 3′ LTR. Several full-length highly similar copies for both SM-Fogey and SM-Diluvium can be found in *S. moellendorffii* WGS. For example, nine putatively intact copies of SM-Fogey presented in WGS share 98.5% average similarity at the DNA level. All of them are flanked by 5-bp TSD. Despite the fact that only one full-length copy was detected for SM-Diluvium, seven full-length copies presented in the genome showed an average 99.1% DNA similarity. Additionally, the 5′ and 3′LTRs of elements have high similarity, 98.7% on average for SM-Fogey and 97.3% on average for SM-Diluvium. Altogether, these features indicate that SM-Fogey and SM-Diluvium recently retrotransposed and may still be active.

The putative Gag-Pol polyproteins of SM-Fogey and SM-Diluvium were compared to those reported for other plant LTR retrotransposons. The aminoacid domains show the highest similarity to the corresponding regions of the Tekay-like retrotransposon SHMIDT that is adjacent to the disease resistance-priming gene NPR1 in *Beta vulgaris* (EF101866; [27]) and retrotransposons from the diverse *Oryza* species, including LTR retrotransposons from *Oryza sativa* and *Oryza australiensis* (e.g., DQ365821, DP000086; [28]). It appears that SM-Fogey and SM-Diluvium are the most closely related to the Tekay clade of CHD-containing Gypsy LTR retrotransposons from plants. They formed a common branch on the phylogenetic tree, but neither SM-Fogey nor SM-Diluvium can be assigned to this clade because of the low bootstrap support for this cluster (bootstrap 77%; Figure 1(a)).

One more CHD-containing Gypsy LTR retrotransposon, SM-Cranky, was found to be present with a few copies per genome. We detected only one full-length SM-Cranky located in scaffold_1 (between position 5651434 and 5657027) and six additional truncated copies. The retrotransposon is 5.6 kbp long with LTRs of 217 bp, terminating in the two-nucleotide inverted repeat (TG···CA). The LTRs are 98.6% identical to each other. A stretch of 12 bp located with two-nucleotide spacers downstream of the 5′LTR is complementary to the 3′ end of tRNA^{Met}, probably providing a primer site for reverse transcription. The 14-bp PPT was found upstream of the 3′LTR. The putative pseudo-ORF is interrupted by three stop codons, but there are no frameshifts. This sequence can be translated to the protein that bears a resemblance to the characteristic PR, RT, Int, and CHD_II motifs (Figure 1(b)).

2.2. Several Types of Retrotransposon Chromodomains in Plants.
The most intriguing finding that concerns the CHD-containing Gypsy LTR retrotransposons from the spike-moss S. moellendorffii is the presence of both types of LTR retrotransposon chromodomains, CHD_I and CHD_II, in the same plant genome. The SM-Tcn1 [18], SM1-Galahad, and SM2-Galahad LTR retrotransposons from Selaginella carry CHD_I, while SM-Diluvium and SM-Cranky contain CHD_II. The distribution of both retrotransposon CHD types among different taxa is considered to be well known. CHD_I is typical for fungal and animal LTR retrotransposons as well as for green algae (Chlamyvir clade), whereas only CHD_II was found in LTR retrotransposons from seed plants (Tekay, Galadriel, and Reina clades [13]). However, it appears that phylogenetic distribution of CHD_I-containing elements extends to most if not all extant green plant lineages. Previously, we reported that CHD_I-containing LTR retrotransposons belonging to the Tcn1 clade can be found in both moss Physcomitrella and spike-moss Selaginella [16, 18].

To expand our understanding of the evolution and diversity of retrotransposon chromodomains in green plants, we implemented a search throughout sequence databases including but not limited to PlantGDB (http://www.plantgdb.org/) and Phytozome (http://www.phytozome.net/). A plant retrotransposon CHD search was performed with (TBLASTN) using aminoacid sequences of known CHDs as queries (see Section 4). Altogether 114 plant species were investigated and results for some of them are presented in Table 2. The full list and primary information can be found in Supporting Information Table 1S and Table 2S. It should be noted that the majority of sequences available were derived from EST databases. In fact, CHD-containing retrotransposons are barely expressed, and as a rule, they are underrepresented in ESTs [13, 16]. Another factor that has an effect on the final result is the presence of a strong bias with respect to the species diversity that is represented in databases. More than 83% of the analyzed species (95 out of a total of 114) were angiosperms, and only a few representatives from other green plant groups are currently available. Among the analyzed species, only 26 did not produce any significant hits. For 80 species, retrotransposon CHDs were detected in ESTs; CHDs were also found in GenomeSurvey Sequences (GSS) and Whole Genomic Sequences (WGS) databases for 46 investigated species. Among those species for which the WGS database is available, a few did not show the presence of retrotransposon CHDs, including red algae (Porphyrayezoensis, Galdieriasulphuraria, and Cyanidioschyzonmerolae) as well as one green algae, Micromonas pusilla CCMP1545. This result can arise from either the limited

number of sources of sequences available or the loss of this type of retrotransposon.

The majority of the CHDs detected belonged to group II. Only the representatives of the Chlamyvir clade showed the presence of CHD_I which was found in green algae Chlorella vulgaris C-169 and Chlamydomonas reinhardtii (Table 2; [13]). The source of ESTs as well as genomic sequences of gymnosperms is very limited, especially in comparison to those for angiosperms. Nevertheless, we were able to identify a few LTR retrotransposon CHDs in all of the species investigated (Table 2; Supporting Information Table 1S).

Comparative analysis indicated that almost all of the chromodomains of type II can be easily classified as Reina-, Tekay-, or Galadriel-like CHDs based on their sequence similarity. Almost all of the plant species that produced hits in our search contain Reina- and Tekay-like CHDs. Galadriel-like CHDs were underrepresented in the analyzed databases. The phylogenetic analysis based on the aminoacid sequences of newly identified CHDs and CHDs from known plant LTR retrotransposons support these findings with several exceptions: (i) Galadriel-like LTR retrotransposons from Selaginella have CHD_I and are grouped with the Tcn1 clade; (ii) CHDs from conifers, which were previously believed to belong to Reina-like LTR retrotransposons from angiosperms, actually form their own branch; (iii) three CHDs identified in green algae grouped together with the Selaginella SM-Cranky LTR retrotransposon and not with CHDs from the Chlamyvir clade (Figure 2; [13]). Additionally, Tekay-like LTR retrotransposon CHDs from gymnosperms formed a common cluster that appears to be distinct from other Tekay-like CHDs. It is worthwhile to note that Tekay-like CHDs retrieved from representatives of the family Poaceae formed their own branch, with fairly high bootstrap support (bootstrap value 77%; Figure 2). Tekay-like LTR retrotransposons are known to be highly repeated in grass genomes. Moreover, retrotransposon activity has been implicated as playing a major role in genome size evolution in angiosperm lineage. This has been especially well-characterized in the Poaceae (e.g., [29, 30]).

2.3. Clade-Characteristic Protein Motifs Can Be Found in Plant CHDs.
Comparative analysis of primary sequences and tertiary structures of retrotransposon CHDs and the CHDs from functional proteins with known function can provide important insights into the possible roles of some of the specific aminoacids and the retrotransposon CHDs as a whole [3, 31–33]. Based on multiple alignments of selected CHDs from different clades as well as "classical" and shadow CHDs from cellular functional proteins, we identified changes in plant CHDs starting with the CHD_I found in LTR retrotransposons from green algae, moss Physcomitrella, and spike moss Selaginella and culminating in the CHD_II that is isolated from genomes of gymnosperms and angiosperms (Figure 3). The CHDs of LTR retrotransposons obtained from the Selaginella genome represent transitional stages between CHD_I and CHD_II. For example, SM2-Galahad CHD is close to the "classical" CHDs, whereas SM1-Galahad CHD contains a few substitutions in the chromo box motif (Y/f)-(L/F/Y)-(L/I/V)-K-(W/y)-(k/r)-g. The chromo-box is one of

TABLE 2: List of some plant species used in this study, their taxonomy (according to NCBI Taxonomy: http://www.ncbi.nlm.nih.gov/taxonomy) and the results of *in silico* mining of chromodomains by LTR retrotransposon clades. The full list is available in Supporting Information Table S1 in Supplementary Material available online at doi: 10.1155/2012/874743.

Class	Order	Family	Species	EST			GSS/WGS		
				Chlamyvir	Other		Chlamyvir	other	
Bangiophyceae	Bangiales	Bangiaceae	*Porphyra yezoensis*	—	—		NA	NA	
	Cyanidiales	Cyanidiaceae	*Galdieria sulphuraria*	—	—		NA	NA	
Trebouxiophyceae	Chlorellales	Chlorellaceae	*Chlorella vulgaris*	—	—		7	—	
Chlorophyceae	Chlamydomonadales	Chlamydomonadaceae	*Chlamydomonas reinhardtii*	—	—		13	3	
		Volvocaceae	*Volvox carteri f. nagariensis*	3	—		241	20	
				Reina	Tekay	Galadriel	Reina	Tekay	Galadriel
Lycopodiopsida	Lycopodiales	Lycopodiaceae	*Huperzia serrata*	—	—	—	NA	NA	NA
Polypodiopsida	Polypodiales	Pteridaceae	*Adiantum capillus-veneris*	—	1	—	NA	NA	NA
Coniferopsida	Coniferales	Pinaceae	*Picea glauca*	6	2	—	—	—	—
			Pinus taeda	4	5	—	18	2	—
			Pinus banksiana	9	3	—	—	—	—
Monocotyledons	Poales	Poaceae	*Triticum aestivum*	37	5	—	10	91	—
			Brachypodium distachyon	11	—	—	46	371	—
			Oryza sativa Indica Group	5	1	—	64	239	—
			Oryza sativa Japonica Group	55	6	—	23	78	—
			Sorghum bicolor	14	—	—	18	>1000	—
			Zea mays	53	105	—	600	>10000	—
Eudicotyledons	Solanales	Solanaceae	*Solanum lycopersicum*	—	11	—	17	>1000	80
			Solanum tuberosum	2	8	1	11	>1000	12
			Nicotiana tabacum	23	39	2	185	2973	297
	Lamiales	Phrymaceae	*Mimulus guttatus*	2	—	1	16	370	13
	Brassicales	Brassicaceae	*Arabidopsis thaliana*	2	6	—	14	23	—
			Brassica napus	2	3	—	5	99	2
	Sapindales	Rutaceae	*Citrus clementina*	—	—	1	11	8	35
	Fabales	Fabaceae	*Glycine max*	1	2	—	22	252	—
			Lotus japonicus	2	3	—	8	62	—
	Vitales	Vitaceae	*Vitis vinifera*	—	4	4	18	59	327

NA: no data available.

the motifs that is essential for hydrophobic core formation [31, 32]. This motif is characteristic for the "classical" CHDs of chromodomain-containing proteins [3]. SM-Diluvium, lacks Y24 (corresponding to position 1 in the multiple alignment represented in Figure 3), but it still has both aromatic aminoacids, W45 and Y48 (positions 28 and 31), which form methyl binding cages [3, 32].

With respect to different clades of LTR retrotransposons, Galadriel-like CHDs lost essential aminoacids (Y24 and Y48) and diverged significantly from CHD_I. Nevertheless,

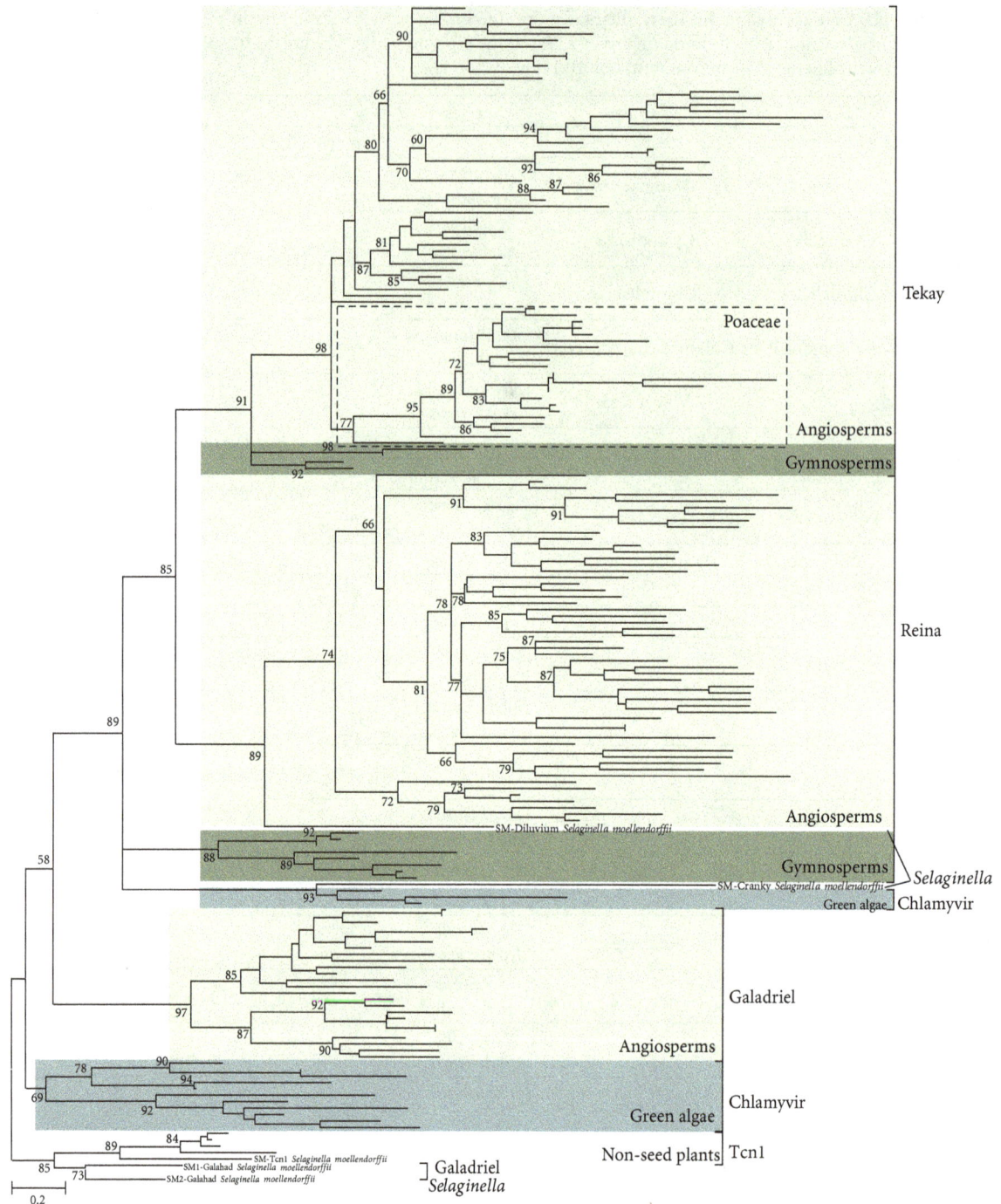

FIGURE 2: Maximum likelihood (ML) phylogenetic tree based on LTR retrotransposon chromodomain (CHD) aminoacid sequences including CHDs from newly identified chromodomain-containing LTR retrotransposons from *Selaginella moellendorffii*. Statistical support was evaluated by using aLTR; nodes with aLTR statistics over 50% are shown. The plant-specific clades Chlamyvir, Reina, CRM, Galadriel, and Tekay, as well as the Tcn1 clade, are indicated. The names of the host species and the accession numbers for the LTR elements are available in Supporting Information Figure 1S. The taxonomic range of the host species is indicated by colored boxes and includes angiosperms, gymnosperms, green algae, and nonseed plants (*Selaginella* and *Physcomitrella*).

Galadriel-like CHDs have the conservative motif (Y/f)-(L/Y)-(I/V)-k-W-k-g (single-letter code, capital letters represent the most prominent aminoacid), which is very close to the chromo box. Reina-like LTR retrotransposons have traces of this motif, but Tekay-like CHDs have lost the motif

with the exception of V43 (position 26 in the multiple alignment) and the highly conservative residue W45 (Figure 3). At the same time, Tekay-like CHDs have a number of conservative characteristic motifs. For example, the protein motif (K/R)-X-(L/T)-R-X-(k/r) is present in all of the investigated

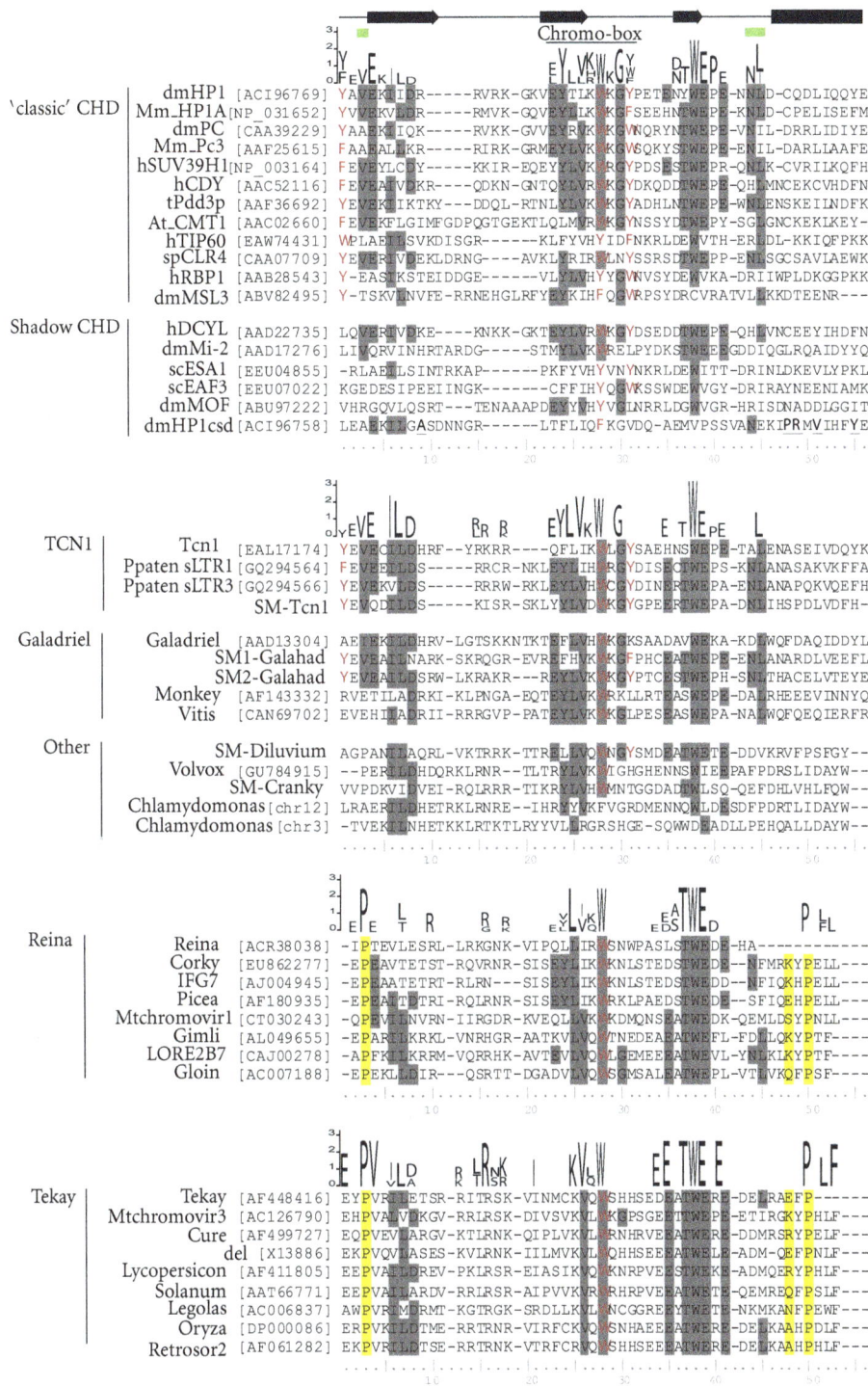

FIGURE 3: Multiple alignment of chromodomains. The following were used in the alignment: the "classical" and "shadow" chromo-like motifs of functional proteins from diverse animal species and *Arabidopsis* (At_CMT1); group I and group II chromodomains from diverse plant LTR retrotransposons belonged to Tcn1, Galadriel, Reina, and Tekay clades as well as unclassified (Other); CHD_I from Tcn1 LTR retrotransposons of fungi *Cryptococcus neoformans*. The GenBank accession numbers are indicated for each sequence. The most conservative domains for each of the groups are shown on the top. The chromo-box is indicated on the very top, which also shows the secondary structure elements (arrows indicate β-strands; rectangle α-helix) and conserved residues that form the complementary surface that is responsible for H3 peptide recognition (green boxes) [3–7]. Three aminoacids that have been shown to be under positive selection are highlighted in yellow.

Tekay-like retrotransposon CHDs but not in Reina- or Galadriel-like CHDs. One more Tekay-characteristic motif, EEXTWEXE, is highly conserved in Tekay CHDs. A similar motif can be found in diverse LTR retrotransposon CHDs; however, this motif is not as conserved in other clades as it is in Tekay. The aminoacid motif TWE is extremely conserved among all of the retrotransposon CHDs, with a few exceptions in green algae and is believed to have important functions. Interestingly, this particular motif is absent in the majority of shadow CHDs and in a few "classical" CHDs of chromodomain-containing proteins. This motif corresponds to the $\beta3$ strand in the tertiary structure [3, 31].

While a majority of plant retrotransposon CHDs lack conserved aromatic residues Y24 and Y48, they still retain high sequence similarity with known "classical" CHDs of functional proteins. This similarity is much higher than the similarity among "classical" and shadow CHDs. Moreover, the analysis of tertiary structure shows the presence of all of the structural features that are characteristic of "classical" CHDs (see below).

2.4. Positive Selection for Retrotransposon CHDs in Plants. For further understanding of processes that lead to the current diversity of plant retrotransposon CHDs, we examined evolutionary constraints that shape the essential domains of plant LTR retrotransposon polyproteins. The presence of conservative motifs in CHD sequences among diverse LTR retrotransposons suggests that evolution has been strongly constrained. At the same time, the presence of conserved clade-specific motifs, as well as the transition from CHD_I to CHD_II can indicate that some aminoacid changes had a selective advantage during diversification between clades and could be accumulated at a rate higher than expected under natural evolution (positive selection). To distinguish between these possibilities and to indicate the positions that evolved under positive selection, we analyzed the nonsynonymous to synonymous substitutions rate ratio (ω, see Section 4 for details).

Only retrotransposons that maintained intact domains were chosen for further analysis. CHD, core integrase (Int), and reverse transcriptase (RT) domains of 39 LTR retrotransposons were used as datasets. Phylogenetic trees based on multiple alignments of nucleotide sequences were reconstructed for each domain separately. As expected, the major difference in tree topologies among datasets included the appearance of common clusters for Galadriel-like LTR retrotransposons from *Selaginella* and the Tcn1 clade on tree reconstructions based on CHD sequences. Overall, RT- and CHD-based phylogenies are similar, whereas the Int-based tree has a different position for the Reina clade (Figure 4).

First, we estimated the ω ratio averaged over all of the sites and all of the lineages using M0 model. This model yields estimates that are close to 0: $\hat{\omega} = 0.022$ for RT, $\hat{\omega} = 0.039$ for Int, and $\hat{\omega} = 0.068$ for the CHD domain. This low rate indicated that there is a dominating role of purifying selection in the evolution of all of the domains. The clade-site test demonstrated that no events of positive selection were inferred for RT, which was expected for a highly conserved

protein domain evolving under strict constraints (e.g., [34–36]). Unexpectedly, strong positive selection was detected by clade-site test on the branch subtending the Reina clade on the Int phylogenetic tree (R branch; Figure 4(b)). This result was confirmed by a branch-site test of positive selection (modified model A—modified model A with $\omega_2 = 1$ fixed comparison) that detected positive selection on the R branch with a significance level of 0.01 (Table 3), taking into consideration multiple testing corrections (see Section 4 for details). Inference of positive selection can be an artifact when the synonymous substitutions reaching saturation. However, it is highly unlikely, taken in consideration the nature of the analyzed sequences and the pattern of substitutions (see Figure 3).

The branch-site tests also revealed several events of positive selection on the CHD phylogenetic tree. The evolutionary changes of CHDs in the Galadriel-Tekay-Reina group most probably occurred under positive selection at the 0.05 significance level based on the mixture distribution by Hommel correction procedure (GTR branch on Figure 4(c)). Nonsynonymous substitutions were also significantly elevated above background on branches subtending the Tekay-Reina group and the Galadriel clade (branch site significance level of 0.1; TR and G branches on Figure 4(c), resp.). We found evidence for the positive selection of CHDs for the GTR and TR branches, with positive signals coming from a few codons (significance level 0.1; Figure 5). Specifically, three codons appeared to be under positive selection for the TR branch at the cutoff posterior probability 95% (corresponding to aminoacid resides in positions 3, 46, and 48 of the multiple alignments presented in Figure 5(a)). Proline residues that are located in positions 3 and 48 of CHDs from Tekay and Reina LTR retrotransposons are highly conserved in these clades (see also Figure 3). Such a high degree of conservation may indicate that these residues are functionally important for both the Reina- and Tekay-like CHDs. P3 is located in a position that corresponds to the residue V3 in "classical" CHDs and is believed to participate in the formation of a complementary surface that is responsible for histone 3 peptide (H3) recognition (based on a study of CHDs from histone protein 1, dmHP1, from *Drosophila melanogaster*; [3, 31–33].

The P48 and E46 residues are located in the area that corresponds to the helical structure in dmHP1 CHD and other CHDs. The presence of P48 in the region, which is expected to form an α-helix, should have significant effects on secondary structure. Prolines are rarely found in α and β structures, because the structure's side chain α-N can form only one hydrogen bond [37], which would reduce the stability of such structures. At the same time, prolines are easily accommodated in a variety of turns; for example, as a Pro-X corner (where X is a variable aminoacid residue) [38]. We reconstructed the tertiary structure of some representatives of plant retrotransposon CHDs using I-TASSER [39]. All of the representatives clearly exhibited the presence of tertiary structure similar to that of dmHP1 (Figure 5(b)). However, as expected due to the presence of P48, CHDs from LTR retrotransposons that belong to the Reina and Tekay clades bear additional helix structures in comparison to dmHP1.

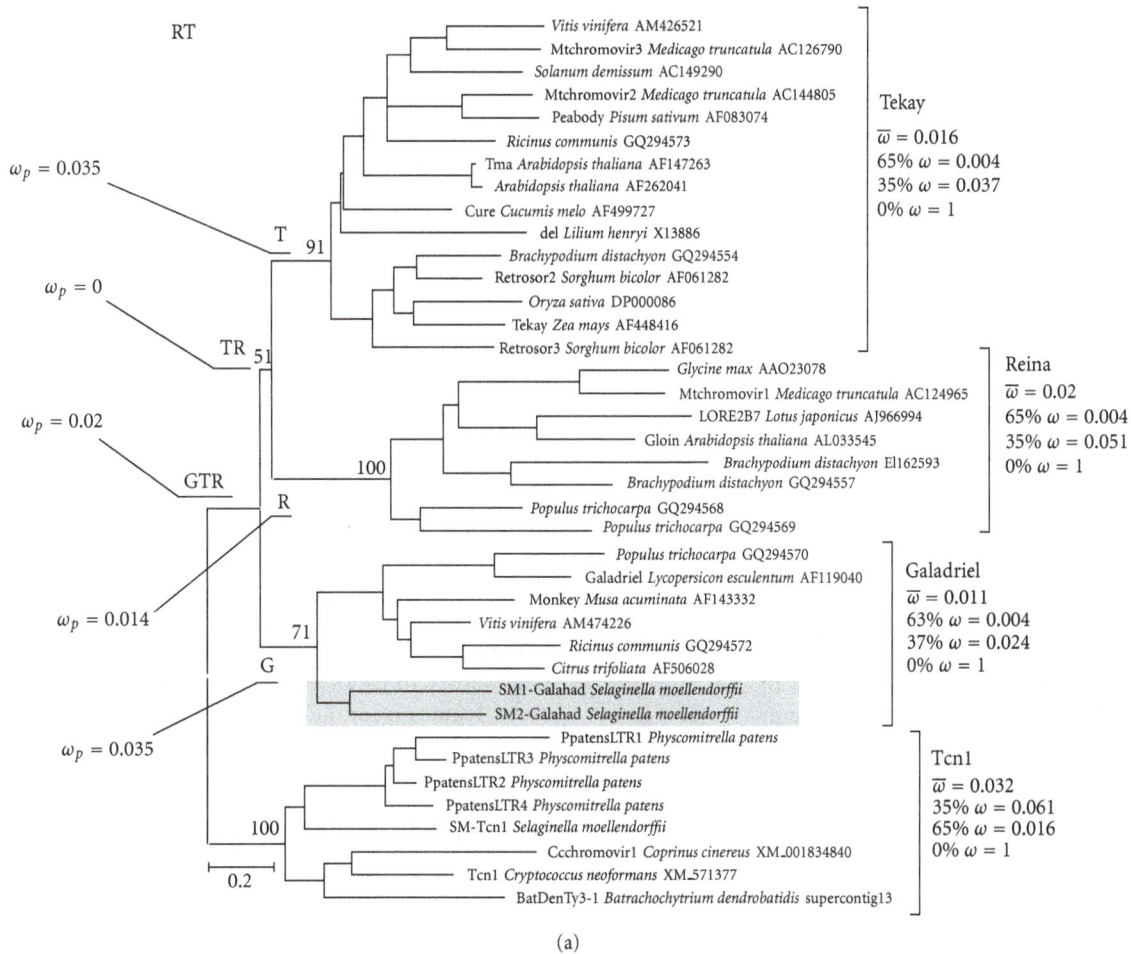

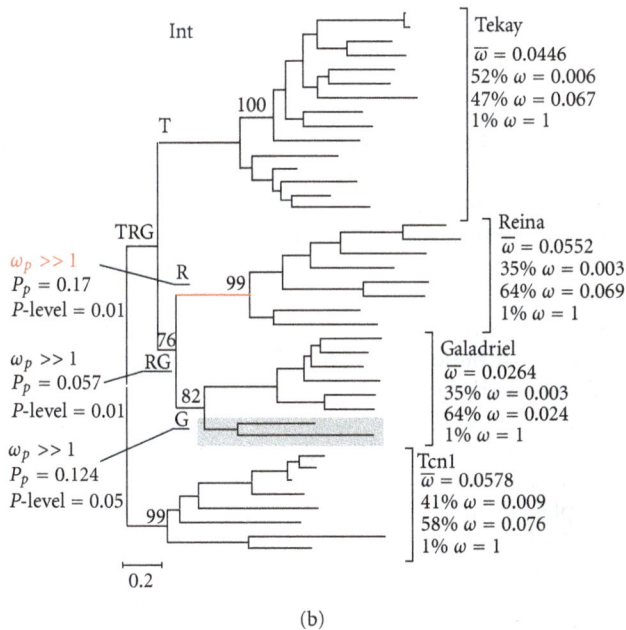

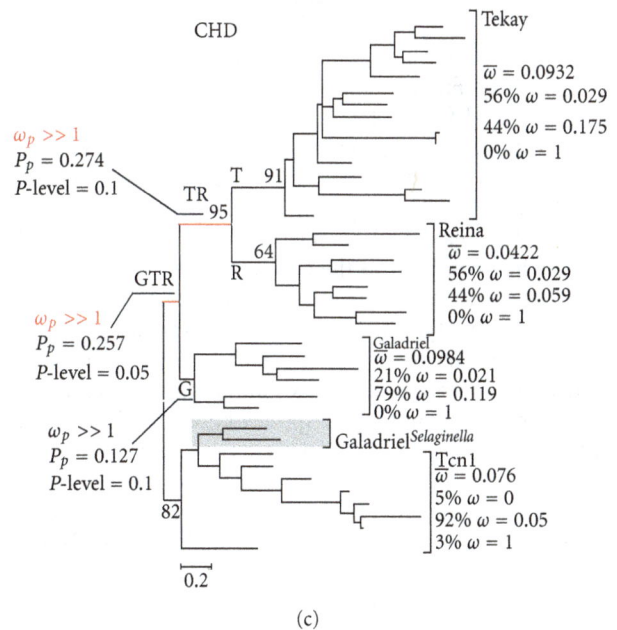

FIGURE 4: Maximum likelihood (ML) phylogenetic trees of sampled reverse transcriptase (a), integrase (b), and chromodomains (c) from 39 LTR retrotransposons. The plant-specific clades: Reina, Galadriel, and Tekay, as well as the Tcn1 clade, are indicated. Statistical support was evaluated by bootstrapping (100 replications); bootstrap values within clades are not shown. The changing in position of SM1-Galahad and SM2-Galahad LTR retrotransposons from *Selaginella* are shown by a gray box for each tree. The results of selection tests are reported for the tested branches, as is the proportion of sites under particular selective regimes for clades/groups. The red color indicates branches where sites under positive selection at the cutoff posterior probability 90% were identified.

TABLE 3: Maximum likelihood estimates and LRT statistics for the chromodomain.

Foreground branch	$2\Delta\ell$	$p_{1/2\chi_0^2+1/2\chi_1^2}$	$\hat{p}_0$	$\hat{p}_1$	$\hat{\omega}_0$	$\hat{\omega}_2$	Positively selected sites (90%)
base T	1.634722	0.1005255	0.71854	0.02675	0.07183	∞	
base R	1.11799	0.145176	0.69307	0.02571	0.07074	19.50879	
base G*	**4.766098**	**0.014513**	**0.84189**	**0.03127**	**0.07093**	∞	none
base TR*	**3.720944**	**0.026867**	**0.70023**	**0.02601**	**0.07199**	∞	3/9/23/46/48
base GTR*	**5.652052**	**0.0087175**	**0.71684**	**0.02660**	**0.07097**	∞	1/30/44/45/46/47/49

* Branches Galadriel (G), Tekay-Reina (TR), Galadriel-Tekay-Reina (GTR) are detected to be under positive selection by Hommel test procedure.

The P48 is located between the two helices and appears to be crucial for the helix-helix structure formation that is specific to plant LTR retrotransposon CHDs.

3. Discussion

The evolutionary history of retrotransposons includes the gain (and loss) of functional enzymatic domains, which allows them to adapt to a constantly changing genomic environment [12, 43–45]. The chromodomain (CHD) is believed to be a comparatively recent acquisition of LTR retrotransposons [12, 45]. The role of the chromodomains most likely is in targeting the insertion of new LTR retrotransposon copies into heterochromatic regions by recognizing specific heterochromatic histone marks and/or other factors [15]. As a consequence, LTR retrotransposons can easily avoid subsequent inactivation and elimination (purifying selection) because the chance to interfere with any coding sequence is small in heterochromatic regions [46].

Comparative and phylogenetic analysis demonstrates that plant LTR retrotransposon CHDs represent heterogeneous groups of enzymatic domains with a complex evolutionary history. First, it appears that plant retrotransposon CHD_II evolved from CHD_I, which can still be found in genomic sequences of green algae (Chlamyvir clade) and nonseed plants such as mosses and lycophytes (Tcn1 clade; Figure 6; and [13, 16, 18]). In addition, the SM1-Galahad and SM2-Galahad found in lycophyte Selaginella carried CHD_I, while belonged to the Galadriel clade of plant LTR retrotransposons. All of the other known representatives of this clade possess typical plant CHD_II domain. Lycophyte Selaginella appears to be a unique model species for the investigation of chromodomains among plants (Figure 6). This species is the only plant species known to have both types of retrotransposon CHDs in its genome. Moreover, LTR retrotransposon CHDs found in Selaginella represent transitional stages between CHD_I (found in fungi and animals) and CHD_II (described from angiosperms). Interestingly, the few CHDs that were found in gymnosperms are also distant from "typical" angiosperm CHD_II domain. We believe that the evolutionary history of plant LTR retrotransposon CHDs and their diversity among different plant groups reflect changes that occurred in chromatin organization (e.g., the distribution of heterochromatin/euchromatin mark; or molecular differences in specific heterochromatin/euchromatin marks) from green algae to higher plants. It was proposed earlier that although the histone methylation marks are conserved among eukaryotes, the distribution of the individual marks and their functional meaning may have diverged as different phyla evolved [47].

The occurrence of heterochromatic marks in plants differs from that of fungi and mammals. For example, histone H3 trimethyl-K9 (H3K9me3) is a heterochromatin-specific mark in Schizosaccharomyces pombe [48] and in mammals [49, 50] but has never been found to be associated with heterochromatin in plants (reviewed in [51]). Moreover, heterochromatin-specific marks have an uneven distribution among plants: histone H3 dimethyl-K27 (H3K27me2) has been shown to be a typical modification in the heterochromatic regions of Arabidopsis thaliana, Vicea faba, Zea mays, and Secale cereale, but it was not detected in species such as Glycine max, Plantago ovate, and Hordeum vulgare [52–58]; only two species so far, S. cereale and in V. faba, showed labeling of heterochromatin with histone H3 trimethyl-K27 (H3K27me3) [53, 55]. Very little is known about chromatin organization in mosses, lycophytes, and ferns [59]. The limited information that is available for gymnosperms indicates that their heterochromatic marks seem to be quite different from those of angiosperms [47]. For example, histone H3 monomethyl-K9 (H3K9me1), H3K9me2, and histone H3 monomethyl-K27 (H3K27me1) modifications, which are believed to be associated with silencing and heterochromatin formation in A. thaliana, are underrepresented in Picea abies and Pinus sylvestris. At the same time, H3K9me3 and H3K27me3 are typical heterochromatic marks in two gymnosperm species but are not present in Arabidopsis [47].

One would expect mobile elements to be very sensitive to any changes in host genome function and organization; they must adapt or be eliminated from the genome. Divergence in the distribution of individual heterochromatin-associated histone methylation marks could trigger the evolutionary changes of LTR retrotransposon CHDs in plants, which would result in a shift from the original CHD_I (still present in green algae, mosses, and lycophytes) to CHD_II (all higher plants) and subsequent subdivision into clade-specific CHDs (Tekay-, Reina-, and Galadriel-like CHDs). The initial plasticity of CHDs provided a wide range of possibilities for evolution. Chromodomains carry diverse functions in cells, from the recognition of specific H3 histone modifications to protein dimerisation as well as DNA and RNA binding (for review [3]). It is believed that considerable diversity of recognition by CHDs is generated within the CHD family through relatively few aminoacid substitutions at the aromatic cage or the peptide-binding sites [31]. While the function of retrotransposon CHDs is generally unknown, it was

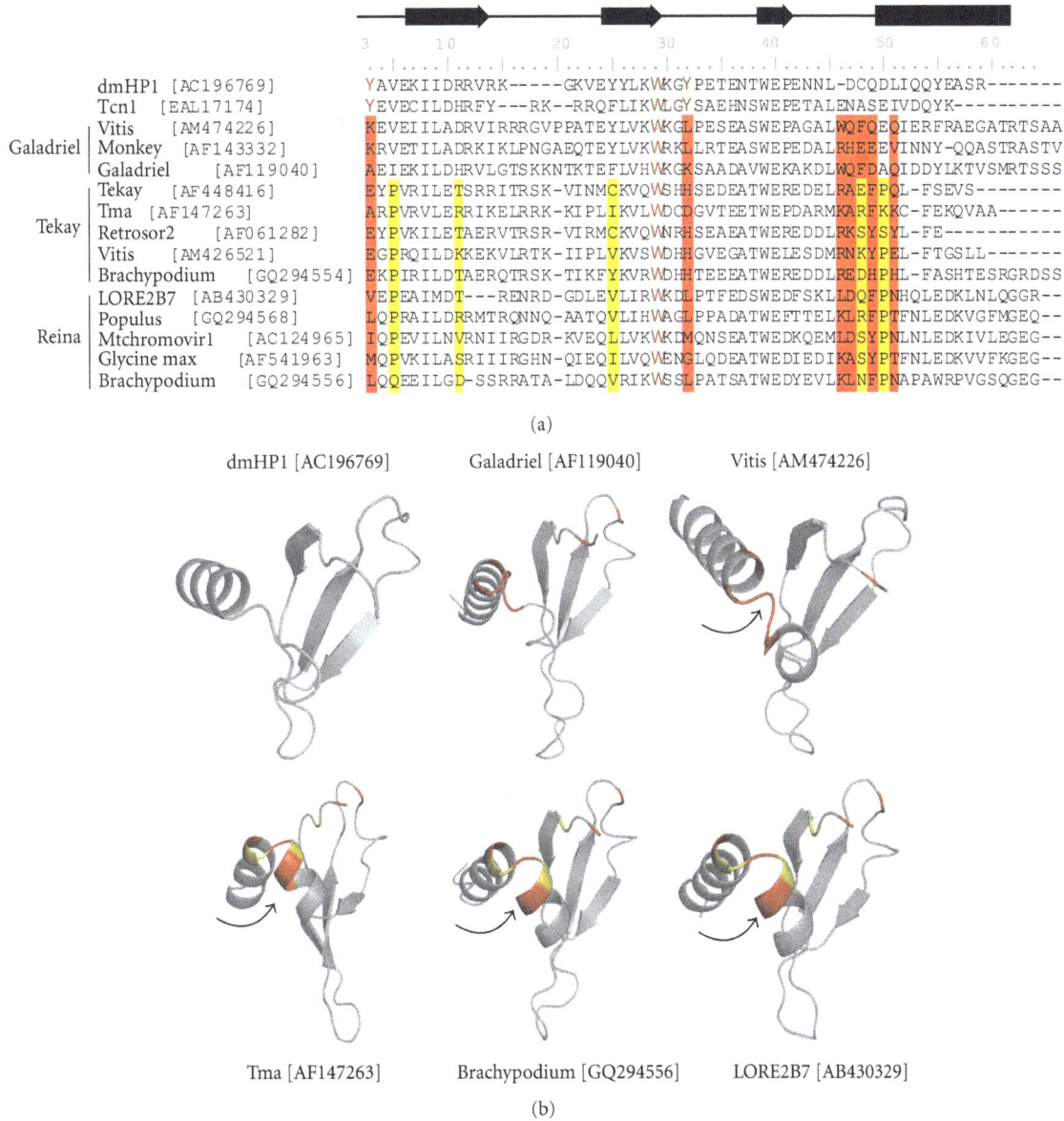

FIGURE 5: Multiple alignment including the "classical" chromodomain (dmHP1 from *Drosophila melanogaster*) and sampled representatives of group I and group II chromodomains from LTR retrotransposons (a); estimated tertiary structure for dmHP1 CHD (pdb : 1q3l; [7]) and predicted tertiary structures for some LTR retrotransposon CHDs (b). The GenBank accession numbers are indicated for each sequence used. The secondary structure elements are shownon the top (arrows indicate β-strands; rectangle α-helix) [3–7]. Three aminoacids that were shown to be under positive selection in the Tekay-Reina cluster are highlighted in yellow. The aminoacids that are potentially under positive selection in the Galadiriel-Reina-Tekay cluster are highlighted in red. An additional helix structure is indicated by an arrow for representatives of the Reina and Tekay clades.

demonstrated that CHD_I of MAGGY LTR retrotransposon from rice-blast fungus *Magnaporthe oryzae* targets the integration of new copies to heterochromatin by recognizing H3K9me2 and H3K9me3 modifications [15]. Although a colocalization of TFL2, a dmHP1-like homolog in *Arabidopsis*, and CHD_II of the Tma LTR retrotransposon from *A. thaliana* were shown in the same study, the actual interacting factor(s) for plant CHD_II was not found.

The sequence divergence between CHD_I and CHD_II is a key to the functional differences, and positive selection appears to be involved in the diversification of CHDs during the evolutionary history of LTR retrotransposons. The presence of positive selection is uncommon among LTR retrotransposon domains [34–36]. It was proposed earlier that LTR retrotransposons rarely undergo substitution events that are driven by positive selection, which allows elements to remain unrecognized by the host genome and to escape silencing [36]. However, the CHD itself provides the possibility of escaping silencing by the specific targeting of heterochromatic regions when LTR retrotransposons integrate [15], and their rapid evolution could be advantageous. Most of the genes for which positive selection has been

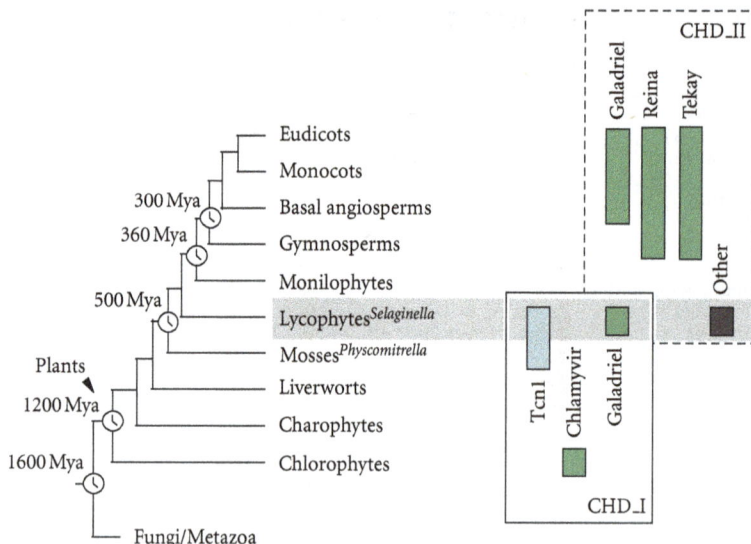

FIGURE 6: Distribution of different clades of chromodomain-containing LTR retrotransposons in plants as well as CHD_I and CHD_II. The evolutionary tree is represented according to Bowman et al., 2007 [40] and Berbee and Taylor, 2001 [41] with minor modifications. Divergence times (Mya: million years ago) are indicated according to Hedges, 2002 [42]. Other: SM-Diluvium and SM-Cranky.

documented are involved in interactions between the organism and the environment (e.g., [60]) and/or are subjected to genetic conflict (e.g., [61–63]). What could be driving the evolution of chromodomains? One of the most attractive explanations is coevolutionary pressures; for example, plant LTR retrotransposon CHDs might evolve after heterochromatin-associated histone methylation marks the host species. This scenario could explain the shift from CHD_I to CHD_II after the divergence of the plant and fungi/metazoa groups as well as the divergence of CHDs between angiosperms and gymnosperms. It is possible that rapid adaptation coupled with subsequent strong selective pressure not only led to the adaptation of LTR retrotransposons to a changing "chromatin" environment in plants in general, but it also may have contributed to a functional diversification of clade-specific LTR retrotransposon CHDs. In other words, while Galadriel-, Reina- and Tekay-like CHDs were still involved in targeted-integration of new LTR retrotransposon copies, they could possibly recognize different chromatin marks or factors.

Mobile elements in general, and LTR retrotransposons in particular, are important parts of plant genomes. While it is still believed that mobile elements are selected against and silenced in host genomes to prevent their harmful effects, increasing numbers of studies indicate that mobile elements have been positively selected as major components of heterochromatin (see [64]). Chromodomain-containing LTR retrotransposons are the most remarkable example of mobile elements that developed the mechanism of targeted integration into heterochromatic regions. In the present study, we inferred that interactions between chromodomain-containing LTR retrotransposons and the host genome resulted in the present diversity of plant LTR retrotransposon CHDs and, most likely, led to the retrotransposon-enriched genome organization in plants. It is necessary to note that chromodomain-containing LTR retrotransposons in plant genomes represent a large pool of diverse chromatin remodeling domains, which also possess high evolutionary plasticity (for a review, see [3]). The potential roles of LTR retrotransposon chromodomains in genome and chromatin organization are poorly understood but should not be underestimated.

4. Materials and Methods

4.1. Genomic Sequence Screening and Sequence and Phylogenetic Analysis. Selaginella moellendorffii genomic sequence is available at the DOE Joint Genome Institute ([65]; http://genome.jgi-psf.org/Selmo1/Selmo1.info.html). We performed BLASTN and TBLASTN searches of the *S. moellendorffii* database (http://genome.jgi-psf.org/Selmo1/Selmo1.info.html; with default parameters) using the SM-Tcn1 retrotransposon [18] and previously described retrotransposons from other species as queries: Osr35—AC068924; rn377-208—AK068625; Reina—U69258; RIRE3—AC119148; Tekay—AF050455; Retrosor2—AF061282; Tma—AF147263; Galadriel—AF119040. The full-length copies of newly identified Gypsy LTR retrotransposons were discovered in genomic sequences and analyzed by UniPro uGENE software (http://ugene.unipro.ru/). The LTR retrotransposon sequences obtained during BLASTN and TBLASTN searches were localized using UniPro uGENE "Find pattern" option with default parameters (both strands, match percent: 100%, whole sequence range). Open reading frames were detected by UniPro uGENE "Find ORFs" option with default parameters. Pseudo-ORFs were manually reconstructed. Putative consensus sequences were reconstructed based on multiple alignments of copies. All DNA alignments were performed by ClustalW [66] with default parameters and were edited manually in UniPro uGENE.

The LTR retrotransposon chromodomain search was carried out using BLAST (BLASTN, TBLASTN, and BLASTP).

BLAST analysis was performed using sequence databases that were accessible from the National Center for Biotechnology Information (NCBI) server (http://blast.ncbi.nlm.nih.gov/Blast.cgi; BLASTP and MegaBLAST with default parameters), the U.S. Department of Energy Joint Genome Institute (http://genome.jgi-psf.org/), PlantGDB (http://www.plantgdb.org/; BLASTN, TBLASTN and BLASTP with default parameters), and Phytozome, a tool for green plant comparative genomics (http://www.phytozome.net/; BLASTN with default parameters). The full list of species investigated and primary information can be found in Supporting Information Table 1S and Table 2S. Aminoacid sequences of the known CHDs were used as queries: MAGGY—L35053; Tcn1—XM_571377; Retrosor2—AF061282; Tma—AF147263; Galadriel—AF119040. To discriminate between functional proteins and retrotransposon CHDs, the next round of BLASTP was performed using newly identified CHDs as queries.

All multiple alignments were performed by ClustalW [66] and were edited manually in UniPro uGENE (http://ugene.unipro.ru/). Phylogenetic analyses were performed using the Maximum Likelihood (ML) method in the PhyML 3.0 program [67]. Neighbor-Joining analysis was performed using the MEGA5 software [68]. Statistical support for the NJ tree was evaluated by bootstrapping (number of replications, 1000) [69]. Statistical support for the ML tree was evaluated by approximate likelihood ratio test (aLRT; Figure 2) and by bootstrapping (number of replications, 100; Figure 4) [69, 70]. The tertiary structures of investigated CHD peptides were predicted using I-TASSER server with default parameters (http://zhanglab.ccmb.med.umich.edu/I-TASSER/) [39]. The tertiary structure of dmHP1 from *Drosophila melanogaster* is available in Protein Data Bank (http://www.rcsb.org/pdb/home/home.do) under ID—1q3l [7].

4.2. Test for Selection.

The multiple alignment of CHD sequences was performed using ClustalW [66] available at the RevTrans 1.4 Server ([71]; http://www.cbs.dtu.dk/services/RevTrans). RevTrans takes a set of DNA sequences, virtually translates them, aligns the peptide sequences, and uses this as a scaffold for constructing the corresponding DNA multiple alignment. The phylogenetic trees of domains (Figure 4) were obtained with the maximum likelihood algorithm implemented in the PhyML 3.0 program [67]. From nucleotide sequence alignments for each domain, we reconstructed the phylogenetic trees under HKY85 + G model. The PhyML tree searching algorithm was chosen as the best of subtree pruning and regrafting (SPR) and nearest neighbor interchange (NNI) for more thorough explorations of the space of topologies. To assess the reliability of the reconstructed phylogenies, we performed 100 bootstrap reconstructions for each domain [69].

The nonsynonymous to synonymous substitution rate ratio ($\omega = d_N/d_S$) provides a measure of natural selection at the protein level, with $\omega = 1$, >1, and <1, indicating neutral evolution, purifying selection, and positive selection, respectively. We use codeml program (the PAML package) to perform lineage and clade-specific analyses of d_N/d_S ratios (ω) [72–74]. The program codeml implements large collec-

tion of codon substitution models. Positive selection is tested using a likelihood ratio test (LRT) comparing a null model that does not allow $\omega > 1$ with an alternative model that does. When two models are nested, twice the log-likelihood difference between the two models can be compared with the χ^2 distribution, with the difference in the number of parameters between the two models as the degrees of freedom (df).

At first, to evaluate whether any of three chosen domains from four diverse clades are undergoing positive selection affecting all sites over prolonged time, the simplest one-ratio site model (M0) was used (codeml parameters used were as follows: model = 0, NSsites = 0). One should keep in mind, since our null hypothesis was always the absence of positive selection failing to reject a null hypothesis and/or providing a null hypothesis were interpreted as an absence of positive selection in either case. At the second stage, we tested each of the clades/groups for a signature of positive selection using clade-site test, this compares the modified clade model C (model = 3, NSsites = 2) with the neutral M1a model (model = 0, NSsites = 0). For this comparison df was set to 3. We conducted one specific test for each domain using extension of clade model C which allows for more than two branch types. Branches leading to appropriate clade/group were labeled: Tekay: T; Reina: R; Galadriel: G; Tekay-Reina: TR; Galadriel-Tekay-Reina: GTR; Reina-Galadriel: RG; Tekay-Reina-Galadriel: TRG, other branches were used as "background". The main purpose of the test was to identify whether there is at least one branch potentially under positive selection. All maximum likelihood estimates for RT branches of the site class 2 ω ratio were close to 0, thus taking in the consideration previous results we conclude that there are not any events of positive selection. For CHD and Int domains there was at least one estimate greater than one. Nevertheless, this clade-based test does not directly examine whether any ω ratio is significantly greater than one.

Finally, we applied the LRT based on two branch-site models with one, where ω was to be estimated and other with ω fixed to 1, to test every of five aforementioned branch of the CHD and Int trees for evidence of positive selection, this test is also known as the branch-site test of positive selection. Branch-site models of codon substitution allow ω to vary both among sites in the protein and across branches on the tree and provide a means to detect short episodes of molecular adaptation affecting just a few sites [75]. In these models it is assumed that the branches are *a priori* divided into foreground and background. Only foreground lineages may have experienced positive selection. One of two branch-site models presented in codeml—modified model A (model = 2, NSsites = 2) was used for comparison with null hypothesis [75]. The model assumes four classes of sites. Site class 0 includes codons that are conserved throughout the tree, with $0 < \omega_0 < 1$ estimated. Site class 1 includes codons that are evolving neutrally throughout the tree with $\omega_1 = 1$. Site classes 2a and 2b include codons that are conserved or neutral on the background branches, but become under positive selection on the foreground branches with $\omega_2 > 1$ estimated. The model involves four parameters in the ω distribution: p_0, p_1, ω_0, and ω_2. The null hypothesis was also modified model A but with $\omega_2 = 1$

fixed (codeml options were switched from fix_omega = 0 to fix_omega = 1 and omega = 1). A likelihood ratio test (LRT) based on models was found to have satisfactory accuracy and reasonable power [75–79]. Branch-site models allow only two types of branches thus the most common approach to test several branches on the tree is to treat every branch as foreground in turn. The probability of rejecting falsely at least one of null hypotheses in such tests can be high. The correction for multiple testing becomes necessary. The Hommel procedure that controls family-wise error rate (FWER) was used as correction method. The results of Bayes empirical Bayes (BEB) approach which accommodates uncertainties in the maximum likelihood estimates were used to identify sites under positive selection if the likelihood ratio test after correction procedure was significant. Table 3 summarizes maximum likelihood estimates and test statistics for LRTs corresponding to the every of 5 branches leading to appropriate clades/groups of the CHD tree (see Supporting Information Protocol S1 for details).

Authors' Contribution

A. Novikov and G. Smyshlyaev contributed equally to this paper.

Acknowledgments

Authors thank Dr. Mark Farman (University of Kentucky, USA) for the helpful comments and stylistic suggestions. The *Selaginella* sequence data were produced by the US Department of Energy Joint Genome Institute (http://www.jgi.doe.gov) in collaboration with the user community.

References

[1] R. Paro and D. S. Hogness, "The Polycomb protein shares a homologous domain with a heterochromatin-associated protein of *Drosophila*," *Proceedings of the National Academy of Sciences of the United States of America*, vol. 88, no. 1, pp. 263–267, 1991.

[2] E. V. Koonin, S. Zhou, and J. C. Lucchesi, "The chromo superfamily: new members, duplication of the chrome domain and possible role in delivering transcription regulators to chromatin," *Nucleic Acids Research*, vol. 23, no. 21, pp. 4229–4233, 1995.

[3] A. Brehm, K. R. Tufteland, R. Aasland, and P. B. Becker, "The many colours of chromodomains," *BioEssays*, vol. 26, no. 2, pp. 133–140, 2004.

[4] F. Aasland and A. F. Stewart, "The chrome shadow domain, a second chrome domain in heterochromatin-binding protein 1, HP1," *Nucleic Acids Research*, vol. 23, no. 16, pp. 3168–3173, 1995.

[5] L. J. Ball, N. V. Murzina, R. W. Broadhurst et al., "Structure of the chromatin binding (chromo) domain from mouse modifier protein 1," *EMBO Journal*, vol. 16, no. 9, pp. 2473–2481, 1997.

[6] S. A. Jacobs, S. D. Taverna, Y. Zhang et al., "Specificity of the HP1 chromo domain for the methylated N-terminus of histone H3," *EMBO Journal*, vol. 20, no. 18, pp. 5232–5241, 2001.

[7] S. A. Jacobs and S. Khorasanizadeh, "Structure of HP1 chromodomain bound to a lysine 9-methylated histone H3 tail," *Science*, vol. 295, no. 5562, pp. 2080–2083, 2002.

[8] S. V. Brasher, B. O. Smith, R. H. Fogh et al., "The structure of mouse HP1 suggests a unique mode of single peptide recognition by the shadow chromo domain dimer," *EMBO Journal*, vol. 19, no. 7, pp. 1587–1597, 2000.

[9] A. J. Bannister, P. Zegerman, J. F. Partridge et al., "Selective recognition of methylated lysine 9 on histone H3 by the HP1 chromo domain," *Nature*, vol. 410, no. 6824, pp. 120–124, 2001.

[10] M. Lachner, D. O'Carroll, S. Rea, K. Mechtler, and T. Jenuwein, "Methylation of histone H3 lysine 9 creates a binding site for HP1 proteins," *Nature*, vol. 410, no. 6824, pp. 116–120, 2001.

[11] J. Nakayama, J. C. Rice, B. D. Strahl, C. D. Allis, and S. I. S. Grewal, "Role of histone H3 lysine 9 methylation in epigenetic control of heterochromatin assembly," *Science*, vol. 292, no. 5514, pp. 110–113, 2001.

[12] H. S. Malik and T. H. Eickbush, "Modular evolution of the integrase domain in the Ty3/Gypsy class of LTR retrotransposons," *Journal of Virology*, vol. 73, no. 6, pp. 5186–5190, 1999.

[13] B. Gorinšek, F. Gubenšek, and D. Kordiš, "Evolutionary genomics of chromoviruses in eukaryotes," *Molecular Biology and Evolution*, vol. 21, no. 5, pp. 781–798, 2004.

[14] I. Marín and C. Lloréns, "Ty3/Gypsy retrotransposons: description of new Arabidopsis thaliana elements and evolutionary perspectives derived from comparative genomic data," *Molecular Biology and Evolution*, vol. 17, no. 7, pp. 1040–1049, 2000.

[15] X. Gao, Y. Hou, H. Ebina, H. L. Levin, and D. F. Voytas, "Chromodomains direct integration of retrotransposons to heterochromatin," *Genome Research*, vol. 18, no. 3, pp. 359–369, 2008.

[16] O. Novikova, V. Mayorov, G. Smyshlyaev et al., "Novel clades of chromodomain-containing Gypsy LTR retrotransposons from mosses (Bryophyta)," *Plant Journal*, vol. 56, no. 4, pp. 562–574, 2008.

[17] O. Novikova, "Chromodomains and LTR retrotransposons in plants," *Communitative and Integrative Biology*, vol. 2, no. 2, pp. 158–162, 2009.

[18] O. Novikova, G. Smyshlyaev, and A. Blinov, "Evolutionary genomics revealed interkingdom distribution of Tcn1-like chromodomain-containing Gypsy LTR retrotransposons among fungi and plants," *BMC Genomics*, vol. 11, no. 1, article 231, 2010.

[19] H. Nakayashiki, T. Awa, Y. Tosa, and S. Mayama, "The C-terminal chromodomain-like module in the integrase domain is crucial for high transposition efficiency of the retrotransposon MAGGY," *FEBS Letters*, vol. 579, no. 2, pp. 488–492, 2005.

[20] A. Hizi and H. L. Levin, "The integrase of the long terminal repeat-retrotransposon Tf1 has a chromodomain that modulates integrase activities," *Journal of Biological Chemistry*, vol. 280, no. 47, pp. 39086–39094, 2005.

[21] T. Wicker, F. Sabot, A. Hua-Van et al., "A unified classification system for eukaryotic transposable elements," *Nature Reviews Genetics*, vol. 8, no. 12, pp. 973–982, 2007.

[22] P. Sanmiguel and J. L. Bennetzen, "Evidence that a recent increase in maize genome size was caused by the massive amplification of intergene retrotransposons," *Annals of Botany*, vol. 82, pp. 37–44, 1998.

[23] W. N. Stewart and G. W. Rothwell, *Paleobotany and the Evolution of Plants*, Cambridge University Press, New York, NY, USA, 1993.

[24] M. Parniske, B. B. H. Wulff, G. Bonnema, C. M. Thomas, D. A. Jones, and J. D. G. Jones, "Homologues of the Cf-9 disease

resistance gene (Hcr9s) are present at multiple loci on the short arm of tomato chromosome 1," *Molecular Plant-Microbe Interactions*, vol. 12, no. 2, pp. 93–102, 1999.

[25] K. J. Dej, T. Gerasimova, V. G. Corces, and J. D. Boeke, "A hotspot for the *Drosophila* gypsy retroelement in the ovo locus," *Nucleic Acids Research*, vol. 26, no. 17, pp. 4019–4024, 1998.

[26] H. M. Temin, "Origin of retroviruses from cellular moveable genetic elements," *Cell*, vol. 21, no. 3, pp. 599–600, 1980.

[27] D. Kuykendall, J. Shao, and K. Trimmer, "A nest of LTR retrotransposons adjacent the disease resistance-priming gene NPR1 in Beta vulgaris L. U.S. Hybrid H20," *International Journal of Plant Genomics*, vol. 2009, Article ID 576742, 2009.

[28] B. Piegu, R. Guyot, N. Picault et al., "Doubling genome size without polyploidization: dynamics of retrotransposition-driven genomic expansions in Oryza australiensis, a wild relative of rice," *Genome Research*, vol. 16, no. 10, pp. 1262–1269, 2006.

[29] G. Caetano-Anollés, "Evolution of genome size in the grasses," *Crop Science*, vol. 45, no. 5, pp. 1809–1816, 2005.

[30] C. Vitte and O. Panaud, "LTR retrotransposons and flowering plant genome size: emergence of the increase/decrease model," *Cytogenetic and Genome Research*, vol. 110, no. 1-4, pp. 91–107, 2005.

[31] K. Tajul-Arifin, R. Teasdale, T. Ravasi et al., "Identification and analysis of chromodomain-containing proteins encoded in the mouse transcriptome," *Genome Research*, vol. 13, no. 6 B, pp. 1416–1429, 2003.

[32] P. R. Nielsen, D. Nietlispach, H. R. Mott et al., "Structure of the HP1 chromodomain bound to histone H3 methylated at lysine 9," *Nature*, vol. 416, no. 6876, pp. 103–107, 2002.

[33] J. F. Flanagan, L. Z. Mi, M. Chruszcz et al., "Double chromodomains cooperate to recognize the methylated histone H3 tail," *Nature*, vol. 438, no. 7071, pp. 1181–1185, 2005.

[34] Y. Xiong and T. H. Eickbush, "Origin and evolution of retroelements based upon their reverse transcriptase sequences," *EMBO Journal*, vol. 9, no. 10, pp. 3353–3362, 1990.

[35] S. Boissinot and A. V. Furano, "Adaptive evolution in LINE-1 retrotransposons," *Molecular Biology and Evolution*, vol. 18, no. 12, pp. 2186–2194, 2001.

[36] R. S. Baucom, J. C. Estill, C. Chaparro et al., "Exceptional diversity, non-random distribution, and rapid evolution of retroelements in the B73 maize genome," *PLoS Genetics*, vol. 5, no. 11, Article ID e1000732, 2009.

[37] S. J. Wood, "Prolines and amyloidogenicity in fragments of the alzheimer's peptide β/A4," *Biochemistry*, vol. 34, no. 3, pp. 724–730, 1995.

[38] P. Y. Chou and G. D. Fasman, "β-Turns in proteins," *Journal of Molecular Biology*, vol. 115, no. 2, pp. 135–175, 1977.

[39] A. Roy, A. Kucukural, and Y. Zhang, "I-TASSER: a unified platform for automated protein structure and function prediction," *Nature protocols*, vol. 5, no. 4, pp. 725–738, 2010.

[40] J. L. Bowman, S. K. Floyd, and K. Sakakibara, "Green genes-comparative genomics of the green branch of life," *Cell*, vol. 129, no. 2, pp. 229–234, 2007.

[41] M. L. Berbee and J. W. Taylor, "Fungal molecular evolution: gene trees and geologic time," in *The Mycota: A Comprehensive Treatise on Fungi as Experimental Systems for Basic and Applied Research*, D. J. McLaughlin, E. G. McLaughlin, and P. A. Lemke, Eds., vol. 7 of *Systematics and Evolution, part B*, pp. 229–246, Springer, New York, NY, USA, 2001.

[42] S. B. Hedges, "The origin and evolution of model organisms," *Nature Reviews Genetics*, vol. 3, no. 11, pp. 838–849, 2002.

[43] H. S. Malik, "Ribonuclease H evolution in retrotransposable elements," *Cytogenetic and Genome Research*, vol. 110, no. 1-4, pp. 392–401, 2005.

[44] H. S. Malik and T. H. Eickbush, "Modular evolution of the integrase domain in the Ty3/Gypsy class of LTR retrotransposons," *Journal of Virology*, vol. 73, no. 6, pp. 5186–5190, 1999.

[45] C. Llorens, M. A. Fares, and A. Moya, "Relationships of gag-pol diversity between Ty3/Gypsy and Retroviridae LTR retroelements and the three kings hypothesis," *BMC Evolutionary Biology*, vol. 8, no. 1, article 276, 2008.

[46] P. Dimitri and N. Junakovic, "Revising the selfish DNA hypothesis: new evidence on accumulation of transposable elements in heterochromatin," *Trends in Genetics*, vol. 15, no. 4, pp. 123–124, 1999.

[47] J. Fuchs, G. Jovtchev, and I. Schubert, "The chromosomal distribution of histone methylation marks in gymnosperms differs from that of angiosperms," *Chromosome Research*, vol. 16, no. 6, pp. 891–898, 2008.

[48] T. Yamada, W. Fischle, T. Sugiyama, C. D. Allis, and S. I. S. Grewal, "The nucleation and maintenance of heterochromatin by a histone deacetylase in fission yeast," *Molecular Cell*, vol. 20, no. 2, pp. 173–185, 2005.

[49] A. H. F. M. Peters, S. Kubicek, K. Mechtler et al., "Partitioning and plasticity of repressive histone methylation states in mammalian chromatin," *Molecular Cell*, vol. 12, no. 6, pp. 1577–1589, 2003.

[50] J. C. Rice, S. D. Briggs, B. Ueberheide et al., "Histone methyltransferases direct different degrees of methylation to define distinct chromatin domains," *Molecular Cell*, vol. 12, no. 6, pp. 1591–1598, 2003.

[51] C. Liu, F. Lu, X. Cui, and X. Cao, "Histone methylation in higher plants," *Annual Review of Plant Biology*, vol. 61, pp. 395–420, 2010.

[52] A. M. Lindroth, D. Shultis, Z. Jasencakova et al., "Dual histone H3 methylation marks at lysines 9 and 27 required for interaction with CHROMOMENTHYLASE3," *EMBO Journal*, vol. 23, no. 21, pp. 4286–4296, 2004.

[53] J. Fuchs, D. Demidov, A. Houben, and I. Schubert, "Chromosomal histone modification patterns—from conservation to diversity," *Trends in Plant Science*, vol. 11, no. 4, pp. 199–208, 2006.

[54] J. Shi and R. K. Dawe, "Partitioning of the maize epigenome by the number of methyl groups on histone H3 lysines 9 and 27," *Genetics*, vol. 173, no. 3, pp. 1571–1583, 2006.

[55] M. Carchilan, M. Delgado, T. Ribeiro et al., "Transcriptionally active heterochromatin in rye B chromosomes," *Plant Cell*, vol. 19, no. 6, pp. 1738–1749, 2007.

[56] S. Marschner, K. Kumke, and A. Houben, "B chromosomes of B. dichromosomatica show a reduced level of euchromatic histone H3 methylation marks," *Chromosome Research*, vol. 15, no. 2, pp. 215–222, 2007.

[57] J. Fuchs and I. Schubert, "Chromosomal distribution and functional interpretation of epigenetic histone marks in plants," in *Plant Cytogenetics, Vol. 1: Genome Structure and Chromosome Function*, H. Bass and J. A. Birchler, Eds., Springer, New York, NY, USA, 2011.

[58] M. K. Dhar, J. Fuchs, and A. Houben, "Distribution of Eu- and heterochromatin in plantagoovata," *Cytogenetic and Genome Research*, vol. 125, no. 3, pp. 235–240, 2009.

[59] S. Spiker, "An evolutionary comparison of plant histones," *Biochimica et Biophysica Acta*, vol. 400, no. 2, pp. 461–467, 1975.

[60] Z. Yang and J. R. Bielawski, "Statistical methods for detecting molecular adaptation," *Trends in Ecology and Evolution*, vol. 15, no. 12, pp. 496–503, 2000.

[61] C. F. Qi, F. Bonhomme, A. Buckler-White et al., "Molecular phylogeny of *Fv1*," *Mammalian Genome*, vol. 9, no. 12, pp. 1049–1055, 1998.

[62] H. S. Malik and S. Henikoff, "Positive selection of iris, a retroviral envelope-derived host gene in *Drosophila melonogaster*," *PLoS Genetics*, vol. 1, no. 4, pp. 0429–0443, 2005.

[63] D. Vermaak, S. Henikoff, and H. S. Malik, "Positive selection drives the evolution of rhino, a member of the heterochromatin protein 1 family in *Drosophila*," *PLoS Genetics*, vol. 1, no. 1, pp. 0096–0108, 2005.

[64] C. Biémont, "Are transposable elements simply silenced or are they under house arrest?" *Trends in Genetics*, vol. 25, no. 8, pp. 333–334, 2009.

[65] J. A. Banks, T. Nishiyama, M. Hasebe et al., "The *Selaginella* genome identifies genetic changes associated with the evolution of vascular plants," *Science*, vol. 332, no. 6032, pp. 960–963, 2011.

[66] J. D. Thompson, D. G. Higgins, and T. J. Gibson, "CLUSTAL W: improving the sensitivity of progressive multiple sequence alignment through sequence weighting, position-specific gap penalties and weight matrix choice," *Nucleic Acids Research*, vol. 22, no. 22, pp. 4673–4680, 1994.

[67] S. Guindon, J. F. Dufayard, V. Lefort, M. Anisimova, W. Hordijk, and O. Gascuel, "New algorithms and methods to estimate maximum-likelihood phylogenies: assessing the performance of PhyML 3.0," *Systematic Biology*, vol. 59, no. 3, pp. 307–321, 2010.

[68] K. Tamura, D. Peterson, N. Peterson, G. Stecher, M. Nei, and S. Kumar, "MEGA5: molecular evolutionary genetics analysis using maximum likelihood, evolutionary distance, and maximum parsimony methods," *Molecular Biology and Evolution*, vol. 28, no. 10, pp. 2731–2739, 2011.

[69] M. Anisimova and O. Gascuel, "Approximate likelihood-ratio test for branches: a fast, accurate, and powerful alternative," *Systematic Biology*, vol. 55, no. 4, pp. 539–552, 2006.

[70] S. Guindon and O. Gascuel, "A simple, fast, and accurate algorithm to estimate large phylogenies by maximum likelihood," *Systematic Biology*, vol. 52, no. 5, pp. 696–704, 2003.

[71] R. Wernersson and A. G. Pedersen, "RevTrans: multiple alignment of coding DNA from aligned amino acid sequences," *Nucleic Acids Research*, vol. 31, no. 13, pp. 3537–3539, 2003.

[72] Z. Yang, "Maximum-likelihood models for combined analyses of multiple sequence data," *Journal of Molecular Evolution*, vol. 42, no. 5, pp. 587–596, 1996.

[73] Z. Yang, "PAML: a program package for phylogenetic analysis by maximum likelihood," *Computer Applications in the Biosciences*, vol. 13, no. 5, pp. 555–556, 1997.

[74] Z. Yang, "PAML 4: phylogenetic analysis by maximum likelihood," *Molecular Biology and Evolution*, vol. 24, no. 8, pp. 1586–1591, 2007.

[75] J. Zhang, R. Nielsen, and Z. Yang, "Evaluation of an improved branch-site likelihood method for detecting positive selection at the molecular level," *Molecular Biology and Evolution*, vol. 22, no. 12, pp. 2472–2479, 2005.

[76] Z. Yang, W. S. W. Wong, and R. Nielsen, "Bayes empirical Bayes inference of amino acid sites under positive selection," *Molecular Biology and Evolution*, vol. 22, no. 4, pp. 1107–1118, 2005.

[77] J. P. Bielawski and Z. Yang, "A maximum likelihood method for detecting functional divergence at individual codon sites, with application to gene family evolution," *Journal of Molecular Evolution*, vol. 59, no. 1, pp. 121–132, 2004.

[78] Z. Yang and R. Nielsent, "Codon-substitution models for detecting molecular adaptation at individual sites along specific lineages," *Molecular Biology and Evolution*, vol. 19, no. 6, pp. 908–917, 2002.

[79] M. Anisimova and Z. Yang, "Multiple hypothesis testing to detect lineages under positive selection that affects only a few sites," *Molecular Biology and Evolution*, vol. 24, no. 5, pp. 1219–1228, 2007.

17

Correlation of Vernalization Loci *VRN-H1* and *VRN-H2* and Growth Habit in Barley Germplasm

Mohsen Mohammadi, Davoud Torkamaneh, and Hamid-Reza Nikkhah

Cereals Research Department, Seed and Plant Improvement Institute (SPII), P.O. Box 4119, Karaj, Iran

Correspondence should be addressed to Mohsen Mohammadi; mohsen@ualberta.ca

Academic Editor: Peter Langridge

Vernalization requirement is a key component in determining the overall fitness of developmental patterns of barley to its environment. We have used previously reported markers and spring-sown growth habit nursery to characterize the genotypes of barley germplasm in an applied barley breeding ground to establish a baseline of information required to understand the relationship between adaptation of autumn-sown barley germplasm in diverse regions with warm (W), moderate (M), or cold climates (C). This study revealed that twenty entries were detected with the presence of the vernalization critical region in *VRN-H1* locus and complete presence of the three geneclusters *ZCCT-Ha*, *-Hb*, and *-Hc* in *VRN-H2* locus represented as genotype *vrn-H1/Vrn-H2 (V1w/V2w)*. Of these genotypes, 17 entries showed winter growth habit whereas the remaining three revealed facultative growth habit indicating reduced vernalization requirements possibly due to *VRN-H3* and photoperiod sensitivity loci as compared to the landmark winter growth habit entries in this group. Twenty-four entries were detected with the lack of vernalization critical region in *VRN-H1* locus but complete presence of the three geneclusters *ZCCT-Ha*, *-Hb*, and *-Hc* in *VRN-H2* locus represented as genotype *Vrn-H1/Vrn-H2 (V1s/V2w)*. However, only half of these germplasms were identified with spring growth habit in spring-sown nursery, and the rest of the germplasms in this group revealed facultative growth habits due to possible variation in the length of deletion in *VRN-H1*. Four germplasms showed vernalization insensitive phenotype due to the lack of a functional *ZCCT-Ha* and/or *ZCCT-Hb* alleles in *VRN-H2* and the deletion in the vernalization critical region of *VRN-H1*. These germplasms revealed a complete spring type growth habit. Only one entry showed reduced vernalization requirement solely due to the deletion in functional *ZCCT-Hb* allele in *VRN-H2* and not due to the deletion in the vernalization critical region of *VRN-H1*.

1. Introduction

Increases in yield and adaptation of cereals rest on life cycle adjustments to environmental constraints explained, in part, by the transition from vegetative to reproductive growth, flowering, and maturity [1, 2]. Optimal fitness of developmental patterns to the environment takes place when the crop utilizes most of inputs from the environment on the one hand and on the other hand escapes from deleterious effects of the adverse environmental conditions (i.e., frost stress and late maturity) [3], [4] leading to superior crop performance and yield potential [3]. Barley is a temperate cereal grown in diverse climatic regions from spring growth habit to winter growth habit [5]. Some barley varieties require prolonged exposure to cold temperatures to initiate flowering, a process referred to as vernalization, whereas others initiate flowering without any obligation for vernalization. Therefore, learning more about genetic variation of loci governing vernalization in barley breeding materials is beneficial for developing new cultivars for target environments.

The concept of vernalization sensitivity in barley is different from that in wheat and the sensitivity to low temperature exposure in barley is not an absolute concept. For a vernalization-sensitive wheat, genotype to induce flowering, the seedlings require exposure to low temperatures (4–6°C) for 4–6 weeks [6]. The vernalization-sensitive wheat plant does not flower if the vernalization requirement is not fulfilled, whereas an unvernalized vernalization-sensitive barley genotype may induce flowering in late cropping season which is not efficient economically [7]. A three-locus epistatic model of vernalization sensitivity has been proposed by Takahashi and Yasuda [8], which then was reduced to two-locus epistatic model on the ground that allelic variations in *VRN-H3* locus is limited [9]. According to this model, vernalization response in barley is mainly determined by *VRN-H1* locus on chromosome 5H and *VRN-H2* locus on chromosome 4H, where genotypes with dominant allele *Vrn-H2* on *VRN-H2* locus and recessive allele *vrn-H1* on *VRN-H1* locus require the greatest vernalization requirement.

The candidate gene responsible for the *VRN-1* locus in wheat and barley is a MADS-box floral meristem identity gene (HvBM5A). High expression of *VRN1* is required for the initiation of reproductive development. In some varieties of barley and wheat, MADS-box gene is expressed at high basal levels without requiring a prolonged exposure to cold temperature. Whereas other varieties require prolonged exposure to cold temperature to induce expression of MADS-box at higher levels of expressions required for transition to reproductive stages [10–13]. Further genetic studies revealed that mutations in the promoter or deletions within the first intron of HvBM5A gene lead to dominant Vrn-H1 spring type alleles (nonvernalization requiring) and are known to be responsible for the high level of expression of HvBM5A and the reductions of vernalization requirements in barley [10, 13, 14]. Investigation of a large collection of barley germplasm narrowed down the allelic variations of *VRN-H1* locus to nine haplotypes (1A, 1B, 2, 3, 4A, 4B, 5A, 5B, and 5C) within the European germplasm [1, 2] of which haplotypes 1A and 5C are typical winter type alleles. The distinction between winter and spring *Vrn-H1* haplotypes is based on the presence or absence of the key regulatory element localized in the first intron of barley, a so-called "critical region" [14, 15]. Further investigation of variation in *Vrn-H1* winter haplotypes revealed a 486-bp discriminative deletion found only within the solo long terminal repeat (LTR) of haplotype 5C but not on 1A resulting in amplification of 344 bp and 830 bp products in winter haplotypes 5C and 1A, respectively [16].

The *VRN-H2* locus is a dominant repressor of flowering. Genotypes lack a dominant (functional) copy of *VRN2* flower early [17]. The candidate gene proposed for Vrn-2 locus in wheat and barley is a ZCCT zinc finger CCT domain transcription factor [17]. Transgenic and RNA interference studies have shown that the ZCCT gene was responsible for the variation in the vernalization requirement in wheat [17] Loss-of-function or complete deletion mutation in the coding region of this gene leads to recessive spring type vrn-H2 growth habit alleles. The ZCCT candidate gene cluster in *VRN-H2* locus consists of three tightly linked genes, ZCCT-Ha, ZCCT-Hb, and ZCCT-Hc [18], and more references. Laurie et al. [19] have mapped *VRN-H2* locus on the distal part of as mentioned by long chromosome 4H in barley. *VRN-H3*, the third vernalization locus in barley, is located on chromosome 1H Takahashi and Yasuda [8]. The candidate gene proposed for this locus is *HvFT1* which is homologous to the *Arabidopsis FT* gene [20, 21]. The elevated expression of *HvFT1* accelerates flower development [20]. Dominant *Vrn-H3* allele results in elevated expression levels of *HvFT1*. In genotypes requiring prolonged exposure to low-temperature treatment for flower development dominant *Vrn-H3* allele provides a bypass of the vernalization requirement [20].

The expression of *VRN-H3* is also mediated by a pathway downstream of *VRN-H1* and *VRN-H2* and interactions with alleles at *VRN-H2* and *PPD-H1* loci. When a genotype possesses recessive *vrn-H2* or vernalization fulfillment represses dominant *Vrn-H2*, *HvFT1* is induced under long-day conditions and accelerates flower development, though this pathway requires active *Ppd-H1* allele [22–24].

The barley germplasms grown in Iran are either developed within the country or they were introduced from the ICARDA barley breeding program. The knowledge of germplasm growth habit is limited to flowering in spring sown nurseries. Neither the allelic composition and distribution of vernalization sensitivity loci have been characterized, nor it is clear as to how such allelic variations contribute to the overall adaptation of the country's barley germplasm through the adjustment of flowering time. In this study, we unravel allelic distribution of vernalization loci in barley. The objective of this investigation is the identification of the allelic combinations of *VRN-H1* and *VRN-H2* loci deployed in a collection of contemporary cultivars by using allele-specific molecular markers and to enhance our understanding of the underlying factors governing overall adaptation of barley to various geographical regions within the barley breeding program.

2. Materials and Methods

2.1. Plant Materials. A total of 51 barley genotypes including cultivars that either developed locally or introduced from external breeding programs, elite experimental lines, and local landraces collected from Iran were selected. The climatic diversity represents regions with moderate temperatures, areas with terminal heat stress, and regions prone to frost stress. In some regions with either moderate temperature or terminal heat stress, the barley production is also under drought and salinity conditions.

2.2. Growth Habit Phenotyping. The growth habit nursery was planted on a single row of one-meter-long plot for each entry during spring 2012 where the minimum temperature did not fall below. Growth habit phenotypes were scored in three categories of W, F, and S representing genotypes with no tiller flowered on the row, some tillers flowered, and all tillers fully flowered, respectively. Genotypes were scored as spring when all tillers of the entire row flowered, as facultative when only few tillers were flowered, and winter when none of the tillers on the entire row flowered by the end of summer 2012.

2.3. DNA Extraction. Small-scale nucleic acid isolation method suitable for marker-assisted selection procedures was used to extract DNA from young leaves of barley genotypes. Fresh leaf tissue of two-week old barley plants were flash frozen in liquid nitrogen and ground into a fine powder. Extraction was performed in 2 mL sterile tubes where a quarter of each 2 mL tube was filled with frozen leaf powder. Then, 0.9 mL of preheated (up to 65°C) CTAB buffer (2% (w/v) cetyltrimethylammonium bromide, 200 mM Tris/HCl pH 8.0, 20 mM EDTA pH 8.0, 1.4 M NaCl, and 1% (v/v) freshly added β-mercaptoethanol) was added to each tube. Extraction mixes were incubated for 45 min at

TABLE 1: The list of primers names, sequences, and expected PCR product bands used in this study with the original names obtained in the references cited.

Vrn-H1			
HvBM5.84F	5′-TGAGGGTATGAGTGGCGCTAG-3′	437 bp ±	Koti et al., 2006 [25]
HvBM5.85R	5′-TCTCATAGGTTCTAGACAAAGCATAG-3′		
HvBM5A-intronI-F3b	5′-CTTGCATGTGTTGTCGGTCT-3′	344/830 bp	Cockram et al., 2009 [16]
HvBM5A-intronI-R3b	5′-GCTGGGACAAGACTCTACGG-3′		
Vrn-H2			
ZCCTH. 14F:	5′-CAAGGAATATCAAGTACATATCTGC-3′	600 bp ±	Szűcs et al., 2007 [26]
ZCCTH. 19R:	5′-CCGTATTTATTGAGTTGGTGGTG-3′		
ZCCTb. 8F:	5′-GCATCAATGCACCCTACCTCTT-3′	600 bp ±	Szűcs et al., 2007 [26]
ZCCTb. 11R:	5′-GGAAAACAATGGTGAGAGTAGTACAG-3′		
ZCCT. HcF:	5′-CACCATCGCATGATGCAC-3′	200 bp ±	Yan et al., 2006 [20]
ZCCT. HcR:	5′-TCATATGGCGAAGCTGGAG-3′		
HvSNF2.01F	5′-CCTGAAGCGAGTATCCATATGC-3′	500–700 bp	von Zitzewitz et al., 2005 [14]
HvSNF2.04R	5′-GCTGCATTATAGAGAAACAACAACG-3′		

$65°C$. Precooled chloroform: isoamyl-alcohol mixture (24 : 1) (0.9 mL) was then added to the heated tubes. The extraction mixes were then centrifuged at max speed (13600 xg) for 15 minutes at $4°C$. The upper layer was then transferred into a new tube and supplemented with 0.7 volumes of precooled ($-20°C$) isopropanol and 0.1 volumes of 3 M ammonium acetate. DNA was allowed to precipitate at $-20°C$ for 45 minutes. The precipitates were pelleted by centrifugation at max speed for 10 minutes at $4°C$. The pellets were washed at least once with 300 μL of precooled 70% (v/v) ethanol. DNA was finally dissolved in 70 μL of nuclease free water.

2.4. DNA Amplifications and Visualization. Deletion-specific primers HvBM5.84F and HvBM5.85R [25] were used to discriminate between recessive and dominant alleles at *VRN-H1* locus based on the presence or absence of the regulatory element in intron I of *HvBM5*. Thermal cycling profile for these primers consisted of an initial denaturing step of $94°C$ for 3 min, followed by 35 cycles of denaturation at $94°C$ for 35 s, annealing at $58°C$ for 45 s, and extension at $72°C$ for 45 s. The PCR reaction was terminated by a final extension of $72°C$ for 10 min. The discrimination between two winter haplotypes was assessed by using primers HvBM5A-intronI-F3b and HvBM5A-intronI-R3b after Cockram et al., [16]. PCR was initiated at $95°C$ for 3 min, followed by 35 cycles of denaturation at $94°C$ for 1 min, annealing at $54°C$ for 45 s, and extension at $72°C$ for 80 s. The PCR reaction was terminated by a final extension of $72°C$ for 5 min. Three deletion-specific primer pairs were used to assess the deletions in three-gene cluster *VRN-H2* locus that is, *ZCCT-Ha*, *ZCCT-Hb*, and *ZCCT-Hc* [20, 25]. The amplification of a physically linked locus HvSNF2 was also used as a positive control as described by Karsai et al., 2005. Thermal cycling profile for these primers consisted of an initial denaturing step of $94°C$ for 4 min, followed by 35 cycles of denaturation at $94°C$ for 45 s, annealing at $54°C$ for 45 s, and extension at $72°C$ for 1 min. The PCR reaction was terminated by a final extension of $72°C$ for 10 min. Full list of primer names and sequences are listed in Table 1. PCR reactions of 25 μL contained 1 μL DNA, 2.5 μL 10X Taq buffer,

TABLE 2: The predicted growth habit for each genotypic class is listed based on allelic combinations of two-locus *VRN-H1/VRN-H2* flowering model. A simplified notation for allelic combinations in relation to growth habit is also provided.

Allelic combinations	Simplified notation	Predicted growth habit
vrn-H1/Vrn-H2	*V1w/V2w*	Winter
vrn-H1/vrn-H2	*V1w/V2s*	Facultative
Vrn-H1/Vrn-H2	*V1s/V2w*	Spring
Vrn-H1/vrn-H2	*V1s/V2s*	Spring

1.5 mM $MgCl_2$, 1 μL dNTP mix (2.5 mM of each dNTP), 0.4 pmol of each primer, and 1 unit of Taq polymerase. PCR reactions were performed in an Eppendorf Mastercycler (Eppendorf AG, Hamburg, Germany) or T100 Cycler (Bio-Rad Laboratories, Inc., USA). DNA amplifications were visualized, after ethidium bromide staining and separation on 1% or 1.5% agarose gel electrophoresis by using UVItec Gel Document system (UVItec Limited, Cambridge, UK).

3. Results and Discussion

We used allele-specific primers to identify allelic variations of vernalization loci *VRN-H1* and *VRN-H2* in barley germplasm. We have also evaluated flowering potentials of the germplasm in the spring-sown nursery to establish whether or not they require vernalization for flowering. Three cultivars were used as positive controls in this study. They included cultivars "Nure" and "Igri" with genotype *vrn-H1/Vrn-H2* (*V1w/V2w*) as typical winter growth habit and cultivar "Morex" with genotype *Vrn-H1/vrn-H2* (*V1s/V2w*) as a typical spring growth habit [14]. Table 2 represents the relationship between the allelic combinations of *VRN-H1* and *VRN-H2* and the predicted growth habit. A simplified notation for spring type versus winter type alleles is also provided to avoid confusion of alleles. PCR amplification using primers HvBM5.84F and HvBM5.85R resulted in PCR products of 437 bp in 21 entries with the same size as observed for Nure and Igri, indicating the presence of the vernalization critical region in intron I region of *VRN-H1* locus suggestive of winter type recessive *vrn-H1* allele (Table 3). Further assessment of

TABLE 3: Germplasm evaluated for allelic variations of *VRN-H1* and *VRN-H2* loci along with their primary cultivation area predicted and observed growth habits are presented. C, M, W-S, and W-N represent cold, moderate, warm-south, and warm-north cultivation regions, respectively. W, F, and S represent winter, facultative, and spring growth habit, respectively.

Genotype/germplasm	Row type	Pedigree	Regions cultivated	VRN-H1 (437 bp)	5C versus 1A	HvSNF	VRN-H2 ZCCT-Ha	VRN-H2 ZCC-Hb	VRN-H2 ZCCT-Hc	Predicted GH	Observed GH
					vrn-H1/Vrn-H2 (V1w/V2w)						
Goharjo	6	Landrace	C	+	830 (1A)	+	+	+	+	W	W
Michailo		Michailo	C (CB, ICARDA)	+	830 (1A)	+	+	+	+	W	W
Radical	6	Radical	C	+	830 (1A)	+	+	+	+	W	W
Makoee	6	Star/FAO	C	+	830 (1A)	+	+	+	+	W	W
Bahman	6	WA 2196-68/NY6005-18, Fl//Scotia I	C	+	830 (1A)	+	+	+	+	W	W
Dasht	2	Probestdwarf (France)	W	+	830 (1A)	+	+	+	+	W	W
L1242	6	L1242	M, C	+	830 (1A)	+	+	+	+	W	W
Karoon	6	Strain205	W-S	+	344 (5C)	+	+	+	+	W	W
Nure	2	Fior 40/Alpha2//Baraka	C	+	830 (1A)	+	+	+	+	W	W
Igri	2	820/1427//Ingrid	C	+	830 (1A)	+	+	+	+	W	W
1-BC-80392	6	Landrace	M, C	+	830 (1A)	+	+	+	+	W	F
1-BC-80453	6	Landrace	M, C	+	830 (1A)	+	+	+	+	W	W
1-BC-80458	6	Landrace	M, C	+	830 (1A)	+	+	+	+	W	W
1-BC-80395	6	Landrace	M, C	+	344 (5C)	+	+	+	+	W	W
Shahr-e-kord	6	Landrace		+	344 (5C)	+	+	+	+	W	W
Arigashar	6	Landrace	W	+	344 (5C)	+	+	+	+	W	W
Ashar	6	Landrace	W	+	344 (5C)	+	+	+	+	W	W
1-BC-80087	6	Landrace	M, C	+	344 (5C)	+	+	+	+	W	W
1-BC-80207	6	Landrace	M, C	+	344 (5C)	+	+	+	+	W	F
1-BC-80017	6	Landrace	M, C	+	344 (5C)	+	+	+	+	W	F
					vrn-H1/vrn-H2 (V1w/V2s)						
1-BC-80318	6	Landrace	M, C	+	344 (5C)	+	+	−	+	F	F
					Vrn-H1/Vrn-H2 (V1s/V2w)						
MB-87-10	6	82S:510/3/Arinar/Aths//DS 29	M	−	NA	+	+	+	+	S	S
Nosrat	6	Karoon/Kavir	M	−	NA	+	+	+	+	S	F
Rihane	6	Rihane	M	−	NA	+	+	+	+	S	F
MBS-87-12	6	Roho/Mazorka//Trompilo	M	−	NA	+	+	+	+	S	F

TABLE 3: Continued.

Genotype/germplasm	Row type	Pedigree	Regions cultivated	VRN-H1 (437 bp)	5C versus 1A	HvSNF	VRN-H2 ZCCT-Ha	VRN-H2 ZCC-Hb	VRN-H2 ZCCT-Hc	Predicted GH	Observed GH
Valfajr	6	CI-108985/Egypt	M, C	−	NA	+	+	+	+	S	F
Torkaman	6	Rihane04/FAO	W-N	−	NA	+	+	+	+	S	F
Gorgan4	2	Herta/Sweden	W-N	−	NA	+	+	+	+	S	S
Jonob	6	Gloria "s"/ Copal "s"/CIMMYT	W-S	−	NA	+	+	+	+	S	F
Aras	2	Arumir/Europe	W-N	−	NA	+	+	+	+	S	S
Yosef	6	L.527/Chn-01//Gostoe/4 /Rhn-08/3/Deir Alla06// DL71/Strain205	M, W	−	NA	+	+	+	+	S	S
20269	6	Kavir/Badia	M	−	NA	+	+	+	+	S	F
20371	6	Legia	C, M	−	NA	+	+	+	+	S	S
1-BC-80162	6	Landrace	M	−	NA	+	+	+	+	S	S
1-BC-80628	6	Landrace	M	−	NA	+	+	+	+	S	S
Nik	6	Lignee 527/NK1272//JLB	M	−	NA	+	+	+	+	S	S
MB-82-12	2	70-63 MB-83-14 Novosadski-444	M	−	NA	+	+	+	+	S	F
EHD-85-8	6	CIRU/3/AGAVE/SUM BARD400//MARCO/4/ PETUNIA 1	M	−	NA	+	+	+	+	S	S
Kavir	6	Arivat (Atlas/Vaughn)	M	−	NA	+	+	+	+	S	S
Fajr30	6	Lignee13l/Gerbel//Alger- Ceres/3/Jonoob	M	−	NA	+	+	+	+	S	F
Teser.93		Teser.93		−	NA	+	+	+	+	S	S
Sararood	2	Landrace	M	−	NA	+	+	+	+	S	F
Badia (20261)	6	Badia	M	−	NA	+	+	+	+	S	S
EH-83-7	6	Congona/Borr	M	−	NA	+	+	+	+	S	S
MBS-82-5	6	Salt tolerant (5 shoori)	M	−	NA	+	+	+	+	S	F
Vrn-H1/vrn-H2 (V1s/V2s)											
Tokak	2	Tokak	C	−	NA	+	+	−	+	S	S
Bereke.54	6	Bereke.54 cold tolerant adapted to southern Kazakstan	C	−	NA	+	−	−	+	S	S
ND-82-10	6	D10	M	−	NA	+	+	−	+	S	S
1-BC-80021	6	Landrace	M	−	NA	+	+	−	+	S	F
Nimrooz	2	Trompillo CMB 74A-432-25B-1Y-1B-1Y-OB	W-S	−	NA	+	+	−	+	S	S
Morex	6	Cree/Bonanza	M	−	NA	+	−	−	−	S	S

these genotypes by means of primer pairs HvBM5A-intronI-F3b and HvBM5A-intronI-R3b resulted in PCR products of 830 bp and 344 bp indicative of winter haplotypes 1A and 5C, respectively (Table 2). Representative gel pictures showing amplifications of 437 bp (vernalization critical region) and discrimination of winter haplotypes 1A versus 5C are shown in Figure 1, respectively. Other genotypes ($n = 30$) did not yield any PCR products when amplified by primers HvBM5.84F and HvBM5.85R suggestive of a deletion in Vrn-H1 leading to a spring type dominant Vrn-H1 allele (Table 3).

PCR amplification using primers HvSNF2.01F and HvSNF2.04R as positive control primers for Vrn-H2 locus resulted in amplification of a PCR band sized 500–700 bp in all genotypes invariably. When assessed by three ZCCT allele-specific primer pairs, only the positive control "Morex" lacked all three ZCCT genes. Most of the genotypes were characterized as winter type dominant Vrn-H2 allele because they yielded the PCR products of coding regions of ZCCT-Ha (600 bp), ZCCT-Hb (600 bp), and ZCCT-Hc (200 bp) (Figure 2). In five genotypes, that is, 1-BC-80318, 1-BC-80021, Tokak, ND-82-10, and Nimrooz, the gene ZCCT-Hb was found deleted. Four of the genotypes were characterized by a spring type dominant Vrn-H1 allele and only 1-BC-80318 possessed the full length winter type recessive vrn-H1 allele. In genotype Bereke.54 both ZCCT-Ha and ZCCT-Hb were absent, and it showed a spring type dominant Vrn-H1 allele (Table 3). Based on a single variety "Fan" observation, Dubcovsky et al. [18] provided preliminary evidence that, of the three tightly linked ZCCT genes in barley, ZCCT-Hb is not sufficiently deterministic for winter growth habit. One year later, when examining the effect of daylength on expression of VRN-H1 and VRN-H2, Trevaskis et al. [22] discovered that in a vernalization requiring winter barley the expression of VRN-H1 is not regulated by daylength. In contrast, daylength was shown to reduce the expression of ZCCT-Ha and ZCCT-Hb but not ZCCT-Hc. This may indicate that both ZCCT-Ha and ZCCT-Hb may be sufficiently deterministic for winter growth habit, and deletion of either of the genes may result in the reduced vernalization requirement. Therefore, for the representation of our results, we have decided to denote recessive vrn-H2 where ZCCT-Hb was absent on the basis of the fact that this may reduce deterministic effect of Vrn-H2 locus for a vernalization requirement (Table 3).

In total, 20 genotypes were identified as vrn-H1/Vrn-H2 (V1w/V2w)—a typical genotype for winter growth habit. Almost all entries in this group revealed winter growth habit phenotype in our growth habit nursery except for 1-BC-80392, 1-BC-80207, and 1-BC-80392 that showed facultative growth habit indicating reduced vernalization requirements as compared with others in this group. In this group, a majority of cultivars grouped with landmarks of winter barley genotypes in our study, that is, "Nure" and "Igri" where the VRN-H1 locus harbored the winter haplotype 1A. Cultivars "Bahman," "Goharjo," and "Makoee" are well-adapted winter barley varieties grown in zone C with acceptable levels of winter survival. In contrast, cultivars Karoon and landrace accessions Arigashar and Ashar from Baluchestan Province

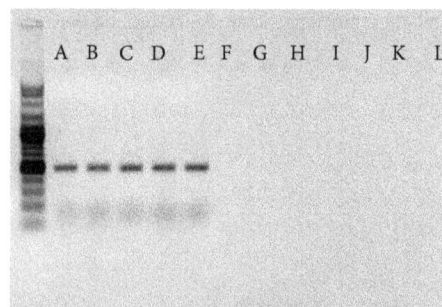

(a) Critical vernalization regulatory region in intron I

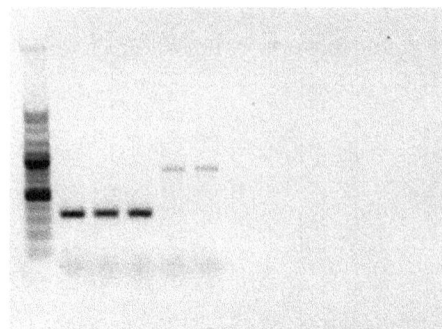

(b) Winter haplotypes 1A versus 5C

FIGURE 1: (a) Amplification of 437 bp PCR product, by using primers HvBM5.84F and HvBM5.85R, indicative of presence of critical vernalization regulatory region in VRN-H1 locus—a winter type recessive vrn-H1 allele in A: Karoon, B: Shahr-e-Kord, C: Arigashar, D: Bahman, and E: Radical, and no amplification suggestive of a deletion in critical vernalization region in VRN-H1 locus leading to a spring type dominant Vrn-H1 allele in F: Torkman, G: Jonoob, H: Kavir, I: Nosrat, J: Nimrooz, K: Yousof, and L: Morex. (b) Discrimination between winter haplotypes 5C and 1A shown by amplification of 344 bp PCR band, by using primers HvBM5A-intronI-F3b and HvBM5A-intronI-R3b, in A: Karoon, B: Shahr-e-Kord, and C: Arigashar and amplification of 830 bp PCR band by using the same primers in D: Bahman and E: Radical.

revealed genotype of vrn-H1/Vrn-H2 (V1w/V2w) where the VRN-H1 locus harbored the winter haplotype 5C.

A total of 24 genotypes were identified as Vrn-H1/Vrn-H2 (V1s/V2w) with expected spring growth habit. These genotypes were grouped with landmarks of spring growth habit barley cultivars in our study, that is, "Yousof" and "Kavir." This group is characterized with a deletion of the vernalization critical region from VRN-H1 intron I. However, in this study we do not report the length of deletion on spring type dominant Vrn-H1 alleles. Therefore, some of these genotypes might represent minimal or reduced vernalization requirements. Various studies have reported that sole deletion observed in the vernalization critical region in VRN-H1 locus may not be sufficient for the removal of the vernalization requirement. Rather, the length of intron I associated with the deletion is more relevant to the reduction in vernalization requirements [27, 28]. Our phenotyping growth habit data revealed that only half of these entries fully flowered in growth habit nursery, and others showed degrees of facultative growth habit. We concluded that there

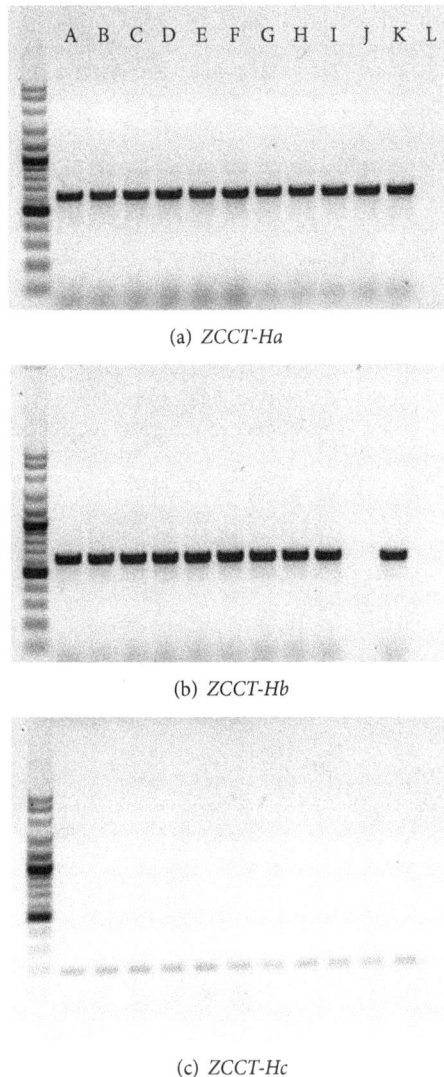

(a) *ZCCT-Ha*

(b) *ZCCT-Hb*

(c) *ZCCT-Hc*

FIGURE 2: (a) Amplification of 600 bp PCR product, by using primers ZCCTH.14F and ZCCTH.19R, indicative of presence of *ZCCT-Ha* in all germplasms tested except L: Morex. (b) Amplification of 600 bp PCR product, by using primers ZCCTb.8F and ZCCTb.11R, indicative of the presence of *ZCCT-Hb* in all germplasms tested except J: Nimrooz and L: Morex (c) Amplification of 200 bp PCR product, by using primers ZCCT.HcF and ZCCT.HcR, indicative of the presence of *ZCCT-Hc* in all germplasms tested except L: Morex. Germplasms included in this figure include A: Karoon, B: Shahr-e-Kord, C: Arigashar, D: Bahman, E: Radical, F: Torkman, G: Jonoob, H: Kavir, I: Nosrat, J: Nimrooz, K: Yousof, and L: Morex.

may be variation in length of deletion in *VRN-H1* locus amongst genotypes in our study, and therefore, some of these genotypes represent reduced vernalization requirements.

Four germplasms showed vernalization insensitive phenotype due to the lack of a functional *ZCCT-Ha* and/or *ZCCT-Hb* alleles in *VRN-H2* and the deletion in the vernalization critical region of *VRN-H1*. These germplasms were Nimrooz, ND-82-10, Tokak, and Bereke-54 which together with Morex revealed complete spring type growth habit. The only entry with the reduced vernalization requirement solely

FIGURE 3: Geographical regions based on the current climatic classification for the barley breeding program. Agricultural areas located in zone W-N (star) are characterized by warm and humid climate during the reproductive stage. Areas located in zone W-S (circle) have hot summers and are prone to heat stress and drought. Zone M (square) does not have an incidence of extreme temperatures. Regions located in zone C (triangle) are characterized by prolonged freezing temperatures and frequent frost damage. The map was generated by using GPS Visualizer server available at (http://www.gpsvisualizer.com/), and the names of locations were added manually in power point. This figure is shown best at this size or smaller in the printed version.

due to the deletion in functional *ZCCT-Hb* allele in *VRN-H2* locus and not due to the deletion of the vernalization critical region was "1-BC-80318". This entry revealed a facultative phenotype in the growth habit nursery.

The barley breeding program in Iran operates on the basis of viewing the country as four mutually distinct climatic characteristics (Figure 3) on the basis of variation in temperature and humidity. This environmental classification has been in place for nearly three decades and includes W-N, W-S, M, and C. The region W-S, characterized by warm and humid climate particularly in reproductive growth stages, is mostly distributed in the north of the country. The regions classified as W-S are mainly in southern Iran and encompass regions with rapidly increasing temperatures in the spring followed by a very hot summer leading to a combined heat and likely drought stress. In region M, the growing season is normally not associated with an extreme temperature or humidity conditions. This region is mostly distributed in the central, western, eastern, and northeastern parts of the country. Regions in zone C are prone to frost damage during the vegetative stage (C), and therefore, winter survival plays a vital role in crop yield.

In Iran, barley is predominantly sown in autumn each year and harvested in the summer of the following year. Spring-sown barley production is not practiced in Iran. What is neglected from this mutual classification of regions—with respect to vernalization requirements—is that nearly all cultivated regions of the country including W-N, W-S, and M are entailed to a mild winter which yet is sufficient for fulfilling vernalization requirements. That explains why some of the cultivars developed or introduced for regions W (Karoon and Jonoob) and M (Nosrat and Rihane) are entailed with full or reduced vernalization requirement. This study drew a baseline understanding of the relationship between genotypes at vernalization loci in barley and growth habit, expected flowering time, and winter survival as related to various geographical regions in an applied barley breeding ground. Genotypic analysis of barley germplasm as represented here increased our awareness of allelic variations in vernalization loci and the adaptation of barley germplasm to various regions of the country. Such information may be used in breeding practices for designing crosses for recombining allelic forms of *VRN* in barley and attempting marker-assisted selection for earliness and winter survival. However, further investigation is required to advance our knowledge about SNP variation and InDel evens in *VRN-H3* locus and allelic variation in barley photoperiod response loci, that is, *PPD-H1* and *PPD-H2*.

Acknowledgments

This study was conducted at the Cereals Research Department. The authors are thankful for the Seed and Plant Improvement Institute (SPII) financial support through Grant No. 02-03-03-9012 made available to M. Mohammadi. Special thanks are due to Mr. Ahmad Yousefi, senior barley breeder at SPII.

References

[1] J. Cockram, E. Chiapparino, S. A. Taylor et al., "Haplotype analysis of vernalization loci in European barley germplasm reveals novel *VRN-H1* alleles and a predominant winter *VRN-H1/VRN-H2* multi-locus haplotype," *Theoretical and Applied Genetics*, vol. 115, no. 7, pp. 993–1001, 2007.

[2] J. Cockram, H. Jones, F. J. Leigh et al., "Control of flowering time in temperate cereals: genes, domestication, and sustainable productivity," *Journal of Experimental Botany*, vol. 58, no. 6, pp. 1231–1244, 2007.

[3] R. A. Richards, "Defining selection criteria to improve yield under drought," *Plant Growth Regulation*, vol. 20, no. 2, pp. 157–166, 1996.

[4] G. A. Slafer, "Genetic basis of yield as viewed from a crop physiologist's perspective," *Annals of Applied Biology*, vol. 142, no. 2, pp. 117–128, 2003.

[5] D. L. Lister, S. Thaw, M. A. Bower et al., "Latitudinal variation in a photoperiod response gene in European barley: insight into the dynamics of agricultural spread from "historic" specimens," *Journal of Archaeological Science*, vol. 36, no. 4, pp. 1092–1098, 2009.

[6] J. R. Porter and M. Gawith, "Temperatures and the growth and development of wheat: a review," *European Journal of Agronomy*, vol. 10, no. 1, pp. 23–36, 1999.

[7] I. Karsai, K. Mészáros, L. Láng, P. M. Hayes, and Z. Bedö, "Multivariate analysis of traits determining adaptation in cultivated barley," *Plant Breeding*, vol. 120, no. 3, pp. 217–222, 2001.

[8] R. Takahashi and S. Yasuda, "Genetics of earliness and growth habit in barley," in *Barley Genetics II*, R. A. Nilan, Ed., pp. 388–408, Washington State University Press, 1971.

[9] S. Yasuda, J. Hayashi, and I. Moriya, "Genetic constitution for spring growth habit and some other characters in barley cultivars in the Mediterranean coastal regions," *Euphytica*, vol. 70, no. 1-2, pp. 77–83, 1993.

[10] J. Danyluk, N. A. Kane, G. Breton, A. E. Limin, D. B. Fowler, and F. Sarhan, "TaVRT-1, a putative transcription factor associated with vegetative to reproductive transition in cereals," *Plant Physiology*, vol. 132, no. 4, pp. 1849–1860, 2003.

[11] B. Trevaskis, D. J. Bagnall, M. H. Ellis, W. J. Peacock, and E. S. Dennis, "MADS box genes control vernalization-induced flowering in cereals," *Proceedings of the National Academy of Sciences of the United States of America*, vol. 100, no. 22, pp. 13099–13104, 2003.

[12] B. Trevaskis, D. J. Bagnall, M. H. Ellis, J. Peacock, and E. J. Dennis, "MADS box genes control vernalization-induced flowering in cereals," *Proceedings of the National Academy of Sciences of the United States of America*, vol. 100, no. 22, pp. 13099–13104, 2003.

[13] L. Yan, A. Loukoianov, G. Tranquilli, M. Helguera, T. Fahima, and J. Dubcovsky, "Positional cloning of the wheat vernalization gene VRN1," *Proceedings of the National Academy of Sciences of the United States of America*, vol. 100, no. 10, pp. 6263–6268, 2003.

[14] J. von Zitzewitz, P. Szucs, J. Dubcovsky et al., "Molecular and structural characterization of barley vernalization genes," *Plant Molecular Biology*, vol. 59, no. 3, pp. 449–467, 2005.

[15] D. Fu, P. Szucs, L. Yan et al., "Large deletions within the first intron in VRN-1 are associated with spring growth habit in barley and wheat," *Molecular Genetics and Genomics*, vol. 273, no. 1, pp. 54–65, 2005.

[16] J. Cockram, C. Norris, and D. M. O'Sullivan, "PCR-based markers diagnostic for spring and winter seasonal growth habit in barley," *Crop Science*, vol. 49, no. 2, pp. 403–410, 2009.

[17] L. Yan, A. Loukoianov, A. Blechl et al., "The wheat VRN2 gene is a flowering repressor down-regulated by vernalization," *Science*, vol. 303, no. 5664, pp. 1640–1644, 2004.

[18] J. Dubcovsky, C. Chen, and L. Yan, "Molecular characterization of the allelic variation at the VRN-H2 vernalization locus in barley," *Molecular Breeding*, vol. 15, no. 4, pp. 395–407, 2005.

[19] D. A. Laurie, N. Pratchett, J. H. Bezant, and J. W. Snape, "RFLP mapping of five major genes and eight quantitative trait loci controlling flowering time in a winter × spring barley (*Hordeum vulgare* L.) cross," *Genome*, vol. 38, no. 3, pp. 575–585, 1995.

[20] L. Yan, D. Fu, C. Li et al., "The wheat and barley vernalization gene VRN3 is an orthologue of FT," *Proceedings of the National Academy of Sciences of the United States of America*, vol. 103, no. 51, pp. 19581–19586, 2006.

[21] S. Faure, J. Higgins, A. Turner, and D. A. Laurie, "The FLOWERING LOCUS T-like gene family in barley (Hordeum vulgare)," *Genetics*, vol. 176, no. 1, pp. 599–609, 2007.

[22] B. Trevaskis, M. N. Hemming, W. J. Peacock, and E. S. Dennis, "HvVRN2 responds to daylength, whereas HvVRN1 is regulated by vernalization and developmental status," *Plant Physiology*, vol. 140, no. 4, pp. 1397–1405, 2006.

[23] A. Turner, J. Beales, S. Faure, R. P. Dunford, and D. A. Laurie, "The pseudo-response regulator Ppd-H1 provides adaptation to photoperiod in barley," *Science*, vol. 310, no. 5750, pp. 1031–1034, 2005.

[24] M. N. Hemming, W. J. Peacock, E. S. Dennis, and B. Trevaskis, "Low-temperature and daylength cues are integrated to regulate *Flowering Locus T* in barley," *Plant Physiology*, vol. 147, no. 1, pp. 355–366, 2008.

[25] K. Koti, I. Karsai, P. Szucs et al., "Validation of the two-gene epistatic model for vernalization response in a Winter Spring barley cross," *Euphytica*, vol. 152, no. 1, pp. 17–24, 2006.

[26] P. Szűcs, J. S. Skinner, I. Karsai et al., "Validation of the VRN-H2/VRN-H1 epistatic model in barley reveals that intron length variation in VRN-H1 may account for a continuum of vernalization sensitivity," *Molecular Genetics and Genomics*, vol. 277, no. 3, pp. 249–261, 2007.

[27] M. N. Hemming, S. Fieg, W. J. Peacock, E. S. Dennis, and B. Trevaskis, "Regions associated with repression of the barley *(Hordeum vulgare) VERNALIZATION1* gene are not required for cold induction," *Molecular Genetics and Genomics*, vol. 282, no. 2, pp. 107–117, 2009.

[28] M. C. Casao, E. Igartua, I. Karsai, J. M. Lasa, M. P. Gracia, and A. M. Casas, "Expression analysis of vernalization and day-length response genes in barley *(Hordeum vulgare* L.) indicates that *VRNH2* is a repressor of *PPDH2* (*HvFT3*) under long days," *Journal of Experimental Botany*, vol. 62, no. 6, pp. 1939–1949, 2011.

18

Molecular Breeding to Improve Salt Tolerance of Rice (*Oryza sativa* L.) in the Red River Delta of Vietnam

Le Hung Linh,[1] Ta Hong Linh,[1] Tran Dang Xuan,[2] Le Huy Ham,[1] Abdelbagi M. Ismail,[3] and Tran Dang Khanh[1]

[1] *Agricultural Genetics Institute, Tu Liem, Hanoi, Vietnam*
[2] *Graduate School for International Development and Cooperation (IDEC), Hiroshima University, Hiroshima 739-8529, Japan*
[3] *International Rice Research Institute, College, Los Baños, Laguna, Philippines*

Correspondence should be addressed to Tran Dang Khanh, khanhkonkuk@gmail.com

Academic Editor: Akhilesh Kumar Tyagi

Rice is a stable food in Vietnam and plays a key role in the economy of the country. However, the production and the cultivating areas are adversely affected from the threats of devastation caused by the rise of sea level. Using marker-assisted backcrossing (MABC) to develop a new salt tolerance rice cultivar is one of the feasible methods to cope with these devastating changes. To improve rice salt tolerance in BT7 cultivar, FL478 was used as a donor parent to introgress the *Saltol* QTL conferring salt tolerance into BT7. Three backcrosses were conducted and successfully transferred positive alleles of *Saltol* from FL478 into BT7. The plants numbers IL-30 and IL-32 in BC_3F_1 population expected recurrent genome recovery of up to 99.2% and 100%, respectively. These selected lines that carried the *Saltol* alleles were screened in field for their agronomic traits. All improved lines had *Saltol* allele similar to the donor parent FL478, whereas their agronomic performances were the same as the original BT7. We show here the success of improving rice salt tolerance by MABC and the high efficiency of selection in early generations. In the present study, MABC has accelerated the development of superior qualities in the genetic background of BT7.

1. Introduction

Salinity is one of the major impediments to enhancing production in rice growing areas worldwide. One-fifth of irrigated arable lands in the world has been reported to be adversely influenced by high soil salinity [1]. As per the report of FAO, 2010 [2], over 800 million ha of worldwide land are severely salt affected and approximately 20% of irrigated areas (about 45 million ha) are estimated to suffer from salinization problems by various degrees. This is more serious since irrigated areas are responsible for one-third of world's food production. In Asia, 21.5 million hectares of land areas are affected by salinity and estimated to cause the loss of up to 50% fertile land by the 21st midcentury [3].

Rice is the most important food crop for over half of the world's population and supplies 20% of daily calories [4]. Rice is a major crop in Vietnam, as the world's second-largest rice exporter after Thailand, together accounting for 50% of the world rice trade. Vast portions of the food producing regions in the country will be inundated by sea water, expected to be at about 19.0%−37.8% of the Mekong River Delta (MRD) and about 1.5%−11.2% of the Red River Delta (RRD). With sea level rise by 1 m, approximately 40,000 km^2 will be inundated, and salinity intrusion is expected to cover about 71% of the MRD and RRD, together with other coastal regions. Vietnam is formidably dealing with salinity intrusion which is causing adverse influence on 1 million ha, equally with 3% of total Vietnam area [5]. The economic loss by salt intrusion in 2005 was up to 45 million USD, which is equivalent to 1.5% of annual rice productivity in the Mekong Delta [6]. It has a salinity threshold of 3 dS/m, with a 12% reduction in yield per dS/m, beyond this threshold. Therefore, rice yields can be reduced by up to 50% when grown under moderate (6 dS/m) salinity levels [7]. The crop yield reduction in salt soils can be overcome by soil reclamation or by improving salt tolerance in target crops. Therefore, the need for enhancement in salt tolerance in rice is well understood.

In the last ten years, a rapid progress has been made towards the development of molecular marker technologies and their application in linkage mapping molecular dissection of the complex agronomical traits and marker-assisted breeding [8]. Rice cultivars grown in saline soil are sensitive at both the vegetative and reproduction stages. However, salinity tolerance at different growth stages seems to be managed by independent genes. *Saltol* is a major quantitative trait locus (QTL) and was identified in the salt-tolerant cultivar Pokkali. Its location was detected on chromosome 1. This QTL confers salinity tolerance at the vegetative stage and explains from 64% to 80% of the phenotypic variance [9]. Several studies reported that this QTL was detected in some other rice varieties [7, 10].

The basis of MABC strategy is to transfer a specific allele at the target locus from a donor line to a recipient line while selecting against donor introgressions across the rest of the genomes [11]. The use of molecular markers, which permit the genetic dissection of the progeny at each generation, increases the speed of the selection process, thus increasing genetic gain per unit time [12]. The main advantages of MABC are (1) efficient foreground selection for the target locus, (2) efficient background selection for the recurrent parent genome, (3) minimization of linkage drag surrounding the locus being introgressed, and (4) rapid breeding of new genotypes with favorable traits. The effectiveness of MABC depends on the availability of closely linked markers and/or franking markers for the target locus, the size of the population, the number of backcrosses, and the position and number of markers for background selection [13]. MABC has previously been used in rice breeding to incorporate the bacterial blight resistance gene *Xa21* [14, 15] and waxy gene [16] into elite cultivars. The availability of the large-effect QTL *Saltol* for salinity tolerance in rice, a theoretical framework for MABC, and the existence of intolerant varieties that are widely accepted by farmers provided an opportunity to develop cultivars that would be suitable for larger areas of submergence-prone rice [17]. Molecular breeding technologies have been widely applied in countries all over the world. It provides powerful tool for development of stress tolerant varieties that can deal with the adverse effects from climate change. However, application of molecular breeding as MABC has just initiated sporadically in Vietnam. Hence, the attempt of this study was to develop a salinity-tolerant version of the widely grown BT7 by applying the MABC method. The improved cultivar may be useful for growing in the soil salinity of the coastal areas of Vietnamese Deltas.

2. Materials and Methods

2.1. Plant Materials and Crossing Scheme. The scheme for constructing the plant materials used in this study is summarized in Figure 1. A highly-salt tolerant FL478 (IR 66946-3R-178-1-1) was used as the donor of *Saltol* QTLs, whereas BT7 (*O. sativa* spp. *indica*), a popular growing Vietnamese elite cultivar with high quality and popularly grown in the Red River Delta of Vietnam, was used as the recipient parent. A total of 477 SSR markers distributed in the 12 chromosomes including foreground, recombinant, and background markers were screened. For the MABC scheme, BT7 was crossed with FL478 to obtain F1 seeds (Figure 1). F1s were backcrossed with BT7 to obtain a large number of BC_1F_1 seeds. In the BC_1F_1 generation, individual plants that were heterozygous at the *Saltol* locus were identified reducing the population size for further screening (foreground selection). From the individual plants that were heterozygous for *Saltol*, those that were homozygous for the recipient allele at one marker locus (RM10825) distally franking the *Saltol* locus (i.e., recombinant) were identified. We termed this as "recombinant selection" [18]. Some used markers are shown in detail in Table 1. From these recombinant plants, individuals with the fewest number of markers from the donor genome were selected (background selection). In the second and third BC generations, the same strategy was followed for selection of individual plants with the desired allele combination at the target loci including selection for recombinants between *Saltol* and the nearest proximal marker locus (RM10694) and suitable genomic composition at the nontarget loci and crossed with the recipient parent to develop the next generation. The selected BC2 and BC3 plants were self-pollinated for further analyses.

2.2. Molecular Marker Analysis. DNA was extracted from juvenile leaves of 2-week-old plants using a modified protocol as described by Zheng et al. (1995) [19]. PCR was performed in $10\,\mu$L reactions containing 5–25 ng of DNA template, $1\,\mu$L 10X TB buffer (containing 200 mM Tris-HCl pH 8.3, 500 mM KCl, 15 mM $MgCl_2$), $1\,\mu$L of 1 mM dNTP, $0.50\,\mu$L each of $5\,\mu$M forward and reverse primers, and $0.25\,\mu$L of *Taq* DNA polymerase ($4\,U/\mu$L) using an MJ Research single or dual 96-well thermal cycler. After initial denaturation for 5 min at 94°C, each cycle comprised 1 min denaturation at 94°C, 1 min annealing at 55°C, and 2 min extension at 72°C with a final extension for 5 min at 72°C at the end of 35 cycles. The PCR products were mixed with bromophenol blue gel loading dye and were analyzed by electrophoresis on 8% polyacrylamide gel using mini vertical polyacrylamide gels for high throughput manual genotyping (CBS Scientific Co. Inc., CA, USA). The gels were stained in 0.5 mg/mL ethidium bromide and were documented using Alpha Imager 1220 (Alpha Innotech, CA, USA). Microsatellite or simple sequence repeat (SSR) markers was used for selection [20].

2.3. Foreground and Recombinant Selection. At the initial stages of the experiment, for selection of the *Saltol* locus (foreground), the reported rice microsatellite (RM) markers RM493 and RM3412b, which were found to be tightly linked to *Saltol*, were used for foreground selection. For franking markers used for recombinant selection, about 5 Mb region of the *Saltol* region was targeted. Four polymorphic microsatellite markers (RM1287, RM10694, RM562, and RM7075) were identified for recombinant selection (Table 1, Figure 2).

TABLE 1: Details of markers for foreground and recombinant selection.

Markers	Mb	Forward primer	Reverse primer	Motif	No. of repeats	SSR start	SSR end
RM10694	11.0	TTTCCCTGGTTTCAAGCTTACG	AGTACGGTACCTTGATGGTAGAAAGG	AC	18	10969040	10969075
AP3206f	11.2	GCAAGAATTAATCCATGTGAAAGA	AGTGCAGGATCTGCCATGA	—	—	—	—
RM3412B	11.6	TGATGGATCTCTGAGGTGTAAAGAGC	TGCACTAATCTTTCTGCCACAGC	—	—	—	—
RM10748	11.8	CATCGGTGACCACCTTCTCC	CCTGTCATCTATCTCCCTCAAGC	AG	14	11758005	11758032
RM493	12.3	GTACGTAAACGCGGAAGGTGACG	CGACGTACGAGATGCCGATCC	AAG	9	12264091	12264117
RM140	12.3	CTTGCACAAGAGATGATGATGAGC	CATGCTGAGAAATAGTACGCTTGG	AG	12	12284725	12284748
RM10825	13.3	GGACACAAGTCCATGATCCTATCC	CTTTCCTTTCCATCCTTGTTGC	AAG	10	13306166	13306195
RM562	14.6	GGAAAGGAAGAATCAGACACAGAGC	GTACCGTTCCTTTCGTCACTTCC	AAG	13	14610402	14610446

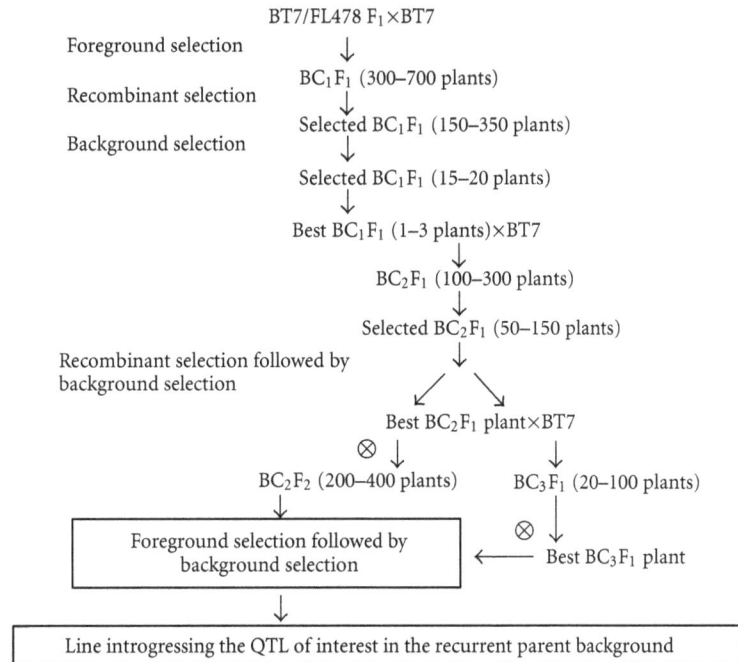

FIGURE 1: The scheme of applying MABC to improve salt tolerance in BT7 cultivar.

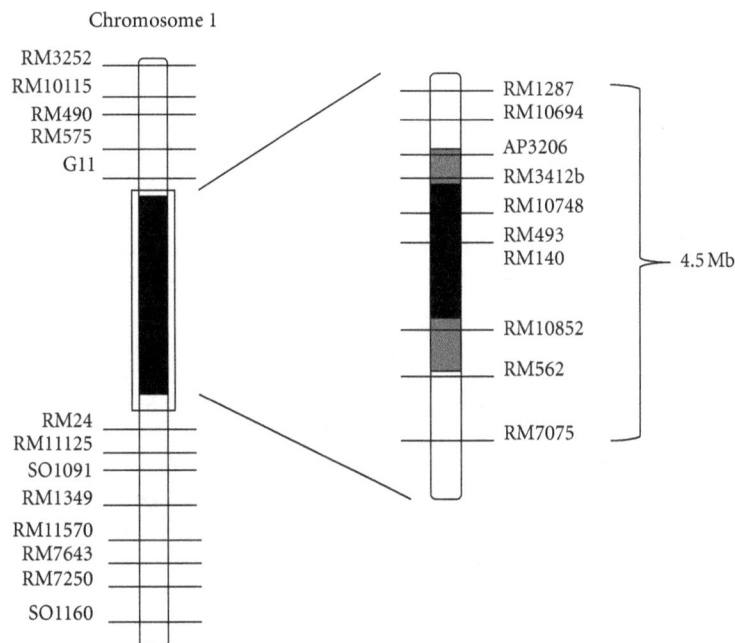

FIGURE 2: Graphical representation of the regions on chromosome 1 containing *Saltol*. White portions of the bar = homozygous BT7 segment, black regions = homozygous *Saltol* segment, and diagonal slashes = regions where crossing over occurred. Markers polymorphics between B-T7 and FL478 are labeled on both sides of the chromosome. The estimated distances in kb between the SSR markers and their orders are available at http://www.gramene.org/ [21].

2.4. Background Selection. Microsatellite markers unlinked to *Saltol* covering all the chromosomes including the *Saltol* carrier chromosome 1, that were polymorphic between the two parents, were used for background selection to recover the recipient genome (Figure 3). Based on the polymorphic information, initially evenly spaced microsatellite markers were selected per chromosome. At least four polymorphic microsatellite markers per chromosome were used. The microsatellite markers that revealed fixed (homozygous) alleles at nontarget loci at one generation were not screened at the next BC generation. Only those markers that were not fixed for the recurrent parent allele were analyzed in

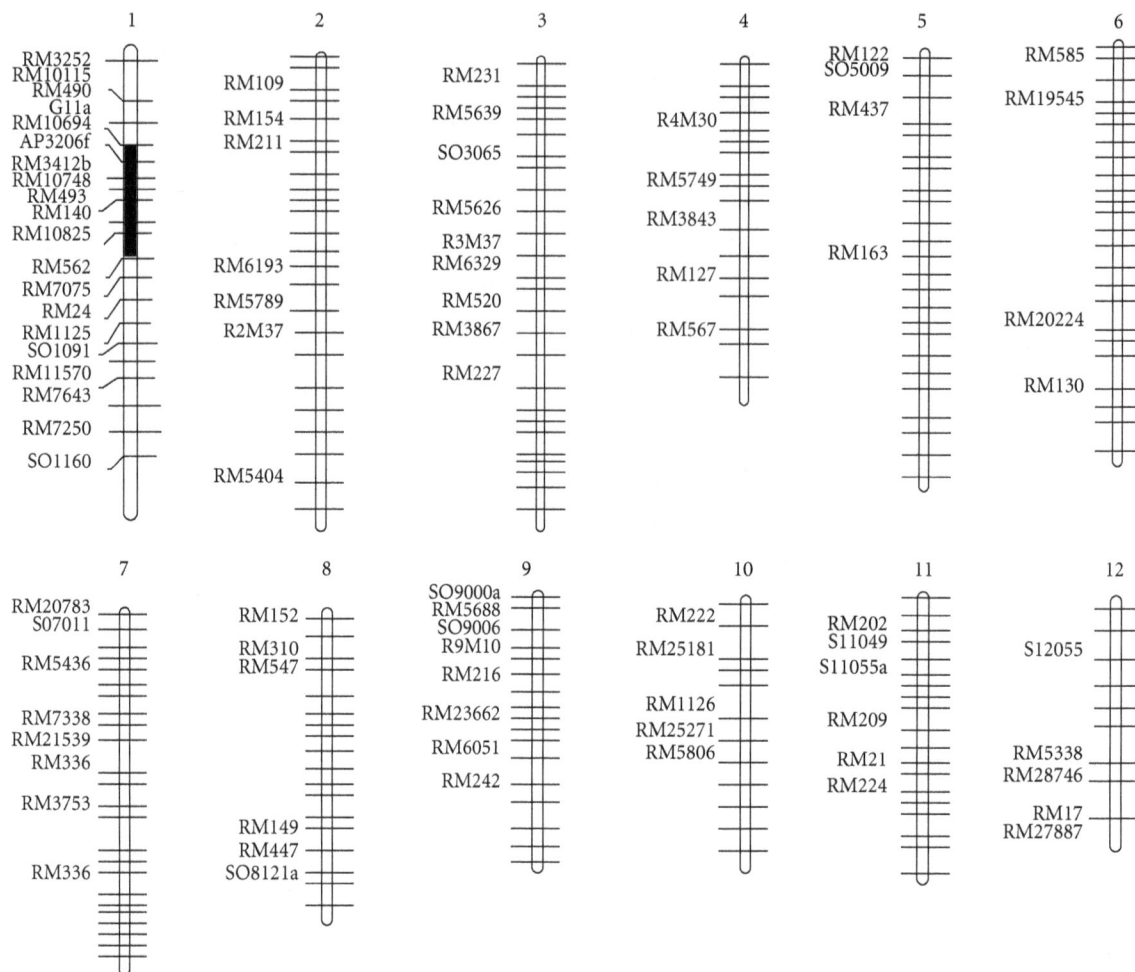

FIGURE 3: Graphical representation of mapping. Chromosome numbers are at the top of the bars. White portions of the bars are derived from BT7 and dark regions dark region is the linkage between the SSR markers and *Saltol*. Markers polymorphics between BT7 and FL478 are labeled on the left of the chromosomes.

the following generations. For the selected plants from BC_2F_1 and BC_3F_1, an additional 84 microsatellite markers were used to check the fixation of the recipient genome.

2.5. Screening for Agronomic Traits. The BC_3F_1 plants with the parents BT7 and FL478 were grown in a field at the Thanh Tri, Hanoi, Vietnam. The plants were laid in a 20×15 cm distance and evaluated for 12 traits: (1) days to heading (dth) were evaluated as the number of days from sowing until the panicle headed; (2) plant height (ph) was measured in centimeters from the soil surface to the tip of the tallest panicle (awns excluded); (3) panicle length (pl) was measured in centimeters from the neck to the panicle tip; (4) panicle number (pn) was calculated as the number of panicles per plant; (5) 1,000 seed weight (tsw) was measured in grams as the weight of 1,000 fully filled seeds per plant; (6) primary branch number (pb) was estimated as the number of primary branches per panicle; (7) secondary branch number (sb) was estimated as the number of secondary branch per panicle; (8) seed per panicle (sp) was calculated as the number of fully filled seed per panicle; (9) spikelets per

panicle (spp) were calculated as the number of spikelets per panicle.

2.6. Statistical Analyses. The molecular weights of the different alleles were calculated by Alpha Ease Fc 5.0 software. The marker data was analyzed using the software Graphical Genotyper [22]. The homozygous recipient allele, homozygous dominant allele, and heterozygous allele were scored as "A," "B," and "H," respectively. The percentage of markers homozygous for recipient parent (%A) and the percent recipient alleles including heterozygous plants (%R) were calculated. All experimental analyses of the agronomic traits were performed in a completely randomized design at least thrice. Data were analyzed with the use of the Duncan's multiple-range test ($P < 0.05$).

3. Results

3.1. Foreground and Recombinant Selections. As, the obtained result from screening of 30 SSR markers at the target region on chromosome 1 for polymorphic markers, ten markers

TABLE 2: Performance of principal agronomic traits and salt tolerance of the plants numbers IL-30 and IL 32, which were selected as the most promising lines.

Cultivar/breeding line	Saltol present	Agronomic traits								
		DTH (d)	PH (cm)	PL (cm)	PN	PB	SB	SP	SPP	TWG (g)
BT7	—	109[a]	107.5[a]	21.0[a]	6.3[a]	8.3[a]	3.5[a]	127[a]	137.6[a]	18.5[a]
FL478	Saltol	110[a]	101.2[b]	22.8[b]	7.8[b]	7.6[b]	3.4[a]	106[b]	125.6[b]	28.7[b]
IL-30	Saltol	110[a]	110.3[c]	21.2[a]	6.5[a]	8.1[a]	3.5[a]	130[a]	140.1[a]	18.7[a]
IL-32	Saltol	110[a]	106.5[a]	21.1[a]	6.5[a]	8.2[a]	3.5[a]	129[a]	139.7[a]	18.8[a]
LSD$_{(0.05)}$		0.27	0.38	0.52	0.08	0.10	0.09	0.61	0.56	0.41

Means with the same letter in a column are not significantly different at $P < 0.05$. Abbreviations present agronomical traits which were presented in Section 2.

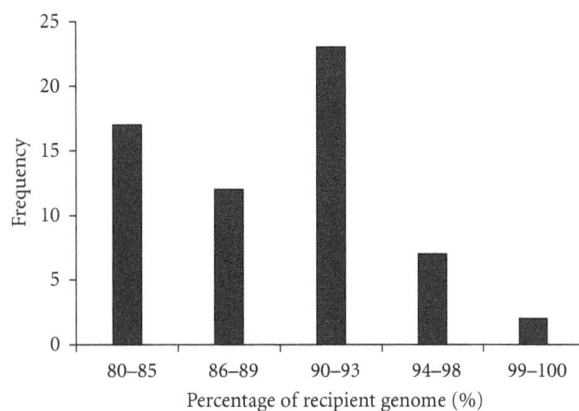

FIGURE 4: Frequency distribution of the percentage of recurrent parent genome (BT7) in the BC_3F_1 population derived from the cross between BT7 and FL478. The vertical axis of each figure represents the relative numbers of BC_3F_1 plants.

showed polymorphics between the parents. Two markers, namely, RM493 and RM3412b, tightly linked to Saltol, and four markers RM1287, RM562, RM3252, and RM490 were detected for foreground and recombinant selection, respectively. In each backcross generation (BC_1F_1-BC_3F_1), the target locus Saltol was monitored by markers linked to the Saltol genes. Individual BCnF$_1$ plants were first selected based on the heterozygous nature of all the target loci at Saltol region. Only a few of such selected individuals that had the least donor alleles of the background markers were chosen to be backcrossed with BT7. In advanced backcrosses and selfed generations, polymorphic markers RM493 and RM3412b tightly linked with Saltol were used to screen.

Four polymorphic markers between BT7 and FL478 at target region were used to screen individual BC_1F_1 plants. In conjunction with background selection, the Saltol is on chromosome 1 of few selected individuals, including plants number 1, 7, 8 and 26 in BC_2F_1, whereas the plants numbers 10, 14, 30, 41, and 359 in BC_3F_1 were characterized with two markers for foreground selection (RM493 and RM3412b). When the selected plants of BC_3F_1 (plants number 10, 30, 32, and 359) were screened with these two markers, the alleles of markers from RM3412 (12597139 bp) through RM493 (13376867 bp) were of the donor (FL478) type, and the alleles of all the remaining markers from RM1287 (11836436 bp) to RM562 (16232926 bp) onwards were of BT7, indicating that these plants were single recombinants.

3.2. Background Selection. A total of 477 SSR markers were screened for polymorphism between BT7 and FL478. Among them, 89 (18.7%) markers showed polymorphisms on 4% polyacrylamide between the parents. The 89 polymorphic markers were used for background selection. The results for polymorphism by SSR marker analysis are diagrammed in Figure 3. In BC_1F_1, A total of 30 microsatellite markers were used for background selection in 25 BC_1F_1 plants resulting from foreground and recombinant selection (Figures 1 and 2). Based on the foreground and background selection, two selected BC_1F_1 plants (nos. 7 and 13) were developed BC_2F_1 populations. In the BC_2F_1 population, 43 polymorphic markers were used for background selection in 19 BC_2F_1 plants resulting from foreground and recombinant selection plants nos. 21 and 41. For plant no. 21 chromosomes 5 and 8 were of complete recipient types. In this experiment, the background analysis of BC_3F_1 revealed the recurrent genome recovery of up to 100% at which individual lines were ranging from 81% to 100% as shown in Figure 4. The recurrent genome recovered in the plants no.s IL-30 and IL-32 is expected to be 99.2% and 100%, respectively (Figures 4 and 5).

Table 2 showed the agronomic traits in field screening of the IL to compare with the BT7. In general, there is no significant difference between the morphological traits of IL and BT7. However, the plant height (PH) of IL-30 and IL-32 was 4-5 cm higher than that of BT7. The agronomic traits including day to heading (DTH) and secondary plant

Ind number: 1 [1]:

	(%)	cM
A	99.3	252.9
H	0.7	1.8

Plant IL-30

Legend: A, B, H

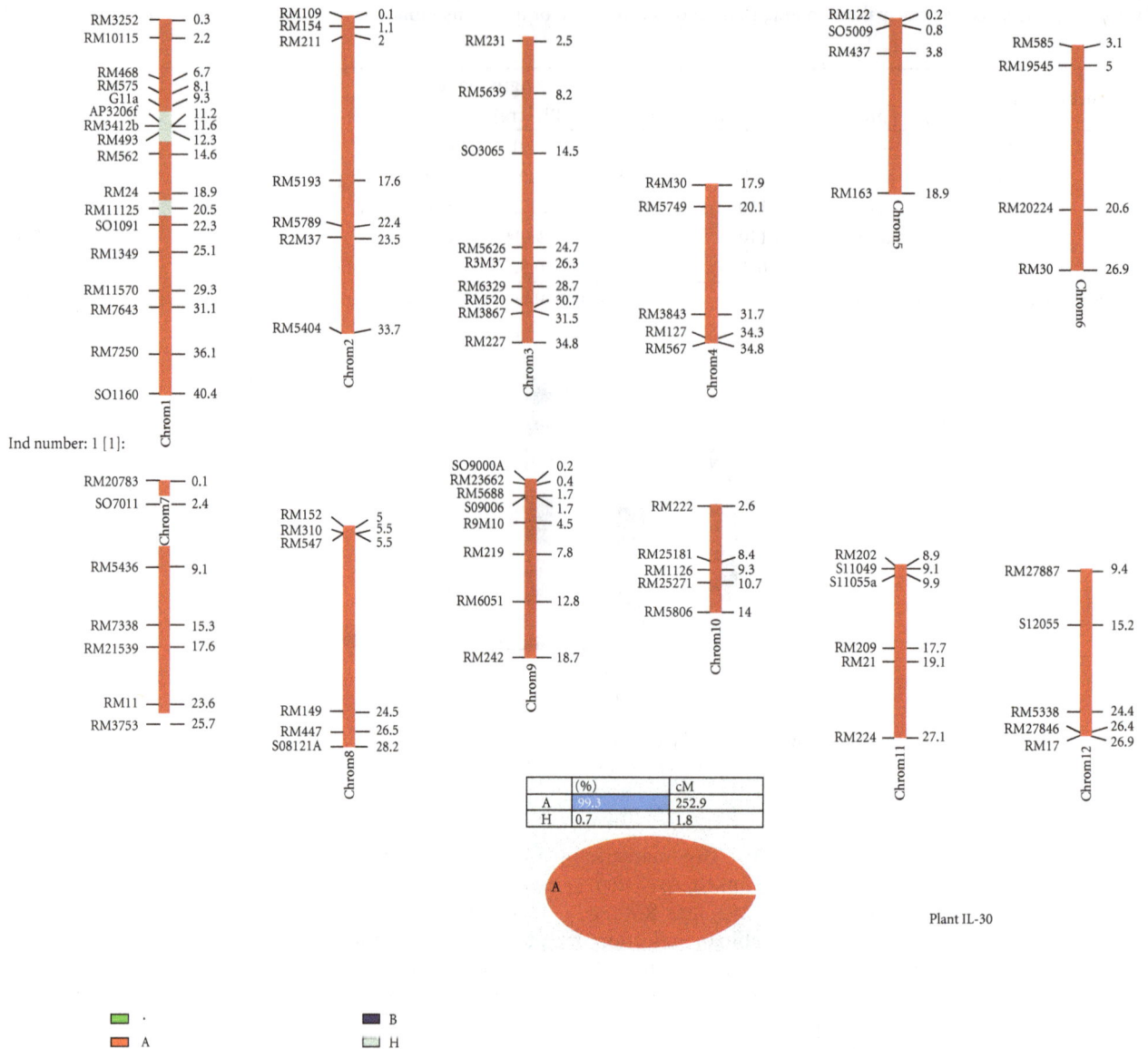

FIGURE 5: Graphical representation of the plant IL-30 genotype. Chromosome numbers are located at the bottom of the bar. Black portions of the bars are derived from BT7 and slash regions indicated the *Saltol* and FL478 introgressions. Markers are labeled on the right side of the chromosomes.

number (SP) were similar to those of the recurrent parent, BT7 (Table 2). Moreover, the other traits such as panicle length (PL), panicle number (PN), primary plant number (PN), seed per panicle (SP), spikelets per panicle (spp.), and grain yield, 1.000-grain weight of each the selected lines was almost the same as those of BT7 (Table 2).

4. Discussion

Climate change is causing negative impacts on rice production, which is the most important crop in Vietnam, and its production is mostly confined to the most vulnerable coastal regions. Climate change is severely aggravating the adverse impacts of abiotic stresses on rice production. Most of the rice production lands in coastal areas are already being affected by the rising sea level, increasing the incidences of salinity. However, salt stress problems in field crops can effectively be mitigated through the use of tolerant rice varieties and proper management and mitigation strategies. It is imperative to develop salt tolerance rice cultivars with high yield potential and grain quality using modern tools of biotechnology. However, it is often difficult to incorporate salt tolerance genes into high yielding varieties by conventional breeding methods due to the unexpected linkage drag encountered in the progenies, which affects yield and grain quality characteristics of rice cultivars [23, 24]. Applying molecular breeding has just initiated in some recent years. Lang et al. [25] applied marker-assisted selection (MAS) to improve salt tolerance in OMCS2000 rice cultivar, a widely grown cultivar in North Vietnam. Current efforts in various institutions in Vietnam are directed towards marker-assisted backcrossing strategy to introgress the favourable alleles for salinity tolerance QTLs into elite rice line which could considerably minimize breeding time as well as make arduous screening unnecessary [25, 26].

It is also challenging to achieve a definite goal of salt tolerance using conventional breeding strategies when the target gene is linked with an unfavorable dominant gene [27]. Nevertheless, since markers have been found as linked to some the specific traits of interest and used as the tools of biotechnology, it is plausible to transfer valuable genes of salt tolerance stresses in rice without linkage drag [28]. In this study, BT7 was selected as the recipient parent because it is good quality rice and always gives high profit for milled rice.

Our study focuses on combining the useful agronomic traits of BT7 with *Saltol* QTL/gen, which attached salt tolerance in backcross breeding lines by conversion to the recurrent parent genotype using molecular genotyping with SSR markers. We successfully transferred the *Saltol* from donor line FL478 into BT7. The *Saltol* gene was identified in an introgression line, highly salt-tolerant FL478 (IR 66946-3R-178-1-1), which inherited the gene from the Pokkali [29].

Here, we used the MABC breeding method to transfer the *Saltol* gene into a popular cultivar by phenotype and genotype selection. Using SSR markers (RM493 and RM3412b) the *Saltol* gene ensured efficient foreground selection. The codominant nature of SSR markers could be very useful in addition to gene-based markers for the introgression of the *Saltol* locus into a wide range of recipient elite cultivars. The selfed progenies or recombinant homozygote plants in the target region were selected from 300 to 478 plants for each backcrossing generation with foreground selection. Our results demonstrate that a major salt tolerance gene (*Saltol*) from the donor parent FL478 was successfully transferred into the BT7 genetic background and expressed similar phenotypic characteristics when compared with BT7.

5. Conclusions

We have developed a salt tolerance of BT7 variety by using marker-assisted backcross, which was controlled by a major *Saltol* QTL. The recovery of the recurrent parent genome by molecular genotyping and selection could increase the efficiency of the MABC strategy, and this was achievable in a short span of time. This study could have a good impact on rice breeding, and it is applicable for the introduction of important agronomic traits into the genomes of popular rice cultivars.

Abbreviation

PCR: Polymerase chain reaction
SSR: Simple sequence repeat marker
MABC: Marker-assisted backcrossing
MAS: Marker-assisted selection
QTL: Quantitative trait loci.

Acknowledgments

This work has been supported by DANIDA (project code: 09-P01-VIE). The authors would like to thank the staff and students at Vietnam Agricultural Genetics Institute for field work. Thanks also go to the staff of the International Rice Research Institute (IRRI) for their technical aids.

References

[1] S. Negrão, B. Courtois, N. Ahmadi, I. Abreu, N. Saibo, and M. M. Oliveira, "Recent updates on salinity stress in rice: from physiological to molecular responses," *Critical Reviews in Plant Sciences*, vol. 30, no. 4, pp. 329–377, 2011.

[2] Food and Agriculture Organization, "Report of salt affected agriculture," 2010, http://www.fao.org/ag/agl/agll/spush/.

[3] R. Nazar, N. Iqbal, A. Masood, S. Syeed, and N. A. Khan, "Understanding the significance of sulfur in improving salinity tolerance in plants," *Environmental and Experimental Botany*, vol. 70, no. 2-3, pp. 80–87, 2011.

[4] World Rice Statistics, http://www.irri.org.

[5] M. H. Nguyen, P. V. Thu, and N. T. Cuong, "Evaluating arable soil composition in order to well manage the soil resource of Vietnam," in *Proceedings of the 9th Workshop of Vietnam Institute of Meteorology Hydrology and Environment*, vol. 9, pp. 437–442, Vietnam, 2006.

[6] Ministry of Agriculture and Rural Development, "Vietnam News Agency," 2005.

[7] Z. H. Ren, J. P. Gao, L. G. Li et al., "A rice quantitative trait locus for salt tolerance encodes a sodium transporter," *Nature Genetics*, vol. 37, no. 10, pp. 1141–1146, 2005.

[8] D. Singh, A. Kumar, A. S. Kumar et al., "Marker assisted selection and crop management for salt tolerance: a review," *African Journal Biotechnology*, vol. 10, no. 66, pp. 14694–14698, 2011.

[9] P. Bonilla, J. Dvorak, D. Mackill, K. Deal, and G. Gregorio, "RFLP and SSLP mapping of salinity tolerance genes in chromosome 1 of rice (*Oryza sativa* L.) using recombinant inbred lines," *Philippine Agricultural Scientist*, vol. 65, no. 1, pp. 68–76, 2002.

[10] H. Takehisa, T. Shimodate, Y. Fukuta et al., "Identification of quantitative trait loci for plant growth of rice in paddy field flooded with salt water," *Field Crops Research*, vol. 89, no. 1, pp. 85–95, 2004.

[11] C. N. Neeraja, R. Maghirang-Rodriguez, A. Pamplona et al., "A marker-assisted backcross approach for developing submergence-tolerant rice cultivars," *Theoretical and Applied Genetics*, vol. 115, no. 6, pp. 767–776, 2007.

[12] F. Hospital, "Marker-assisted breeding," in *Plant Molecular Breeding*, H. J. Newbury, Ed., pp. 30–59, Blackwell Publishing, Oxford, UK, 2003.

[13] M. Frisch and A. E. Melchinger, "Selection theory for marker-assisted backcrossing," *Genetics*, vol. 170, no. 2, pp. 909–917, 2005.

[14] S. Chen, X. H. Lin, C. G. Xu, and Q. Zhang, "Improvement of bacterial blight resistance of 'Minghui 63', an elite restorer line of hybrid rice, by molecular marker-assisted selection," *Crop Science*, vol. 40, no. 1, pp. 239–244, 2000.

[15] S. Chen, C. G. Xu, X. H. Lin, and Q. Zhang, "Improving bacterial blight resistance of '6078', an elite restorer line of hybrid rice, by molecular marker-assisted selection," *Plant Breeding*, vol. 120, no. 2, pp. 133–137, 2001.

[16] P. H. Zhou, Y. F. Tan, Y. Q. He, C. G. Xu, and Q. Zhang, "Simultaneous improvement for four quality traits of Zhenshan 97, an elite parent of hybrid rice, by molecular marker-assisted selection," *Theoretical and Applied Genetics*, vol. 106, no. 2, pp. 326–331, 2003.

[17] D. J. Mackill, "Breeding for resistance to abiotic stresses in rice: the value of quantitative trait loci," in *Plant Breeding*, K. R. Lamkey and M. Lee, Eds., pp. 201–212, Blackwell Publishing, Ames, Iowa, 2006.

[18] B. C. Y. Collard and D. J. Mackill, "Marker-assisted selection: an approach for precision plant breeding in the twenty-first century," *Philosophical Transactions of the Royal Society B*, vol. 363, no. 1491, pp. 557–572, 2008.

[19] K. Zheng, P. K. Subudhi, J. Domingo, G. Magpantay, and N. Huang, "Rapid DNA isolation for marker assisted selection in rice breeding," *Rice Genetics News Letter*, vol. 12, pp. 255–258, 1995.

[20] S. R. McCouch, L. Teytelman, Y. Xu et al., "Development and mapping of 2240 new SSR markers for rice (*Oryza sativa* L.)," *DNA Research*, vol. 9, no. 6, pp. 199–207, 2002.

[21] SSR Marker, 2011, http://www.gramene.org/.

[22] R. Van Berloo, "GGT 2.0: versatile software for visualization and analysis of genetic data," *Journal of Heredity*, vol. 99, no. 2, pp. 232–236, 2008.

[23] J. U. Jeung, H. G. Hwang, H. P. Moon, and K. K. Jena, "Fingerprinting temperate japonica and tropical indica rice genotypes by comparative analysis of DNA markers," *Euphytica*, vol. 146, no. 3, pp. 239–251, 2005.

[24] U. S. Yeo and J. K. Shon, "Linkage analysis between some agronomic traits and resistance gene to brown planthopper in rice," *Korean Journal in Plant Breeding*, vol. 33, no. 4, pp. 287–293, 2001.

[25] N. T. Lang, B. C. Buu, and A. M. Ismail, "Molecular mapping and marker assisted selection for salt tolerance in rice (*Oryza sativa* L.)," *Omonrice*, vol. 16, pp. 50–56, 2008.

[26] C. M. F. Elahi, Z. I. Seraj, N. M. Rasul et al., "Breeding rice for salinity tolerance using the Pokkali allele: finding a linked marker," in *In Vitro Culture, TransFormation and Molecular Markers for Crop Improvement*, A. S. Islam, Ed., pp. 157–169, Science Publisher, NH, USA, 2004.

[27] J. Jairin, S. Teangdeerith, P. Leelagud et al., "Development of rice introgression lines with brown planthopper resistance and KDML105 grain quality characteristics through marker-assisted selection," *Field Crops Research*, vol. 110, no. 3, pp. 263–271, 2009.

[28] D. J. Mackill, "Molecular markers and marker-assisted selection in rice," in *Genomic Assisted Crop Improvement, Genomics Applications in Crops*, R. K. Varshney and R. Tuberosa, Eds., vol. 2, pp. 147–169, Springer, New York, NY, USA, 2007.

[29] M. J. Thomson, M. de Ocampo, J. Egdane et al., "Characterizing the saltol quantitative trait locus for salinity tolerance in rice," *Rice*, pp. 1–13, 2010.

Permissions

The contributors of this book come from diverse backgrounds, making this book a truly international effort. This book will bring forth new frontiers with its revolutionizing research information and detailed analysis of the nascent developments around the world.

We would like to thank all the contributing authors for lending their expertise to make the book truly unique. They have played a crucial role in the development of this book. Without their invaluable contributions this book wouldn't have been possible. They have made vital efforts to compile up to date information on the varied aspects of this subject to make this book a valuable addition to the collection of many professionals and students.

This book was conceptualized with the vision of imparting up-to-date information and advanced data in this field. To ensure the same, a matchless editorial board was set up. Every individual on the board went through rigorous rounds of assessment to prove their worth. After which they invested a large part of their time researching and compiling the most relevant data for our readers. Conferences and sessions were held from time to time between the editorial board and the contributing authors to present the data in the most comprehensible form. The editorial team has worked tirelessly to provide valuable and valid information to help people across the globe.

Every chapter published in this book has been scrutinized by our experts. Their significance has been extensively debated. The topics covered herein carry significant findings which will fuel the growth of the discipline. They may even be implemented as practical applications or may be referred to as a beginning point for another development. Chapters in this book were first published by Hindawi Publishing Corporation; hereby published with permission under the Creative Commons Attribution License or equivalent.

The editorial board has been involved in producing this book since its inception. They have spent rigorous hours researching and exploring the diverse topics which have resulted in the successful publishing of this book. They have passed on their knowledge of decades through this book. To expedite this challenging task, the publisher supported the team at every step. A small team of assistant editors was also appointed to further simplify the editing procedure and attain best results for the readers.

Our editorial team has been hand-picked from every corner of the world. Their multi-ethnicity adds dynamic inputs to the discussions which result in innovative outcomes. These outcomes are then further discussed with the researchers and contributors who give their valuable feedback and opinion regarding the same. The feedback is then collaborated with the researches and they are edited in a comprehensive manner to aid the understanding of the subject.

Apart from the editorial board, the designing team has also invested a significant amount of their time in understanding the subject and creating the most relevant covers. They scrutinized every image to scout for the most suitable representation of the subject and create an appropriate cover for the book.

The publishing team has been involved in this book since its early stages. They were actively engaged in every process, be it collecting the data, connecting with the contributors or procuring relevant information. The team has been an ardent support to the editorial, designing and production team. Their endless efforts to recruit the best for this project, has resulted in the accomplishment of this book. They are a veteran in the field of academics and their pool of knowledge is as vast as their experience in printing. Their expertise and guidance has proved useful at every step. Their uncompromising quality standards have made this book an exceptional effort. Their encouragement from time to time has been an inspiration for everyone.

The publisher and the editorial board hope that this book will prove to be a valuable piece of knowledge for researchers, students, practitioners and scholars across the globe.

List of Contributors

K. R. Gedye, J. L. Gonzalez-Hernandez, V. Owens and A. Boe
Department of Plant Sciences, South Dakota State University, Brookings, SD 57007, USA

Andrea Kunova, Elena Zubko and Peter Meyer
Centre for Plant Sciences, University of Leeds, Leeds LS2 9JT, UK

Stephane Deschamps
Du Pont Agricultural Biotechnology, P.O. Box 80353, Wilmington, DE 19880, USA

Kishore Nannapaneni, Yun Zhang and Kevin Hayes
Du Pont Pioneer, P.O. Box 1004, Johnston, IA 50131, USA

Yogesh T. Jasrai and Himanshu A. Pandya
Department of Bioinformatics, Applied Botany Center, University School of Sciences, Gujarat University, Ahmedabad 380 009, India
Department of Botany, University School of Sciences, Gujarat University, Ahmedabad 380 009, India

Sivakumar Prasanth Kumar, Saumya K. Patel and Ravi G. Kapopara
Department of Bioinformatics, Applied Botany Center, University School of Sciences, Gujarat University, Ahmedabad 380 009, India

Yangfan Luo, Jennifer Bragg, Olin Anderson, John Vogel and Yong Q. Gu
Western Regional Research Center, USDA-ARS, 800 Buchanan Street, Albany, CA 94710, USA

Xianting Wu
Western Regional Research Center, USDA-ARS, 800 Buchanan Street, Albany, CA 94710, USA
Department of Plant Sciences, University of California, Davis, CA 95616, USA

Jiajie Wu
State Key Laboratory of Crop Biology, Shandong Agricultural University, 61 Daizong Avenue, Taian, Shandong 271018, China

Erika Asamizu, Hiroshi Ezura and Tohru Ariizumi
Faculty of Life and Environmental Sciences, University of Tsukuba, 1-1-1 Tennodai, Tsukuba 305-8572, Japan

Kenta Shirasawa, Hideki Hirakawa, Shusei Sato, Satoshi Tabata and Daisuke Shibata
Kazusa DNA Research Institute, 2-6-7 Kazusa-kamatari, Kisarazu 292-0818, Japan

Kentaro Yano
School of Agriculture, Meiji University, 1-1-1 Higashi-mita, Tama-ku, Kawasaki 214-8571, Japan

Cosette Abdallah
Environment and Agro-Biotechnologies Department, Centre de Recherche Public-Gabriel Lippmann, 41 rue du Brill, 4422 Belvaux, Luxembourg
UMR Agroecologie INRA 1347/Agrosup/Universite de Bourgogne, Pole Interactions Plantes Microorganismes ERL 6300 CNRS, Boite Postal 86510, 21065 Dijon Cedex, France

Jenny Renaut and Kjell Sergeant
Environment and Agro-Biotechnologies Department, Centre de Recherche Public-Gabriel Lippmann, 41 rue du Brill, 4422 Belvaux, Luxembourg

Eliane Dumas-Gaudot
UMR Agroecologie INRA 1347/Agrosup/Universite de Bourgogne, Pole Interactions Plantes Microorganismes ERL 6300 CNRS, Boite Postal 86510, 21065 Dijon Cedex, France

Kelvin H. P. Khoo, Amanda J. Able and Jason A. Able
School of Agriculture, Food & Wine, Waite Research Institute, The University of Adelaide, Waite Campus, PMB1, Glen Osmond, SA, 5064, Australia

Bhupendra Chaudhary
School of Biotechnology, Gautam Buddha University, Greater Noida 201 308, India

Jens Boesger, Volker Wagner, Wolfram Weisheit and Maria Mittag
Institute of General Botany and Plant Physiology, Friedrich Schiller University Jena, Am Planetarium 1, 07743 Jena, Germany

Weifan Gao and Din-Pow Ma
Department of Biochemistry, Molecular Biology, Entomology, and Plant Pathology, Mississippi State University, Mississippi State, MS 39762, USA

Johnie N. Jenkins and Sukumar Saha
USDA/ARS Crop Science Research Laboratory, P.O. Box 5367, Mississippi State, MS 39762, USA

Yufang Guo
Department of Plant and Soil Sciences, Mississippi State University, Mississippi State, MS 39762, USA

David M. Stelly
Department of Soil and Crop Sciences, Texas A&M University, College Station, TX 77845, USA

Katsuyuki Ichitani, Yuma Takemoto, Kotaro Iiyama and Muneharu Sato
Faculty of Agriculture, Kagoshima University, 1-21-24 Korimoto, Kagoshima, Kagoshima 890-0065, Japan

Satoru Taura
Institute of Gene Research, Kagoshima University, 1-21-24 Korimoto, Kagoshima, Kagoshima 890-0065, Japan

Runze Li
Department of Statistics, The Pennsylvania State University, University Park, PA 16802, USA

Kiranmoy Das
Department of Statistics, Temple University, Philadelphia, PA 19122, USA
Department of Statistics, The Pennsylvania State University, University Park, PA 16802, USA
Center for Statistical Genetics, The Pennsylvania State University, Hershey, PA 17033, USA

Zhongwen Huang and Junyi Gai
Department of Agronomy, Henan Institute of Science and Technology, Xinxiang 453003, China
National Center for Soybean Improvement, National Key Laboratory of Crop Genetics and Germplasm Enhancement, Soybean Research Institute, Nanjing Agricultural University, Nanjing 210095, China

Rongling Wu
Department of Statistics, The Pennsylvania State University, University Park, PA 16802, USA
Center for Statistical Genetics, The Pennsylvania State University, Hershey, PA 17033, USA

Jafar Mammadov, Rajat Aggarwal and Ramesh Buyyarapu
Department of Trait Genetics and Technologies, Dow AgroSciences LLC, 9330 Zionsville Road, Indianapolis, IN 46268, USA

Siva Kumpatla
Department of Biotechnology Regulatory Sciences, Dow AgroSciences LLC, 9330 Zionsville Road, Indianapolis, IN 46268, USA

Santosh Kumar
Department of Plant Science, University of Manitoba, Winnipeg, MB, Canada R3T 2N2

Travis W. Banks
Department of Applied Genomics, Vineland Research and Innovation Centre, Vineland Station, ON, Canada L0R 2E0

Sylvie Cloutier
Department of Plant Science, University of Manitoba, Winnipeg, MB, Canada R3T 2N2
Cereal Research Centre, Agriculture and Agri-Food Canada, Winnipeg, MB, Canada R3T 2M9

Anton Novikov
Laboratory of Molecular Genetic Systems, Institute of Cytology and Genetics, Novosibirsk, 630090, Russia

Georgiy Smyshlyaev
Department of Natural Sciences, Novosibirsk State University, Novosibirsk, 630090, Russia

Olga Novikova
Department of Plant Pathology, University of Kentucky, Lexington, KY 40546, USA
Department of Biological Sciences, University at Albany, Life Sciences Building 2061, 1400 Washington Avenue, Albany, NY 12222, USA

Mohsen Mohammadi, Davoud Torkamaneh and Hamid-Reza Nikkhah
Cereals Research Department, Seed and Plant Improvement Institute (SPII), P.O. Box 4119, Karaj, Iran

Le Hung Linh, Ta Hong Linh, Le Huy Ham and Tran Dang Khanh
Agricultural Genetics Institute, Tu Liem, Hanoi, Vietnam

Tran Dang Xuan
Graduate School for International Development and Cooperation (IDEC), Hiroshima University, Hiroshima 739-8529, Japan

Abdelbagi M. Ismail
International Rice Research Institute, College, Los Banos, Laguna, Philippines

www.ingramcontent.com/pod-product-compliance
Lightning Source LLC
Chambersburg PA
CBHW080257230326

41458CB00097B/5080